Seasonality Revisited

The Use of Irregular Seasonality in Quantitative Time Series Analysis and Forecasting

Also by Kevin B. Burk

Astrology: Understanding the Birth Chart

The Complete Node Book

The Relationship Handbook:
How to Understand and Improve Every Relationship in Your Life

The Relationship Workbook:
How to Understand and Improve Every Relationship in Your Life

Astrology Math Made Easy

The Relationship Workbook:
How to Design and Create Your Ideal Romantic Relationship

The Relationship Workbook:
The Secrets of Successful Team Building

Astrological Relationship Handbook:
How to Use Astrology to Understand Every Relationship in Your Life

Astrological Relationship Workbook:
How to Use Astrology to Understand Every Relationship in Your Life

Anger Mastery: Get Angry, Get Happy

Principles of Practical Natal Astrology:
Talented Astrologer Training Book 1

Seasonality Revisited

The Use of Irregular Seasonality in Quantitative Time Series Analysis and Forecasting

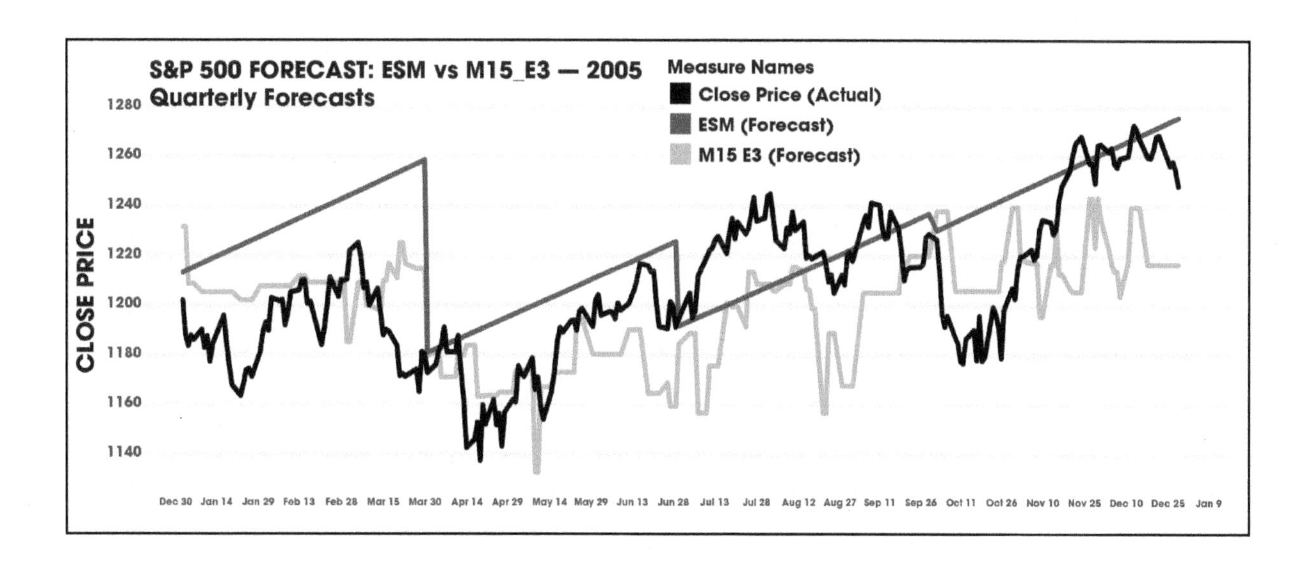

Kevin B. Burk

ISBN 978-0-9764496-9-7

Printed in the United States of America

Published by
Serendipity Press
Houston, Texas

Source files, raw data, R scripts and Excel files used to generate the results of this study are available by request at TheScienceofAstrology.com.

Book design and cover design by Kevin B. Burk

Abstract

This monograph revisits the theory of seasonality by asserting that **seasonality is a quality of time**, and that **the calendar and clock are not the only ways to measure time**.

If seasonality is a quality of time, then the seasons of a time series are defined by the divisions of time within the seasonal model and not by patterns of peaks and troughs in the data itself. Every season has an effect that can be quantified. **Cohen's d** provides an objective metric to quantify the magnitude of the effect of any season in relation to the mean value of the calendar year. This allows for comparison of seasonal effects within a seasonal model, between different seasonal models, and even between seasonal models applied to different data sets.

The historic variance of the effect sizes can be quantified and used to anticipate the predictive value of a season. The more consistent the historic effect size is, the lower the variance, and the greater the expectation that the seasonal pattern will continue in the future.

Each historic season has a mean value, and seasonal forecasts that predict the mean value for future seasons can be generated using a moving average of non-contiguous historical data, considering historical values of the season itself. These seasonal forecast signals can be combined with traditional forecasts to create a hybrid forecast that includes the seasonal influences.

Because the seasonal forecast is separate from the traditional forecast, it is also possible to incorporate only the seasonal data of seasons with the lowest variance and the greatest expectation that they will improve the accuracy of the final forecast. This is called **Targeted Seasonality**.

The seasonal models of the calendar and the clock are based on the cycles of the Sun as observed from the Earth. These are examples of **regular seasonality**. This study introduces seasonal models based on the observed speed and direction of the planet Mercury. These are examples of **irregular seasonality**.

The quantified significance of **Calendar Month** seasonality is used as the baseline of comparison to evaluate the potential significance of the irregular Mercury-based seasonal models. The seasonal models are compared across three extensive and unrelated data sets: **Transportation On-Time Performance**, **Car Crash**, and **Financial Market** data. The Mercury-based irregular seasonal models showed a far greater percentage of significant seasons than the **Calendar Month** model. The ability to compare and contrast seasonal effects in this way demonstrates the practical value of the effect- and variance-based approach to quantifying seasonal influences. It also confirms that irregular seasonal models can reveal patterns in time series data that are otherwise undetectable.

To test the value of incorporating irregular seasonal influences in time series forecasting, 76 quarterly forecasts covering a 19-year period from 2000–2018 were generated for each of **430 individual financial data sets** (10 stock market indexes, 379 individual stocks, 21 commodities, 10 interest rates and bonds, and 10 currency exchange rates). The aggregate accuracy of twelve different forecast models was then ranked and compared with both mean absolute percentage error (**MAPE**) and root mean square error (**RMSE**) The forecast models include five non-seasonal traditional forecast models (**ARIMA**, **ESM**, **HOLT**, **MEAN**, and **NAÏVE**), the seasonal forecast generated using the **M15 Sign + Speed** seasons, and six hybrid seasonal forecasts that combine the seasonal forecast data with the forecast data of each of the traditional forecast models.

If the forecasts that include the irregular seasonal data are more accurate than the non-seasonal forecasts by a significant percentage (60% or greater), then these irregular seasonal models have clear value in quantitative time series forecasting.

Two different methods, designated, **E3** and **M3,** were used to generate the seasonal forecasts for the entire data set. For the forecasts that used the **E3** model, **the hybrid seasonal forecasts were more accurate than their non-seasonal counterparts 71.30% of the time** (**MAPE**) **and 66.93% of the time** (**RMSE**). For the forecasts that used the **M3** model, **the hybrid seasonal forecasts were more accurate than their non-seasonal counterparts 74.37% of the time** (**MAPE**) and **72.33% of the time** (**RMSE**). This clearly demonstrates that including the irregular seasonal influences has a significant chance of improving the accuracy of the forecast.

Finally, to test the hypothesis that the lower the historical variance of the effects of a season, the greater its value in improving the accuracy of forecasts, the forecast accuracy for both forecast methods were evaluated considering only the seasons in each data set that met the threshold of significance for variance. In every instance, the targeted results were more accurate than their non-targeted counterparts. This strongly suggests that the historical variance of the effects of a season correlates to the value of using that season in time series forecasting.

The conclusions of this study are that there is clear and consistent value in this new approach to seasonality, both with quantitative time series analysis and with quantitative time series forecasting. These discoveries are worth further serious consideration and exploration.

Contents

List of Figures

List of Tables

List of Tables

Acknowledgments

This work would not be possible without the support and contributions of **Stephen Jaye** and **Bill Gold**. Stephen Jaye created the original R scripts to analyze the raw data and generate and aggregate the effect sizes of the seasonal influences. He also suggested the use of Cohen's d to evaluate the significance of the results when the standard significance tests proved to be impractical.

Bill Gold developed the R scripts to generate the different forecast models.

1. Overview

Seasonality promises the ability to see patterns in time series data that can improve the accuracy of time series forecasts. In practice, however, it never quite delivers. Seasonality has become a minor consideration at the beginning of any time series analysis. After a perfunctory decomposition of the time series data, any seasonal signal is either kept or discarded.

The theory of seasonality is sound. It states that a pattern in the historical time series data can be expected to continue in the future, and therefore it could be used to improve the accuracy of time series forecast for that data set. To incorporate seasonality, one must first identify the patterns in the historical data, and next evaluate if those patterns are significant and therefore likely to improve the accuracy of the forecast.

The current approach to seasonality does not adequately address either of these objectives. It treats seasonality as if it were a quality of the data rather than a quality of time, and it assumes that seasonal models are limited to subsets of the calendar or the clock. Neither of these assumptions is correct.

When we approach seasonality as a quality of time, it becomes possible to quantify the expected significance of a season, both in terms of the magnitude of the effect, and also in terms of the expected predictive value of the season. And when we begin to view time series data through seasonal filters that are not limited to divisions of the calendar or the clock, we discover a universe of previously invisible patterns in the time series data. By revisiting how we understand and apply seasonality, we can realize more of the promise of seasonality with more accurate forecasts.

This new approach to seasonality introduces a number of intriguing options that offer significant advantages both in quantitative time series analysis and in quantitative time series forecasting. Each of these ideas will be explored individually before outlining the overall methodology used to test the practical value and potential applications of this new approach to seasonality.

1.1. Quantifying Seasonal Effect Size with Cohen's d

A season is defined as a discrete unit of time that exists and reoccurs as a subset of a larger, recurring unit of time. Therefore, seasonality is a quality of time itself, not of data. This is a critical distinction because in practice, the current approach to seasonality treats seasonality as a quality of data. Seasons are detected by the presence of repeating peaks and troughs in the data, and then those fluctuations are associated with a seasonal model based on equal divisions of the calendar or the clock.

Analysis of seasonal influences is primarily accomplished by considering the results of the time series decomposition. The option exists to plot the seasonal data itself, and even to view the sub-series plots of individual seasons, but this approach does not include the ability to quantify or compare the effects of individual seasons.

When seasonality is viewed as a quality of the time series data, the question is whether or not a seasonal influence exists. When seasonality is viewed as a quality of time itself, the existence of the seasonal influence is no longer the question. Seasons exist in the time series data the moment you apply a specific seasonal model to that data. The effect of each season can then be quantified in relation to the mean values of the calendar year. The key questions now are whether the effect of a season is significant and whether that seasonal influence can be expected to improve the accuracy of a forecast.

Cohen's d parameter (Cohen, 1998), commonly used for t-tests, is a standardized measure that makes it possible to compare effect sizes between data sets or studies. **Cohen's d** considers the difference between the means (in this case, the mean values of a specific season and the reference mean of the calendar year) and the variance to produce a number that describes the magnitude of the effect.

Not only is it possible to quantify the magnitude of a seasonal effect size, but it's also possible to evaluate the direction of the effect, i.e., higher or lower than the reference mean. Negative effect sizes, where the mean values of the season are lower than the reference mean are indicated with a minus sign (e.g., **-M** for a Medium effect in a negative direction), and positive effect sizes, where the values of the season are higher than the reference mean are indicated with a plus sign (e.g., **L+** for a large effect in a positive direction).

1.1.1. Advantages of Cohen's d in Quantitative Time Series Analysis

Using **Cohen's d** to quantify the effect sizes of individual seasons provides significant advantages for quantitative time series analysis.

Because the effect size is an objective metric, this approach makes it possible to compare seasonal influences across multiple dimensions. It can be used to **compare the seasonal influences within a seasonal model** for a single data set. It can be used to **compare different models of seasonality** for a single data set to determine which seasonal models reveal the most significant seasonal influences for that data. And it can be used to **compare seasonal influences between unrelated data sets**.

Because the seasonal effect sizes are relative to the mean values of the referenced calendar year, it's easy to create a detailed narrative of the seasonal fluctuations for any given year. Additionally, the magnitude of the effect is not a primary consideration. Effect size doesn't matter. A small effect simply indicates an average season, where the mean values of the season adhere closely to the mean values of the year.

The overall seasonal effects of any quantity of historical data can be quantified by aggregating the historical seasonal effects and using **Cohen's d** to calculate the historic effect size using weighted means. This is a simple, effective, and quite powerful way to analyze and quantify any long-term seasonal influences in a data set, and of course to be able to compare those long-term seasonal influences within the data set, between different seasonal models, and between unrelated data sets.

1.2. Generating Seasonal Forecasts from Non-Contiguous Data

While the traditional model of forecasting can incorporate seasonal influences in forecasts, the forecasts are not based on the historic seasonal values. Seasonal influences are identified when the time series is decomposed, and then any seasonal influence is fitted to a waveform where the signal of seasonal influence is standardized and plotted so that it fits within the calendar year.

The seasonal naïve (**SNAIVE**) forecast model comes the closest to considering actual seasonal influences, because it takes the value of the forecasted season to be the same as the value of the prior instance of that season. For example, the forecast for next February would be identical to the observed values from last February. More complex forecast methods, including neural network models, can be programmed to consider select historical seasonal influences as well. But none of these applications can be truly called a seasonal forecast because these forecasts are not based on the actual quantified historical seasonal values.

These harmonic seasonal signals have obvious value in forecasting, but they make small adjustments to the contiguous historical observations that are used to compute the individual forecast values. Seasonal forecasts that are based on the actual seasonal data have the potential to provide season-based adjustments to the completed forecast.

One of the unusual attributes of these seasonal forecasts is that they can be based on a moving average of non-contiguous data. Using **Calendar Month** seasonality as an example, the forecast for January 2021 would be based on the average values of January 2020, January 2019, and January 2018. Using traditional forecast methods, three-year-old historical data would never be considered; for daily aggregated data, even three-month-old data has questionable predictive value.

The quantified effect sizes of each season show the relationship between the mean values of the season and the mean values of the calendar year. A forecast based on the historical seasonal effects can be viewed in the same way. The forecasted values of each season can be considered in relation to the forecasted reference mean. The seasonal forecast can be combined with the results of a traditional forecast, which will adjust the forecasted values for each season, potentially improving the overall accuracy of the forecast.

Consider that a traditional forecast, even one that incorporates traditional seasonal influences, either resolves to a virtual trend line or adheres to the shape of the underlying cycles. There are no random, unexpected variances in the forecast values. But when a seasonal forecast is combined with a traditional forecast, each season of the forecast introduces a flex point in the hybrid (combined) forecast where the forecast values can adjust, the direction can change, or the trend can normalize.

1.3. Incorporating Variance-Based Targeted Seasonality

When applied to forecasting, the theory of seasonality suggests that historic patterns are likely to continue in the future. It's reasonable to expect that the more consistent a pattern is, the more predictive value it will have, but this is difficult to quantify using the current approach to seasonality.

When seasonal influences are quantified using **Cohen's d**, it becomes possible to evaluate the expected forecast value of a season by considering the historic effect sizes of that season. This can be quantified by measuring the variance of the percent difference between the mean values of the season and the mean values of the calendar year across every historic instance of the season. If the variance is low, it means that the effect of the season is historically consistent, which suggests that the historic values of that season will improve the accuracy of forecasts for future values of that season. If the variance is high, it means that the effect of the season is inconsistent, which suggests that the historic values of that season will not improve the accuracy of forecasts for that season.

These variance scores are not remotely objective. A historical variance of 40 for a season might be quite low for one time series and unacceptably high for another. For the purposes of this study, a the historical variance of a season is considered to be significant (i.e., low enough to have expected forecast value) if that variance is less than half of the mean variance of all of the seasons in that seasonal model.

The ability to identify individual seasons with the highest expected forecast value is important because when working with non-contiguous seasonal forecasts, seasonal adjustments can be included on a season-by-season basis. Forecast adjustments can be made only for seasons that are expected to improve the forecast accuracy, and seasons without that expectation can be left unchanged. This kind of **Targeted Seasonality** is not possible with the current approach to seasonality, which requires that the entire seasonal influence either be included in or excluded from the forecast.

1.4. Irregular Seasonal Models

Seasonality is a quality of time. Human beings have a limited perception of the dimension of time. We have objective references for everything that can be defined in the three dimensions of the physical world. We understand length, width, height, and weight without having to measure them. You can easily appreciate the difference between the distance from your house to the deli and the difference from your house to New Dehli.

We can't do the same thing with time.

Our perception of time is entirely subjective. We can perceive the entirety of a ruler, but we can't perceive the entirety of an hour. Our experience of time is limited to the present moment.

The only correct answer to the question, "What time is it?" is, "Now."

It's *always* now.

We rely on external references to orient ourselves in the dimension of time. We can only understand time when it's expressed in terms of the calendar or the clock, and this is why traditional seasonal models in statistics are based on divisions of the calendar and the clock. However, the calendar and the clock are not the only ways to measure time.

We rarely remember that the calendar and the clock require objective, external references to mark the start of each cycle, and those objective, external references are celestial. A day may be defined as the time that it takes the Earth to revolve once on its axis, but the only way that we can measure that is with an external, objective reference. A day is the time from one sunrise to the next, or from one sunset to the next. A year is understood to be the time that it takes the Earth to orbit the Sun, but we require an external, objective reference to measure where we are in that cycle. That reference is the Vernal Equinox, the moment when the Sun is exactly overhead when viewed from the equator. The Vernal equinox is also what defines 0° Aries on the great circle of the zodiac, and that brings us to the subject of astrology.

Astrology has been a part of every human civilization for more than 5,000 years. Whatever astrology has come to represent today, it retains its original purpose, which is to enable human beings to measure time.

Until the 17th century astrology and astronomy were synonymous, but today they represent entirely unrelated fields of study. Astronomy uses a heliocentric model of the Universe and is concerned with the physical properties of the cosmos. Astrology uses a geocentric model of the Universe and is concerned with how the cycles of the planets—observed from the Earth—measure time.

The seasonal models currently in use, divisions of the calendar or the clock, are based on the astrological cycles of the Sun (day, year) and the Moon (month). These cycles are examples of **regular seasonality**.

Regular seasonality is regular because **the duration of the individual seasons is uniform**, and **the recurrence of the seasons within the calendar year is also uniform**. The current application of seasonality in statistics assumes regular seasonality, and the mathematical models that identify seasonal influences via decomposition assume and require regular seasonality.

Irregular seasonality is based on the observed cycles of planets other than the Sun or Moon. Those planetary cycles do not correspond with the solar-based calendar and clock system of time measurement. Because we can only understand time when it's expressed in terms of the calendar or the clock, working with irregular seasonal models requires a matrix that maps the irregular seasons on the calendar so the time series data can be analyzed.

Irregular seasonality is irregular because **the duration of the individual seasons is not uniform from instance to instance**, and **the rate of recurrence of the seasons within the calendar year is also not uniform**. Not every season will occur in every calendar year.

This study explores irregular seasonal cycles based on the speed and direction of the planet Mercury. Of all of the planets, Mercury's orbit is the most eccentric. From a heliocentric (astronomical) perspective, Mercury orbits the Sun once every 88 days. From a geocentric (astrological) perspective, Mercury completes a cycle of the entire zodiac on roughly an annual basis, with an average of three retrograde cycles each year, each lasting approximately three weeks. During the year, the average daily motion of Mercury varies from 0°00 to 2°12 degrees in direct motion, or from 0°00 to 1°23 degrees in retrograde motion.

Here's where things get really interesting.

The **Calendar Month** seasonal model consists of only 12 seasons, each of which lasts approximately 30 days. A quarterly forecast using **Calendar Month** seasonality would include three forecast values, and would introduce three potential flex points in the hybrid forecast. The **M15 Sign + Speed** seasonal model consists of **202 individual seasons**, which can last from 1 to 27 days, with no obvious sequence or annual rate of recurrence. A quarterly forecast using the **M15 Sign + Speed** seasonal model often contains **20 or more seasonal adjustments within the forecast period**. (The Mercury-based irregular seasonal models are explored in detail in **Section 3**.)

Figure 1 is a graph of four quarterly forecasts for the S&P 500 for the year 2005. The black line shows the actual close price. The red line is the quarterly forecasts using the exponential smoothing model (**ESM**) method. As is expected with these types of forecasts, the forecast signal is effectively a trend line with no adjustments at all within the forecast period. The gold line is the forecast generated using the **E3** forecast method with the **M15 Sign + Speed** model (detailed in **Section 5**). It includes a significant adjustment for each individual season within each quarterly forecast period. **More importantly, the shape of these adjusted forecasts closely mirrors the shape of the price fluctuations of the actual close price.**

To reiterate: the gold line is an **irregular seasonal forecast** generated using historical close prices of the S&P 500 prior to January 1, 2005. **No existing forecast methodology can generate a forecast signal that looks like this.** And the seasonal forecast hasn't even been combined with the non-seasonal **ESM** forecast yet.

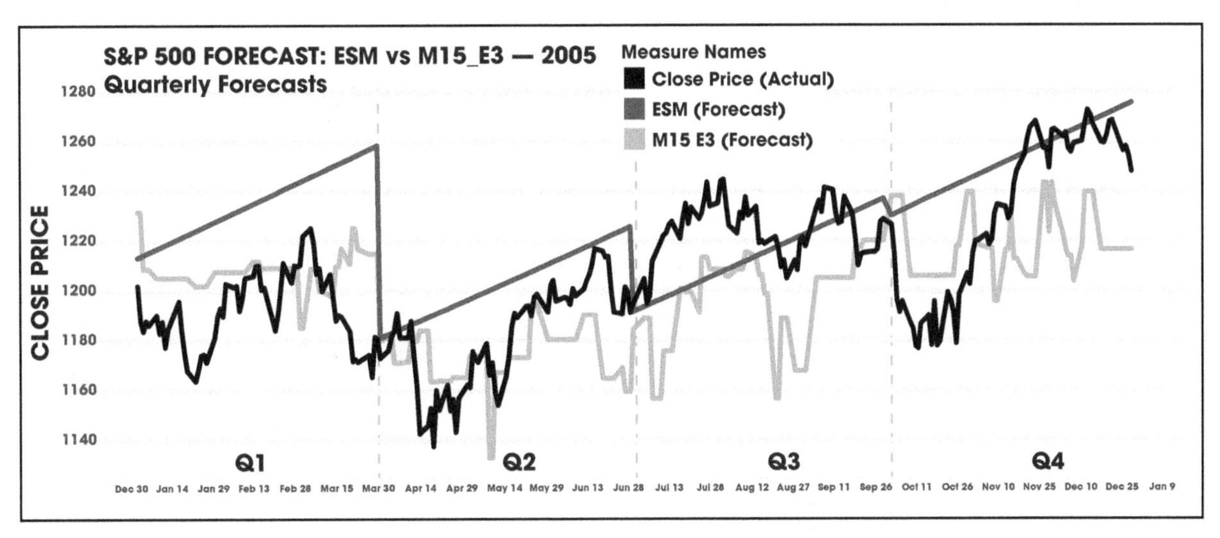

Figure 1: Comparison of Forecast Signals for the S&P 500

1.5. How to Quantify and Evaluate the Practical Value of Irregular Seasonality

This monograph consists of two related studies. The various, extensive data sets used for these studies are detailed in **Section 2**, and **Section 3** provides a detailed exploration of the nine different regular and irregular seasonal models used in the studies. The first study (**Section 4**) explores the use of irregular seasonality in quantitative time series analysis. The second study (**Section 5**) explores the use of irregular seasonality in quantitative time series forecasting.

1.5.1. The Use of Irregular Seasonality in Quantitative Time Series Analysis

This study uses the effect- and variance-based approach to quantify seasonal influences using **Cohen's d** to compare nine different seasonal models. **Calendar Month** seasonality, which is a familiar model used in the traditional approach to seasonality, serves as the baseline for comparison, and the metric being considered is the percentage of seasons within each seasonal model that can be classified as significant. When considering effect size, seasons with an effect magnitude of Medium or greater are significant. When considering variance, seasons with a variance that is less than half of the mean variance of the entire seasonal model are considered significant and expected to have predictive value. (This idea will be tested as a part of the forecast study in **Section 5**.)

To support the hypothesis that seasonality is a quality of time and not of data, the study is repeated three times with three distinct collections of data. Two data collections have some expectations of a **Calendar Month** seasonal influence (**Transportation**

On-Time Performance and **Car Crash**), and the third data collection has no expectation of a **Calendar Month** seasonal influence (sample **Financial Market** data).

The percentage of significant seasons for **Calendar Month** seasonality for each data set within each category of data are compared to the percentage of significant seasons of two regular seasonal models (**Lunar Month**, **Moon Sign**) and six irregular seasonal models based on the cycles of the planet Mercury (**M30 Speed**, **M15 Speed**, **M30 Month + Speed**, **M15 Month + Speed**, **M30 Sign + Speed**, **M15 Sign + Speed**).

First, the three regular seasonal models for each data set, **Calendar Month**, **Lunar Month**, and **Moon Sign**, are compared. This demonstrates the practical value of the analysis that is possible when **Cohen's d** is used to quantify the effect size of seasonal influences. Next, the regular seasonal models are compared to the irregular seasonal models to evaluate which seasonal models have the greatest percentage of significant seasons in terms of both effect and variance. Finally, the average significance for each seasonal model and each data set is calculated to compare the relative significance of each seasonal model across the entire collection of data.

The irregular, Mercury-based seasonal models consistently showed greater significance than the Calendar Month model, both in terms of effect size and variance (expected forecast value). This is a compelling demonstration of how irregular seasonal models can reveal significant patterns in time series data that are otherwise invisible.

1.5.2. The Use of Irregular Seasonality in Quantitative Time Series Forecasting

This study, presented in **Section 5**, evaluates the potential value of irregular seasonal models. In the simplest terms, it compares the accuracy of forecasts that include the irregular seasonal influences with the accuracy of forecasts that do not. If the forecasts that include the irregular seasonal data are more accurate than the non-seasonal forecasts, then this new approach to seasonality is clearly worth further investigation.

The study evaluates and compares the aggregate accuracy of forecasts generated for a total of **430 individual financial data sets**, including 10 stock market indexes, 379 individual stocks, 21 commodities, 10 interest rates and bonds, and 10 currency exchange rates. The stocks were selected primarily because they had the required 39 years of historical data: 20 years to be able to generate the seasonal forecasts using the **M15 Sign + Speed** seasonal model, and 19 years to forecast and evaluate the forecast accuracy.

For each data set, a total of **76 quarterly forecasts** from Q1 2000 through Q4 2019 were generated using **12 different forecast models.** The reference forecasts are generated using standard non-seasonal forecast methods: autoregressive integrated moving average (**ARIMA**), **ESM**, **HOLT**, **MEAN**, and **NAÏVE**. The hybrid seasonal forecasts combine the irregular seasonal forecast for the **M15 Sign + Speed** seasonal model with each of the traditional forecasts. The entire study was repeated twice using two different methods of generating the irregular seasonal forecast, **E3** and **M3**. The forecast methodology is explored in detail in Section 5.

Each of the 430 data sets has a total of 12 different forecasts: five non-seasonal traditional forecasts, six hybrid forecasts that combine the irregular seasonal forecast with the traditional forecast, and the irregular seasonal forecast itself. The aggregate accuracy of each forecast method is computed both by mean absolute percentage error (**MAPE**) and root mean square error (**RMSE**). The results are then analyzed to determine how often the forecasts that include the irregular seasonal data are more accurate than the forecasts that do not include the irregular seasonal data. This can be measured in two different dimensions.

The first dimension ranks the accuracy of each of the 12 forecasts and considers the number and percentage of the 430 data sets where one of the seasonal forecasts is ranked #1 or is ranked either #1 or #2 in overall accuracy by both **MAPE** and **RMSE**.

The second dimension compares the accuracy of each of the traditional forecast methods to the accuracy of the hybrid versions of those forecasts methods that include the seasonal influences, and considers the number and percentage of the 430 data sets where the hybrid seasonal forecast is more accurate than the traditional, non-seasonal counterpart.

There is no objective metric for this kind of study, so it requires an arbitrary threshold of significance to interpret the results. For the purposes of this study, any result that falls within 20 percentage points of 50% (i.e., in the range of 40% to 60%) is not significant. Should the seasonal forecasts be more accurate than the non-seasonal forecasts more than 60% of the time, it would establish the clear expectation that these irregular seasonal models have the potential to improve the accuracy of quantitative time series forecasts.

For the forecasts that used the **E3** model, **the hybrid seasonal forecasts were more accurate than their non-seasonal counterparts 71.30% of the time** (**MAPE**) **and 66.93% of the time** (**RMSE**). For the forecasts that used the **M3** model, **the hybrid seasonal forecasts were more accurate than their non-seasonal counterparts 74.37% of the time** (**MAPE**) and **72.33% of the time** (**RMSE**). This clearly demonstrates that including the irregular seasonal influences has a significant chance of improving the accuracy of the forecast.

The final portion of this study tests the hypothesis that seasons with lower historical variance have a higher expectation of improving forecast accuracy and considers the potential value of **Targeted Seasonality**. The accuracy of both the **E3** and **M3** forecasts are recalculated considering only the seasons for each data set that met the threshold of significance of variance. The targeted seasonal forecasts were significantly more accurate than the targeted non-seasonal forecasts, and also more accurate than the non-targeted seasonal forecasts.

Not only does this support the idea that the forecast value of a season can be quantified in advance by evaluating the variance of the historic effect sizes of the season, but it also demonstrates the potential value of **Targeted Seasonality**, where only the seasonal influences of selected seasons are used to adjust the forecast values.

2. Data Sets and Data Integrity

The two studies in this monograph utilize three separate and extensive categories of time series data. Section 4, the study that explores the use of irregular seasonality in quantitative time series analysis, uses the **Transportation On-Time Performance** data set, the **Car Crash** data set, and a small subset of the **Financial Market** data set. Section 5, the study that explores the use of irregular seasonality in quantitative time series forecasting, uses the full **Financial Market** data set.

The models of seasonal influence considered all require daily aggregated data. Confidence in data integrity is based both on the consistency of the results across the data set and also the presence of data for all or most of the individual dates within the data set. When data is missing or clearly inaccurate for groups of sequential data, it has a negative effect on the confidence of the data integrity.

2.1. Transportation On-Time Performance

The **Transportation On-Time Performance** data set evaluates the on-time performance of flights, Amtrak trains, city buses, and light rail. Each data set has a different criteria for a "late" departure: more than 15 minutes for flights and Amtrak trains, and more than 5 minutes for city bus and light rail. The percentage of late departures for each season is compared to the percentage of late departures for the calendar year.

The airline flight delay data comes from the **United States Department of Transportation Statistics** website (**http://www.transtats.bts.gov**), from the **Airline On-Time Performance Database**.

The Amtrak performance data comes from the **Amtrak Status Maps** website (**http://statusmaps.com**), with the assistance of Chris Juckins, who compiled the Amtrak data into a database.

The local bus and light rail data was obtained from the individual cities via Freedom of Information Law (FOIL) requests.

The nine individual data sets included in the **Transportation On-Time Performance** studies, and the confidence in the data integrity are listed in Table 1.

Table 1: On-Time Performance Data Sets

DATA	DATES	RECORDS	CONF.
Flights	Jan 1988–Dec 2017	172,960,056	Medium
Amtrak	July 4, 2006–May 10, 2018	10,668,374	High
Chicago (Bus)	Jan 2009–Dec 2017	418,962,950	High
Dallas (DART) (Bus)	Jan 2012–Aug 2018	49,787,039	High
King County (Bus)	Feb 18, 2012–Mar 9, 2018	7,176,709	Medium
Philadelphia (Bus)	Oct 31, 2011–Jun 16, 2018	68,466,941	Low
San Francisco (Bus)	Oct 15, 2011–May 22, 2018	100,252,351	Low
Chicago (Rail)	Jan 2011–May 2018	191,075	High
Dallas (DART) (Rail)	Mar 2011–Aug 2018	45,671,313	Medium

The **Philadelphia Bus** data set is missing data for multiple non-sequential individual dates, as well as for the following sequential dates: February 6–9, 2012; February 11–20, 2012; April 20–May 13, 2012; August 16–September 5, 2015.

The **San Francisco Bus** data set is missing data for multiple non-sequential individual dates, as well as the entire month of January in 2017.

2.2. Car Crash

The **Car Crash** data set evaluates seasonal influences present in the rates of car crashes.

Fatal highway crash data was obtained from the **Fatality Analysis Reporting System (FARS)** data maintained by the **National Highway Traffic Safety Administration** at **www-fars.nhtsa.dot.gov/.**

California crash data was obtained from the **Statewide Integrated Traffic Records System (SWITRS)** database at **iswitrs.chp.ca.gov**.

Iowa crash data was obtained from the **Motor Vehicle Division** of the **Iowa DOT** website, at **Iowadot.gov/mvd/factsandstats**.

Michigan crash data was obtained from the **Michigan Traffic Crash Facts (MTCF)** website at **MichiganTrafficCrashFacts.org**.

New Jersey crash data was obtained from the State of New Jersey Department of Transportation website at **state.nj.us/transportation/refdata/accident/**.

New York City crash data was obtained from the **NYC Open Data** website at **https://data.cityofnewyork.us/Public-Safety/NYPD-Motor-Vehicle-Collisions/h9gi-nx95**.

New York State crash data was obtained from the New York State Open Data portal at **https://data.ny.gov/Transportation/Motor-Vehicle-Crashes-Case-Information-Three-Year-/e8ky-4vqe.**

Texas crash data was obtained from the **Texas Department of Transportation Crash Records Information System (CRIS)** at **txdot.gov/government/enforcement-crash-statistics.html**.

The car crash data and confidence in data integrity is presented in Table 2.

Table 2: Car Crash Data Sets

DATA	DATES	RECORDS	CONF.
FARS	Jan 1980–Dec 2016	1,369,970	High
California	Jan 2001–Dec 2016	6,002,314	Medium
Iowa	Jan 1, 2006–Sep 21, 2016	234,123	Medium
Michigan	Jan 2004–Dec 2016	4,007,367	High
New Jersey	Jan 2001–Dec 2016	3,644,021	Medium
New York City	July 1, 2012–Jun 4, 2019	1,509,269	High
New York State	Jan 2014–Dec 2016	895,917	High
Texas	Jan 2010–Dec 2017	4,351,088	High

California: Dates with fewer than 500 reported crashes removed (808 dates).
Iowa: Dates with fewer than 40 reported crashes removed (366 dates).
New Jersey: Dates with fewer than 300 reported crashes removed (1,093 dates).

2.3. Financial Market

The **Financial Market** data set includes a total of 430 examples of financial data including 10 stock market indexes, 379 individual stocks, 21 commodities, 10 interest rates and bonds, and 10 currency exchange rates. The individual stocks are grouped by sector and category. The stocks were selected based on the amount of historical data available. The forecast methodology used in this study optimally requires 20 years of historical data. Forecasts were generated for the years 2000–2018, so virtually any stock with historical prices dating back to 1980 was selected for this study. To maximize representation within each stock category and sector, a small number of stocks with less extensive historical data were included. A complete list of all of the financial data sets included in this study is provided in **Appendix A**.

For the study exploring the use of irregular seasonality in quantitative time series analysis (**Section 4**) a subset of this data was selected. The entire financial data set is used for the study exploring the use of irregular seasonality in quantitative time series forecasting (**Section 5**).

Most of the data was obtained from the **Yahoo Finance** website at (**http://finance.yahoo.com**). Commodity future and interest rate data was obtained from **Macrotrends** (**http://macrotrends.net**) and **Investing.com** (**http://investing.com**).

In many cases, data was missing for the close price for random days in a given year. While this had no effect on the Mercury-based forecasts because the missing data was simply ignored when calculating the mean values for each season, the missing data threw off the **ARIMA** and **ESM** forecasts, making them invalid. To insure the overall accuracy of the **ARIMA** and **ESM** forecasts, the close price for any missing date has been replaced with the prior close price. Overall the confidence level of the integrity of this data set is high.

3. Regular and Irregular Seasonal Models

A season can be defined as a discrete unit of time that exists and reoccurs as a subset of a larger recurring unit of time. The larger unit of time is usually linked to the cycle of a reference planet through the zodiac.

The current applications of seasonality in statistics are based on various divisions of the calendar year, which is based on apparent cycle of the Sun through the zodiac (the orbit of the Earth around the Sun). Months, weeks, days, and hours are all subdivisions of the calendar year, and can all be considered seasons. The uniformity and consistency of these subdivisions of the year constitutes **regular seasonality**.

When the cycles of planets other than the Earth are used to measure time, the seasonal divisions are examples of **irregular seasonality**. The larger cycle that contains the seasons does not correspond with a solar calendar year. To make things even more complicated, when viewed from the Earth, the planets appear to slow down and change direction and this sequence of retrograde periods is its own cycle, independent of the cycle of the planet through the zodiac. The only way to understand these irregular, astrology-based seasonal models is to create a matrix that can translate them into calendar and/or clock units of time.

This study considers and compares examples of both **regular seasonality** and **irregular seasonality.**

3.1. Regular Seasonality

Regular seasonality is regular because **the duration of the individual seasons** is uniform, and **the recurrence of the seasons within the calendar year** is also uniform. The current application of seasonality in statistics assumes regular seasonality. The mathematical models designed to identify seasonal influences and to incorporate those influences in forecasts both assume and require regular seasonality.

This study explores three different regular seasonality models: **Calendar Month**, **Lunar Month**, and **Moon Sign**.

3.1.1. Calendar Month

Throughout this study, **Calendar Month** seasonality will be used as the standard of comparison. A version of **Calendar Month** seasonality is a familiar and established part of the traditional approach to seasonality. The **Calendar Month** seasons for each data set can be quantified using **Cohen's d** to evaluate the effect size and variance of historical effects for each season. The percentage of significant **Calendar Month** seasons can then be used as the minimum threshold of significance for every other seasonal model. The methodology used to quantify effect sizes is described in detail in **Section 4**.

It should be noted that the **Calendar Month** model of seasonality used in this study is not identical to the monthly seasonal influence that can result from the decomposition of a time series. When decomposing a time series, monthly seasonality is approximated as the calendar year is divided into 12 equal seasons, independent of the actual calendar. The evaluation of **Calendar Month** seasonality in this study strictly adheres to the actual calendar months such that actual duration of each **Calendar Month** season is, in fact, irregular, ranging from 28 to 31 days.

3.1.2. Lunar Month

The word "month" shares the same linguistic roots as the word "moon." Before the adoption of the solar-based Julian and Gregorian calendars, a month was a unit of time defined by a complete lunar cycle, beginning and ending at the new moon phase. The Moon orbits the Earth and makes a complete cycle of the zodiac once every 27.32 days. However, a lunar month is a synodic period based on the relationship between the Moon and the Sun. A lunar month lasts approximately 29.5 days, although the precise duration varies throughout the year. While there are 12 lunar months in a year, they do not track with the calendar months, because 12 lunar months is approximately 354 days.

If we identify individual **Lunar Month** seasons based on the sign in which the new moon occurs, it's possible to have a rough idea of how a particular **Lunar Month** season relates to a **Calendar Month** season. The approximate start and end date ranges for **Lunar Month** seasons is shown in Table 3.

If the start date ranges look familiar, they should. They are the approximate dates associated with one's "Sun Sign" or "Horoscope Sign." What these dates actually represent are the dates when the Sun occupies those signs. The new moon occurs when the Moon is conjunct the Sun (at the same longitudinal degree of a sign).

The solar and lunar calendars align on a 19-year cycle; that is, every 19 years, the **Lunar Month** seasons (new moons) begin on approximately the same dates of the Gregorian calendar. It is relatively common to have two consecutive **Lunar Month** seasons in a single sign in the same calendar year. For example, if the New Moon in Aries occurs on March 21st, there will be a second New Moon in Aries on April 18th, marking a **Lunar Month** season that continues through May 16th. The total duration of that year's Aries **Lunar Month** season would be approximately 56 days rather than the traditional 28 to 29 days.

Table 3: Approximate Calendar Date Ranges of Lunar Months

LUNAR MONTH	START DATE RANGE	END DATE RANGE
ARIES	March 21st – April 19th	April 18th – May 17th
TAURUS	April 20th – May 20th	May 18th – June 17th
GEMINI	May 21st – June 20th	June 18th – July 18th
CANCER	June 21st – July 22nd	July 19th– August 19th
LEO	July 23rd – August 22nd	August 20th – September 19th
VIRGO	August 23rd – September 22nd	September 20th – October 20th
LIBRA	September 23rd – October 22nd	October 21st – November 19th
SCORPIO	October 23rd – November 21st	November 20th – December 19th
SAGITTARIUS	November 22nd – December 21st	December 20th – January 18th
CAPRICORN	December 22nd – January 19th	January 19th – February 16th
AQUARIUS	January 20th – February 18th	February 17th – March 18th
PISCES	February 19th – March 20th	March 19th – April 17th

3.1.3. Moon Sign

The Moon makes a complete cycle of the zodiac once every 27.32 days. The periods of time that the Moon occupies each of the zodiac signs can be defined as a season. **Moon Sign** seasonality is regular seasonality and each season has a fixed position in the sequence, but the larger containing cycle is a month, not a year. There are approximately 12 non-contiguous instances of each **Moon Sign** season in each calendar year, each lasting approximately 2.5 days. For the purpose of this study, the **Moon Sign** season for a given calendar date is determined by the sign of the Moon at noon Eastern Standard Time on that date. The **Moon Sign** seasons will be either 2 or 3 days long.

3.2. Irregular Seasonality

Irregular seasonality is irregular because **the duration of the individual seasons is not uniform** from instance to instance, and **the rate of recurrence of the seasons within the calendar year** is also not uniform. Not every season will occur in every calendar year.

Irregular seasonality is based on planetary cycles that do not correspond with the solar-based calendar and clock system of time measurement. Because we can only *understand* time when it's expressed in terms of the calendar or the clock, working with irregular seasonal models requires a matrix that maps the irregular seasons on the calendar so that the time series data can be analyzed.

This study explores irregular seasonal cycles based on the speed and direction of the planet Mercury. This section describes the methodology for generating those seasons, and explores the extent of the irregular nature of the Mercury-based seasons.

Of all of the planets, Mercury's orbit is the most eccentric. From a heliocentric (astronomical) perspective, Mercury orbits the Sun once every 88 days. From a geocentric (astrological) perspective, however, Mercury can never be more than 28° from the Sun. Mercury completes a cycle of the entire zodiac on a roughly annual basis, with an average of three retrograde cycles each year, each lasting approximately three weeks.

During the year, the average daily motion of Mercury varies from 0°00 to 2°12 degrees in direct motion, or from 0°00 to 1°23 degrees in retrograde motion.

3.2.1. Definitions of Mercury Seasons

A daily ephemeris for the planet Mercury was generated using **Solar Fire** astrology software. Because all of the data sets used in this study are based in the United States, and many of the data sets involve events that occur during regular business hours (stock market trades), the ephemeris lists the daily position of Mercury at noon Eastern Standard Time. The ephemeris includes both the longitudinal position of Mercury for each day (sign and degree), the direction of motion, and the average daily motion (speed) of Mercury.

3.2.1.1. Speed-Based Seasons

Based on the speed and direction of Mercury, each calendar date can be associated with a speed category. The **M15** speeds are based on 15-minute increments of average daily motion, and the **M30** speeds are based on 30-minute increments of average daily motion. Direct motion speeds are noted with a "**D**" and retrograde motion speeds are noted with an "**R**."

The slowest speeds, which occur as Mercury changes direction, are individually categorized. For the direct speeds, the **D/R** indication means that Mercury is slowing down and preparing to turn retrograde; the **D/D** indication means that Mercury is speeding up, having just finished a retrograde cycle. For the retrograde speeds, the **R/R** indication means that Mercury is speeding up, having just turned retrograde; the **R/D** indication means that Mercury is slowing down, preparing to turn direct.

The **M15** series contains 19 unique speeds (11 direct and 8 retrograde), and the **M30** series contains 10 unique speeds (6 direct and 4 retrograde). The Mercury speeds for the **M15** and **M30** seasonal periods are shown in Table 4.

Table 4: Mercury Seasonal Periods: Speed and Direction

15-Minute (M15)		30-Minute (M30)	
D1.1	D/R 0°00 to 0°15	D3.1	D/R 0°00 to 0°30
D1.2	D/R 0°16 to 0°30		
D1.3	D/D 0°00 to 0°15	D3.2	D/D 0°00 to 0°30
D1.4	D/D 0°16 to 0°30		
D1.5	D 0°31 to 0°45	D3.3	D 0°31 to 1°00
D1.6	D 0°46 to 1°00		
D1.7	D 1°01 to 1°15	D3.4	D 1°01 to 1°30
D1.8	D 1°16 to 1°30		
D1.9	D 1°31 to 1°45	D3.5	D 1°31 to 2°00
D1.10	D 1°46 to 2°00		
D1.11	D 2°01 to 2°12	D3.6	D 2°01 to 2°12
R1.1	R/R 0°00 to 0°15	R3.1	R/R 0°00 to 0°30
R1.2	R/R 0°16 to 0°30		
R1.3	R/D 0°00 to 0°15	R3.2	R/D 0°00 to 0°30
R1.4	R/D 0°16 to 0°30		
R1.5	R 0°31 to 0°45	R3.3	R 0°31 to 1°00
R1.6	R 0°46 to 1°00		
R1.7	R 1°01 to 1°15	R3.4	R 1°00 to 1°23
R1.8	R 1°16 to 1°23		

3.2.1.2. Sign-Related Seasonality

Because Mercury can never be more than 28° from the Sun, the potential sign position of Mercury depends on the calendar month. In any given month, Mercury can only be in three or four different signs. The correlation between Mercury signs and calendar months is shown in Table 5.

3.2.1.3. Mercury Sign and Speed Combinations

The fastest Mercury speeds, either in direct or retrograde motion, only occur in certain signs. Table 6 illustrates the sign-dependent speeds for both the **M15** and **M30** series. When these are taken into consideration, the result is a total of 202 **unique sign + speed seasons** for the **M15** series and 108 **unique sign + speed seasons** for the **M30** series.

Table 5: Mercury Possible Sign Positions by Calendar Month

MONTH	SIGNS	MONTH	SIGNS	MONTH	SIGNS
JANUARY	Sagittarius	FEBRUARY	Capricorn	MARCH	Aquarius
	Capricorn		Aquarius		Pisces
	Aquarius		Pisces		Aries
	Pisces				
APRIL	Pisces	MAY	Aries	JUNE	Taurus
	Aries		Taurus		Cancer
	Taurus		Cancer		Gemini
	Gemini		Gemini		Leo
JULY	Gemini	AUGUST	Cancer	SEPTEMBER	Leo
	Cancer		Leo		Virgo
	Leo		Virgo		Libra
	Virgo		Libra		Scorpio
OCTOBER	Virgo	NOVEMBER	Libra	DECEMBER	Scorpio
	Libra		Scorpio		Sagittarius
	Scorpio		Sagittarius		Capricorn
	Sagittarius				

Table 6: Mercury Sign-Dependent Speeds

SPEED	SIGNS	SPEED	SIGNS	SPEED	SIGNS
M15 D1.10	Aries	M15 R1.6	Aries	M30 D3.6	Aries
	Taurus		Leo		Taurus
	Gemini		Virgo		Gemini
	Cancer		Libra		Cancer
	Leo		Scorpio		Leo
	Virgo		Sagittarius		
	Libra		Capricorn		
	Aquarius		Aquarius		
	Pisces		Pisces		
M15 D1.11	Aries	M15 R1.7	Virgo	M30 R3.4	Virgo
	Taurus		Libra		Libra
	Gemini		Scorpio		Scorpio
	Cancer		Sagittarius		Sagittarius
	Leo		Capricorn		Capricorn
			Aquarius		Aquarius
			Pisces		Pisces
		M15 R1.8	Scorpio		
			Sagittarius		
			Capricorn		
			Aquarius		

3.2.2. Duration of Individual Mercury Sign + Speed Seasons

The first dimension of irregularity with the Mercury seasons is the duration of each season. When considering the cycles from 1975 through 2030, the **M15** seasons can last from 1 to 27 days (59.2% of the seasons are 1 to 4 days), and the **M30** seasons can last from 1 to 32 days (55.7% of the seasons are 1 to 7 days). The distribution of the duration of the calculated **M15/M30 Sign + Speed** seasons is shown in Table 7 and Figure 2.

Table 7: Distribution of Duration of Mercury Seasons (Sign + Speed) from 1975-2030

# of DAYS	M30 (2427 Total)	M30 %	M15 (4030 Total)	M15 %
1 DAY	77	3.17%	428	10.62%
2 DAYS	93	3.83%	816	20.25%
3 DAYS	297	12.24%	676	16.77%
4 DAYS	249	10.26%	466	11.56%
5 DAYS	204	8.41%	290	7.20%
6 DAYS	206	8.49%	288	7.15%
7 DAYS	216	8.90%	252	6.25%
8 DAYS	141	5.81%	179	4.44%
9 DAYS	121	4.99%	119	2.95%
10 DAYS	81	3.34%	103	2.56%
11 DAYS	74	3.05%	81	2.01%
12 DAYS	94	3.87%	68	1.69%
13 DAYS	119	4.90%	48	1.19%
14 DAYS	81	3.34%	54	1.34%
15 DAYS	54	2.22%	25	0.62%
16 DAYS	77	3.17%	38	0.94%
17 DAYS	60	2.47%	18	0.45%
18 DAYS	52	2.14%	18	0.45%
19 DAYS	45	1.85%	35	0.87%
20 DAYS	13	0.54%	7	0.17%
21 DAYS	17	0.70%	1	0.02%
22 DAYS	7	0.29%	1	0.02%
23 DAYS	7	0.29%	2	0.05%
24 DAYS	6	0.25%	3	0.07%
25 DAYS	10	0.41%	5	0.12%
26 DAYS	12	0.49%	7	0.17%
27 DAYS	10	0.41%	2	0.05%
30 DAYS	1	0.04%		
31 DAYS	1	0.04%		
32 DAYS	2	0.08%		

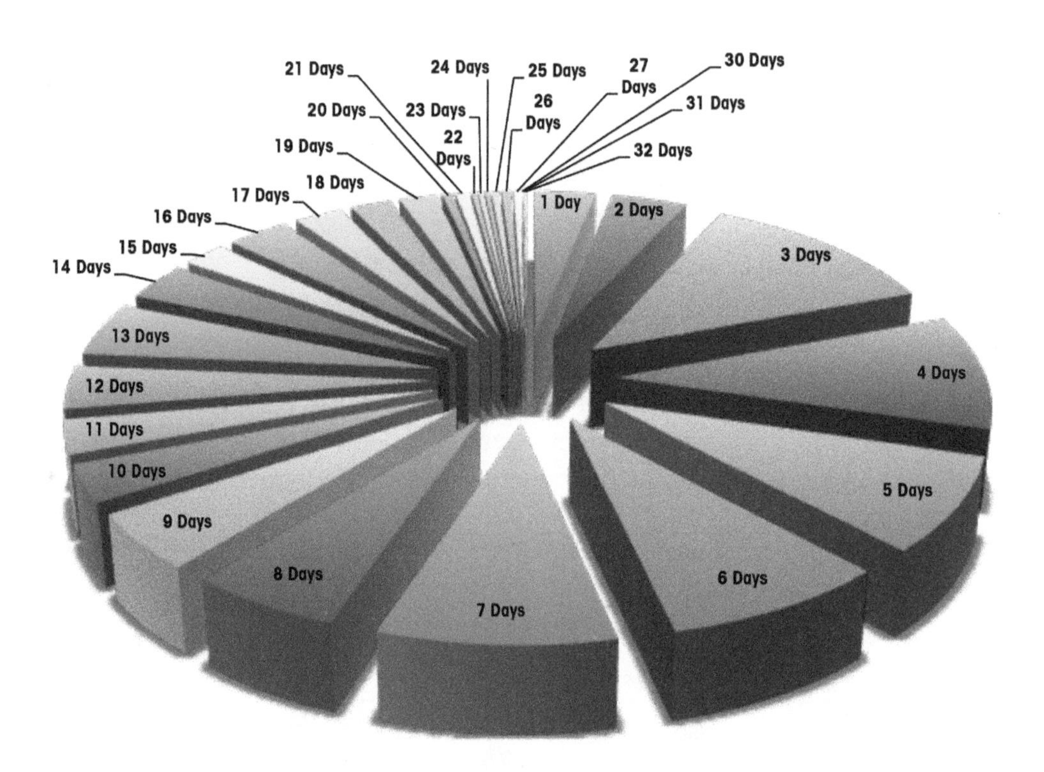

M30 Sign + Speed Season Durations

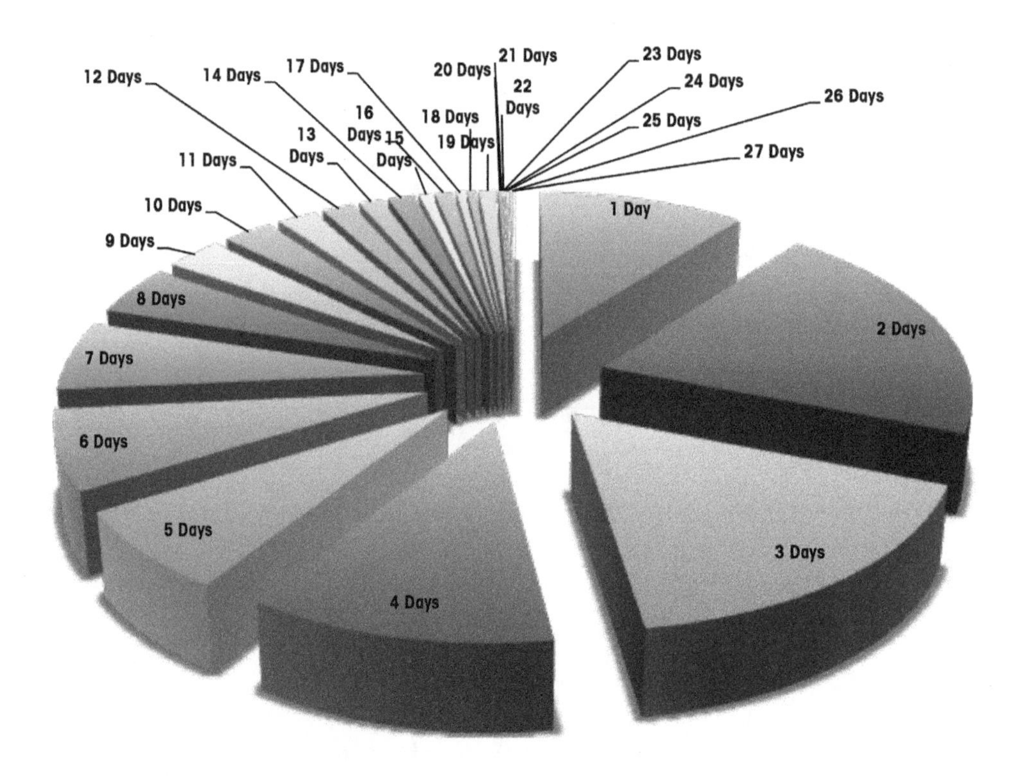

M15 Sign + Speed Season Durations

Figure 2: Distribution of Duration of M30 and M15 Mercury Seasons (Sign + Speed) from 1975-2030

While there is a wide range of possible durations for the Mercury seasons, the maximum duration of a season is limited by the speed of Mercury. When the Mercury seasons combine both the sign and speed of Mercury, every Mercury season has the potential to last only a single day because Mercury can change signs while maintaining the same speed category. The maximum duration of each Mercury season by speed is shown in Table 8.

Table 8: Maximum Duration of Mercury Seasons (Speed) from 1975-2030

M15 Speed	Maximum Days	M30 Speed	Maximum Days
D1.1	4	D3.1	7
D1.2	4	D3.2	8
D1.3	4	D3.3	19
D1.4	5	D3.4	32
D1.5	8	D3.5	27
D1.6	11	D3.6	14
D1.7	19	R3.1	9
D1.8	26	R3.2	9
D1.9	27	R3.3	15
D1.10	16	R3.4	16
D1.11	14		
R1.1	4		
R1.2	6		
R1.3	4		
R1.4	6		
R1.5	13		
R1.6	10		
R1.7	8		
R1.8	8		

3.2.3. Annual Recurrence of Mercury Sign-Speed Seasons

The second dimension of irregularity with the Mercury seasons is with the cyclical nature of the specific seasons. While some sign-speed combinations occur virtually every year, others are quite rare and infrequent. In general, the rarest seasons are also the seasons of the shortest duration, which makes it difficult to evaluate their significance.

Figure 3 on page 25 illustrates the recurrence of Mercury Sign + Speed seasons by year. Table 9 on page 26 summarizes the recurrence of each Sign + Speed Mercury season. Table 10 on page 27 analyzes each of the 202 **M15** Mercury seasons and the number of years the occurs, and Table 11 on page 28 analyzes each of the 108

M30 Mercury seasons. A more detailed analysis of the extremes will provide a greater understanding of these seasonal cycles.

3.2.3.1. M15 Outliers

Aquarius R1.8: During the 56 years under consideration, the R1.8 Aquarius season occurred a total of 4 times: 2 days in 1976; 1 day in 1989; 1 day in 2009; and 2 days in 2022.

Virgo R1.7: The R1.7 Virgo season occurred a total of 8 times: 2 days in 1976; 3 days in 1983; 1 day in 1989; 3 days in 1996; 4 days in 2009; 2 days in 2016; 2 days in 2022; and 2 days in 2029.

Pisces R1.7: The R1.7 Pisces season occurred a total of 8 times: 1 day in 1981; 4 days in 1987; 3 days in 1994; 3 days in 2000; 5 days in 2007; 2 days in 2013, 5 days in 2020, and 2 days in 2027.

Libra D1.9: The D1.9 Libra season occurred every year except for 2014 (55 out of 56 years). Most years, the duration of this season was quite substantial. Every year except for 1975, 1981, 1988, 1994, 2001, 2007, 2021 and 2027, the season lasted at least 5 days, with the longest duration being 18 days in 1996, 2016, and 2029.

Aquarius D1.9: The D1.9 Aquarius season occurred every year except for 2002 (55 out of 56 years). Most years, the duration of this season was quite substantial. Every year except for 1976, 1982, 1989, 1995, 2009, 2015, 2022 and 2028, the season lasted at least 5 days, with the longest duration being 17 days in 1991, 2000, 2004, 2007, 2013, 2020, and 2024, and 18 days in 1987.

3.2.3.2. M30 Outliers

Virgo R3.4: During the 56 years under consideration, the R3.4 Virgo season occurred a total of 8 times: 2 days in 1976; 3 days in 1983; 1 day in 1989; 3 days in 1996; 4 days in 2009; 2 days in 2016; 2 days in 2022; and 2 days in 2029.

Pisces R3.4: The R3.4 Pisces season occurred a total of 8 times: 1 day in 1981; 4 days in 1987; 3 days in 1994; 3 days in 2000; 5 days in 2007; 2 days in 2013, 5 days in 2020, and 2 days in 2027.

Leo D3.5, Virgo D3.5, Pisces D3.5: These three seasons occur every year. The shortest duration of any occurrence is 3 days, and the longest is 18 days.

Libra D3.5: The Libra D3.5 season occurred every year except for 2014 (55 out of 56 years). Most years, the duration of this season was quite substantial. Every year except for 1975, 1981, 1988, 1994, 2001, 2007, 2021 and 2027, the season lasted at least 5 days, with the longest duration being 19 days in 2005.

Aquarius D3.5: The D3.5 Aquarius season occurred every year except for 2002 (55 out of 56 years). Most years, the duration of this season was quite substantial. Every year except for 1976, 1982, 1989, 1995, 2009, 2015, 2022 and 2028, the season lasted at least 5 days, with 11 years where the duration was 18 days.

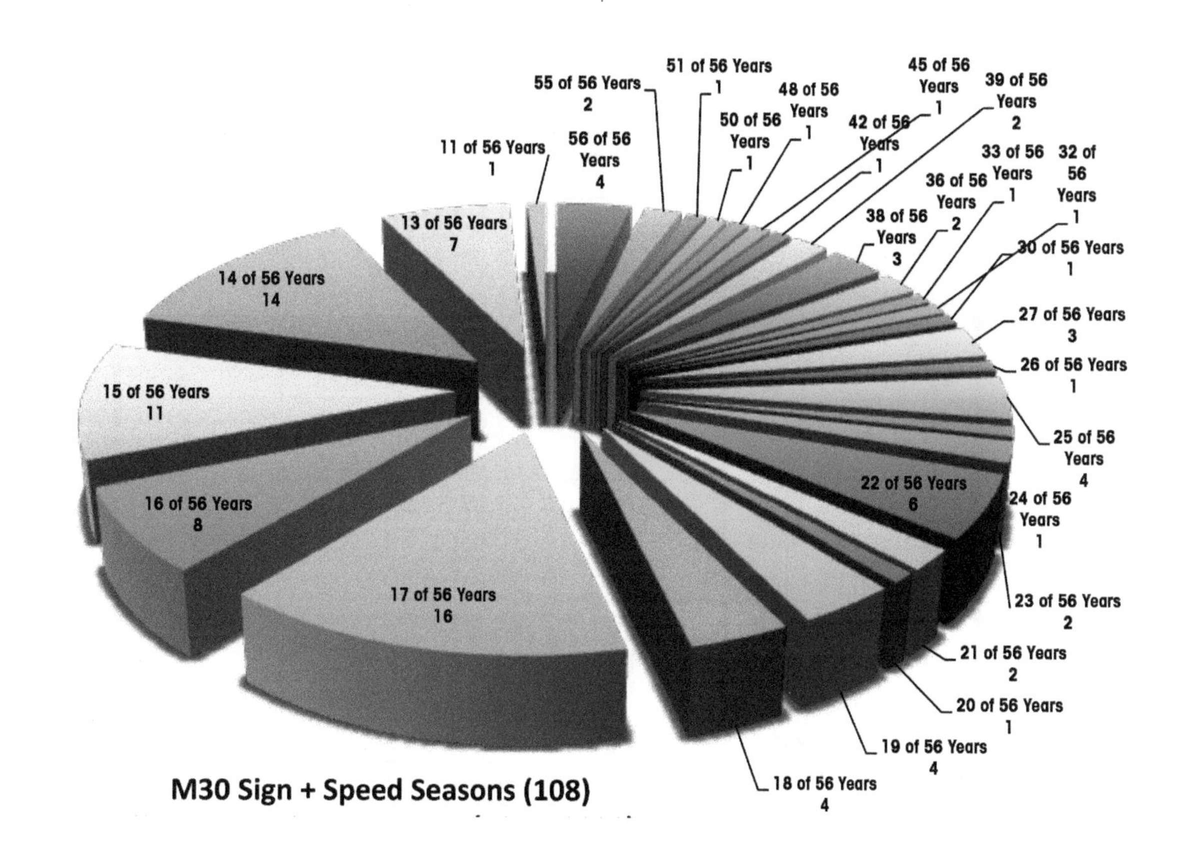

M30 Sign + Speed Seasons (108)

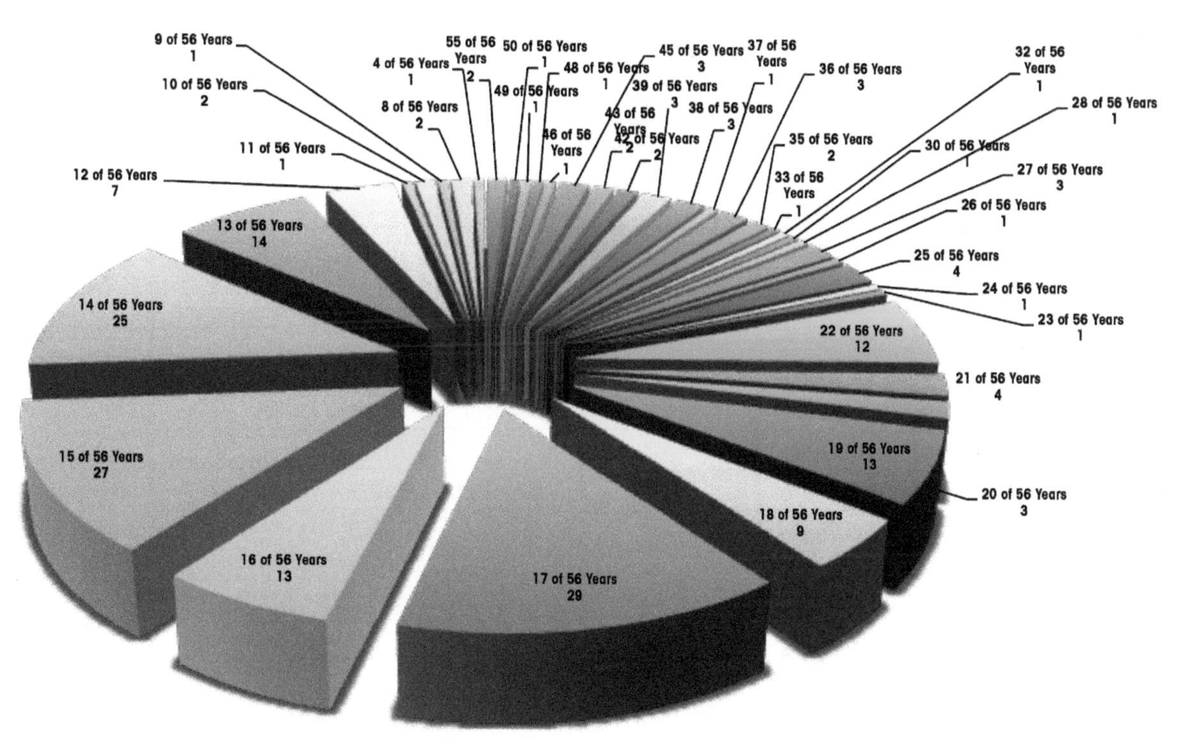

M15 Sign + Speed Seasons (202)

Figure 3: Recurrence of Mercury Seasons by Year (Sign + Speed) from 1975-2030

Table 9: Recurrence of Mercury Seasons by Year (Sign + Speed) from 1975-2030

YEARS	M15	YEARS	M30
4 out of 56 Years	1	8 out of 56 Years	2
8 out of 56 Years	2	11 out of 56 Years	1
9 out of 56 Years	1	13 out of 56 Years	7
10 out of 56 Years	2	14 out of 56 Years	14
11 out of 56 Years	1	15 out of 56 Years	11
12 out of 56 Years	7	16 out of 56 Years	8
13 out of 56 Years	14	17 out of 56 Years	16
14 out of 56 Years	25	18 out of 56 Years	4
15 out of 56 Years	27	19 out of 56 Years	4
16 out of 56 Years	13	20 out of 56 Years	1
17 out of 56 Years	29	21 out of 56 Years	2
18 out of 56 Years	9	22 out of 56 Years	6
19 out of 56 Years	13	23 out of 56 Years	2
20 out of 56 Years	3	24 out of 56 Years	1
21 out of 56 Years	4	25 out of 56 Years	4
22 out of 56 Years	12	26 out of 56 Years	1
23 out of 56 Years	1	27 out of 56 Years	3
24 out of 56 Years	1	30 out of 56 Years	1
25 out of 56 Years	4	32 out of 56 Years	1
26 out of 56 Years	1	33 out of 56 Years	1
27 out of 56 Years	3	36 out of 56 Years	2
28 out of 56 Years	1	38 out of 56 Years	3
30 out of 56 Years	1	39 out of 56 Years	2
32 out of 56 Years	1	42 out of 56 Years	1
33 out of 56 Years	1	45 out of 56 Years	1
35 out of 56 Years	2	48 out of 56 Years	1
36 out of 56 Years	3	50 out of 56 Years	1
37 out of 56 Years	1	51 out of 56 Years	1
38 out of 56 Years	3	55 out of 56 Years	2
39 out of 56 Years	3	56 out of 56 Years	4
42 out of 56 Years	2		
43 out of 56 Years	2		
45 out of 56 Years	3		
46 out of 56 Years	1		
48 out of 56 Years	1		
49 out of 56 Years	1		
50 out of 56 Years	1		
55 out of 56 Years	2		

Table 10: M15 Sign + Speed Seasons, Yearly Occurrences 1975-2030 (56 Years)

SIGN	SPEED	YEARS	%
ARIES	D1.1	14	25.00%
	D1.2	14	25.00%
	D1.3	13	23.21%
	D1.4	12	21.43%
	D1.5	18	32.14%
	D1.6	15	26.79%
	D1.7	18	32.14%
	D1.8	25	44.64%
	D1.9	37	66.07%
	D1.10	49	87.50%
	D1.11	27	48.21%
	R1.1	14	25.00%
	R1.2	14	25.00%
	R1.3	13	23.21%
	R1.4	13	23.21%
	R1.5	18	32.14%
	R1.6	10	17.86%
TAURUS	D1.1	14	25.00%
	D1.2	14	25.00%
	D1.3	13	23.21%
	D1.4	14	25.00%
	D1.5	17	30.36%
	D1.6	16	28.57%
	D1.7	19	33.93%
	D1.8	25	44.64%
	D1.9	36	64.29%
	D1.10	45	80.36%
	D1.11	38	67.86%
	R1.1	14	25.00%
	R1.2	14	25.00%
	R1.3	13	23.21%
	R1.4	14	25.00%
	R1.5	17	30.36%
GEMINI	D1.1	14	25.00%
	D1.2	14	25.00%
	D1.3	14	25.00%
	D1.4	13	23.21%
	D1.5	16	28.57%
	D1.6	17	30.36%
	D1.7	21	37.50%
	D1.8	26	46.43%
	D1.9	35	62.50%
	D1.10	43	76.79%
	D1.11	39	69.64%
	R1.1	14	25.00%
	R1.2	14	25.00%
	R1.3	14	25.00%
	R1.4	14	25.00%
	R1.5	15	26.79%
CANCER	D1.1	12	21.43%
	D1.2	13	23.21%
	D1.3	12	21.43%
	D1.4	13	23.21%
	D1.5	15	26.79%
	D1.6	15	26.79%
	D1.7	18	32.14%
	D1.8	25	44.64%
	D1.9	35	62.50%
	D1.10	45	80.36%
	D1.11	39	69.64%
	R1.1	12	21.43%
	R1.2	13	23.21%
	R1.3	12	21.43%
	R1.4	13	23.21%
	R1.5	15	26.79%
LEO	D1.1	12	21.43%
	D1.2	13	23.21%
	D1.3	14	25.00%
	D1.4	14	25.00%
	D1.5	17	30.36%
	D1.6	15	26.79%
	D1.7	19	33.93%
	D1.8	25	44.64%
	D1.9	39	69.64%
	D1.10	50	89.29%
	D1.11	27	48.21%
	R1.1	12	21.43%
	R1.2	13	23.21%
	R1.3	14	25.00%
	R1.4	15	26.79%
	R1.5	17	30.36%
	R1.6	9	16.07%
VIRGO	D1.1	15	26.79%
	D1.2	16	28.57%
	D1.3	17	30.36%
	D1.4	17	30.36%
	D1.5	22	39.29%
	D1.6	19	33.93%
	D1.7	19	33.93%
	D1.8	30	53.57%
	D1.9	48	85.71%
	D1.10	42	75.00%
	R1.1	15	26.79%
	R1.2	16	28.57%
	R1.3	17	30.36%
	R1.4	17	30.36%
	R1.5	22	39.29%
	R1.6	19	33.93%
	R1.7	8	14.29%
LIBRA	D1.1	17	30.36%
	D1.2	17	30.36%
	D1.3	17	30.36%
	D1.4	16	28.57%
	D1.5	22	39.29%
	D1.6	21	37.50%
	D1.7	19	33.93%
	D1.8	32	57.14%
	D1.9	55	98.21%
	D1.10	18	32.14%
	R1.1	17	30.36%
	R1.2	17	30.36%
	R1.3	17	30.36%
	R1.4	16	28.57%
	R1.5	23	41.07%
	R1.6	22	39.29%
	R1.7	20	35.71%
SCORPIO	D1.1	15	26.79%
	D1.2	15	26.79%
	D1.3	15	26.79%
	D1.4	16	28.57%
	D1.5	22	39.29%
	D1.6	20	35.71%
	D1.7	18	32.14%
	D1.8	36	64.29%
	D1.9	45	80.36%
	R1.1	15	26.79%
	R1.2	15	26.79%
	R1.3	15	26.79%
	R1.4	16	28.57%
	R1.5	22	39.29%
	R1.6	22	39.29%
	R1.7	19	33.93%
	R1.8	15	26.79%
SAGITTARIUS	D1.1	15	26.79%
	D1.2	15	26.79%
	D1.3	11	19.64%
	D1.4	10	17.86%
	D1.5	18	32.14%
	D1.6	17	30.36%
	D1.7	19	33.93%
	D1.8	36	64.29%
	D1.9	38	67.86%
	R1.1	15	26.79%
	R1.2	15	26.79%
	R1.3	13	23.21%
	R1.4	14	25.00%
	R1.5	22	39.29%
	R1.6	22	39.29%
	R1.7	21	37.50%
	R1.8	18	32.14%
CAPRICORN	D1.1	16	28.57%
	D1.2	17	30.36%
	D1.3	17	30.36%
	D1.4	17	30.36%
	D1.5	19	33.93%
	D1.6	21	37.50%
	D1.7	28	50.00%
	D1.8	42	75.00%
	D1.9	38	67.86%
	R1.1	16	28.57%
	R1.2	15	26.79%
	R1.3	17	30.36%
	R1.4	17	30.36%
	R1.5	17	30.36%
	R1.6	17	30.36%
	R1.7	16	28.57%
	R1.8	13	23.21%
AQUARIUS	D1.1	17	30.36%
	D1.2	17	30.36%
	D1.3	16	28.57%
	D1.4	15	26.79%
	D1.5	22	39.29%
	D1.6	22	39.29%
	D1.7	20	35.71%
	D1.8	33	58.93%
	D1.9	55	98.21%
	D1.10	19	33.93%
	R1.1	17	30.36%
	R1.2	17	30.36%
	R1.3	16	28.57%
	R1.4	15	26.79%
	R1.5	24	42.86%
	R1.6	22	39.29%
	R1.7	19	33.93%
	R1.8	4	7.14%
PISCES	D1.1	15	26.79%
	D1.2	15	26.79%
	D1.3	14	25.00%
	D1.4	14	25.00%
	D1.5	19	33.93%
	D1.6	17	30.36%
	D1.7	17	30.36%
	D1.8	27	48.21%
	D1.9	46	82.14%
	D1.10	43	76.79%
	R1.1	15	26.79%
	R1.2	15	26.79%
	R1.3	14	25.00%
	R1.4	14	25.00%
	R1.5	19	33.93%
	R1.6	18	32.14%
	R1.7	8	14.29%

Table 11: M30 Sign + Speed Seasons, Yearly Occurrences 1975-2030 (56 Years)

	SPEED	YEARS	%		SPEED	YEARS	%		SPEED	YEARS	%		SPEED	YEARS	%
ARIES	D3.1	14	25.00%	TAURUS	D3.1	14	25.00%	GEMINI	D3.1	14	25.00%	CANCER	D3.1	13	23.21%
	D3.2	13	23.21%		D3.2	14	25.00%		D3.2	14	25.00%		D3.2	13	23.21%
	D3.3	18	32.14%		D3.3	17	30.36%		D3.3	17	30.36%		D3.3	15	26.79%
	D3.4	25	44.64%		D3.4	25	44.64%		D3.4	26	46.43%		D3.4	25	44.64%
	D3.5	56	100.00%		D3.5	51	91.07%		D3.5	48	85.71%		D3.5	50	89.29%
	D3.6	27	48.21%		D3.6	38	67.86%		D3.6	39	69.64%		D3.6	39	69.64%
	R3.1	14	25.00%		R3.1	14	25.00%		R3.1	14	25.00%		R3.1	13	23.21%
	R3.2	14	25.00%		R3.2	14	25.00%		R3.2	15	26.79%		R3.2	13	23.21%
	R3.3	18	32.14%		R3.3	17	30.36%		R3.3	15	26.79%		R3.3	15	26.79%
LEO	D3.1	13	23.21%	VIRGO	D3.1	16	28.57%	LIBRA	D3.1	17	30.36%	SCORPIO	D3.1	15	26.79%
	D3.2	14	25.00%		D3.2	17	30.36%		D3.2	17	30.36%		D3.2	16	28.57%
	D3.3	17	30.36%		D3.3	22	39.29%		D3.3	22	39.29%		D3.3	22	39.29%
	D3.4	25	44.64%		D3.4	30	53.57%		D3.4	32	57.14%		D3.4	36	64.29%
	D3.5	56	100.00%		D3.5	56	100.00%		D3.5	55	98.21%		D3.5	45	80.36%
	D3.6	27	48.21%		R3.1	16	28.57%		R3.1	17	30.36%		R3.1	15	26.79%
	R3.1	13	23.21%		R3.2	17	30.36%		R3.2	17	30.36%		R3.2	16	28.57%
	R3.2	15	26.79%		R3.3	22	39.29%		R3.3	23	41.07%		R3.3	22	39.29%
	R3.3	17	30.36%		R3.4	8	14.29%		R3.4	20	35.71%		R3.4	19	33.93%
SAGITTARIUS	D3.1	15	26.79%	CAPRICORN	D3.1	17	30.36%	AQUARIUS	D3.1	17	30.36%	PISCES	D3.1	15	26.79%
	D3.2	11	19.64%		D3.2	17	30.36%		D3.2	16	28.57%		D3.2	14	25.00%
	D3.3	18	32.14%		D3.3	21	37.50%		D3.3	22	39.29%		D3.3	19	33.93%
	D3.4	36	64.29%		D3.4	42	75.00%		D3.4	33	58.93%		D3.4	27	48.21%
	D3.5	38	67.86%		D3.5	38	67.86%		D3.5	55	98.21%		D3.5	56	100.00%
	R3.1	15	26.79%		R3.1	16	28.57%		R3.1	17	30.36%		R3.1	15	26.79%
	R3.2	14	25.00%		R3.2	17	30.36%		R3.2	16	28.57%		R3.2	14	25.00%
	R3.3	23	41.07%		R3.3	18	32.14%		R3.3	24	42.86%		R3.3	19	33.93%
	R3.4	21	37.50%		R3.4	16	28.57%		R3.4	19	33.93%		R3.4	8	14.29%

4. The Use of Irregular Seasonality in Quantitative Time Series Analysis

This study explores the application and potential value of the effect- and variance-based approach to seasonality in quantitative time series analysis. The **Calendar Month** seasonal model is used as a comparative baseline to explore the relative significance of other seasonal models, and specifically to explore the potential value of the seasonal influences that can only be seen through the lens of the irregular, Mercury-based seasonal models. Both the magnitude of the seasonal effects and the historical variance of the effect sizes will be considered.

Before addressing the methodology of the study itself, we must first explore the methodology that will be used to quantify the effects of seasonal influences to make it possible to evaluate and compare the potential value of different seasonal models.

4.1. Methodologies Used to Quantify the Effect of Seasonal Influences

This study uses **Cohen's d** parameter (Cohen, 1988) to quantify the magnitude of the effect for each season. **Cohen's d**, commonly used for t-tests, is a standardized measure that allows for comparison of effect sizes between data sets or studies. The calculation of **Cohen's d** parameter takes into account the difference between the means (the mean for a specific season and the reference mean of the calendar year) and the variance to produce a number that describes the magnitude of the effect. Table 12 defines the **Cohen's d** effect sizes and ranges used in this study.

The effect size is an objective metric, so this approach makes it possible to compare seasonal influences across multiple dimensions. It can be used to **compare the seasonal influences within a seasonal model** for a single data set. It can be used to **compare different models of seasonality** for a single data set to determine which seasonal models reveal the most significant seasonal influences for that data. And it can be used to **compare seasonal influences between unrelated data sets.**

Table 12: Cohen's d Effect Sizes

Effect (d) SIZE	d	REFERENCE	Study Range
Very Small (VS)	0.01	Sawilowsky, 2009	< 0.15
Small (S)	0.20	Cohen, 1988	0.15 – 0.35
Medium (M)	0.50	Cohen, 1988	0.35 – 0.65
Large (L)	0.80	Cohen, 1988	0.65 – 1.00
Very Large (VL)	1.20	Sawilowsky, 2009	1.00 – 1.85
Huge (H)	2.0	Sawilowsky, 2009	> 1.85

Because this method of evaluating seasonal effect size is relative to a reference mean, it is possible not only to evaluate the **magnitude of the effect**, but also the **direction of the effect**, i.e., higher or lower than the reference mean. Negative effect sizes, where the mean for the season is lower than the reference mean, are indicated with a minus sign (e.g., **-M** for a Medium effect in a negative direction), and positive effect sizes, where the mean for the season is higher than the reference mean, are indicated with a plus sign (e.g., **L+** for a Large effect in a positive direction).

The direction of the effect has the greatest practical value when analyzing the **On-Time Performance** (**OTP**) data, because on-time performance is commonly measured against a reference mean. A certain percentage of trips will depart late. The direction of the effect indicates whether the on-time performance was better or worse than the reference mean. The direction of the effect has less practical value when analyzing data sets without a clear expectation of a mean value, such as the stock market data. A negative effect size for a season does not necessarily represent a drop in the stock price.

For the purpose of this study, effect sizes of Medium or greater magnitude are considered to be significant.

4.1.1. Methodology to Quantify Individual Historical Seasonal Effect Sizes

The historic time series data is analyzed to generate effect sizes for each historic instance of each season. Daily averages are used to determine the mean and standard deviation for each individual season and for the calendar year. The initial analysis of each data set generates variance and effect size for each instance of each season by comparing it to the mean and standard deviation of the calendar year that contains the instance of the season.

To illustrate, let's consider flight on-time performance, which has a strong monthly seasonal component. Any flight that pushes back from the gate more than 15 minutes after the scheduled departure time is classified as late. When considering

30 years of flight departure data from the Department of Transportation, historically, the greatest percentage of late flights occurs in the month of December. And some of the best on-time performance occurs in September. The expectation is that the historic seasonal effect sizes should tell the same story.

Table 13 on page 32 lists the flight delays for the months of December and September, from 1988 through 2017. The **Actual Difference** column is the difference between the percent of late departures during the season and the percent of late departures for that year. The **Percent Difference** column is the percent difference between the monthly mean and the annual reference mean.

The percentage of late flights in December was greater than the annual percentage of late flights 28 out of 30 years. There are 8 Very Large effects, 9 Large effects, and 7 Medium effects, all in a positive direction, meaning a greater than average percentage of late departures.

The percentage of late flights in September was lower than the annual percentage of late flights every single year. There are 15 Large effects, 12 Medium Effects, and 3 Small effects, all in a negative direction.

4.1.2. Methodology to Aggregate Historical Seasonal Effect Sizes for Analysis

While the historical seasonal data is necessary to generate seasonal forecasts, it's impractical for analysis. To get an overall picture of the patterns of on-time performance over the past 30 years using the **Calendar Month** seasonal model would require that we consider 360 individual seasons, each with its own effect size. To simplify this, the individual seasonal results can be aggregated using weighted averages.

A weighted mean is calculated for each season's mean and standard deviation, and a weighted reference mean and standard deviation are also calculated. These derived weighted means and standard deviations are then used to determine the average effect size for each season.

Table 14 on page 33 shows the results for the **Calendar Month** seasonality for the flight on-time performance data set. In this table, the **Actual Difference** and **Percent Difference** refer to the weighted means. Significant results (Medium or greater) are highlighted.

We can clearly see that December has both the largest effect size for flight delays and the greatest percent difference between the seasonal means and the reference means at 29.31%. The next worst month to travel is June. It has only a Small effect, but there is a 14.81% greater than average rate of flight delays overall. Both September and October show significant effect sizes with a negative Medium effect, but September shows a 26.68% lower rate of delays, while October shows only a 17.56% lower rate of delays.

Table 13: Annual Seasonal Flight Delays for December and September, 1988-2017

SEASON	ACTUAL DIFFERENCE	PERCENT DIFFERENCE	EFFECT SIZE	SEASON	ACTUAL DIFFERENCE	PERCENT DIFFERENCE	EFFECT SIZE
1988 December	3.12%	23.62%	M+	1988 September	-4.88%	-36.94%	–L
1989 December	6.26%	38.43%	L+	1989 September	-3.09%	-18.97%	–M
1990 December	8.55%	63.66%	L+	1990 September	-4.18%	-31.12%	–M
1991 December	3.14%	27.19%	M+	1991 September	-4.57%	-39.57%	–L
1992 December	6.92%	61.79%	VL+	1992 September	-2.21%	-19.73%	–M
1993 December	3.60%	30.00%	M+	1993 September	-2.68%	–22.33	–M
1994 December	3.44%	26.52%	M+	1994 September	-5.20%	-40.09%	–L
1995 December	10.46%	66.16%	VL+	1995 September	-5.89%	-37.25%	–L:
1996 December	8.00%	42.90%	VL+	1996 September	-4.18%	-22.41%	–M
1997 December	3.96%	24.84%	M+	1997 September	-5.91%	-37.08%	–L
1998 December	4.79%	28.84%	L+	1998 September	-4.56%	-27.45%	–M
1999 December	-0.46%	-2.64%	–S	1999 September	-3.57%	-20.47%	–M
2000 December	8.69%	42.20%	VL+	2000 September	-4.50%	-21.86%	–M
2001 December	0.24%	1.44%	S+	2001 September	-2.56%	-15.39%	–S
2002 December	3.72%	27.00%	M+	2002 September	-4.55%	-33.02%	–L
2003 December	4.90%	37.52%	L+	2003 September	-3.28%	-25.11%	–M
2004 December	6.42%	37.85%	L+	2004 September	-5.52%	–32.55%	–L
2005 December	6.06%	33.19%	L+	2005 September	-4.91%	-26.89%	–L
2006 December	4.16%	20.49%	M+	2006 September	-2.10%	-10.34%	–S
2007 December	6.94%	32.19%	L+	2007 September	-6.83%	-31.68%	–L
2008 December	8.77%	45.39%	VL+	2008 September	-7.78%	-40.27%	–L
2009 December	6.71%	39.33%	VL+	2009 September	-5.60%	-32.83%	–L
2010 December	6.53%	37.19%	VL+	2010 September	-4.83%	-27.51%	–L
2011 December	-3.02%	-17.28%	–M	2011 September	-3.62%	-20.71%	–M
2012 December	5.18%	31.41%	L+	2012 September	-1.66%	-10.07%	–S
2013 December	8.57%	43.70%	VL+	2013 September	-4.69%	-23.92%	–M
2014 December	1.50%	7.15%	S+	2014 September	-5.03%	-23.98%	–L
2015 December	2.36%	12.79%	S+	2015 September	-5.82%	-31.54%	–L
2016 December	5.22%	30.40%	L+	2016 September	-3.81%	-22.19%	–M
2017 December	1.41%	7.75%	S+	2017 September	-4.92%	-27.03%	–L

Table 14: Aggregate Calendar Month Seasonality for Flight On-Time Performance, 1998 - 2017

SEASON	ACTUAL DIFFERENCE	PERCENT DIFFERENCE	EFFECT SIZE
January	1.95%	11.73%	S+
February	0.82%	4.93%	S+
March	0.62%	3.72%	S+
April	-1.94%	-11.67%	–S
May	-1.79%	-10.75%	–S
June	2.46%	14.81%	S+
July	1.92%	11.53%	S+
August	0.60%	3.64%	S+
September	-4.43%	-26.68%	–M
October	-2.92%	-17.56%	–M
November	-2.36%	-14.17%	–S
December	4.87%	29.31%	L+

4.1.3. Methodology to Quantify Variance as a Measure of Expected Seasonal Forecast Value

These average effect sizes are quite useful for historical time series analysis, and they make it easy to identify significant patterns in the flight data. But to identify which seasons are the most likely to have value in forecasting, we need to consider the pattern behind the average effect sizes. The more consistently a pattern repeats in the historical data, the more likely that pattern will continue in the future. We need to refer back to the original data and evaluate the variance within each season's effect sizes. If the effect sizes of the historical iterations of a season are consistent—if the variance is low—then it's reasonable to expect that the pattern will continue and the season could improve the accuracy of forecasts. If the effect sizes of the historical seasons are inconsistent—if the variance is high—then there is no reason to expect that the pattern will continue or that the season will improve the forecast accuracy.

We know that over a 30 year period, the month of December represents the season with the greatest percentage of late flight departures, with a Large positive effect overall (Table 14). But the percent difference between the historical means (shown in

Table 13) varies from -17.28% to +66.16%, a total range of 83.44%. The effect sizes range from Medium negative to Very Large positive. Over 30 years, they average out to a Large positive effect, but the variance is substantial. The actual percentage of late flights varies too much between historical instances to have confidence in the forecast value of this seasonal influence.

The month of September has a Medium negative effect overall and represents a season with better than average on-time performance (Table 14). When we consider the historical effect sizes in Table 13, the variance is extremely low. Every effect is in a negative direction, so the delays during this season are consistently lower than the mean. The percent difference between the means varies from –10.07% to –40.27%, a total range of only 30.2% The percentage of late flights during the month of September is quite consistent across historical seasonal instances, so incorporating this data into a forecast will probably improve the forecast accuracy for the month of September.

The metric used to evaluate the variance within instances of a season is the percent difference between the seasonal mean and the reference mean (the calendar year). This metric was selected in an effort to normalize the results across unrelated data sets. The percentages were converted into whole numbers by multiplying them by 100, such that 4.06% would become 4.06, rather than 0.0406, and the variance of the historical differences was calculated for each season.

The hope was that using the percent difference as the metric would produce some kind of objective scale of variance that could be used to determine significance, but that was not the case. Therefore, an arbitrary method to determine the relative threshold of significance was needed.

The mean of all of the historical variances of all of the seasons was calculated for each seasonal model for each data set. **For the purposes of this study, any season with a variance of less than half of the mean variance of the entire data set is considered to be significant, and expected to have value when forecasting for that data set.**

4.1.4. Applications and Limitations of Seasonal Effect Size and Variance

The effect sizes are standardized, which allows for an objective comparison of the effect of a single variable across different data sets. For example, the effect sizes allow for a direct comparison of the aggregate seasonal influence of the month of January on flight delays (**S+**), car crashes in the state of Texas (**-M**), and the Dow Jones Industrial Average (**-M**).

When considering aggregated seasonal influences, the effect sizes operate in multiple dimensions because each data set includes related results with individual effect sizes. Within a data set, the effect size can be combined with the percent difference between the weighted mean for the season and the weighted reference mean to get a more accurate and specific evaluation of the relative size of each seasonal effect. For example, when considering the **Calendar Month** seasonal effects for on-time performance of flight departures, **January**, **February**, and **March** each have the same effect size, **S+**. However, the **January S+** represents a 11.73% difference from the reference mean, the **February S+** represents a 4.93% difference from the reference mean, and the **March S+** represents a 3.72% difference from the reference mean. The actual effect size for **January** is far greater than the effect in either **February** or **March**.

Because the weighted reference mean is constant within a single data set and is not tied to any one seasonal model, both the aggregate effect size and the percent difference between the weighted means can be used to compare the relative significance of different seasonal models within a single data set. This is what makes it possible to compare the effect sizes and significance of the irregular, Mercury-based seasonal models to the effect size and significance of the **Calendar Month** seasonal model.

Although the effect sizes are objective and standardized such that they can be used for seasonal comparisons across different data sets, the percent difference metric is tied to the specific data set; it is not objective, and cannot be used to compare results between two unrelated data sets. The **-M** result for Texas car crashes in **January** represents a -9.59% difference from the mean, while the **-M** result for the Dow Jones in **January** represents a -3.05% difference from the mean, but because these differences refer to unrelated means, the numbers cannot be compared.

The direction of the effect provides useful information for historical analysis, but it does not translate to forecasting. A negative effect size for a season does not mean that the values for the forecasted season will go down; it does not even mean that the values for the forecasted season will be lower than the values of the preceding season. It merely suggests that the values of the forecasted season will be lower than the mean values for the year. This expectation does not have a practical application in forecasting.

When the annual reference means are known, the combination of the percent difference between the weighted means and the effect size and direction can provide a fascinating level of detail and reveal small, yet significant patterns in the time series. Analyzing the historical seasonal effects, significance, and variance can help to determine if a particular seasonal model is likely to improve forecast accuracy.

4.2. Study Objectives and Methodology

By its very definition, seasonality is a quality of time. The question is whether or not that definition has practical value that merits further exploration when applied to quantitative time series analysis and forecasting.

Because seasonal models are not limited to the calendar or the clock, there is technically no limit to the number of different seasonal models. Each seasonal model is like a lens that filters the underlying time series data and reveals hidden patterns and influences. Not every seasonal model will reveal significant patterns, and not every hidden pattern will be useful. For this reason, it's necessary to have some kind of methodology that can evaluate the anticipated value of a particular seasonal model for a given data set.

There has never been a need to quantify or compare seasonal influences before, so we need a baseline reference to adjust expectations of significance. In this study, **Calendar Month** seasonality will be used as the baseline. It's important to note that the **Calendar Month** model is not precisely the same as the approach to monthly seasonality currently used in time series forecasting. Monthly seasonal signals in traditional forecasting (via time series decomposition) do not account for the irregular duration of the calendar months and instead look for patterns in the data that correspond to equal divisions of 12. The quantified effects of the **Calendar Month** seasons in this study adhere to the actual calendar dates.

The percentage of significant results, both in terms of effect size and variance for the **Calendar Month** seasonal model will be used as a benchmark to compare the seasonal effects of two models of regular seasonality (**Lunar Month**, **Moon Sign**), and six models of irregular seasonality (**M30 Speed**, **M15 Speed**, **M30 Month + Speed**, **M15 Month + Speed**, **M30 Sign + Speed**, **M15 Sign + Speed**). The seasonal models are defined in **Section 3**.

The comparison is straightforward. The percentage of significant seasons for the **Calendar Month**, both by effect size and by variance, is taken as the absolute minimum threshold of interest. The greater the difference in percentage of significant seasons when compared to the **Calendar Month** seasonality, the more potential value that seasonal model has for that data set.

To support the hypothesis that seasonality is a quality of time and not of data, this study is repeated three times, with three distinct data sets. Two data sets have some expectation of the presence of a **Calendar Month** seasonal influence (**Transportation On-Time Performance** and **Car Crash**), and the third data set has no expectation of a **Calendar Month** seasonal influence (sample **Financial Market** data).

4.3. On-Time Performance Seasonality

The **Transportation On-Time Performance** data sets include records of flight delays from U.S. airports, Amtrak train delays, city bus delays, and light rail delays. Particularly with regards to the flight delays, strong evidence of a monthly seasonal influence exists—we've already established that December is the worst month to fly, with the greatest percentage of late departures, and September is the best month to fly, with the smallest percentage of late departures. The data sets and confidence in the data integrity can be found in **Section 2**.

4.3.1. Analysis of Regular Seasonal Influences

Table 15 on page 38 shows the seasonality by **Calendar Month** for each of the transportation data sets, Table 16 on page 39 shows the seasonality by **Lunar Month**, and Table 17 on page 40 shows the seasonality by **Moon Sign**. The data sets are ordered based on the total number of records in each data set, from the most data (**Flights**, 172,960,056 departures) to the least (**Chicago Rail**, 191,075 departures).

Each table is divided into two sections. The upper section evaluates the **aggregate effect size with direction** (indicating whether the result is higher or lower than the weighed reference mean for the season), and **the percent difference between the weighted mean for the season and the weighted reference mean for the season**. This number is a consistent metric within each data set and allows for a more granular comparison of the actual effect size between different models of seasonality (e.g., between **Calendar Month** and the **Lunar Month**). Significant effects (Medium or greater) are highlighted, and the number and percentage of seasons with significant effects is presented at the bottom of each column.

The bottom section shows the **aggregate effect size and the variance score** for the historical effects of that season. The lower the historical variance of the effect sizes of a season, the greater the expectation that the season can be used to improve the accuracy of forecasts. The **VAR** row shows the minimum significant threshold number for each data set. **A season is considered to be significant if its variance is less than 50% of the mean variance of all of the seasons for that data set and seasonal model.** Significant seasons are highlighted, and the number and percentage of seasons with significant variance is presented at the bottom of each column.

Table 15: On-Time Performance Seasonality: Calendar Month

		FLIGHTS 172,960,056	CHI (BUS) 418,962,950	SF (BUS) 100,252,351	PHIL (BUS) 68,466,941	DART (BUS) 49,787,039	DART (RAIL) 45,671,313	AMTRAK 10,668,374	KING (BUS) 7,176,709	CHI (RAIL) 191,075
CALENDAR MONTH — EFFECT & MEAN % DIF	JAN	S+ 11.73%	–M -20.83%	–M -9.16%	–L -15.74%	–M -12.15%	–S -23.74%	–S -3.40%	S+ 1.86%	S+ 0.53%
	FEB	S+ 4.93%	–S -1.73%	–S -2.45%	–M -10.85%	S+ 1.10%	–S -19.77%	VS+ 0.14%	–S -0.41%	VS+ 0.03%
	MAR	S+ 3.72%	–S -10.83%	–S -5.35%	–M -10.82%	S+ 0.28%	–S -10.80%	–S -1.35%	–S -1.34%	–S -3.78%
	APR	-S -11.67%	–M -11.63%	–S -2.45%	S+ 1.12%	–S -1.94%	–S -12.21%	–S -3.09%	–M -6.95%	–S -5.63%
	MAY	-S -10.75%	S+ 9.21%	–VS -0.16%	S+ 6.74%	S+ 1.42%	–S -11.23%	S+ 2.56%	VS+ 0.02%	–S -0.95%
	JUN	S+ 14.81%	M+ 17.69%	–S -0.36%	S+ 3.24%	–S -7.88%	–S -2.76%	L+ 16.66%	M+ 5.87%	S+ 1.16%
	JUL	S+ 11.53%	S+ 6.33%	–S -3.33%	S+ 0.56%	–M -14.85%	–S -2.42%	L+ 18.20%	S+ 3.50%	S+ 1.75%
	AUG	S+ 3.64%	S+ 1.67%	S+ 5.87%	S+ 5.41%	–S -1.00%	–S -5.00%	M+ 11.99%	S+ 0.75%	S+ 5.08%
	SEP	-M -26.68%	M+ 18.66%	M+ 10.59%	L+ 16.41%	M+ 12.74%	–S -22.75%	–M -9.50%	S+ 4.16%	–S -2.63%
	OCT	-M -17.56%	S+ 10.03%	M+ 8.26%	M+ 10.55%	L+ 17.77%	M+ 93.76%	–M -7.69%	S+ 0.64%	–S -4.10%
	NOV	-S -14.17%	–S -4.37%	–S -2.46%	M+ 8.13%	S+ 5.83%	–S -26.86%	–M -8.52%	–S -0.64%	S+ 1.70%
	DEC	L+ 29.31%	–S -11.00%	S+ 1.18%	–S -6.77%	S+ 1.59%	S+ 27.83%	–S -5.16%	–M -6.40%	S+ 5.36%
	SIG	3 25.00%	4 33.33%	3 25.00%	6 50.00%	4 33.33%	1 8.33%	6 50.00%	3 25.00%	0 0.00%
CALENDAR MONTH — EFFECT & VARIANCE	JAN	S+ 463.47	–M 224.12	–M 62.34	–L 34.66	–M 36.18	–S 421.29	–S 271.80	S+ 217.86	S+ 66.69
	FEB	S+ 315.92	–S 328.99	–S 67.17	–M 89.50	S+ 29.94	–S 353.37	VS+ 299.70	–S 173.12	VS+ 166.89
	MAR	S+ 158.22	–S 56.25	–S 64.14	–M 61.94	S+ 48.90	–S 283.05	–S 82.33	–S 159.65	–S 66.84
	APR	-S 131.67	–M 43.57	–S 15.70	S+ 94.78	–S 12.63	–S 692.10	–S 110.50	–M 63.72	–S 52.52
	MAY	-S 152.54	S+ 32.58	–VS 10.25	S+ 13.50	S+ 37.30	–S 1240.28	S+ 124.24	VS+ 27.64	–S 113.95
	JUN	S+ 202.36	M+ 58.49	–S 16.86	S+ 24.85	–S 8.15	–S 657.37	L+ 29.27	M+ 54.12	S+ 21.24
	JUL	S+ 211.34	S+ 42.07	–S 1.89	S+ 28.60	–M 18.42	–S 887.24	L+ 58.12	S+ 41.41	S+ 43.80
	AUG	S+ 80.91	S+ 56.97	S+ 27.03	S+ 12.79	–S 87.29	–S 2596.97	M+ 26.82	S+ 22.94	S+ 92.72
	SEP	-M 68.48	M+ 17.26	M+ 27.23	L+ 61.49	M+ 25.42	–S 240.61	–M 42.23	S+ 29.41	–S 244.90
	OCT	-M 96.08	S+ 22.00	M+ 67.63	M+ 3.62	L+ 6.75	M+ 4174.84	–M 57.40	S+ 25.24	–S 140.65
	NOV	-S 188.20	–S 44.12	–S 64.72	M+ 278.78	S+ 57.17	–S 340.94	–M 113.54	–S 54.49	S+ 155.31
	DEC	L+ 349.95	–S 56.86	S+ 77.07	–S 18.20	S+ 182.01	S+ 6012.89	–S 226.14	–M 116.01	S+ 93.86
	VAR	100.80	40.97	20.92	30.11	22.92	745.87	60.09	41.07	52.47
	SIG	3 25.00%	3 25.00%	4 33.33%	6 50.00%	4 33.33%	7 58.33%	5 41.67%	4 33.33%	2 16.67%

Table 16: On-Time Performance Seasonality: Lunar Month

		FLIGHTS 172,960,056	CHI (BUS) 418,962,950	SF (BUS) 100,252,351	PHIL (BUS) 68,466,941	DART (BUS) 49,787,039	DART (RAIL) 45,671,313	AMTRAK 10,668,374	KING (BUS) 7,176,709	CHI (RAIL) 191,075
LUNAR MONTH — EFFECT & MEAN % DIF	ARIES	-S -12.18%	-S -10.41%	-S -3.84%	S+ 1.53%	S+ 1.37%	-S -5.26%	-S -2.59%	-M -5.98%	-S -5.81%
	TAURUS	-S -8.42%	S+ 10.60%	-S -0.57%	S+ 5.37%	-S -0.61%	-S -9.66%	S+ 3.88%	-VS -0.12%	S+ 0.43%
	GEMINI	S+ 12.32%	M+ 14.97%	-S -1.70%	S+ 0.64%	-M -10.17%	-S -9.80%	L+ 15.05%	S+ 4.74%	VS+ 0.22%
	CANCER	S+ 13.53%	S+ 6.57%	-VS -0.12%	S+ 1.83%	-M -13.12%	S+ 3.32%	L+ 17.96%	S+ 3.57%	S+ 1.24%
	LEO	-S -2.79%	S+ 5.03%	S+ 6.09%	S+ 5.64%	S+ 0.77%	-S -12.58%	S+ 7.34%	S+ 1.34%	S+ 3.17%
	VIRGO	-M -23.38%	M+ 15.79%	M+ 10.15%	L+ 16.68%	M+ 15.60%	S+ 18.37%	-M -8.03%	S+ 3.60%	-S -2.33%
	LIBRA	-M -19.44%	S+ 8.39%	S+ 7.47%	M+ 10.12%	L+ 18.27%	M+ 62.79%	-M -8.17%	S+ 0.41%	-S -1.21%
	SCO.	-S -11.49%	-S -4.68%	-S -1.93%	M+ 7.69%	S+ 5.99%	-S -16.94%	-M -8.11%	-VS -0.12%	S+ 0.35%
	SAG.	L+ 37.59%	-M -13.17%	-VS -0.17%	-M -8.41%	-S -4.71%	S+ 19.07%	-S -2.66%	-M -6.58%	S+ 4.68%
	CAP.	S+ 8.81%	-M -18.78%	-S -7.20%	-L -15.82%	-M -12.63%	-S -17.98%	-S -3.60%	S+ 0.62%	S+ 1.62%
	AQU.	S+ 2.73%	-S -3.71%	-S -4.12%	-M -10.52%	S+ 2.93%	-S -21.29%	S+ 0.72%	-VS -0.10%	VS+ 0.12%
	PISCES	S+ 2.47%	-S -10.84%	-S -3.43%	-M -9.62%	-S -2.65%	-S -17.97%	-S -2.01%	-S -2.10%	-S -4.03%
	SIG	3 25.00%	4 33.33%	1 8.33%	7 58.33%	5 41.67%	1 8.33%	5 41.67%	2 16.67%	0 0.00%
LUNAR MONTH — EFFECT & VARIANCE	ARIES	-S 177.17	-S 76.67	-S 22.87	S+ 101.23	S+ 22.22	-S 756.34	-S 186.24	-M 67.51	-S 70.67
	TAURUS	-S 215.85	S+ 58.47	-S 8.85	S+ 1878.04	-S 21.43	-S 1554.89	S+ 80.15	-VS 47.93	S+ 170.57
	GEMINI	S+ 230.32	M+ 77.92	-S 38.80	S+ 23.98	-M 27.35	-S 956.48	L+ 47.52	S+ 44.03	VS+ 14.86
	CANCER	S+ 149.47	S+ 54.91	-VS 35.32	S+ 21.51	-M 36.12	S+ 555.78	L+ 48.08	S+ 45.58	S+ 59.55
	LEO	-S 190.03	S+ 83.67	S+ 34.73	S+ 5.18	S+ 67.54	-S 2160.98	S+ 82.01	S+ 23.28	S+ 138.05
	VIRGO	-M 103.93	M+ 28.76	M+ 21.53	L+ 71.52	M+ 17.04	S+ 2265.42	-M 52.00	S+ 17.01	-S 231.86
	LIBRA	-M 91.87	S+ 56.94	S+ 97.73	M+ 3.53	L+ 9.63	M+ 6832.49	-M 53.86	S+ 19.39	-S 57.53
	SCO.	-S 192.53	-S 61.74	-S 79.89	M+ 129.88	S+ 41.70	-S 630.18	-M 84.22	-VS 55.42	S+ 121.50
	SAG.	L+ 957.56	-M 41.99	-VS 146.75	-M 29.54	-S 417.98	S+ 12576	-S 293.73	-M 165.72	S+ 122.69
	CAP.	S+ 363.35	-M 122.02	-S 109.29	-L 41.79	-M 60.61	-S 442.64	-S 190.86	S+ 128.02	S+ 83.61
	AQU.	S+ 213.84	-S 245.91	-S 104.78	-M 129.78	S+ 49.52	-S 229.24	S+ 211.44	-VS 208.99	VS+ 86.60
	PISCES	S+ 219.29	-S 41.80	-S 67.33	-M 137.38	-S 21.16	-S 286.24	-S 80.47	-S 158.24	-S 67.35
	VAR	129.38	39.62	31.99	107.22	33.01	1218.62	58.77	40.88	51.03
	SIG	2 16.67%	1 8.33%	3 25.00%	8 66.67%	6 50.00%	7 58.33%	4 33.33%	3 25.00%	1 8.33%

Table 17: On-Time Performance Seasonality: Moon Sign

		FLIGHTS 172,960,056	CHI (BUS) 418,962,950	SF (BUS) 100,252,351	PHIL (BUS) 68,466,941	DART (BUS) 49,787,039	DART (RAIL) 45,671,313	AMTRAK 10,668,374	KING (BUS) 7,176,709	CHI (RAIL) 191,075
MOON SIGN — EFFECT & MEAN % DIF	ARIES	-S -0.77%	-S -1.59%	-S -1.81%	-S -0.66%	-VS -0.21%	-S -10.60%	-S -0.29%	-VS -0.03%	S+ 0.35%
	TAURUS	-S -0.68%	S+ 2.36%	S+ 0.97%	S+ 0.53%	S+ 0.36%	-S -10.08%	S+ 0.80%	S+ 1.33%	S+ 3.95%
	GEMINI	S+ 1.39%	-S -1.40%	-S -0.65%	-S -1.68%	-S -0.37%	S+ 5.17%	-S -0.51%	-S -1.35%	-S -2.95%
	CANCER	VS+ 0.20%	S+ 1.50%	S+ 1.01%	S+ 1.02%	S+ 0.91%	-S -7.72%	-S -2.65%	VS+ 0.10%	S+ 0.52%
	LEO	S+ 1.83%	-VS -0.02%	S+ 1.55%	S+ 0.42%	-VS -0.14%	S+ 9.06%	-S -0.26%	S+ 0.82%	S+ 0.86%
	VIRGO	S+ 1.29%	S+ 2.49%	S+ 2.29%	S+ 0.43%	S+ 0.80%	S+ 20.61%	S+ 0.96%	-S -0.24%	-S -0.95%
	LIBRA	-S -0.89%	-S -1.68%	-VS -0.18%	-S -1.01%	-S -1.38%	-S -8.29%	S+ 0.92%	-S -0.79%	-VS -0.22%
	SCO.	-S -1.07%	-S -3.28%	-S -0.94%	-S -1.66%	-S -1.52%	-S -18.90%	S+ 1.54%	-S -0.96%	-S -0.72%
	SAG.	-S -2.34%	S+ 1.17%	S+ 1.46%	S+ 0.47%	S+ 0.83%	-S -12.87%	S+ 0.53%	S+ 0.36%	-S -2.73%
	CAP.	-S -0.93%	S+ 0.54%	-S -1.35%	S+ 0.32%	S+ 0.64%	S+ 18.84%	-S -0.35%	-S -1.21%	-S -0.32%
	AQU.	S+ 1.56%	-S -1.23%	-S -2.23%	S+ 1.60%	-S -0.41%	S+ 7.05%	-S -0.31%	S+ 0.18%	S+ 0.58%
	PISCES	S+ 0.65%	S+ 1.01%	-S -1.23%	S+ 0.77%	-S -0.54%	S+ 10.90%	-S -0.26%	S+ 1.44%	S+ 1.68%
	SIG	0 0.00%	0 0.00%	0 0.00%	0 0.00%	0 0.00%	0 0.00%	0 0.00%	0 0.00%	0 0.00%
MOON SIGN — EFFECT & VARIANCE	ARIES	-S 74.19	-S 24.61	-S 47.30	-S 44.33	-VS 35.94	-S 392.54	-S 17.72	-VS 71.22	S+ 20.94
	TAURUS	-S 56.31	S+ 50.72	S+ 5.89	S+ 17.80	S+ 8.83	-S 253.13	S+ 10.83	S+ 3.83	S+ 20.10
	GEMINI	S+ 79.83	-S 14.48	-S 64.05	-S 25.20	-S 8.44	S+ 1174.52	-S 6.61	-S 54.00	-S 27.24
	CANCER	VS+ 72.07	S+ 12.28	S+ 55.69	S+ 22.11	S+ 21.94	-S 268.87	-S 23.34	VS+ 31.66	S+ 18.28
	LEO	S+ 49.34	VS+ 60.05	S+ 30.08	S+ 40.12	-VS 25.04	S+ 1034.94	-S 9.47	S+ 37.04	S+ 43.89
	VIRGO	S+ 69.48	S+ 13.37	S+ 29.25	S+ 24.65	S+ 30.49	S+ 1346.68	S+ 15.53	-S 40.40	-S 20.27
	LIBRA	-S 34.77	-S 18.69	-VS 23.45	-S 55.89	-S 14.32	-S 548.99	S+ 26.41	-S 12.92	-VS 26.37
	SCO.	-S 85.90	-S 8.81	-S 33.95	-S 18.43	-S 17.04	-S 584.76	S+ 14.68	-S 4.42	-S 31.96
	SAG.	-S 73.11	S+ 33.17	S+ 38.04	S+ 42.26	S+ 15.13	-S 369.00	S+ 23.11	S+ 12.92	-S 4.28
	CAP.	-S 100.46	S+ 23.42	-S 102.72	S+ 83.65	S+ 40.01	S+ 494.26	-S 6.38	-S 48.76	-S 62.03
	AQU.	S+ 52.42	-S 11.16	-S 10.76	S+ 69.33	-S 27.11	S+ 4113.92	-S 6.03	S+ 21.48	S+ 14.02
	PISCES	S+ 59.23	S+ 47.52	-S 56.34	S+ 15.25	-S 18.54	S+ 1044.35	-S 5.39	S+ 42.91	S+ 22.07
	VAR	33.63	13.26	20.73	19.13	10.95	484.42	6.90	15.90	12.98
	SIG	0 0.00%	3 25.00%	2 16.67%	3 25.00%	2 16.67%	4 33.33%	4 33.33%	4 33.33%	1 8.33%

We'll begin by considering the **Flights** data set (column 1 of each table). An expectation of a **Calendar Month** seasonal influence exists for the on-time performance of flights, and the popular belief is that the greatest percentage of flight delays occurs during the winter months. Only three months have significant effect sizes: **September**, **October**, and **December**, and of those, **December** is the only month where there is a greater than average rate of late departures.

We see similar results when considering the **Lunar Month** for the **Flights** data set. As with the **Calendar Month**, there are three **Lunar Month** seasons with significant results: **-M** for **Virgo** and **Libra**, and **L+** for **Sagittarius**. These seasons roughly correspond with the months of **September**, **October**, and **December**, which tracks with the **Calendar Month** seasonality; however, the **Sagittarius** season could mark a 28-day period that begins anywhere from November 23rd to December 21st. This means it could extend well into the calendar month of **January**.

However, the effect sizes don't tell the complete story. During the **Calendar Month** of **December**, the **L+** effect size represents 29.31% increase in delays over the reference mean, but the **L+** effect for the **Lunar Month** of **Sagittarius** represents a 37.59% increase in delays over the same reference mean. This suggests that historical flight delays correlate more with the **Lunar Month** than with the **Calendar Month**. While the effect sizes are comparable when the lunar months are aligned with the calendar months, the percent difference with the **Lunar Month** model tells a more significant story than the percent difference with the **Calendar Month** model.

There are **no significant effect size results** when considering the **Moon Sign** for the **Flights** data set—or indeed for any of the on-time performance data sets.

The variance numbers tell quite a different story. While there are also three **Calendar Month** seasons with significant variance scores for the flight data, those months are **August**, **September**, and **October**. As significant as the effect size is for the **December** season, the variance of the historical effects is far too high to expect the historical delays in **December** to be of any use in forecasting future **December** flight delays.

While the **Lunar Month** effect sizes seem to provide more insight into historical patterns of flight delays than the **Calendar Month** model does, only two **Lunar Month** seasons, **Virgo** and **Libra**, have a low enough variance to improve forecast accuracy. No **Moon Sign** seasons had significance variance for the Flights data set.

Next, let's consider the Philadelphia city bus data set (column 4 in each table). Philadelphia city buses seem to experience significantly higher delays in **September** (**L+**), **October** (**M+**), and **November** (**M+**). While it might be easy to blame this on weather conditions, **December** shows a lower than average rate of delays (**-S**), and the other three significant results are **January** (**-L**), **February** (**-M**), and **March** (**-M**), so weather does not seem to be a deciding factor. The months with the lowest variance and the most significance for forecasting are **May, June, July, August, October**, and **December**. The only overlap is **July** and **October**, both of which have both significant effect sizes and low variance.

When considering the **Lunar Month** for the Philadelphia city bus data set, there are seven significant effect results, with the greatest number of delays occurring during

Virgo (**L+**), which could represent a 28-day period that begins between August 23rd and September 23rd. There are eight significant variance results, and once again, a minimum of overlap between the significant effect sizes and the significant variance. Only **Virgo**, **Libra**, **Sagittarius**, and **Capricorn** qualify as significant in both categories.

While the **Philadelphia** city bus data shows impressive seasonality with the **Calendar Month** model, the **Lunar Month** seasonal model has a truly exceptional expectation of improving forecast accuracy with eight out of twelve seasons (66.67%) registering significantly low historic variance. Even the **Moon Sign** seasonal model has three significant variance seasons for Philadelphia.

The results of the **DART (Rail)** seasonal models (column 6 in each table) must be considered (and questioned) because of the unusually high variance of these historic results compared to the variance of the other data sets. When considering the mean (50%) variance threshold for the **Calendar Month**, the **VAR** row at the bottom of each table) the other data sets range from 20.92 to 100, but the mean threshold for **DART (Rail)** is 745.87. With the **Lunar Month** seasonality, the highest variance threshold (**Flights**) is 129.38, while the **DART (Rail)** is 1218.62. And even in the **Moon Sign** seasonality, the **Flights** variance is 33.63, and the **DART (Rail)** is 484.42.

While there is no objective standard to compare variance between data sets, this extreme discrepancy raises questions. It's possible that missing, inconsistent, or inaccurate values in the daily data could cause these unusually high variance scores.

4.3.2. Analysis of Irregular Seasonal Influences

Table 18 through Table 26 compare the three regular seasonality models (**Calendar Month**, **Lunar Month**, **Moon Sign**) with six Mercury-based irregular seasonality models (**M30 Speed**, **M15 Speed**, **M30 Month + Speed**, **M15 Month + Speed**, **M30 Sign + Speed**, **M15 Sign + Speed**) for each individual data set.

Each table is divided into two sections. The top section is a straight count of the aggregate effect sizes. Results with a effect size of Medium or greater are considered to be significant (and are highlighted), and the number and percentage of seasons with significant effect sizes is highlighted at the bottom of the section. The bottom section (the last two rows of each table) evaluates the variance. The **VAR** row shows the minimum significant threshold number for each data set, and the **SIG** row below it shows the number and percentage of seasons with a significant variance.

Table 27 on page 47 provides a summary that compares the significance of the **Calendar Month** seasonal model with the significance of the **M30 Sign + Speed** and the **M15 Sign + Speed** seasonal models in terms of effect size, across the entire data set. Table 28 on page 48 provides a summary that compares the significance of the **Calendar Month** seasonal model with the significance of the **M30 Sign + Speed** and the **M15 Sign + Speed** seasonal models in terms of variance, across the entire data set.

Table 18: Flight OTP Seasonality Results (172,960,056 Records)

		CALENDAR MONTH	LUNAR MONTH	MOON SIGN	M30 SPEED	M15 SPEED	M30 MONTH + SPEED	M15 MONTH + SPEED	M30 SIGN + SPEED	M15 SIGN + SPEED
EFFECT SIZE + MEAN % DIF	#	12	12	12	10	19	109	203	108	202
	VS	0 0.00%	0 0.00%	1 8.33%	2 20.00%	3 15.79%	6 5.50%	8 3.94%	0 0.00%	3 1.49%
	S	9 75.00%	9 75.00%	11 91.67%	8 80.00%	16 84.21%	96 88.07%	168 82.76%	72 66.67%	122 60.40%
	M	2 16.67%	2 16.67%	0 0.00%	0 0.00%	0 0.00%	7 6.42%	25 12.32%	24 22.22%	50 24.75%
	L	1 8.33%	1 8.33%	0 0.00%	0 0.00%	0 0.00%	0 0.00%	2 0.99%	9 8.33%	21 10.40%
	VL	0 0.00%	0 0.00%	0 0.00%	0 0.00%	0 0.00%	0 0.00%	0 0.00%	3 2.78%	6 2.97%
	H	0 0.00%	0 0.00%	0 0.00%	0 0.00%	0 0.00%	0 0.00%	0 0.00%	0 0.00%	0 0.00%
	SIG	3 25.00%	3 25.00%	0 0.00%	0 0.00%	0 0.00%	7 6.42%	27 13.30%	36 33.33%	77 38.12%
VAR	VAR	100.80	129.38	33.63	126.43	182.12	294.56	422.09	425.80	559.08
	SIG	3 25.00%	2 16.67%	0 0.00%	3 30.00%	7 36.84%	31 28.44%	59 29.06%	40 37.04%	72 35.64%

Table 19: Chicago Bus OTP Seasonality Results (418,962,950 Records)

		CALENDAR MONTH	LUNAR MONTH	MOON SIGN	M30 SPEED	M15 SPEED	M30 MONTH + SPEED	M15 MONTH + SPEED	M30 SIGN + SPEED	M15 SIGN + SPEED
EFFECT SIZE	#	12	12	12	10	19	109	203	108	202
	VS	0 0.00%	0 0.00%	1 8.33%	1 10.00%	3 15.79%	7 6.42%	1 0.49%	3 2.78%	4 1.98%
	S	8 66.67%	8 66.67%	11 91.67%	9 90.00%	15 78.95%	77 70.64%	139 68.47%	48 44.44%	94 46.53%
	M	4 33.33%	4 33.33%	0 0.00%	0 0.00%	1 5.26%	20 18.35%	39 19.21%	45 41.67%	67 33.17%
	L	0 0.00%	0 0.00%	0 0.00%	0 0.00%	0 0.00%	4 3.67%	14 6.90%	11 10.19%	29 14.36%
	VL	0 0.00%	0 0.00%	0 0.00%	0 0.00%	0 0.00%	1 0.92%	7 3.45%	1 0.93%	7 3.47%
	H	0 0.00%	0 0.00%	0 0.00%	0 0.00%	0 0.00%	0 0.00%	3 1.48%	0 0.00%	1 0.50%
	SIG	4 33.33%	4 33.33%	0 0.00%	0 0.00%	1 5.26%	25 22.94%	63 31.03%	57 52.78%	104 51.49%
VAR	VAR	40.97	39.62	13.26	77.96	103.13	175.36	597.28	213.84	548.33
	SIG	3 25.00%	1 8.33%	3 25.00%	3 30.00%	5 26.32%	59 54.13%	144 71.29%	59 54.63%	132 65.35%

Table 20: San Francisco Bus OTP Seasonality Results (100,252,351 Records)

		CALENDAR MONTH	LUNAR MONTH	MOON SIGN	M30 SPEED	M15 SPEED	M30 MONTH + SPEED	M15 MONTH + SPEED	M30 SIGN + SPEED	M15 SIGN + SPEED
EFFECT SIZE	#	12	12	12	10	19	109	203	108	200
EFFECT SIZE	VS	1 8.33%	2 16.67%	1 8.33%	2 20.00%	2 10.53%	4 3.67%	7 3.45%	2 1.85%	5 2.50%
EFFECT SIZE	S	8 66.67%	9 75.00%	11 91.67%	8 80.00%	17 89.47%	84 77.06%	122 60.10%	68 62.96%	120 60.00%
EFFECT SIZE	M	3 25.00%	1 8.33%	0 0.00%	0 0.00%	0 0.00%	16 14.68%	41 20.20%	23 21.30%	34 17.00%
EFFECT SIZE	L	0 0.00%	0 0.00%	0 0.00%	0 0.00%	0 0.00%	2 1.83%	21 10.34%	11 10.19%	22 11.00%
EFFECT SIZE	VL	0 0.00%	0 0.00%	0 0.00%	0 0.00%	0 0.00%	3 2.75%	12 5.91%	4 3.70%	17 8.50%
EFFECT SIZE	H	0 0.00%	0 0.00%	0 0.00%	0 0.00%	0 0.00%	0 0.00%	0 0.00%	0 0.00%	2 1.00%
EFFECT SIZE	SIG	3 25.00%	1 8.33%	0 0.00%	0 0.00%	0 0.00%	21 19.27%	74 36.45%	38 35.19%	75 37.50%
VAR	VAR	20.92	31.99	20.73	29.59	55.74	85.84	149.42	85.68	131.92
VAR	SIG	4 33.33%	3 25.00%	2 16.67%	3 30.00%	7 36.84%	55 50.46%	87 43.07%	48 44.44%	89 44.50%

Table 21: Philadelphia Bus OTP Seasonality Results (68,466,941 Records)

		CALENDAR MONTH	LUNAR MONTH	MOON SIGN	M30 SPEED	M15 SPEED	M30 MONTH + SPEED	M15 MONTH + SPEED	M30 SIGN + SPEED	M15 SIGN + SPEED
EFFECT SIZE	#	12	12	12	10	19	106	196	101	184
EFFECT SIZE	VS	0 0.00%	0 0.00%	0 0.00%	0 0.00%	0 0.00%	1 0.94%	3 1.53%	2 1.98%	4 2.17%
EFFECT SIZE	S	6 50.00%	5 41.67%	12 100%	10 100%	19 100%	81 76.42%	114 58.16%	47 46.53%	84 45.65%
EFFECT SIZE	M	4 33.33%	5 41.67%	0 0.00%	0 0.00%	0 0.00%	12 11.32%	39 19.90%	28 27.72%	44 23.91%
EFFECT SIZE	L	2 16.67%	2 16.67%	0 0.00%	0 0.00%	0 0.00%	9 8.49%	21 10.71%	14 13.86%	31 16.85%
EFFECT SIZE	VL	0 0.00%	0 0.00%	0 0.00%	0 0.00%	0 0.00%	3 2.83%	19 9.69%	9 8.91%	18 9.78%
EFFECT SIZE	H	0 0.00%	0 0.00%	0 0.00%	0 0.00%	0 0.00%	0 0.00%	0 0.00%	1 0.99%	3 1.63%
EFFECT SIZE	SIG	6 50.00%	7 58.33%	0 0.00%	0 0.00%	0 0.00%	24 22.64%	79 40.31%	52 51.49%	96 52.17%
VAR	VAR	30.11	107.22	19.13	27.91	43.96	112.02	139.57	122.31	149.34
VAR	SIG	6 50.00%	8 66.67%	3 25.00%	3 30.00%	9 47.37%	57 53.77%	95 48.47%	50 49.50%	80 43.48%

Table 22: DART Bus OTP Seasonality Results (49,787,039 Records)

		CALENDAR MONTH	LUNAR MONTH	MOON SIGN	M30 SPEED	M15 SPEED	M30 MONTH + SPEED	M15 MONTH + SPEED	M30 SIGN + SPEED	M15 SIGN + SPEED
EFFECT SIZE	#	12	12	12	10	19	108	201	108	201
	VS	0 0.00%	0 0.00%	2 16.67%	0 0.00%	2 10.53%	8 7.41%	9 4.48%	3 2.78%	3 1.49%
	S	8 66.67%	7 58.33%	10 83.33%	10 100%	17 89.47%	74 68.52%	118 58.71%	61 56.48%	100 49.75%
	M	3 25.00%	4 33.33%	0 0.00%	0 0.00%	0 0.00%	20 18.52%	50 24.88%	29 26.85%	57 28.36%
	L	1 8.33%	1 8.33%	0 0.00%	0 0.00%	0 0.00%	5 4.63%	19 9.45%	10 9.26%	26 12.94%
	VL	0 0.00%	0 0.00%	0 0.00%	0 0.00%	0 0.00%	1 0.93%	5 2.49%	5 4.63%	12 5.97%
	H	0 0.00%	0 0.00%	0 0.00%	0 0.00%	0 0.00%	0 0.00%	0 0.00%	0 0.00%	3 1.49%
	SIG	4 33.33%	5 41.67%	0 0.00%	0 0.00%	0 0.00%	26 24.07%	74 36.82%	44 40.74%	98 48.76%
VAR	VAR	22.92	33.01	10.95	29.98	53.70	105.55	148.04	101.48	130.70
	SIG	4 33.33%	6 50.00%	2 16.67%	5 50.00%	10 52.63%	54 50.00%	92 45.77%	46 42.59%	87 43.28%

Table 23: DART Rail OTP Seasonality Results (45,671,313 Records)

		CALENDAR MONTH	LUNAR MONTH	MOON SIGN	M30 SPEED	M15 SPEED	M30 MONTH + SPEED	M15 MONTH + SPEED	M30 SIGN + SPEED	M15 SIGN + SPEED
EFFECT SIZE	#	12	12	12	10	19	109	203	108	201
	VS	0 0.00%	0 0.00%	0 0.00%	2 20.00%	2 10.53%	4 3.67%	3 1.48%	2 1.85%	9 4.48%
	S	11 91.67%	11 91.67%	12 100%	8 80.00%	16 84.21%	84 77.06%	141 69.46%	90 83.33%	150 74.63%
	M	1 8.33%	1 8.33%	0 0.00%	0 0.00%	1 5.26%	17 15.60%	44 21.67%	10 9.26%	29 14.43%
	L	0 0.00%	0 0.00%	0 0.00%	0 0.00%	0 0.00%	3 2.75%	8 3.94%	2 1.85%	6 2.99%
	VL	0 0.00%	0 0.00%	0 0.00%	0 0.00%	0 0.00%	1 0.92%	6 2.96%	2 1.85%	4 1.99%
	H	0 0.00%	0 0.00%	0 0.00%	0 0.00%	0 0.00%	0 0.00%	1 0.49%	2 1.85%	3 1.49%
	SIG	1 8.33%	1 8.33%	0 0.00%	0 0.00%	1 5.26%	21 19.27%	59 29.06%	16 14.81%	42 20.90%
VAR	VAR	745.87	1218.62	484.42	1246.56	2302.30	3217.46	3167.44	4092.76	6886.42
	SIG	7 58.33%	7 58.33%	4 33.33%	5 50.00%	14 73.68%	84 77.06%	132 65.35%	74 68.52%	141 70.15%

Table 24: Amtrak OTP Seasonality Results (10,668,374 Records)

		CALENDAR MONTH	LUNAR MONTH	MOON SIGN	M30 SPEED	M15 SPEED	M30 MONTH + SPEED	M15 MONTH + SPEED	M30 SIGN + SPEED	M15 SIGN + SPEED
EFFECT SIZE	#	12	12	12	10	19	109	203	108	202
	VS	1 8.33%	0 0.00%	0 0.00%	1 10.00%	1 5.26%	8 7.34%	5 2.46%	1 0.93%	0 0.00%
	S	5 41.67%	7 58.33%	12 100%	8 80.00%	17 89.47%	76 69.72%	128 63.05%	40 37.04%	84 41.58%
	M	4 33.33%	3 25.00%	0 0.00%	1 10.00%	1 5.26%	18 16.51%	49 24.14%	33 30.56%	63 31.19%
	L	2 16.67%	2 16.67%	0 0.00%	0 0.00%	0 0.00%	7 6.42%	19 9.36%	29 26.85%	38 18.81%
	VL	0 0.00%	0 0.00%	0 0.00%	0 0.00%	0 0.00%	0 0.00%	2 0.99%	5 4.63%	16 7.92%
	H	0 0.00%	0 0.00%	0 0.00%	0 0.00%	0 0.00%	0 0.00%	0 0.00%	0 0.00%	1 0.50%
	SIG	6 50.00%	5 41.67%	0 0.00%	1 10.00%	1 5.26%	25 22.94%	70 34.48%	67 62.04%	118 58.42%
VAR	VAR	60.09	58.77	6.90	34.28	50.97	70.08	97.38	128.42	146.13
	SIG	5 41.67%	4 33.33%	4 33.33%	3 30.00%	4 21.05%	57 52.29%	99 49.01%	47 43.52%	91 45.05%

Table 25: King County Bus OTP Seasonality Results (7,176,709 Records)

		CALENDAR MONTH	LUNAR MONTH	MOON SIGN	M30 SPEED	M15 SPEED	M30 MONTH + SPEED	M15 MONTH + SPEED	M30 SIGN + SPEED	M15 SIGN + SPEED
EFFECT SIZE	#	12	12	12	10	19	109	202	108	200
	VS	1 8.33%	3 25.00%	2 16.67%	2 20.00%	2 10.53%	1 0.92%	4 1.98%	2 1.85%	2 1.00%
	S	8 66.67%	7 58.33%	10 83.33%	8 80.00%	17 89.47%	79 72.48%	124 61.39%	62 57.41%	100 50.00%
	M	3 25.00%	2 16.67%	0 0.00%	0 0.00%	0 0.00%	21 19.27%	37 18.32%	22 20.37%	51 25.50%
	L	0 0.00%	0 0.00%	0 0.00%	0 0.00%	0 0.00%	6 5.50%	28 13.86%	17 15.74%	26 13.00%
	VL	0 0.00%	0 0.00%	0 0.00%	0 0.00%	0 0.00%	2 1.83%	9 4.46%	5 4.63%	21 10.50%
	H	0 0.00%	0 0.00%	0 0.00%	0 0.00%	0 0.00%	0 0.00%	0 0.00%	0 0.00%	0 0.00%
	SIG	3 25.00%	2 16.67%	0 0.00%	0 0.00%	0 0.00%	29 26.61%	74 36.63%	44 40.74%	98 49.00%
VAR	VAR	41.07	40.88	15.90	17.22	26.39	54.79	74.74	69.33	88.57
	SIG	4 33.33%	3 25.00%	4 33.33%	3 30.00%	5 26.32%	57 52.29%	100 49.50%	40 37.04%	70 35.00%

Table 26: Chicago Rail OTP Seasonality Results (191,075 Records)

		CALENDAR MONTH	LUNAR MONTH	MOON SIGN	M30 SPEED	M15 SPEED	M30 MONTH + SPEED	M15 MONTH + SPEED	M30 SIGN + SPEED	M15 SIGN + SPEED
EFFECT SIZE	#	12	12	12	10	19	109	203	108	201
	VS	1 8.33%	2 16.67%	1 8.33%	2 20.00%	1 5.26%	2 1.83%	11 5.42%	4 3.70%	7 3.48%
	S	11 91.67%	10 83.33%	11 91.67%	8 80.00%	18 94.74%	88 80.73%	128 63.05%	79 73.15%	125 62.19%
	M	0 0.00%	0 0.00%	0 0.00%	0 0.00%	0 0.00%	17 15.60%	39 19.21%	20 18.52%	53 26.37%
	L	0 0.00%	0 0.00%	0 0.00%	0 0.00%	0 0.00%	2 1.83%	22 10.84%	4 3.70%	12 5.97%
	VL	0 0.00%	0 0.00%	0 0.00%	0 0.00%	0 0.00%	0 0.00%	3 1.48%	1 0.93%	4 1.99%
	H	0 0.00%	0 0.00%	0 0.00%	0 0.00%	0 0.00%	0 0.00%	0 0.00%	0 0.00%	0 0.00%
	SIG	0 0.00%	0 0.00%	0 0.00%	0 0.00%	0 0.00%	19 17.43%	64 31.53%	25 23.15%	69 34.33%
VAR	VAR	52.47	51.03	12.98	21.00	36.08	71.44	105.32	86.94	125.73
	SIG	2 16.67%	1 8.33%	1 8.33%	4 40.00%	6 31.58%	58 53.21%	96 47.52%	38 35.19%	71 35.32%

Table 27: OTP Seasonality Summary: Effect Size

ON TIME PERFORMANCE DATA SET	CALENDAR MONTH (12)		M30 SIGN + SPEED (108)		% DIFF. MONTH / M30	M15 SIGN + SPEED (202)		% DIFF. MONTH / M15
FLIGHTS	3	25.00%	36	33.33%	33.32%	77	38.12%	52.48%
CHICAGO (BUS)	4	33.33%	57	52.78%	58.36%	104	51.49%	54.49%
SAN FRANCISCO (BUS)	3	25.00%	38	35.19%	40.76%	75	37.50%	50.00%
PHILADELPHIA (BUS)	6	50.00%	52	51.49%	2.98%	96	52.17%	4.34%
DART (BUS)	4	33.33%	44	40.74%	22.23%	98	48.76%	46.29%
DART (RAIL)	1	8.33%	16	14.81%	77.79%	42	20.90%	150.90%
AMTRAK	6	50.00%	67	62.04%	24.08%	118	58.42%	16.84%
KING COUNTY (BUS)	3	25.00%	44	40.74%	62.96%	98	49.00%	96.00%
CHICAGO (RAIL)	0	00.00%	23	23.15%	N/A	69	34.33%	N/A
AVERAGE SIGNIFICANCE	27.78%		39.36%		41.68%	43.41%		56.26%

Table 28: OTP Seasonality Summary: Variance

ON TIME PERFORMANCE DATA SET	CALENDAR MONTH (12)		M30 SIGN + SPEED (108)		% DIFF. MONTH / M30	M15 SIGN + SPEED (202)		% DIFF. MONTH / M15
FLIGHTS	3	25.00%	40	37.04%	48.16%	72	35.64%	42.56%
CHICAGO (BUS)	3	25.00%	59	54.63%	118.52%	132	65.35%	161.40%
SAN FRANCISCO (BUS)	4	33.33%	48	44.44%	33.33%	89	44.50%	33.51%
PHILADELPHIA (BUS)	6	50%	50	49.50%	-1.00%	80	43.48%	-13.04%
DART (BUS)	4	33.33%	46	42.59%	27.78%	87	43.28%	29.85%
DART (RAIL)	7	58.33%	74	68.52%	17.40%	141	70.15%	11.82%
AMTRAK	5	41.87%	47	43.52%	3.94%	91	45.05%	7.59%
KING COUNTY (BUS)	4	33.33%	40	37.04%	11.13%	70	35.00%	5.01%
CHICAGO (RAIL)	2	16.67%	38	35.19%	111.10%	71	35.32%	111.88%
AVERAGE SIGNIFICANCE	35.21%		45.83%		30.16%	46.42%		31.84%

4.3.3. On-Time Performance Seasonality Conclusions

A reasonable degree of **Calendar Month** seasonality is assumed for these data sets, and this approach, which measures the effect of seasonality as a function of effect size and variance, has quantified the monthly seasonality for each data set and identified the number and percentage of significant seasons.

The results for the **Calendar Month** seasonality can be used as a baseline to compare the presence and significance of other types of seasonality, both regular and irregular. With the exception of the **Philadelphia Bus** data, the **M15** and **M30 Sign + Speed** models showed greater significance for effect and variance than the **Calendar Month** model, and with the exception of Amtrak, the increase in significance is considerable.

When considering the entire on-time performance data set, **Calendar Month** seasonality had a 27.8% average rate of significance for effect size and a 35.21% average rate of significance for variance. The **M30 Sign + Speed** model had a 39.36% average rate of significance for effect size (**a 41.68% increase**) and a 45.83% average rate of significance for variance (**a 30.16% increase**). The **M15 Sign + Speed** model had a 43.41% average rate of significance for effect size (**a 56.26% increase**) and a 46.52% average rate of significance for variance (**a 31.84% increase**).

That these irregular, Mercury-based seasonal models appear to have so much more significance than the **Calendar Month** model of seasonality strongly suggests that these Mercury-based irregular seasonal models are worthy of further study and consideration, at least when considering transportation on-time performance data sets where **Calendar Month** seasonality is often evaluated.

4.4. Car Crash Seasonality Study

The **Car Crash** data sets evaluate two different metrics. The **Fatality Analysis Reporting System (FARS)** data tracks the numbers of fatal highway crashes, and contains data from January 1980 through December 2016. The other seven data sets contain data on car crashes that were serious enough to require a police report, whether or not there were any fatalities. The **FARS** data has been evaluated extensively in other studies, and includes a strong expectation of **Calendar Month** seasonal effects (the greatest numbers of fatal highway crashes occur in the summer months, when the greatest number of cars are on the highway system). It's reasonable to expect some degree of **Calendar Month** seasonal influence in the other crash data sets, although it's perhaps not reasonable to expect the same seasonal influence, with an increase in crashes during the summer months.

Note that the results for the **New York** data are questionable. Only three years of data were available for **New York State**. Confidence in the effect sizes is strong, but because the population is so limited, confidence in the seasonal variance—and the forecast potential—is low.

4.4.1. Analysis of Regular Seasonal Influences

Table 29 on page 50 shows the seasonality by **Calendar Month** for each of the data sets, Table 30 on page 51 shows the seasonality by **Lunar Month**, and Table 31 on page 52 shows the seasonality by **Moon Sign**. Once again, there are no significant effect sizes for the **Moon Sign** seasonality.

We'll begin with the **FARS** data set because it encompasses the greatest period of time, and because it has a high expectation of a **Calendar Month** seasonal influence. When considering the effect size, there are five significant seasons in both the **Calendar Month** and **Lunar Month** seasons. The highest rates of crashes occur in **July** (M+, 9.56%) and **August** (M+, 11.69%), which roughly correspond to the **Lunar Month** seasons of **Cancer** (M+, 10.07%) and **Leo** (M+, 10.82%). The lowest rates of crashes occur in **January** (-L, -15.86%), **February** (-M, -15.25%), and **March** (-M, -11.47%), which correspond to **Capricorn** (-M, 14.86%), **Aquarius**, (-M, 14.76%), and **Pisces** (-M, 10.55%).

However, when considering the variance, there are no significant seasons for either the **Calendar Month** or **Lunar Month** for **FARS**. The seasonal effect sizes may be useful in forecasting the relative number of crashes in each season, but it does not appear that either the **Calendar Month** or **Lunar Month** seasonality will be of use when forecasting the precise number of fatal highway car crashes for any given season.

Table 29: Crash Seasonality: Calendar Month

		FARS 1,369,970	CALIFORNIA 6,002,314	IOWA 234,123	MICHIGAN 4,007,367	NEW JERSEY 3,644,021	NYC 1,509,269	NEW YORK 895,917	TEXAS 4,351,088
CALENDAR MONTH — EFFECT & MEAN % DIF	JAN	–L -15.86%	–S -2.98%	M+ 16.81%	M+ 20.84%	–S -1.67%	–M -8.76%	M+ 9.31%	–M -9.59%
	FEB	–M -15.25%	S+ 2.69%	M+ 17.26%	M+ 13.32%	–S -4.24%	–M -8.01%	S+ 4.90%	S+ 1.20%
	MAR	–M -11.47%	S+ 0.68%	–M -13.65%	–M -14.00%	–S -8.06%	–S -3.86%	–M -11.12%	S+ 2.23%
	APR	–S -5.05%	–S -0.55%	–M -16.08%	–M -20.38%	–S -6.96%	–S -3.89%	–L -13.55%	S+ 1.90%
	MAY	S+ 1.64%	S+ 0.77%	–S -10.04%	–S -10.58%	S+ 0.50%	S+ 5.39%	–S -4.41%	S+ 1.60%
	JUN	S+ 7.53%	–S -0.58%	–S -7.72%	–S -5.09%	S+ 5.98%	M+ 9.11%	S+ 4.16%	–S -1.53%
	JUL	M+ 9.56%	–S -4.78%	–M -14.58%	–M -15.78%	–S -1.18%	S+ 0.83%	–S -2.86%	–M -7.70%
	AUG	M+ 11.69%	–S -5.18%	–M -14.27%	–M -17.02%	–S -3.72%	–S -0.42%	–S -6.41%	–S -3.36%
	SEP	S+ 7.61%	–S -5.37%	–S -7.30%	–S -8.03%	–S -1.25%	S+ 2.69%	–S -2.18%	S+ 1.34%
	OCT	S+ 7.42%	S+ 3.19%	S+ 2.69%	S+ 11.05%	S+ 5.42%	S+ 3.40%	S+ 5.77%	M+ 6.21%
	NOV	S+ 2.51%	S+ 5.42%	M+ 18.41%	L+ 24.86%	S+ 5.29%	S+ 1.86%	M+ 11.57%	S+ 4.24%
	DEC	–S -1.26%	S+ 5.24%	L+ 25.92%	L+ 21.69%	M+ 9.36%	VS+ 0.01%	S+ 5.24%	S+ 3.74%
	SIG	5 41.67%	0 0.00%	8 66.67%	8 66.67%	1 8.33%	3 25.00%	4 33.33%	3 25.00%
CALENDAR MONTH — EFFECT & VARIANCE	JAN	–L 19.23	–S 29.11	M+ 167.78	M+ 207.91	–S 33.83	–M 8.04	M+ 141.17	–M 3.61
	FEB	–M 17.48	S+ 22.43	M+ 264.02	M+ 132.19	–S 97.50	–M 18.00	S+ 117.74	S+ 11.06
	MAR	–M 9.61	S+ 51.63	–M 39.28	–M 34.06	–S 29.16	–S 8.24	–M 26.76	S+ 12.31
	APR	–S 11.91	–S 42.07	–M 14.24	–M 19.57	–S 11.71	–S 4.19	–L 40.91	S+ 5.31
	MAY	S+ 7.61	S+ 13.82	–S 33.69	–S 17.40	S+ 5.67	S+ 2.09	–S 10.88	S+ 0.80
	JUN	S+ 12.72	–S 12.81	–S 16.47	–S 15.96	S+ 18.55	M+ 5.61	S+ 9.73	–S 1.43
	JUL	M+ 12.01	–S 125.96	–M 23.77	–M 13.38	–S 2.84	S+ 1.15	–S 10.32	–M 3.58
	AUG	M+ 19.42	–S 131.35	–M 37.90	–M 12.11	–S 6.36	–S 5.02	–S 8.78	–S 6.58
	SEP	S+ 10.00	–S 267.59	–S 51.82	–S 18.56	–S 7.18	S+ 2.76	–S 9.19	S+ 1.17
	OCT	S+ 13.79	S+ 191.06	S+ 20.32	S+ 23.25	S+ 16.64	S+ 7.16	S+ 0.17	M+ 10.03
	NOV	S+ 11.44	S+ 28.00	M+ 74.10	L+ 44.79	S+ 7.49	S+ 8.39	M+ 20.38	S+ 8.74
	DEC	–S 15.43	S+ 49.13	L+ 310.93	L+ 191.04	M+ 36.40	VS+ 1.61	S+ 13.64	S+ 10.12
	VAR	6.69	40.21	43.93	30.43	11.39	3.01	17.07	3.11
	SIG	0 0.00%	5 41.67%	7 58.33%	7 58.33%	5 41.67%	4 33.33%	7 58.33%	3 25.00%

Table 30: Crash Seasonality: Lunar Month

		FARS 1,369,970	CALIFORNIA 6,002,314	IOWA 234,123	MICHIGAN 4,007,367	NEW JERSEY 3,644,021	NYC 1,509,269	NEW YORK 895,917	TEXAS 4,351,088
LUNAR MONTH — EFFECT & MEAN % DIF	ARIES	-S -4.63%	VS+ 0.10%	-M -15.52%	-M -18.98%	-S -5.57%	-S -0.87%	-M -12.35%	S+ 2.25%
	TAURUS	S+ 2.36%	VS+ 0.04%	-S -9.01%	-S -8.90%	S+ 1.70%	M+ 5.92%	-S -2.57%	S+ 2.14%
	GEMINI	S+ 8.00%	-S -2.33%	-S -8.64%	-S -7.75%	S+ 3.92%	M+ 6.36%	S+ 2.83%	-S -3.21%
	CANCER	M+ 10.07%	-S -4.01%	-M -14.58%	-M -15.73%	-S -0.81%	S+ 2.74%	-S -2.20%	-M -6.92%
	LEO	M+ 10.82%	-S -4.42%	-M -13.20%	-M -16.54%	-S -4.08%	-S -0.31%	-S -6.37%	-S -3.00%
	VIRGO	S+ 7.27%	-S -4.54%	-S -4.99%	-S -4.12%	S+ 0.51%	S+ 2.41%	-S -2.29%	S+ 1.55%
	LIBRA	S+ 6.53%	S+ 4.29%	S+ 6.67%	M+ 14.45%	S+ 5.72%	S+ 3.06%	S+ 5.76%	M+ 7.24%
	SCO.	S+ 2.24%	M+ 7.01%	M+ 19.15%	L+ 24.98%	S+ 6.91%	S+ 3.01%	M+ 11.35%	S+ 5.23%
	SAG.	-S -4.31%	S+ 2.08%	L+ 25.51%	M+ 20.29%	S+ 6.02%	-S -1.81%	M+ 7.35%	S+ 0.90%
	CAP.	-M -14.86%	-S -3.07%	M+ 20.34%	M+ 20.83%	VS+ 0.21%	-M -7.89%	S+ 6.30%	-M -8.05%
	AQU.	-M -14.76%	S+ 3.34%	S+ 7.89%	S+ 9.82%	-S -3.08%	-M -6.93%	S+ 3.33%	S+ 0.81%
	PISCES	-M -10.55%	S+ 0.45%	-M -15.35%	-M -15.11%	-M -10.95%	-M -6.46%	-M -12.67%	S+ 1.63%
	SIG	5 41.67%	1 8.33%	7 58.33%	8 66.67%	1 8.33%	5 41.67%	4 33.33%	3 25.00%
LUNAR MONTH — EFFECT & VARIANCE	ARIES	-S 10.77	VS+ 30.54	-M 16.32	-M 22.34	-S 15.57	-S 17.23	-M 14.25	S+ 2.63
	TAURUS	S+ 10.21	VS+ 165.42	-S 23.35	-S 30.19	S+ 6.62	M+ 9.18	-S 25.02	S+ 1.98
	GEMINI	S+ 11.08	-S 26.72	-S 20.00	-S 14.20	S+ 15.29	M+ 6.45	S+ 13.62	-S 4.11
	CANCER	M+ 14.55	-S 379.15	-M 24.07	-M 13.33	-S 2.49	S+ 2.43	-S 5.78	-M 3.45
	LEO	M+ 17.46	-S 164.13	-M 37.31	-M 12.21	-S 3.86	-S 6.62	-S 10.32	-S 9.30
	VIRGO	S+ 8.07	-S 171.79	-S 65.46	-S 43.92	S+ 9.66	S+ 3.08	-S 6.04	S+ 7.17
	LIBRA	S+ 16.02	S+ 400.15	S+ 110.96	M+ 76.16	S+ 17.23	S+ 18.44	S+ 13.03	M+ 4.43
	SCO.	S+ 12.54	M+ 27.17	M+ 93.50	L+ 54.16	S+ 26.52	S+ 2.03	M+ 29.30	S+ 22.32
	SAG.	-S 32.98	S+ 78.12	L+ 76.92	M+ 167.80	S+ 57.40	-S 8.37	M+ 58.54	S+ 29.53
	CAP.	-M 18.46	-S 37.76	M+ 220.76	M+ 161.40	VS+ 40.06	-M 3.35	S+ 198.37	-M 25.05
	AQU.	-M 16.63	S+ 11.25	S+ 146.67	S+ 100.15	-S 86.34	-M 18.35	S+ 79.32	S+ 6.66
	PISCES	-M 12.13	S+ 38.75	-M 55.42	-M 44.42	-M 8.31	-M 9.32	-M 7.16	S+ 4.76
	VAR	7.54	63.79	37.11	30.84	12.06	4.37	19.20	5.06
	SIG	0 0.00%	6 50.00%	4 33.33%	5 41.67%	5 41.67%	4 33.33%	7 58.33%	6 50.00%

Table 31: Crash Seasonality: Moon Sign

		FARS 1,369,970	CALIFORNIA 6,002,314	IOWA 234,123	MICHIGAN 4,007,367	NEW JERSEY 3,644,021	NYC 1,509,269	NEW YORK 895,917	TEXAS 4,351,088
LUNAR MONTH — EFFECT & MEAN % DIF	ARIES	-S -0.48%	S+ 1.00%	-S -0.97%	S+ 2.25%	VS+ 0.11%	S+ 0.56%	-S -0.29%	S+ 0.49%
	TAURUS	-VS -0.22%	S+ 0.33%	-S -1.55%	S+ 1.10%	S+ 1.39%	-S -0.84%	S+ 0.55%	S+ 0.59%
	GEMINI	S+ 0.24%	-S -1.44%	S+ 0.42%	-S -0.47%	-S -0.64%	-S -1.23%	-S -0.43%	-S -0.17%
	CANCER	-S -0.69%	S+ 0.66%	S+ 1.87%	S+ 1.05%	S+ 0.80%	-S -0.68%	-S -0.41%	S+ 1.17%
	LEO	-VS -0.12%	S+ 0.38%	S+ 2.88%	-S -0.66%	-S -0.56%	S+ 1.11%	S+ 5.02%	-S -0.44%
	VIRGO	-VS -0.10%	S+ 0.39%	S+ 0.84%	S+ 1.29%	VS+ 0.22%	S+ 2.42%	S+ 0.48%	S+ 0.79%
	LIBRA	-S -0.25%	-S -1.04%	S+ 1.10%	S+ 0.78%	-S -1.04%	-S -1.54%	-S -1.13%	-S -0.73%
	SCO.	-VS -0.24%	S+ 0.28%	-S -2.09%	S+ 0.85%	-VS -0.15%	VS+ 0.09%	-S -0.90%	-S -1.33%
	SAG.	S+ 0.34%	S+ 0.32%	-S -1.23%	-S -2.30%	-S -1.08%	VS+ 0.10%	-S -2.64%	-S -0.91%
	CAP.	S+ 0.80%	-S -0.32%	-VS -0.19%	-S -3.13%	S+ 1.10%	-S -0.29%	S+ 0.21%	S+ 0.26%
	AQU.	VS+ 0.16%	-S -0.29%	-S -0.56%	-S -1.41%	S+ 0.83%	S+ 0.70%	-S -1.52%	S+ 0.17%
	PISCES	S+ 0.55%	-S -0.26%	-S -0.59%	S+ 0.56%	-S -1.04%	-S -0.46%	S+ 0.85%	VS+ 0.14%
	SIG	0 0.00%	0 0.00%	0 0.00%	0 0.00%	0 0.00%	0 0.00%	0 0.00%	0 0.00%
LUNAR MONTH — EFFECT & VARIANCE	ARIES	-S 11.13	S+ 29.39	-S 26.12	S+ 30.16	VS+ 14.45	S+ 9.44	-S 5.83	S+ 6.68
	TAURUS	-VS 9.41	S+ 51.55	-S 28.84	S+ 19.06	S+ 11.10	-S 6.78	S+ 0.81	S+ 6.73
	GEMINI	S+ 6.94	-S 27.06	S+ 65.32	-S 28.81	-S 7.69	-S 4.22	-S 20.41	-S 4.76
	CANCER	-S 10.58	S+ 11.91	S+ 34.92	S+ 22.30	S+ 11.83	-S 21.70	-S 29.83	S+ 6.78
	LEO	-VS 11.30	S+ 20.13	S+ 41.43	-S 32.77	-S 12.51	S+ 10.65	S+ 23.53	-S 5.49
	VIRGO	-VS 9.61	S+ 24.36	S+ 52.50	S+ 41.04	VS+ 14.43	S+ 9.21	S+ 27.57	S+ 6.33
	LIBRA	-S 10.12	-S 18.49	S+ 73.06	S+ 37.60	-S 28.29	-S 11.37	-S 12.61	-S 5.37
	SCO.	-VS 8.58	S+ 16.18	-S 75.56	S+ 30.91	-VS 14.03	VS+ 17.29	-S 14.32	-S 11.97
	SAG.	S+ 8.36	S+ 12.54	-S 44.89	-S 13.72	-S 12.79	VS+ 5.13	-S 18.04	-S 6.02
	CAP.	S+ 7.94	-S 15.73	-VS 32.21	-S 21.94	S+ 13.29	-S 4.39	S+ 20.27	S+ 5.10
	AQU.	VS+ 11.38	-S 18.23	-S 30.58	-S 37.72	S+ 36.16	S+ 7.20	-S 1.07	S+ 10.09
	PISCES	S+ 11.44	-S 8.94	-S 48.45	S+ 21.38	-S 3.65	-S 11.26	S+ 25.94	VS+ 21.34
	VAR	4.87	10.60	23.08	14.06	7.51	4.94	8.34	4.03
	SIG	0 0.00%	1 8.33%	0 0.00%	1 8.33%	1 8.33%	2 16.67%	3 25.00%	0 0.00%

When considering the other (non-fatal) crash data sets, the greatest number of significant seasonal effects occurs in the **Iowa** and **Michigan** data sets. **Iowa** has eight significant results for the **Calendar Month** and seven significant results for the **Lunar Month**, and **Michigan** has eight significant results for both **Calendar Month** and **Lunar Month**. By contrast, **California** shows no significant effect sizes for **Calendar**

Month and only a single significant effect size for **Lunar Month**, and **New Jersey** shows only a single significant effect size for each model.

The variance tells a different story. Although the effect sizes are small for **California**, the variance is quite low, so five **Calendar Month** seasons and six **Lunar Month** seasons could be expected to improve the accuracy of forecasts for **California** car crashes. **Iowa**, **Michigan**, and **New York** each have seven **Calendar Month** seasons that have a low enough variance to potentially improve forecast accuracy.

4.4.2. Analysis of Irregular Seasonal Influences

Table 32 through Table 39 compare the three regular seasonality models (**Calendar Month**, **Lunar Month**, **Moon Sign**) with six Mercury-based irregular seasonality models (**M30 Speed**, **M15 Speed**, **M30 Month + Speed**, **M15 Month + Speed**, **M30 Sign + Speed**, **M15 Sign + Speed**) for each individual data set.

Each table is divided into two sections. The top section is a straight count of the effect sizes. Results with a effect size of Medium or greater are considered to be significant, and the number and percentage of seasons with significant effect sizes is highlighted at the bottom of the section. The bottom section (the last two rows of each table) evaluates the variance. The second section evaluates the variance. The **VAR** row shows the minimum significant threshold number for each data set, and the **SIG** row below it shows the number and percentage of seasons with a significant variance.

These results are similar to the results of the on-time performance irregular seasonality. Both the **M30 Sign + Speed** and the **M15 Sign + Speed** seasons show high percentages of significant seasonal effects. The most dramatic results can be found in the **California** and **New Jersey** data sets. These had the lowest percentage of significant effect sizes when considering **Calendar Month** seasonal effects (0% for **California** and 8.33% for **New Jersey**), but for the **M15 Sign + Speed**, 29.21% of the seasonal effect sizes are significant for **California**, and 30.20% of the seasonal effect sizes are significant for **New Jersey**.

When considering the seasonal variance, the **FARS** data shows the most dramatic results. Neither the **Calendar Month** nor the **Lunar Month** models had any forecast value, but 20.79% of the **M30 Sign + Speed** seasons and 26.73% of the **M15 Sign + Speed** seasons can be expected to improve the forecast accuracy of this data set.

The lack of sufficient historical data is most problematic when considering the irregular seasonality for the **New York State** data (Table 38 on page 57). Only 79 out of 108 **M30 Sign + Speed** seasons and 139 out of 202 **M15 Sign + Speed** seasons are represented, and a majority of those seasons will have only one or two historical instances.

Table 40 on page 58 provides a summary that compares the significance of the **Calendar Month** seasonal model with the significance of the **M30 Sign + Speed** and the **M15 Sign + Speed** seasonal models in terms of effect size, across the entire data set. Table 41 on page 58 provides a summary that compares the significance of the

Calendar Month seasonal model with the significance of the **M30 Sign + Speed** and the **M15 Sign + Speed** seasonal models in terms of variance, across the entire data set.

Table 32: FARS Crashes Jan 1980-Dec 2016 (1,369,970 Records)

		CALENDAR MONTH	LUNAR MONTH	MOON SIGN	M30 SPEED	M15 SPEED	M30 MONTH + SPEED	M15 MONTH + SPEED	M30 SIGN + SPEED	M15 SIGN + SPEED
EFFECT SIZE	#	12	12	12	10	19	109	203	108	202
	VS	0 0.00%	0 0.00%	5 41.67%	2 20.00%	3 15.79%	11 10.09%	15 7.39%	0 0.00%	4 1.98%
	S	7 58.33%	7 58.33%	7 58.33%	8 80.00%	16 84.21%	97 88.99%	184 90.64%	57 52.78%	99 49.01%
	M	4 33.33%	5 41.67%	0 0.00%	0 0.00%	0 0.00%	1 0.92%	4 1.97%	48 44.44%	88 43.56%
	L	1 8.33%	0 0.00%	0 0.00%	0 0.00%	0 0.00%	0 0.00%	0 0.00%	3 2.78%	11 5.45%
	VL	0 0.00%	0 0.00%	0 0.00%	0 0.00%	0 0.00%	0 0.00%	0 0.00%	0 0.00%	0 0.00%
	H	0 0.00%	0 0.00%	0 0.00%	0 0.00%	0 0.00%	0 0.00%	0 0.00%	0 0.00%	0 0.00%
	SIG	5 41.67%	5 41.67%	0 0.00%	0 0.00%	0 0.00%	1 0.92%	4 1.97%	51 47.22%	99 49.01%
VAR	VAR	6.69	7.54	4.87	11.82	36.45	60.18	118.23	118.23	116.27
	SIG	0 0.00%	0 0.00%	0 0.00%	4 40.00%	8 42.11%	38 34.86%	42 20.79%	42 20.79%	54 26.73%

Table 33: California Crashes Jan 2001-Dec 2016 (6,002,314 Records)

		CALENDAR MONTH	LUNAR MONTH	MOON SIGN	M30 SPEED	M15 SPEED	M30 MONTH + SPEED	M15 MONTH + SPEED	M30 SIGN + SPEED	M15 SIGN + SPEED
EFFECT SIZE	#	12	12	12	10	19	109	203	108	202
	VS	0 0.00%	2 16.67%	0 0.00%	1 10.00%	2 10.53%	5 4.59%	8 3.94%	4 3.70%	13 6.44%
	S	12 100%	9 75.00%	12 100%	9 90.00%	16 84.21%	98 89.91%	158 77.83%	82 75.93%	130 64.36%
	M	0 0.00%	1 8.33%	0 0.00%	0 0.00%	1 5.26%	4 3.67%	28 13.79%	15 13.89%	42 20.79%
	L	0 0.00%	0 0.00%	0 0.00%	0 0.00%	0 0.00%	2 1.83%	9 4.43%	7 6.48%	17 8.42%
	VL	0 0.00%	0 0.00%	0 0.00%	0 0.00%	0 0.00%	0 0.00%	0 0.00%	0 0.00%	0 0.00%
	H	0 0.00%	0 0.00%	0 0.00%	0 0.00%	0 0.00%	0 0.00%	0 0.00%	0 0.00%	0 0.00%
	SIG	0 0.00%	1 8.33%	0 0.00%	0 0.00%	1 5.26%	6 5.50%	37 18.23%	22 20.37%	59 29.21%
VAR	VAR	40.21	63.79	10.60	31.72	44.03	46.37	70.63	83.18	104.97
	SIG	5 41.67%	6 50.00%	1 8.33%	5 50.00%	6 31.58%	61 55.96%	99 49.01%	56 51.85%	97 48.02%

Table 34: Iowa Crashes Jan 1, 2006–Sep 21, 2016 (234,123 Records)

		CALENDAR MONTH	LUNAR MONTH	MOON SIGN	M30 SPEED	M15 SPEED	M30 MONTH + SPEED	M15 MONTH + SPEED	M30 SIGN + SPEED	M15 SIGN + SPEED
EFFECT SIZE	#	12	12	12	10	19	109	203	108	202
	VS	0 0.00%	0 0.00%	1 8.33%	0 0.00%	2 10.53%	7 6.42%	5 2.46%	0 0.00%	1 0.50%
	S	4 33.33%	5 41.67%	11 91.67%	8 80.00%	14 73.68%	87 79.82%	155 76.35%	59 54.63%	106 52.48%
	M	7 58.33%	6 50.00%	0 0.00%	2 20.00%	2 10.53%	12 11.01%	32 15.76%	41 37.96%	79 39.11%
	L	1 8.33%	1 8.33%	0 0.00%	0 0.00%	1 5.26%	3 2.75%	11 5.42%	8 7.41%	16 7.92%
	VL	0 0.00%	0 0.00%	0 0.00%	0 0.00%	0 0.00%	0 0.00%	0 0.00%	0 0.00%	0 0.00%
	H	0 0.00%	0 0.00%	0 0.00%	0 0.00%	0 0.00%	0 0.00%	0 0.00%	0 0.00%	0 0.00%
	SIG	8 66.67%	7 58.33%	0 0.00%	2 20.00%	3 15.79%	15 13.76%	43 21.18%	49 45.37%	95 47.03%
VAR	VAR	43.93	37.11	23.08	49.89	90.85	110.85	160.02	149.31	225.42
	SIG	7 58.33%	4 33.33%	0 0.00%	4 40.00%	8 42.11%	63 57.80%	108 53.47%	69 63.89%	116 57.43%

Table 35: Michigan Crashes Jan 2004–Dec 2016 (4,007,367 Records)

		CALENDAR MONTH	LUNAR MONTH	MOON SIGN	M30 SPEED	M15 SPEED	M30 MONTH + SPEED	M15 MONTH + SPEED	M30 SIGN + SPEED	M15 SIGN + SPEED
EFFECT SIZE	#	12	12	12	10	19	109	203	108	202
	VS	0 0.00%	0 0.00%	0 0.00%	1 10.00%	1 5.26%	8 7.34%	9 4.43%	1 0.93%	3 1.49%
	S	4 33.33%	4 33.33%	12 100%	7 70.00%	15 78.95%	97 88.99%	153 75.37%	36 33.33%	70 34.65%
	M	6 50.00%	7 58.33%	0 0.00%	2 20.00%	2 10.53%	4 3.67%	30 14.78%	52 48.15%	88 43.56%
	L	2 16.67%	1 8.33%	0 0.00%	0 0.00%	1 5.26%	0 0.00%	8 3.94%	13 12.04%	29 14.36%
	VL	0 0.00%	0 0.00%	0 0.00%	0 0.00%	0 0.00%	0 0.00%	2 0.99%	5 4.63%	10 4.95%
	H	0 0.00%	0 0.00%	0 0.00%	0 0.00%	0 0.00%	0 0.00%	1 0.49%	1 0.93%	2 0.99%
	SIG	8 66.67%	8 66.67%	0 0.00%	2 20.00%	3 15.79%	4 3.67%	41 20.20%	71 65.74%	129 63.86%
VAR	VAR	30.43	30.84	14.06	30.45	59.39	98.91	156.94	128.00	219.00
	SIG	7 58.33%	5 41.67%	1 8.33%	4 40.00%	7 36.84%	59 54.13%	101 50.00%	63 58.33%	122 60.40%

Table 36: New Jersey Crashes Jan 2001-Dec 2016 (3,644,021 Records)

		CALENDAR MONTH	LUNAR MONTH	MOON SIGN	M30 SPEED	M15 SPEED	M30 MONTH + SPEED	M15 MONTH + SPEED	M30 SIGN + SPEED	M15 SIGN + SPEED
EFFECT SIZE	#	12	12	12	10	19	109	203	108	202
	VS	0 0.00%	1 8.33%	3 25.00%	4 40.00%	3 15.79%	8 7.34%	7 3.45%	2 1.85%	3 1.49%
	S	11 91.67%	10 83.33%	9 75.00%	6 60.00%	16 84.21%	90 82.57%	157 77.34%	84 77.78%	138 68.32%
	M	1 8.33%	1 8.33%	0 0.00%	0 0.00%	0 0.00%	9 8.26%	34 16.75%	19 17.59%	43 21.29%
	L	0 0.00%	0 0.00%	0 0.00%	0 0.00%	0 0.00%	2 1.83%	5 2.46%	3 2.78%	17 8.42%
	VL	0 0.00%	0 0.00%	0 0.00%	0 0.00%	0 0.00%	0 0.00%	0 0.00%	0 0.00%	0 0.00%
	H	0 0.00%	0 0.00%	0 0.00%	0 0.00%	0 0.00%	0 0.00%	0 0.00%	0 0.00%	1 0.50%
	SIG	1 8.33%	1 8.33%	0 0.00%	0 0.00%	0 0.00%	11 10.09%	39 19.21%	22 20.37%	61 30.20%
VAR	VAR	11.39	12.06	7.51	16.87	42.71	71.24	137.58	69.21	139.77
	SIG	5 41.67%	5 41.67%	1 8.33%	3 30.00%	8 42.11%	66 60.55%	120 59.41%	66 61.11%	121 59.90%

Table 37: New York City Crashes July 1, 2012-Jun 4, 2019 (1,509,269 Records)

		CALENDAR MONTH	LUNAR MONTH	MOON SIGN	M30 SPEED	M15 SPEED	M30 MONTH + SPEED	M15 MONTH + SPEED	M30 SIGN + SPEED	M15 SIGN + SPEED
EFFECT SIZE	#	12	12	12	10	19	109	203	108	201
	VS	1 8.33%	0 0.00%	2 16.67%	2 20.00%	1 5.26%	6 5.50%	4 1.97%	3 2.78%	5 2.49%
	S	8 66.67%	7 58.33%	10 83.33%	8 80.00%	18 94.74%	84 77.06%	129 63.55%	63 58.33%	101 50.25%
	M	3 25.00%	5 41.67%	0 0.00%	0 0.00%	0 0.00%	17 15.60%	47 23.15%	30 27.78%	59 29.35%
	L	0 0.00%	0 0.00%	0 0.00%	0 0.00%	0 0.00%	2 1.83%	17 8.37%	10 9.26%	25 12.44%
	VL	0 0.00%	0 0.00%	0 0.00%	0 0.00%	0 0.00%	0 0.00%	6 2.96%	2 1.85%	11 5.47%
	H	0 0.00%	0 0.00%	0 0.00%	0 0.00%	0 0.00%	0 0.00%	0 0.00%	0 0.00%	0 0.00%
	SIG	3 25.00%	5 41.67%	0 0.00%	0 0.00%	0 0.00%	19 17.43%	70 34.48%	42 38.89%	95 47.26%
VAR	VAR	3.01	4.37	4.94	12.97	16.98	31.43	60.83	26.16	47.96
	SIG	4 33.33%	4 33.33%	2 16.67%	4 40.00%	7 36.84%	55 50.46%	102 50.50%	44 40.74%	79 39.30%

Table 38: New York State Crashes Jan 2014-Dec 2016 (895,917 Records)

		CALENDAR MONTH	LUNAR MONTH	MOON SIGN	M30 SPEED	M15 SPEED	M30 MONTH + SPEED	M15 MONTH + SPEED	M30 SIGN + SPEED	M15 SIGN + SPEED
EFFECT SIZE	#	12	12	12	10	19	85	151	79	139
	VS	0 0.00%	0 0.00%	0 0.00%	1 10.00%	1 5.26%	4 4.71%	2 1.32%	1 1.27%	5 3.60%
	S	8 66.67%	8 66.67%	12 100%	8 80.00%	17 89.47%	59 69.41%	90 59.60%	49 62.03%	68 48.92%
	M	3 25.00%	4 33.33%	0 0.00%	1 10.00%	1 5.26%	14 16.47%	39 25.83%	21 26.58%	42 30.22%
	L	1 8.33%	0 0.00%	0 0.00%	0 0.00%	0 0.00%	7 8.24%	19 12.58%	8 10.13%	23 16.55%
	VL	0 0.00%	0 0.00%	0 0.00%	0 0.00%	0 0.00%	0 0.00%	0 0.00%	0 0.00%	0 0.00%
	H	0 0.00%	0 0.00%	0 0.00%	0 0.00%	0 0.00%	1 1.18%	1 0.66%	0 0.00%	1 0.72%
	SIG	4 33.33%	4 33.33%	0 0.00%	1 10.00%	1 5.26%	22 25.88%	59 39.07%	29 36.71%	66 47.48%
VAR	VAR	17.07	19.20	8.34	10.99	16.35	30.59	69.51	29.94	59.18
	SIG	7 58.33%	7 58.33%	3 25.00%	4 40.00%	7 36.84%	28 32.94%	50 33.11%	22 27.85%	36 25.90%

Table 39: Texas Crashes Jan 2010-Dec 2017 (4,351,088 Records)

		CALENDAR MONTH	LUNAR MONTH	MOON SIGN	M30 SPEED	M15 SPEED	M30 MONTH + SPEED	M15 MONTH + SPEED	M30 SIGN + SPEED	M15 SIGN + SPEED
EFFECT SIZE	#	12	12	12	10	19	109	203	108	201
	VS	0 0.00%	0 0.00%	1 8.33%	0 0.00%	3 15.79%	3 2.75%	5 2.46%	1 0.93%	1 0.50%
	S	9 75.00%	9 75.00%	11 91.67%	10 100%	16 84.21%	87 79.82%	138 67.98%	71 65.74%	116 57.71%
	M	3 25.00%	3 25.00%	0 0.00%	0 0.00%	0 0.00%	16 14.68%	37 18.23%	31 28.70%	64 31.84%
	L	0 0.00%	0 0.00%	0 0.00%	0 0.00%	0 0.00%	3 2.75%	22 10.84%	5 4.63%	20 9.95%
	VL	0 0.00%	0 0.00%	0 0.00%	0 0.00%	0 0.00%	0 0.00%	0 0.00%	0 0.00%	0 0.00%
	H	0 0.00%	0 0.00%	0 0.00%	0 0.00%	0 0.00%	0 0.00%	1 0.49%	0 0.00%	0 0.00%
	SIG	3 25.00%	3 25.00%	0 0.00%	0 0.00%	0 0.00%	19 17.43%	60 29.56%	36 33.33%	84 41.79%
VAR	VAR	3.11	5.06	4.03	9.91	22.71	31.88	62.51	28.99	59.90
	SIG	3 25.00%	6 50.00%	0 0.00%	8 80.00%	12 63.16%	54 49.54%	96 47.52%	52 48.15%	106 52.74%

Table 40: Car Crash Seasonality Summary: Effect Size

CAR CRASH DATA SET	CALENDAR MONTH (12)		M30 SIGN + SPEED (108)		% DIFF. MONTH / M30	M15 SIGN + SPEED (202)		% DIFF. MONTH / M15
FARS	5	41.67%	51	47.22%	13.32%	99	49.01%	17.61%
CALIFORNIA	0	0.00%	22	20.37%	N/A	59	29.21%	N/A
IOWA	8	66.67%	49	45.37%	-31.95%	95	47.03%	-29.46%
MICHIGAN	8	66.67%	71	65.74%	-1.39%	129	63.86%	-4.21%
NEW JERSEY	1	8.33%	22	20.37%	144.54%	61	30.20%	262.55%
NEW YORK CITY	3	25.00%	42	38.89%	55.56%	95	47.26%	89.04%
NEW YORK STATE	4	33.33%	29	36.71%	10.14%	66	47.48%	42.45%
TEXAS	3	25.00%	36	33.33%	33.32%	84	41.79%	67.16%
AVERAGE SIGNIFICANCE	33.33%		38.50%		15.50%	44.48%		33.44%

Table 41: Car Crash Seasonality Summary: Variance

CAR CRASH DATA SET	CALENDAR MONTH (12)		M30 SIGN + SPEED (108)		% DIFF. MONTH / M30	M15 SIGN + SPEED (202)		% DIFF. MONTH / M15
FARS	0	0.00%	42	20.79%	N/A	54	26.73%	N/A
CALIFORNIA	5	41.67%	56	51.85%	24.43%	97	48.02%	15.24%
IOWA	7	58.33%	69	63.89%	9.53%	116	57.43%	-1.54%
MICHIGAN	7	58.33%	63	58.33%	0.00%	122	60.40%	3.55%
NEW JERSEY	5	41.67%	66	61.11%	46.65%	121	59.90%	43.75%
NEW YORK CITY	4	33.33%	44	40.74%	22.23%	79	39.30%	17.91%
NEW YORK STATE	7	58.33%	22	27.85%	-52.25%	36	25.90%	-55.60%
TEXAS	3	25.00%	52	48.15%	92.60%	106	52.74%	110.96%
AVERAGE SIGNIFICANCE	39.58%		46.59%		17.70%	46.30%		16.98%

4.4.3. Car Crash Seasonality Conclusions

Only the **FARS** data had been previously analyzed, so it was the only data set with an established expectation of a **Calendar Month** seasonal influence. A reasonable expectation of a similar seasonal influence for the other car crash data sets existed, and the presence of that seasonal influence has been confirmed using this approach. Once again, the results of the **Calendar Month** seasonality will be used as a baseline to compare the presence and significance of other seasonal influences.

As summarized in Table 40, with the exception of the **Iowa** and **Michigan** data sets, which have an unusually high percentage of significant results for the **Calendar Month** seasonality (66.67%), the percentage of seasons with significant effect sizes was greater for the irregular seasons that combine the Mercury sign and speed than it was for the **Calendar Month**. Even in these cases, the actual percentage of significant results for the irregular seasonality was quite high: 45.37% or greater. As summarized in Table 41, the percentage of seasons with significant variance was greater for the irregular seasons, except in the case of **Iowa** and **New York State**. (As noted in the previous section, the **New York State** data set had insufficient historical data to provide an accurate analysis of the irregular seasonal models, so these results should, perhaps, be isolated.)

When considering the entire car crash data set, **Calendar Month** seasonality had a 33.33% average rate of significance for effect size and a 39.58% average rate of significance for variance. The **M30 Sign + Speed** model had a 38.50% average rate of significance for effect size (**a 15.50% increase**) and a 46.59% average rate of significance for variance (**a 17.70% increase**). The **M15 Sign + Speed** model had a 44.48% average rate of significance for effect size (**a 33.44% increase**) and a 46.30% average rate of significance for variance (**a 16.98% increase**).

That these irregular, Mercury-based seasonal models appear to have so much more significance than the **Calendar Month** model of seasonality strongly suggests that these Mercury-based irregular seasonal models are worthy of further study and consideration. These influences have now been documented in two different and unrelated categories of data where **Calendar Month** seasonality traditionally merits consideration.

4.5. Financial Market Seasonality

This part of the study concludes by considering the seasonal influences found in a small sample of **Financial Market** data sets, including three stock market indexes (**Dow Jones Industrial**, **S&P 500**, **NASDAQ Composite**), three individual stocks from three different sectors (**Apple Computer**, **JP Morgan Chase**, **Walmart**), two commodities (**Gold** futures and **Soybean** futures), and the **Federal Funds Interest Rate**.

The general consensus is that financial markets, and the stock market in particular, involve such completely random patterns of data that any kind of forecasting is a waste of time. Market performance is subject to an unknown quantity of outside influences that cannot be modeled or adequately explained. With the possible exception of the retail sector (represented by **Walmart** in this subset), **Financial Market** data has no expectation **Calendar Month** seasonal influences.

4.5.1. Analysis of Regular Seasonal Influences

Table 42 on page 61 shows the seasonality by **Calendar Month** for each of the data sets. Table 43 on page 62 shows the seasonality by **Lunar Month**, and Table 44 on page 63 shows the seasonality by **Moon Sign**. Yet again, the **Moon Sign** seasonal model produced no significant results.

The only data set in this group that has a possible expectation of a **Calendar Month** seasonal influence is **Walmart**. The retail sector relies on holiday shopping each year, and it's not surprising that **Walmart** has two seasons with significant effect sizes for the **Calendar Month** (**November** and **December**) and also for the **Lunar Month** (**Scorpio** and **Sagittarius**, which roughly correspond to **November** and **December**). It's therefore surprising to see that there are five **Calendar Month** seasons with significant effect sizes for the **Dow Jones**, four for the **S&P 500**, and three for the **NASDAQ**, and frankly shocking that **Apple** stock has significant seasonal effect sizes for *nine* Calendar Month seasons. **JP Morgan Chase**, **Gold Futures**, and the **Federal Funds Rate** perform as expected, with virtually no significant effect sizes for either **Calendar Month** or **Lunar Month.**

When we consider the variance to determine whether these historic seasonal patterns can be expected to improve the accuracy of forecasts, once again, we get an entirely different story. Those wild monthly fluctuations in **Apple's** stock price have little predictive value; only a single season for either **Calendar Month** or **Lunar Month** has a low enough variance to be of use. The regular seasonal models are probably not worth considering for the **Dow Jones** or the **S&P 500**, but **May**, **June**, **July**, and **August** show potential when forecasting for the **NASDAQ**. And even though there are no significant seasonal effect sizes for the **Federal Funds** rate, six **Calendar Month** seasons and five **Lunar Month** seasons have a low enough variance that they could be expected to improve the accuracy of forecasts for that data set.

Table 42: Financial Market Seasonality: Calendar Month

CALENDAR MONTH — EFFECT & MEAN % DIF

	DOW JONES (^DJI)	S&P 500 (^GSPC)	NASDAQ (^IXIC)	APPLE (AAPL)	JP MORGAN CHASE (JPM)	WALMART (WMT)	GOLD FUTURES	SOYBEAN FUTURES	FEDERAL FUNDS RATE
JAN	-M -3.05%	-M -3.02%	-M -3.50%	-VL -11.74%	-M -3.87%	-S -2.17%	-S -0.45%	-S -1.27%	S+ 2.62%
FEB	-M -2.80%	-M -2.76%	-M -2.83%	-L -9.98%	-S -2.10%	-S -2.34%	S+ 0.50%	-S -1.05%	S+ 0.98%
MAR	-M -2.10%	-M -1.94%	-S -1.68%	-M -6.11%	VS+ 0.04%	-S -1.78%	-S -0.68%	S+ 0.88%	S+ 2.38%
APR	-S -0.67%	-S -0.84%	-S -2.26%	-M -4.80%	S+ 0.97%	-S -0.36%	-S -0.62%	S+ 2.10%	S+ 3.56%
MAY	S+ 0.44%	S+ 0.13%	-S -1.33%	-S -1.92%	S+ 1.01%	-S -1.23%	-S -0.79%	M+ 3.85%	S+ 1.10%
JUN	S+ 0.28%	S+ 0.43%	-S -0.18%	-S -1.51%	VS+ 0.06%	-S -0.71%	-S -1.16%	M+ 4.29%	S+ 1.49%
JUL	S+ 0.83%	S+ 1.08%	S+ 1.29%	S+ 0.20%	S+ 0.84%	S+ 0.44%	-S -1.22%	S+ 2.34%	S+ 0.41%
AUG	S+ 0.75%	S+ 0.82%	S+ 1.08%	M+ 5.42%	S+ 1.11%	-VS -0.06%	-S -0.13%	-S -1.39%	-S -0.93%
SEP	S+ 0.45%	S+ 0.77%	S+ 1.89%	L+ 7.51%	-S -0.33%	-S -0.34%	S+ 1.46%	-S -2.21%	-S -1.25%
OCT	S+ 0.21%	S+ 0.32%	S+ 0.82%	L+ 8.33%	-S -1.07%	S+ 0.58%	S+ 1.26%	-M -3.97%	-S -3.24%
NOV	M+ 2.02%	S+ 1.83%	S+ 2.58%	L+ 7.66%	S+ 0.84%	M+ 3.55%	S+ 1.14%	-S -1.80%	-S -4.10%
DEC	L+ 3.40%	M+ 2.91%	M+ 3.85%	M+ 5.65%	S+ 2.20%	M+ 4.28%	S+ 0.72%	-S -1.83%	-S -2.91%
SIG	5 41.67%	4 33.33%	3 25.00%	9 75.00%	1 8.33%	2 16.67%	0 0.00%	3 25.00%	0 0.00%

CALENDAR MONTH — EFFECT & VARIANCE

	DOW JONES (^DJI)	S&P 500 (^GSPC)	NASDAQ (^IXIC)	APPLE (AAPL)	JP MORGAN CHASE (JPM)	WALMART (WMT)	GOLD FUTURES	SOYBEAN FUTURES	FEDERAL FUNDS RATE
JAN	-M 39.41	-M 48.74	-M 128.31	-VL 479.20	-M 231.92	-S 193.04	-S 67.73	-S 94.22	S+ 772.99
FEB	-M 30.32	-M 31.82	-M 77.18	-L 372.45	-S 147.62	-S 142.62	S+ 46.56	-S 85.52	S+ 468.63
MAR	-M 30.89	-M 34.92	-S 83.04	-M 308.30	VS+ 139.29	-S 108.63	-S 36.48	S+ 104.74	S+ 281.01
APR	-S 23.10	-S 27.30	-S 47.57	-M 242.65	S+ 91.50	-S 56.19	-S 24.32	S+ 93.36	S+ 142.94
MAY	S+ 18.04	S+ 17.26	-S 31.78	-S 246.69	S+ 72.72	-S 29.52	-S 22.53	M+ 76.86	S+ 74.11
JUN	S+ 8.61	S+ 11.67	-S 28.24	-S 224.82	VS+ 64.61	-S 38.09	-S 15.86	M+ 78.19	S+ 98.61
JUL	S+ 12.80	S+ 15.16	S+ 29.19	S+ 157.23	S+ 55.57	S+ 58.04	-S 20.56	S+ 49.79	S+ 91.49
AUG	S+ 15.94	S+ 18.11	S+ 33.48	M+ 194.00	S+ 57.91	-VS 66.31	-S 16.82	-S 67.87	-S 76.93
SEP	S+ 27.35	S+ 29.01	S+ 63.42	L+ 260.46	-S 118.60	-S 77.67	S+ 29.28	-S 103.20	-S 109.15
OCT	S+ 36.87	S+ 44.20	S+ 87.19	L+ 444.25	-S 197.80	S+ 80.77	S+ 31.55	-M 112.85	-S 229.14
NOV	M+ 50.47	S+ 61.51	S+ 116.10	L+ 643.78	S+ 211.13	M+ 129.38	S+ 48.92	-S 146.49	-S 431.45
DEC	L+ 69.19	M+ 77.65	M+ 174.80	M+ 923.51	S+ 242.86	M+ 248.38	S+ 69.50	-S 166.51	-S 890.82
VAR	15.13	17.39	37.51	187.39	67.98	51.19	17.92	49.15	152.80
SIG	2 16.67%	3 25.00%	4 33.33%	1 8.33%	3 25.00%	2 16.67%	2 16.67%	0 0.00%	6 50.00%

Table 43: Financial Market Seasonality: Lunar Month

		DOW JONES (^DJI)	S&P 500 (^GSPC)	NASDAQ (^IXIC)	APPLE (AAPL)	JP MORGAN CHASE (JPM)	WALMART (WMT)	GOLD FUTURES	SOYBEAN FUTURES	FEDERAL FUNDS RATE
LUNAR MONTH — EFFECT & MEAN % DIF	ARIES	-S -0.61%	-S -0.82%	-S -2.16%	-M -4.31%	S+ 0.89%	-S -0.55%	-S -0.87%	S+ 2.85%	S+ 1.96%
	TAURUS	S+ 0.42%	S+ 0.15%	-S -1.10%	-S -1.41%	S+ 1.25%	-S -1.68%	-S -0.77%	M+ 3.90%	S+ 1.07%
	GEMINI	S+ 0.15%	S+ 0.29%	-S -0.29%	-S -1.47%	-S -0.57%	-S -0.62%	-S -1.49%	M+ 4.41%	S+ 2.17%
	CANCER	S+ 0.92%	S+ 1.08%	S+ 1.32%	S+ 0.71%	S+ 1.16%	S+ 0.32%	-S -1.42%	S+ 2.25%	-S -0.31%
	LEO	S+ 0.47%	S+ 0.66%	S+ 1.02%	M+ 5.71%	S+ 0.62%	-VS -0.07%	S+ 0.60%	-S -1.66%	-S -0.72%
	VIRGO	S+ 0.90%	S+ 1.16%	S+ 2.11%	L+ 7.75%	-S -0.17%	S+ 0.20%	S+ 1.86%	-S -2.42%	-S -0.91%
	LIBRA	S+ 0.60%	S+ 0.69%	S+ 1.47%	L+ 8.91%	-S -0.51%	S+ 1.28%	S+ 1.50%	-M -3.53%	-S -2.99%
	SCO.	M+ 2.39%	M+ 2.25%	M+ 3.27%	M+ 6.66%	S+ 1.57%	M+ 3.51%	S+ 1.23%	-S -1.48%	-S -3.91%
	SAG.	S+ 1.60%	S+ 1.30%	S+ 1.68%	S+ 3.01%	S+ 0.39%	M+ 2.68%	S+ 0.23%	-S -3.10%	-S -1.95%
	CAP.	-M -2.68%	-M -2.78%	-M -3.58%	-L -11.29%	-S -2.95%	-S -1.53%	-S -0.68%	-S -1.84%	S+ 0.74%
	AQU.	-M -2.32%	-M -2.27%	-S -2.17%	-L -8.32%	-S -1.69%	-S -1.63%	S+ 0.31%	-S -0.30%	S+ 2.05%
	PISCES	-M -1.95%	-S -1.85%	-S -1.77%	-M -6.65%	-S -0.17%	-S -1.82%	-S -0.48%	S+ 0.49%	S+ 2.87%
	SIG	4 33.33%	3 25.00%	2 16.67%	8 66.67%	0 0.00%	2 16.67%	0 0.00%	3 25.00%	0 0.00%
LUNAR MONTH — EFFECT & VARIANCE	ARIES	-S 22.89	-S 25.64	-S 45.45	-M 230.91	S+ 93.44	-S 52.44	-S 23.17	S+ 92.18	S+ 120.11
	TAURUS	S+ 16.25	S+ 14.88	-S 31.42	-S 252.28	S+ 58.66	-S 31.97	-S 18.42	M+ 63.74	S+ 82.93
	GEMINI	S+ 9.32	S+ 12.09	-S 28.64	-S 201.24	-S 67.42	-S 34.24	-S 15.46	M+ 79.78	S+ 82.23
	CANCER	S+ 11.55	S+ 16.25	S+ 29.62	S+ 150.43	S+ 63.28	S+ 55.81	-S 17.82	S+ 49.93	-S 93.00
	LEO	S+ 17.48	S+ 17.50	S+ 35.69	M+ 181.27	S+ 58.76	-VS 64.54	S+ 20.26	-S 75.35	-S 79.18
	VIRGO	S+ 29.63	S+ 31.57	S+ 72.52	L+ 288.57	-S 157.60	S+ 78.33	S+ 30.38	-S 92.67	-S 145.01
	LIBRA	S+ 42.19	S+ 49.91	S+ 93.22	L+ 484.90	-S 222.35	S+ 102.79	S+ 30.52	-M 123.59	-S 215.82
	SCO.	M+ 55.06	M+ 64.95	M+ 120.03	M+ 719.16	S+ 206.14	M+ 185.62	S+ 54.86	-S 150.09	-S 475.37
	SAG.	S+ 50.44	S+ 54.48	S+ 116.49	S+ 626.05	S+ 225.45	M+ 208.80	S+ 41.20	-S 108.97	-S 845.56
	CAP.	-M 29.66	-M 35.13	-M 95.72	-L 432.21	-S 178.83	-S 175.54	-S 59.67	-S 77.82	S+ 519.16
	AQU.	-M 30.14	-M 32.93	-S 89.84	-L 418.89	-S 138.95	-S 137.25	S+ 36.28	-S 84.61	S+ 478.64
	PISCES	-M 30.68	-S 37.33	-S 82.12	-M 319.04	-S 149.67	-S 99.90	-S 40.62	S+ 93.88	S+ 274.54
	VAR	14.39	16.36	35.03	179.37	67.52	51.13	16.19	45.52	142.15
	SIG	2 16.67%	3 25.00%	3 25.00%	1 8.33%	4 33.33%	2 16.67%	1 8.33%	0 0.00%	5 41.67%

Table 44: Financial Market Seasonality: Moon Sign

MOON SIGN — EFFECT & MEAN % DIF

	DOW JONES (^DJI)	S&P 500 (^GSPC)	NASDAQ (^IXIC)	APPLE (AAPL)	JP MORGAN CHASE (JPM)	WALMART (WMT)	GOLD FUTURES	SOYBEAN FUTURES	FEDERAL FUNDS RATE
ARIES	-S -0.13%	-S -0.11%	VS+ 0.02%	S+ 0.95%	-S -0.20%	-S -0.14%	-VS -0.05%	S+ 0.35%	S+ 1.38%
TAURUS	-S -0.19%	-S -0.21%	-S -0.09%	-S -0.80%	-S -0.13%	S+ 0.22%	-VS -0.02%	-S -0.22%	S+ 0.22%
GEMINI	S+ 0.16%	S+ 0.19%	S+ 0.19%	-S -0.12%	-S -0.18%	-S -0.42%	S+ 0.43%	-S -0.10%	-S -0.75%
CANCER	S+ 0.06%	S+ 0.09%	S+ 0.27%	S+ 1.44%	-VS -0.07%	S+ 0.30%	-S -0.14%	S+ 0.16%	S+ 1.04%
LEO	-S -0.28%	-S -0.28%	-S -0.47%	-S -1.47%	-VS -0.02%	VS+ 0.04%	-S -0.28%	-VS -0.03%	-S -0.64%
VIRGO	S+ 0.29%	S+ 0.34%	S+ 0.29%	-S -0.20%	VS+ 0.07%	-S -0.23%	S+ 0.18%	-S -0.20%	-S -0.70%
LIBRA	-S -0.08%	-S -0.05%	VS+ 0.06%	S+ 0.65%	VS+ 0.00%	VS+ 0.06%	-S -0.13%	S+ 0.37%	S+ 0.95%
SCO.	-VS -0.01%	-S -0.06%	-S -0.22%	-S -0.22%	S+ 0.19%	S+ 0.45%	-S -0.42%	-S -0.17%	-S -0.85%
SAG.	S+ 0.23%	S+ 0.28%	S+ 0.32%	-S -0.16%	S+ 0.40%	-S -0.29%	S+ 0.35%	VS+ 0.03%	-S -0.50%
CAP.	-VS -0.04%	-VS -0.05%	-VS -0.01%	S+ 0.91%	VS+ 0.03%	-S -0.24%	VS+ 0.06%	VS+ 0.01%	S+ 2.07%
AQU.	-S -0.39%	-S -0.45%	-S -0.64%	-S -0.94%	-S -0.32%	S+ 0.24%	-S -0.29%	-S -0.57%	-S -1.19%
PISCES	S+ 0.39%	S+ 0.34%	S+ 0.31%	VS+ 0.08%	S+ 0.23%	VS+ 0.01%	S+ 0.33%	S+ 0.39%	-S -0.98%
SIG	0 0.00%	0 0.00%	0 0.00%	0 0.00%	0 0.00%	0 0.00%	0 0.00%	0 0.00%	0 0.00%

MOON SIGN — EFFECT & VARIANCE

	DOW JONES (^DJI)	S&P 500 (^GSPC)	NASDAQ (^IXIC)	APPLE (AAPL)	JP MORGAN CHASE (JPM)	WALMART (WMT)	GOLD FUTURES	SOYBEAN FUTURES	FEDERAL FUNDS RATE
ARIES	-S 1.43	-S 1.70	VS+ 2.64	S+ 18.96	-S 4.68	-S 6.26	-VS 1.63	S+ 5.32	S+ 9.01
TAURUS	-S 1.44	-S 1.35	-S 3.40	-S 22.28	-S 6.17	S+ 4.35	-VS 1.25	-S 4.00	S+ 7.37
GEMINI	S+ 1.02	S+ 1.41	S+ 2.74	-S 17.57	-S 5.77	-S 5.47	S+ 1.82	-S 5.07	-S 14.88
CANCER	S+ 0.97	S+ 1.24	S+ 2.52	S+ 15.08	-VS 5.27	S+ 4.48	-S 1.17	S+ 6.06	S+ 9.56
LEO	-S 1.38	-S 1.78	-S 3.63	-S 23.54	-VS 5.90	VS+ 3.86	-S 1.44	-VS 5.31	-S 11.65
VIRGO	S+ 1.33	S+ 1.35	S+ 3.40	-S 19.73	VS+ 6.74	-S 4.27	S+ 1.31	-S 2.99	-S 9.29
LIBRA	-S 0.76	-S 0.92	VS+ 1.75	S+ 20.07	VS+ 5.77	VS+ 4.68	-S 1.71	S+ 6.77	S+ 12.13
SCO.	-VS 1.51	-S 1.64	-S 3.24	-S 18.90	S+ 7.07	S+ 3.74	-S 1.29	-S 5.22	-S 8.68
SAG.	S+ 2.14	S+ 1.76	S+ 3.00	-S 14.39	S+ 4.15	-S 5.80	S+ 1.11	VS+ 3.03	-S 7.94
CAP.	-VS 1.03	-VS 0.94	-VS 2.23	S+ 13.01	VS+ 8.39	-S 4.20	VS+ 1.23	VS+ 6.79	S+ 14.89
AQU.	-S 1.32	-S 1.47	-S 3.53	-S 18.87	-S 3.74	S+ 5.18	-S 1.26	-S 5.69	-S 9.24
PISCES	S+ 1.74	S+ 1.63	S+ 2.35	VS+ 15.44	S+ 4.13	VS+ 4.62	S+ 1.25	S+ 5.35	-S 15.06
VAR	0.67	0.72	1.43	9.08	2.82	2.37	0.69	2.57	5.40
SIG	0 0.00%	0 0.00%	0 0.00%	0 0.00%	0 0.00%	0 0.00%	0 0.00%	0 0.00%	0 0.00%

4.5.2. Analysis of Irregular Seasonal Influences

Table 45 through Table 53 compare the three regular seasonality models (**Calendar Month, Lunar Month, Moon Sign**) with six Mercury-based irregular seasonality models (**M30 Speed, M15 Speed, M30 Month + Speed, M15 Month + Speed, M30 Sign + Speed, M15 Sign + Speed**) for each individual data set.

Each table is divided into two sections. The top section is a straight count of the effect sizes. Results with a effect size of Medium or greater are considered to be significant, and the number and percentage of seasons with significant effect sizes is highlighted at the bottom of the section. The bottom section (the last two rows of each table) evaluates the variance. The **VAR** row shows the minimum significant threshold number for each data set, and the **SIG** row below it shows the number and percentage of seasons with a significant variance.

Table 54 on page 69 provides a summary that compares the significance of the **Calendar Month** seasonal model with the significance of the **M30 Sign + Speed** and the **M15 Sign + Speed** seasonal models in terms of effect size, across the entire data set. Table 55 on page 69 provides a summary that compares the significance of the **Calendar Month** seasonal model with the significance of the **M30 Sign + Speed** and the **M15 Sign + Speed** seasonal models in terms of variance, across the entire data set.

Table 45: Dow Jones Industrial Average (^DJI) Seasonality (1985-2018)

		CALENDAR MONTH	LUNAR MONTH	MOON SIGN	M30 SPEED	M15 SPEED	M30 MONTH + SPEED	M15 MONTH + SPEED	M30 SIGN + SPEED	M15 SIGN + SPEED
EFFECT SIZE	#	12	12	12	10	19	109	203	108	202
	VS	0 0.00%	0 0.00%	2 16.67%	3 30.00%	2 10.53%	5 4.59%	5 2.46%	1 0.93%	5 2.48%
	S	7 58.33%	8 66.67%	10 83.33%	7 70.00%	16 84.21%	78 71.56%	144 70.94%	53 49.07%	96 47.52%
	M	4 33.33%	4 33.33%	0 0.00%	0 0.00%	1 5.26%	23 21.10%	43 21.18%	37 34.26%	62 30.69%
	L	1 8.33%	0 0.00%	0 0.00%	0 0.00%	0 0.00%	3 2.75%	10 4.93%	15 13.89%	34 16.83%
	VL	0 0.00%	0 0.00%	0 0.00%	0 0.00%	0 0.00%	0 0.00%	1 0.49%	2 1.85%	5 2.48%
	H	0 0.00%	0 0.00%	0 0.00%	0 0.00%	0 0.00%	0 0.00%	0 0.00%	0 0.00%	0 0.00%
	SIG	5 41.67%	4 33.33%	0 0.00%	0 0.00%	1 5.26%	26 23.85%	54 26.60%	54 50.00%	101 50.00%
VAR	VAR	15.13	14.39	0.67	2.70	4.27	1.77	2.22	14.28	14.08
	SIG	2 16.67%	2 16.67%	0 0.00%	3 30.00%	4 21.05%	33 30.28%	84 41.58%	40 37.04%	78 38.61%

Table 46: S&P 500 (^GSPC) Seasonality (1980-2018)

		CALENDAR MONTH	LUNAR MONTH	MOON SIGN	M30 SPEED	M15 SPEED	M30 MONTH + SPEED	M15 MONTH + SPEED	M30 SIGN + SPEED	M15 SIGN + SPEED
EFFECT SIZE	#	12	12	12	10	19	109	203	108	202
	VS	0 0.00%	0 0.00%	1 8.33%	1 10.00%	2 10.53%	3 2.75%	3 1.48%	1 0.93%	2 0.99%
	S	8 66.67%	9 75.00%	11 91.67%	9 90.00%	16 84.21%	88 80.73%	152 74.88%	56 51.85%	108 53.47%
	M	4 33.33%	3 25.00%	0 0.00%	0 0.00%	1 5.26%	14 12.84%	41 20.20%	35 32.41%	58 28.71%
	L	0 0.00%	0 0.00%	0 0.00%	0 0.00%	0 0.00%	4 3.67%	6 2.96%	15 13.89%	28 13.86%
	VL	0 0.00%	0 0.00%	0 0.00%	0 0.00%	0 0.00%	0 0.00%	1 0.49%	1 0.93%	6 2.97%
	H	0 0.00%	0 0.00%	0 0.00%	0 0.00%	0 0.00%	0 0.00%	0 0.00%	0 0.00%	0 0.00%
	SIG	4 33.33%	3 25.00%	0 0.00%	0 0.00%	1 5.26%	18 16.51%	48 23.65%	51 47.22%	92 45.54%
VAR	VAR	17.39	16.36	0.72	2.71	4.42	1.67	2.08	16.14	15.86
	SIG	3 25.00%	3 25.00%	0 0.00%	3 30.00%	4 21.05%	32 29.36%	77 38.12%	32 29.63%	68 33.66%

Table 47: NADAQ Composite (^IXIC) Seasonality (1980-2018)

		CALENDAR MONTH	LUNAR MONTH	MOON SIGN	M30 SPEED	M15 SPEED	M30 MONTH + SPEED	M15 MONTH + SPEED	M30 SIGN + SPEED	M15 SIGN + SPEED
EFFECT SIZE	#	12	12	12	10	19	109	203	108	202
	VS	0 0.00%	0 0.00%	3 25.00%	2 20.00%	1 5.26%	3 2.75%	5 2.46%	4 3.70%	7 3.47%
	S	9 75.00%	10 83.33%	9 75.00%	8 80.00%	17 89.47%	84 77.06%	138 67.98%	62 57.41%	114 56.44%
	M	3 25.00%	2 16.67%	0 0.00%	0 0.00%	1 5.26%	17 15.60%	45 22.17%	33 30.56%	59 29.21%
	L	0 0.00%	0 0.00%	0 0.00%	0 0.00%	0 0.00%	4 3.67%	13 6.40%	8 7.41%	16 7.92%
	VL	0 0.00%	0 0.00%	0 0.00%	0 0.00%	0 0.00%	1 0.92%	2 0.99%	1 0.93%	6 2.97%
	H	0 0.00%	0 0.00%	0 0.00%	0 0.00%	0 0.00%	0 0.00%	0 0.00%	0 0.00%	0 0.00%
	SIG	3 25.00%	2 16.67%	0 0.00%	0 0.00%	1 5.26%	22 20.18%	60 29.56%	42 38.89%	81 40.10%
VAR	VAR	37.51	35.03	1.43	5.82	2.90	36.66	9.53	3.48	36.25
	SIG	4 33.33%	3 25.00%	0 0.00%	4 40.00%	27 24.77%	30 27.78%	5 26.32%	72 35.64%	66 32.67%

Table 48: Apple Computer (AAPL) Seasonality (1981-2018)

		CALENDAR MONTH	LUNAR MONTH	MOON SIGN	M30 SPEED	M15 SPEED	M30 MONTH + SPEED	M15 MONTH + SPEED	M30 SIGN + SPEED	M15 SIGN + SPEED
EFFECT SIZE	#	12	12	12	10	19	109	203	108	202
	VS	0 0.00%	0 0.00%	1 8.33%	2 20.00%	2 10.53%	5 4.59%	3 1.48%	0 0.00%	0 0.00%
	S	3 25.00%	4 33.33%	11 91.67%	8 80.00%	17 89.47%	65 59.63%	111 54.68%	38 35.19%	71 35.15%
	M	4 33.33%	4 33.33%	0 0.00%	0 0.00%	0 0.00%	28 25.69%	49 24.14%	35 32.41%	63 31.19%
	L	4 33.33%	4 33.33%	0 0.00%	0 0.00%	0 0.00%	9 8.26%	34 16.75%	28 25.93%	51 25.25%
	VL	1 8.33%	0 0.00%	0 0.00%	0 0.00%	0 0.00%	2 1.83%	6 2.96%	7 6.48%	17 8.42%
	H	0 0.00%	0 0.00%	0 0.00%	0 0.00%	0 0.00%	0 0.00%	0 0.00%	0 0.00%	0 0.00%
	SIG	9 75.00%	8 66.67%	0 0.00%	0 0.00%	0 0.00%	39 35.78%	89 43.84%	70 64.81%	131 64.85%
VAR	VAR	187.39	179.37	9.08	42.54	62.32	13.48	15.07	188.82	192.12
	SIG	1 8.33%	1 8.33%	0 0.00%	6 60.00%	7 36.84%	24 22.02%	58 28.71%	19 17.59%	38 18.81%

Table 49: JP Morgan Chase (JPM) Seasonality (1980-2018)

		CALENDAR MONTH	LUNAR MONTH	MOON SIGN	M30 SPEED	M15 SPEED	M30 MONTH + SPEED	M15 MONTH + SPEED	M30 SIGN + SPEED	M15 SIGN + SPEED
EFFECT SIZE	#	12	12	12	10	19	109	203	108	202
	VS	2 16.67%	0 0.00%	5 41.67%	1 10.00%	0 0.00%	2 1.83%	2 0.99%	1 0.93%	6 2.97%
	S	9 75.00%	12 100%	7 58.33%	9 90.00%	19 100%	88 80.73%	157 77.34%	80 74.07%	136 67.33%
	M	1 8.33%	0 0.00%	0 0.00%	0 0.00%	0 0.00%	17 15.60%	39 19.21%	21 19.44%	46 22.77%
	L	0 0.00%	0 0.00%	0 0.00%	0 0.00%	0 0.00%	2 1.83%	4 1.97%	5 4.63%	10 4.95%
	VL	0 0.00%	0 0.00%	0 0.00%	0 0.00%	0 0.00%	0 0.00%	1 0.49%	1 0.93%	3 1.49%
	H	0 0.00%	0 0.00%	0 0.00%	0 0.00%	0 0.00%	0 0.00%	0 0.00%	0 0.00%	1 0.50%
	SIG	1 8.33%	0 0.00%	0 0.00%	0 0.00%	0 0.00%	19 17.43%	44 21.67%	27 25.00%	60 29.70%
VAR	VAR	67.98	67.52	2.82	9.92	16.42	6.88	8.00	66.26	66.64
	SIG	3 25.00%	4 33.33%	0 0.00%	3 30.00%	6 31.58%	20 18.35%	54 26.73%	23 21.30%	51 25.25%

Table 50: Walmart (WMT) Seasonality (1980-2018)

		CALENDAR MONTH	LUNAR MONTH	MOON SIGN	M30 SPEED	M15 SPEED	M30 MONTH + SPEED	M15 MONTH + SPEED	M30 SIGN + SPEED	M15 SIGN + SPEED
EFFECT SIZE	#	12	12	12	10	19	109	203	108	202
	VS	1 8.33%	1 8.33%	3 25.00%	1 10.00%	2 10.53%	2 1.83%	4 1.97%	2 1.85%	7 3.47%
	S	9 75.00%	9 75.00%	9 75.00%	9 90.00%	16 84.21%	91 83.49%	150 73.89%	67 62.04%	114 56.44%
	M	2 16.67%	2 16.67%	0 0.00%	0 0.00%	1 5.26%	15 13.76%	40 19.70%	29 26.85%	51 25.25%
	L	0 0.00%	0 0.00%	0 0.00%	0 0.00%	0 0.00%	0 0.00%	7 3.45%	7 6.48%	23 11.39%
	VL	0 0.00%	0 0.00%	0 0.00%	0 0.00%	0 0.00%	1 0.92%	2 0.99%	3 2.78%	7 3.47%
	H	0 0.00%	0 0.00%	0 0.00%	0 0.00%	0 0.00%	0 0.00%	0 0.00%	0 0.00%	0 0.00%
	SIG	2 16.67%	2 16.67%	0 0.00%	0 0.00%	1 5.26%	16 14.68%	49 24.14%	39 36.11%	81 40.10%
VAR	VAR	51.19	51.13	2.37	10.76	15.06	3.70	4.33	46.00	46.16
	SIG	2 16.67%	2 16.67%	0 0.00%	4 40.00%	7 36.84%	22 20.18%	48 23.76%	28 25.93%	54 26.73%

Table 51: Gold Futures Seasonality (1980-2018)

		CALENDAR MONTH	LUNAR MONTH	MOON SIGN	M30 SPEED	M15 SPEED	M30 MONTH + SPEED	M15 MONTH + SPEED	M30 SIGN + SPEED	M15 SIGN + SPEED
EFFECT SIZE	#	12	12	12	10	19	109	203	108	202
	VS	0 0.00%	0 0.00%	3 25.00%	0 0.00%	2 10.53%	1 0.92%	1 0.49%	4 3.70%	4 1.98%
	S	12 100%	12 100%	9 75.00%	10 100%	17 89.47%	80 73.39%	140 68.97%	88 81.48%	160 79.21%
	M	0 0.00%	0 0.00%	0 0.00%	0 0.00%	0 0.00%	24 22.02%	46 22.66%	15 13.89%	31 15.35%
	L	0 0.00%	0 0.00%	0 0.00%	0 0.00%	0 0.00%	4 3.67%	14 6.90%	1 0.93%	5 2.48%
	VL	0 0.00%	0 0.00%	0 0.00%	0 0.00%	0 0.00%	0 0.00%	2 0.99%	0 0.00%	2 0.99%
	H	0 0.00%	0 0.00%	0 0.00%	0 0.00%	0 0.00%	0 0.00%	0 0.00%	0 0.00%	0 0.00%
	SIG	0 0.00%	0 0.00%	0 0.00%	0 0.00%	0 0.00%	28 25.69%	62 30.54%	16 14.81%	38 18.81%
VAR	VAR	17.92	16.19	0.69	3.60	4.95	1.91	1.99	18.43	18.99
	SIG	2 16.67%	1 8.33%	0 0.00%	5 50.00%	6 31.58%	17 15.60%	41 20.30%	26 24.07%	49 24.26%

Table 52: Soybean Futures Seasonality (1980-2018)

		CALENDAR MONTH	LUNAR MONTH	MOON SIGN	M30 SPEED	M15 SPEED	M30 MONTH + SPEED	M15 MONTH + SPEED	M30 SIGN + SPEED	M15 SIGN + SPEED
EFFECT SIZE	#	12	12	12	10	19	109	203	108	202
	VS	0 0.00%	0 0.00%	3 25.00%	1 10.00%	2 10.53%	5 4.59%	2 0.99%	0 0.00%	4 1.98%
	S	9 75.00%	9 75.00%	9 75.00%	9 90.00%	17 89.47%	77 70.64%	138 67.98%	65 60.19%	113 55.94%
	M	3 25.00%	3 25.00%	0 0.00%	0 0.00%	0 0.00%	23 21.10%	47 23.15%	31 28.70%	57 28.22%
	L	0 0.00%	0 0.00%	0 0.00%	0 0.00%	0 0.00%	3 2.75%	12 5.91%	10 9.26%	25 12.38%
	VL	0 0.00%	0 0.00%	0 0.00%	0 0.00%	0 0.00%	1 0.92%	4 1.97%	2 1.85%	3 1.49%
	H	0 0.00%	0 0.00%	0 0.00%	0 0.00%	0 0.00%	0 0.00%	0 0.00%	0 0.00%	0 0.00%
	SIG	3 25.00%	3 25.00%	0 0.00%	0 0.00%	0 0.00%	27 24.77%	63 31.03%	43 39.81%	85 42.08%
VAR	VAR	49.15	45.52	2.57	8.85	12.34	3.06	3.31	49.97	50.02
	SIG	0 0.00%	0 0.00%	0 0.00%	5 50.00%	7 36.84%	22 20.18%	51 25.25%	15 13.89%	34 16.83%

Table 53: Federal Funds Interest Rate Seasonality (1980-2018)

		CALENDAR MONTH	LUNAR MONTH	MOON SIGN	M30 SPEED	M15 SPEED	M30 MONTH + SPEED	M15 MONTH + SPEED	M30 SIGN + SPEED	M15 SIGN + SPEED
EFFECT SIZE	#	12	12	12	10	19	109	203	108	202
	VS	0 0.00%	0 0.00%	0 0.00%	1 10.00%	1 5.26%	1 0.92%	5 2.46%	2 1.85%	5 2.48%
	S	12 100%	12 100%	12 100%	9 90.00%	18 94.74%	82 75.23%	135 66.50%	67 62.04%	111 54.95%
	M	0 0.00%	0 0.00%	0 0.00%	0 0.00%	0 0.00%	21 19.27%	46 22.66%	32 29.63%	69 34.16%
	L	0 0.00%	0 0.00%	0 0.00%	0 0.00%	0 0.00%	4 3.67%	16 7.88%	7 6.48%	15 7.43%
	VL	0 0.00%	0 0.00%	0 0.00%	0 0.00%	0 0.00%	1 0.92%	1 0.49%	0 0.00%	2 0.99%
	H	0 0.00%	0 0.00%	0 0.00%	0 0.00%	0 0.00%	0 0.00%	0 0.00%	0 0.00%	0 0.00%
	SIG	0 0.00%	0 0.00%	0 0.00%	0 0.00%	0 0.00%	26 23.85%	63 31.03%	39 36.11%	86 42.57%
VAR	VAR	152.80	142.15	5.40	23.68	35.06	25.26	24.01	135.20	142.11
	SIG	6 50.00%	5 41.67%	0 0.00%	3 30.00%	4 21.05%	57 52.29%	100 49.50%	41 37.96%	74 36.63%

Table 54: Financial Market Seasonality Summary: Effect Size

STOCK MARKET DATA SET	CALENDAR MONTH (12)		M30 SIGN + SPEED (108)		% DIFF. MONTH / M30	M15 SIGN + SPEED (202)		% DIFF. MONTH / M15
DOW JONES INDUSTRIAL (^DJI)	5	41.67%	54	50.00%	19.99%	101	50.00%	19.99%
S&P 500 (^GSPC)	4	33.33%	51	47.22%	41.67%	92	45.54%	36.63%
NASDAQ COMPOSITE (^IXIC)	3	25.00%	42	38.39%	53.56%	81	40.10%	60.40%
APPLE COMPUTER (AAPL)	9	75.00%	70	64.81%	-13.59%	131	64.85%	-13.53%
JP MORGAN CHASE (JPM)	1	8.33%	27	25.00%	200.12%	60	29.70%	256.54%
WALMART (WMT)	2	16.67%	39	36.11%	116.62%	81	40.10%	140.55%
GOLD FUTURES	0	0.00%	16	14.81%	N/A	38	18.81%	N/A
SOYBEAN FUTURES	3	25.00%	43	39.81%	59.24%	85	42.08%	68.32%
FEDERAL FUNDS RATE	0	0.00%	39	36.11%	N/A	86	42.57%	N/A
AVERAGE SIGNIFICANCE	25.00%		39.14%		56.56%	41.53%		66.12%

Table 55: Financial Market Seasonality Summary: Variance

STOCK MARKET DATA SET	CALENDAR MONTH (12)		M30 SIGN + SPEED (108)		% DIFF. MONTH / M30	M15 SIGN + SPEED (202)		% DIFF. MONTH / M15
DOW JONES INDUSTRIAL (^DJI)	2	16.67%	40	37.04%	122.20%	78	38.61%	131.61%
S&P 500 (^GSPC)	3	25.00%	32	29.63%	18.52%	68	33.66%	34.64%
NASDAQ COMPOSITE (^IXIC)	4	33.33%	72	35.64%	6.93%	66	32.67%	-1.98%
APPLE COMPUTER (AAPL)	1	8.33%	19	17.59%	111.16%	38	18.81%	125.81%
JP MORGAN CHASE (JPM)	3	25.00%	23	21.30%	-14.80%	51	25.25%	1.00%
WALMART (WMT)	2	16.67%	28	25.93%	55.55%	54	26.73%	60.35%
GOLD FUTURES	2	16.67%	26	24.07%	44.39%	49	24.26%	45.53%
SOYBEAN FUTURES	0	0.00%	15	13.89%	N/A	34	16.83%	N/A
FEDERAL FUNDS RATE	6	50.00%	41	37.96%	-24.08%	74	36.63%	-26.74%
AVERAGE SIGNIFICANCE	21.30%		27.01%		26.81%	28.16%		32.21%

4.5.3. Financial Market Seasonality Conclusions

As before, the **Calendar Month** seasonality will be used as a baseline to compare the presence and significance of other seasonal influences. As shown in Table 54, with the exception of **Apple Computer**, which has the highest percentage of significant results for the **Calendar Month** seasonality across this entire study (75%), the percentage of seasons with significant effect sizes was greater for the irregular seasons that combine the Mercury sign and speed than it was for the **Calendar Month**. As shown in Table 55, the percentage of seasons with significant variance was greater in most cases as well, with the notable exception of the **Federal Funds Rate**.

When considering the entire financial market data set used in this study, **Calendar Month** seasonality had a 25% average rate of significance for effect size and a 21.30% average rate of significance for variance. The **M30 Sign + Speed** model had a 39.14% average rate of significance for effect size (**a 56.56% increase**) and a 27.01% average rate of significance for variance (**a 26.81% increase**). The **M15 Sign + Speed** model had a 41.53% average rate of significance for effect size (**a 66.12% increase**) and a 28.16% average rate of significance for variance (**a 32.21% increase**).

This specific category of data had no expectation of any significant seasonal influence and yet the irregular, Mercury-based seasonal models strongly suggest the existence of significant seasonal influences. **As this is the third unrelated category of data that contains these influences, we now have compelling evidence of the presence and potential significance of irregular seasonal influences.**

4.6. Conclusions

This study has clearly demonstrated the advantages of using **Cohen's d** to quantify the effect size of seasonal influences. We have seen the value of using these objective metrics to compare the influence of seasons within a seasonal model, between seasonal models, and even between different data sets. By establishing a baseline threshold of significance, it's possible to quantify and compare the significance of different seasonal models using the percentage of significant seasons as the metric.

This ability to quantify and compare different seasonal models made it possible to consider the potential value of irregular models, specifically six irregular models based on the cycles of the planet Mercury. The irregular seasonal models consistently revealed seasonal patterns that were otherwise undetectable. Most notably, the irregular seasonal models showed significant seasonal influences in the **Financial Market** data, which showed little or no significant seasonal influences with the **Calendar Month** model. This provides compelling evidence of the potential value of irregular seasonal models.

The results of this study also suggest that while the magnitude of the seasonal effect is a useful metric when analyzing the historical seasonal influences, the size of the effect has little correlation with the anticipated forecast value of the season. The lower the variance of the historic effect sizes of a season, the greater the expected forecast value is. The seasons with the most significant effect sizes were rarely the seasons with the lowest historical variance.

The correlation between variance and forecast value is explored as a part of the forecasting study presented in Section 5.

5. The Use of Irregular Seasonality in Quantitative Time Series Forecasting

This study explores whether irregular seasonality can be used to improve the accuracy of quantitative time series forecasts. The scope of this question is vast, and this new effect- and variance-based method of quantifying seasonal influences potentially represents an entirely new and unexplored universe of statistical discovery. All that this study can do is explore how the application of a single irregular seasonal model, **M15 Sign + Speed** can be used to improve forecast accuracy. If the practical value of this single irregular seasonal model is clear, it should justify further research into the potential value of irregular seasonal models.

5.1. Forecast Study Overview and Objectives

The question being considered is whether forecasts that include the irregular seasonal influence of the **M15 Sign + Speed** model are more accurate than forecasts that do not. To answer this question multiple forecasts must be generated for each data set using different forecast methods to evaluate how often the forecasts that include the seasonal influences are more accurate than those that don't. If the seasonal forecasts are consistently more accurate than the non-seasonal forecasts, we can conclude that this irregular, Mercury-based seasonal model has clear value in quantitative time series forecasting.

Because this new effect- and variance-based approach to seasonality allows for selective incorporation of the seasonal influences (as opposed to the traditional application of seasonality which is an all-or-nothing proposition), each set of forecasts can be evaluated twice. The first evaluation considers and compares the accuracy of forecasts for the entire forecast period, while the second evaluation considers and compares only the accuracy of forecasts for dates that correspond to the seasons with significantly low variance for the data set (**Targeted Seasonality**). This allows us to test the hypothesis that the lower the historical variance of the effect sizes of a season is, the more likely that season will have value when forecasting future instances of that season.

Because we are considering only a single seasonal model, the scope of the data being evaluated must be as vast as possible for the results of the study to be considered significant. The difference in forecast accuracy must be demonstrated across hundreds of different forecasts, and the duration of those forecasts must be substantial enough that it accurately reflects the effectiveness of the methodology by minimizing the impact of short-term fluctuations.

The forecast methodology presents another challenge with respect to the data. The seasonal forecasts are based on a moving average of three historical instances of a season. But the irregular nature of the **M15 Sign + Speed** seasonal model requires at least 20 years of historical data, and even that only guarantees sufficient historical references for 200 out of the 202 individual seasons. This means that to be able to generate 19 years of quarterly forecasts (from 2000 through 2018), each data set for each forecast required at least 39 years of data, aggregated on a daily basis.

The only publicly available data that meets these criteria are financial market data. This means that on the face, at least, this study appears to be about how to forecast the stock market. This is not the intention, nor is it the objective of this study.

True, accurately predicting the stock market is the Holy Grail of quantitative time series forecasting. It may be theoretically impossible, but it still represents a substantial portion of the practical applications of time series analysis.

Ironically, the expectation that it is impossible to forecast the stock market makes using stock market data ideal for this study. The random nature of financial data and the inconsistent expectations of accuracy within financial forecasts create a neutral background. Any seasonal patterns will show up in stark contrast, making them easier to identify and evaluate. Moreover, what's being considered is the aggregate accuracy of 19 years of quarterly forecasts, a perspective with no practical application for investors.

This study evaluates forecasts generated for **a total of 430 individual financial data sets**, including **10 stock market indexes**, **379 individual stocks**, **21 commodities**, **10 interest rates and bonds**, and **10 currency exchange rates**. The stocks were selected primarily because they had the requisite 39 years of historical data. They are grouped both by general sector and specific category. A small percentage of the stocks have less than 39 years of historical data. These stocks were included to maximize the representation of each sector and category. For the early forecast years, they may not have three historical references for every season, but this did not have a significant impact on the results.

The complete list of the financial source data, including the start date of the historical data, is included in **Appendix A**.

5.2. Forecast Study Methodology

The methodology of the forecast study is quite involved. First, the Mercury forecast signal must be generated from the historical time series data. Next, that signal must be combined with each of the five traditional forecast signals: autoregressive integrated moving average (**ARIMA**), exponential smoothing model (**ESM**), **MEAN**, **NAÏVE**, and **HOLT**. The accuracy of each forecast is then measured, compared, and ranked. Finally, the entire set of forecasts is re-evaluated using **Targeted Seasonality**, quantifying and comparing the accuracy of only the seasons with significant (low) variance. The methodology to quantify variance is addressed in **Section 4.1.3** on page 33.

5.2.1. Methodology to Generate the Mercury Forecast Signal

The principle that the forecast value of a season is based on a moving average of the historical values of that season is quite simple. Creating a working forecast model was far more complicated than expected.

The initial forecast model was named **E3** because it based the forecast on an equal average of the mean values of the three prior instances of the season. The problem with this approach, at least when forecasting for financial instruments, is that the stock market, and stock prices, rise significantly over time. The mean stock prices used in the moving average could be up to 20 years old. While the shape and inflections of the **E3** seasonal forecast often mirrored the actual ups and downs of the stock price, the values of the forecast were so much lower than the current stock prices that quantifying and comparing the accuracy was futile.

The **E3** forecast model used in this study is based on an equal moving average of the **adjusted mean stock prices** from the three most recent historical periods of the Mercury season. Where only two historical periods exist, the forecast is the average of the adjusted mean stock price for those two periods. Where only one historical period exists, the forecast is the adjusted stock price for that period.

The mean seasonal stock prices are adjusted based on the ratio between the **historical mean**, which is connected to the historical season, and the **reference mean**, which is connected to the current market values. For this study, the **historical mean** is the mean close price of the stock for a 30-day period centered on the historical season. The **reference mean** is the mean stock price of the 10-calendar-day period immediately prior to the quarter being forecasted. The actual mean stock price of each historical season is multiplied by the ratio of these two means to produce the **adjusted mean stock price**. The moving average of the adjusted mean stock price for the three most recent historical instances of the forecast season is the forecast value for the season.

Keep in mind that the seasonal forecast does not generate daily forecast values; it generates a single forecast value that remains constant for the duration of the season. As detailed in **Section 3.2.2 on page 21**, although a season in the **M15 Sign + Speed** model could last as long as 27 days, 59.2% of the seasons last from 1 to 4 days. Technically what's being forecasted is the mean stock price for the season. Changes to the forecasted values occur only between seasons.

Figure 4 on page 75 shows the **M15_E3** forecast plotted against the actual close price of the S&P 500 for the years 2004 through 2006. The forecasts are quarterly, which is why there is often a stark adjustment of the forecast price at the start of each quarter. **Remember that the forecast for each year was generated using only the historical stock data.** Note how the shape of the forecast so closely follows the shape of the actual close price.

Figure 5 on page 76 shows the same **M15_E3** forecast, but this time it's plotted against the mean close price of the S&P500 for each of the **M15** Mercury seasons. This makes it easier to visualize the relationship between the seasonal forecast and the seasonal timescale for the forecast years.

Figure 6 on page 77 compares the quarterly **ESM** forecasts with the quarterly **M15_E3** forecasts. Note the difference in the forecast signals: the daily variations in the **ESM** signal are so small that when viewed from this scale the **ESM** forecasts are effectively trend lines, while the **M15_E3** forecast signal has peaks and troughs similar to the actual close prices.

The closer a forecast line is to the line of the close price, the more accurate the forecast. There are a few quarters where the **M15_E3** forecast appears to be more accurate than the **ESM** forecast, most notably Q1 and Q2 of 2005. In most of the other quarters, however, the trend line of the **ESM** forecast is closer to the actual values of the close signal.

After completing the initial study and considering the advantages and disadvantages of the **E3** forecast methodology, a new forecast model was developed, designated **M3**. Rather than working with the actual close prices and then adjusting those values for inflation and trends, the **M3** forecast considers the relationship between the mean stock price of the historic season and the mean stock price of the quarter that contains the historic season. The percent difference between the seasonal mean and the quarterly reference mean is used to create the **M3** forecast.

Each historic season has an associated percentage, and the mean of the three most recent seasonal percentages is the forecast percentage. The actual forecast for the season is the mean forecast percentage multiplied by the mean stock price of the 10 calendar day period immediately prior to the quarter being forecasted. Figure 7 on page 78 compares the **M15_E3** and the **M15_M3** signals for the S&P 500 for 2004 through 2006. The **M15_M3** signal appears to have a comparable degree of accuracy to the **M15_E3** signal, which is why it warranted a second iteration of the forecast study.

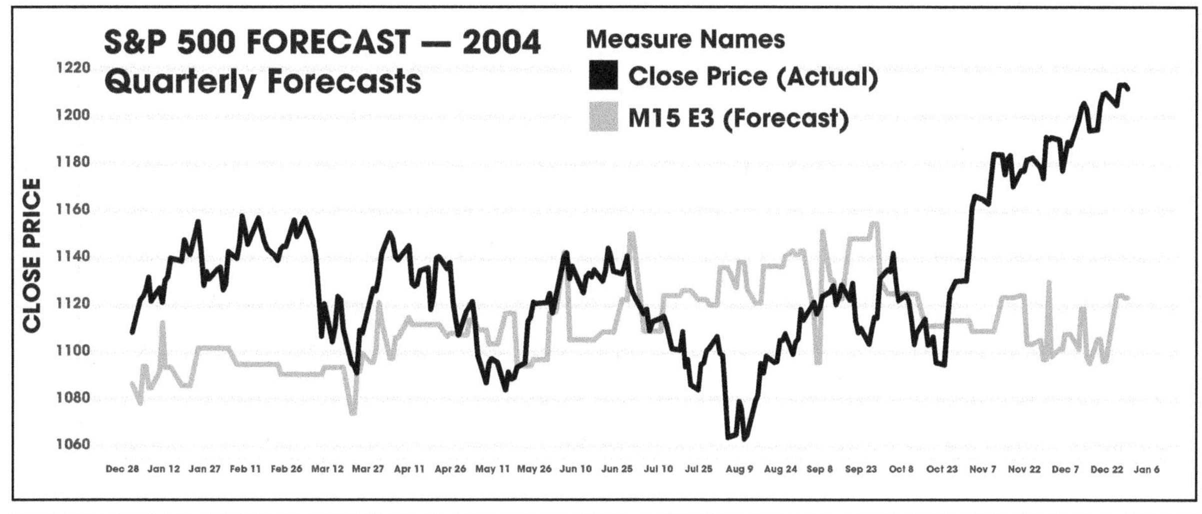

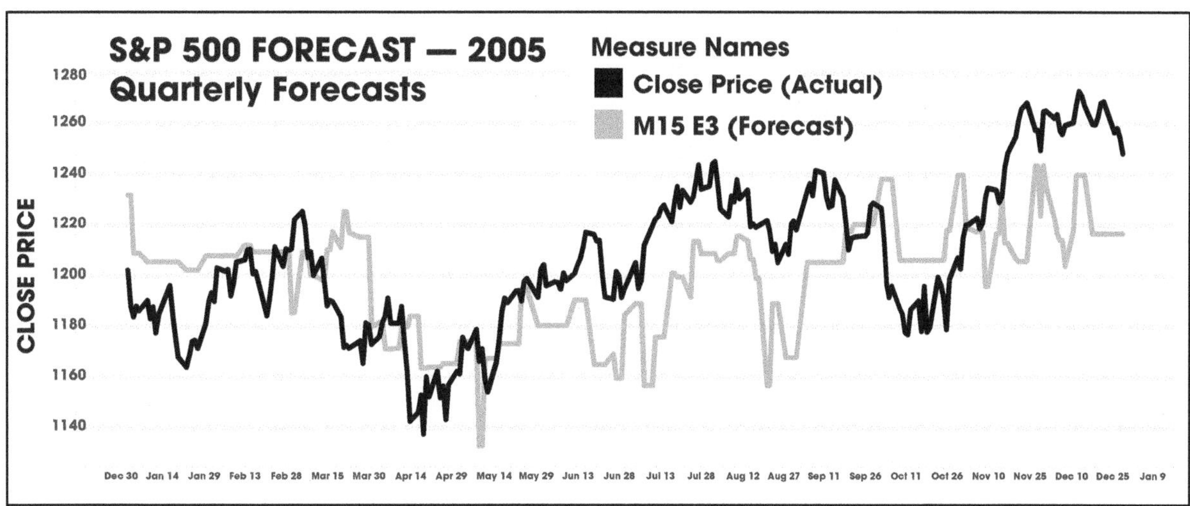

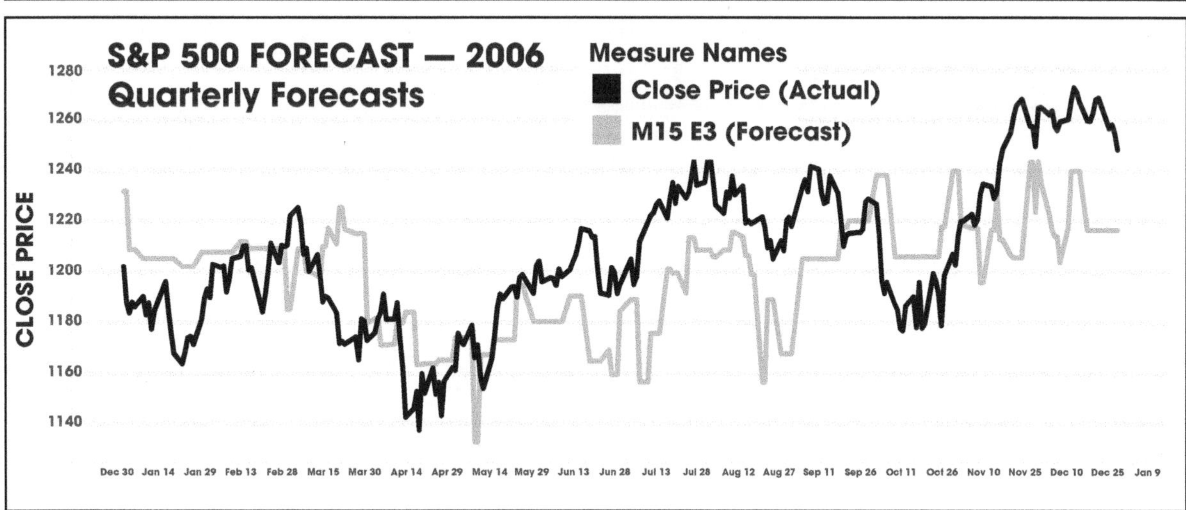

Figure 4: S&P 500 M15_E3 Quarterly Forecasts vs. Actual Close Price, 2004-2006

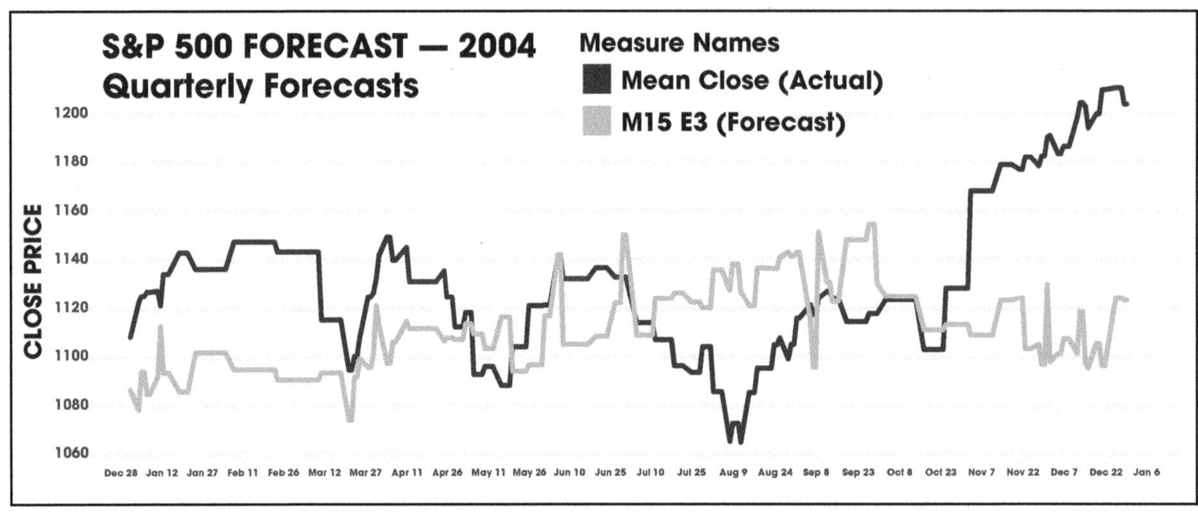

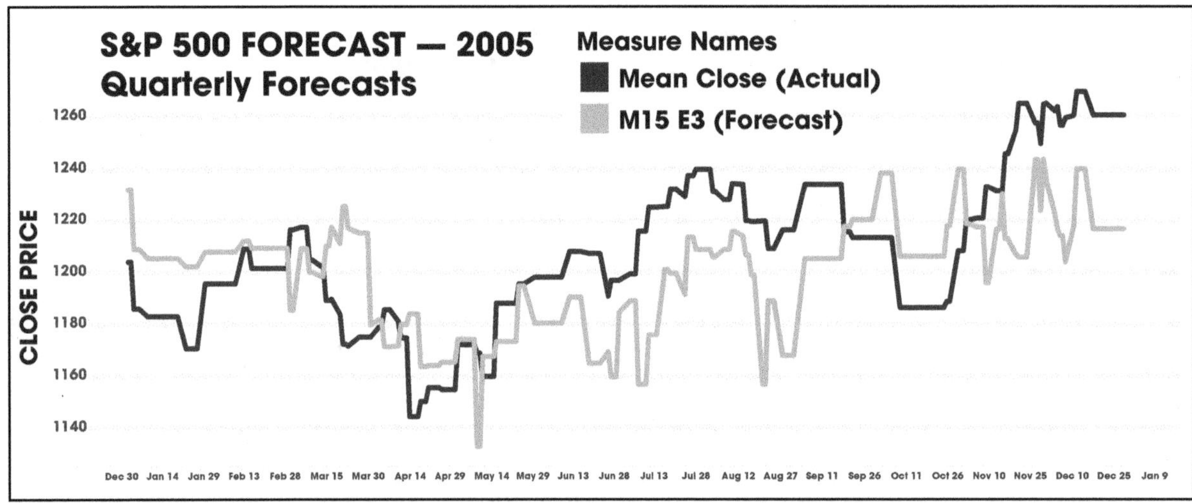

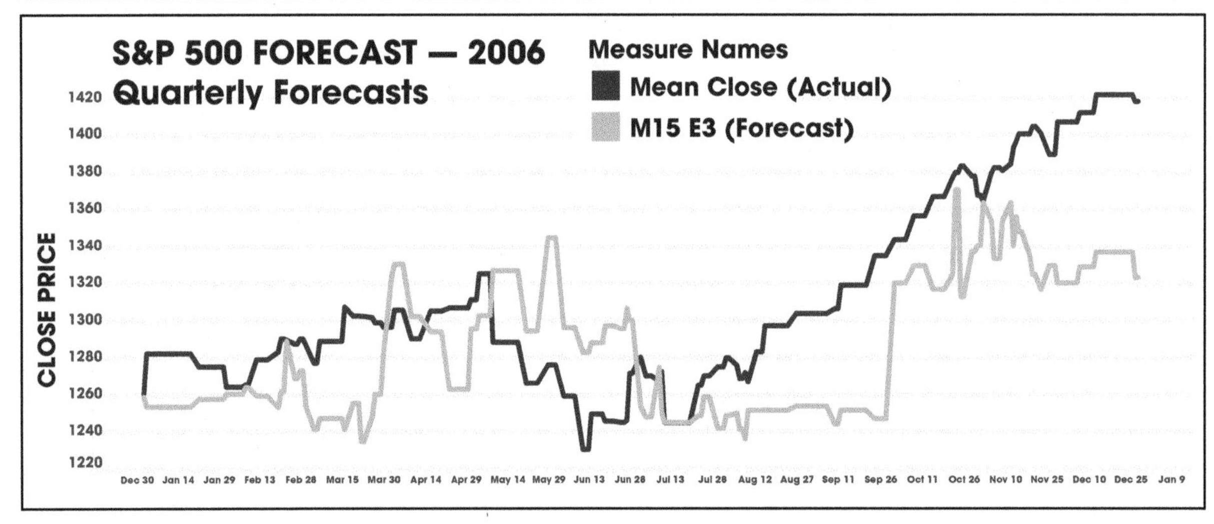

Figure 5: S&P 500 M15_E3 Quarterly Forecasts vs. Actual Mean Close Price, 2004-2006

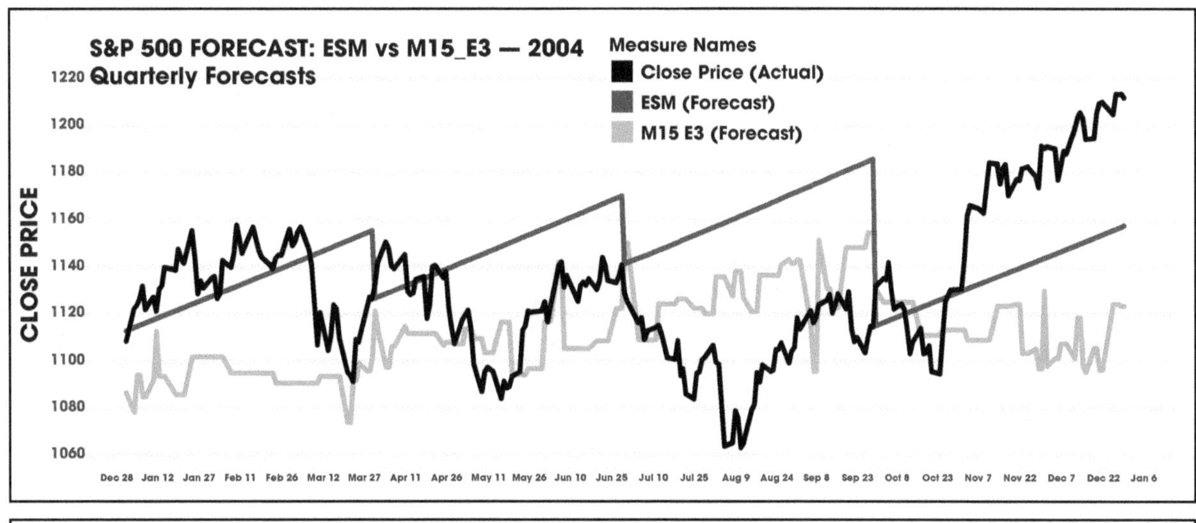

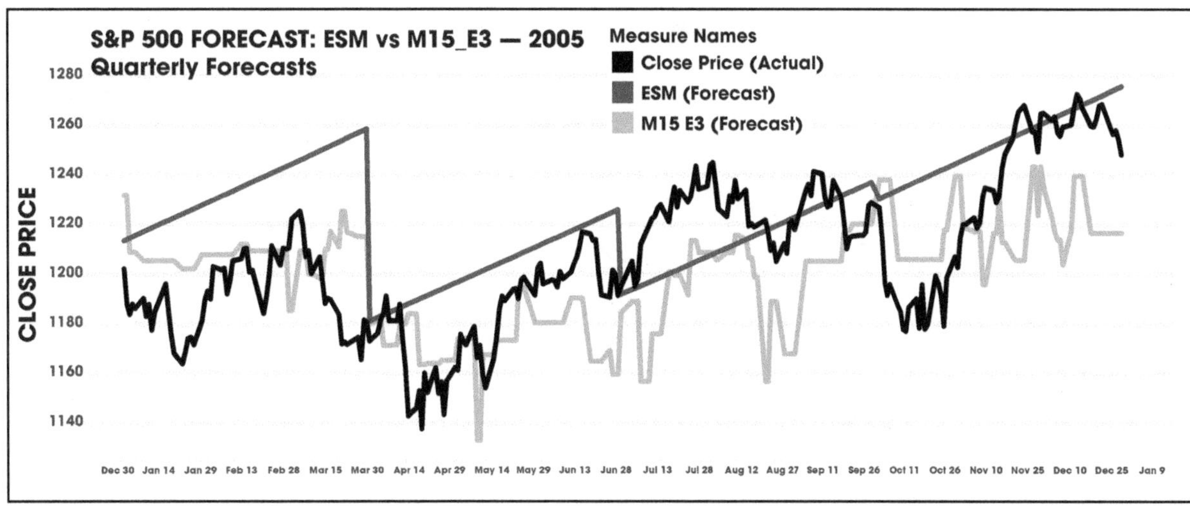

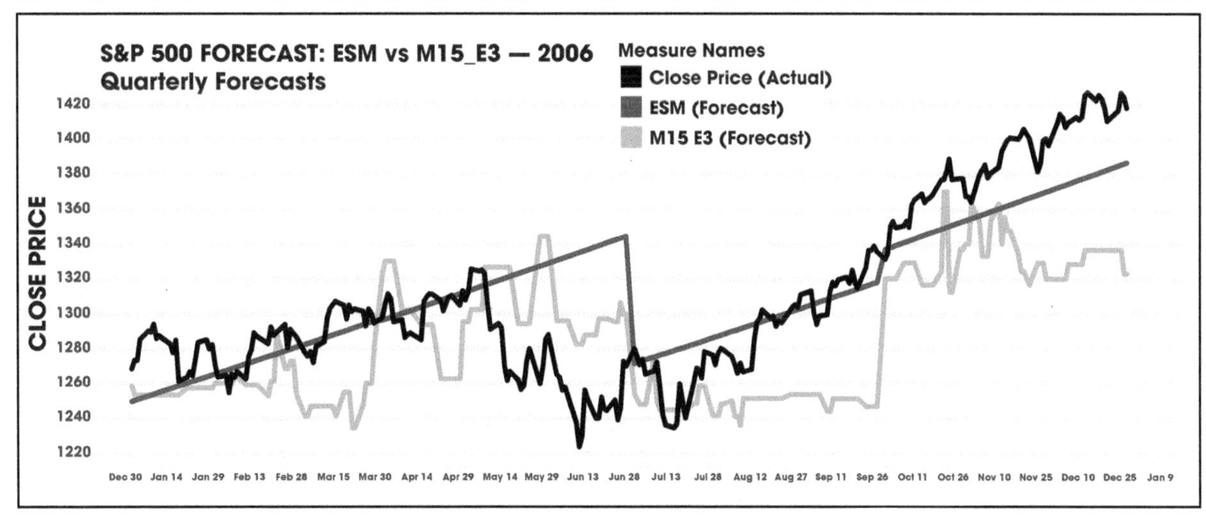

Figure 6: S&P 500 ESM vs M15_E3 Quarterly Forecasts vs. Actual Close Price, 2004-2006

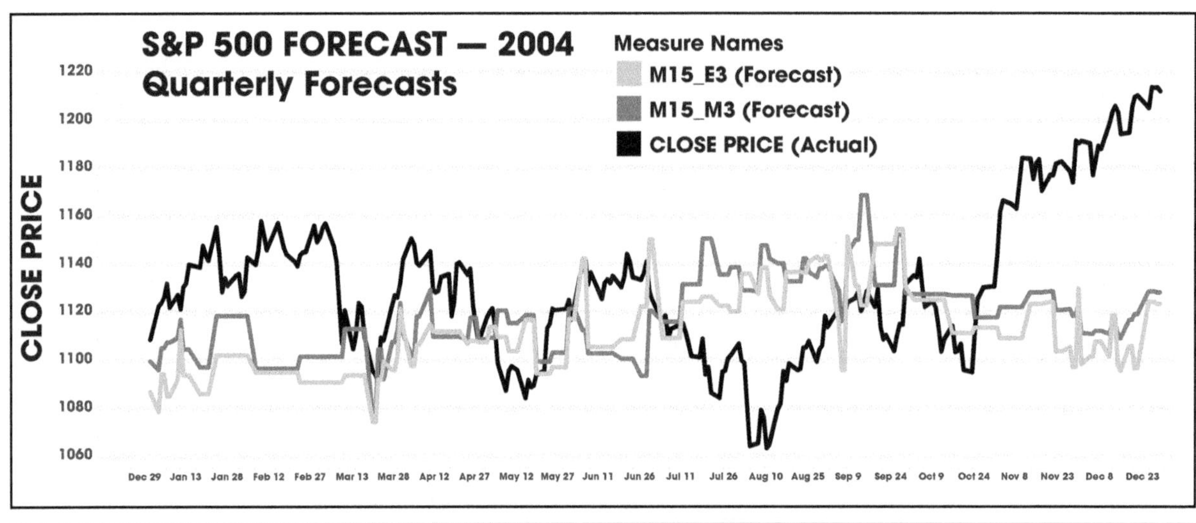

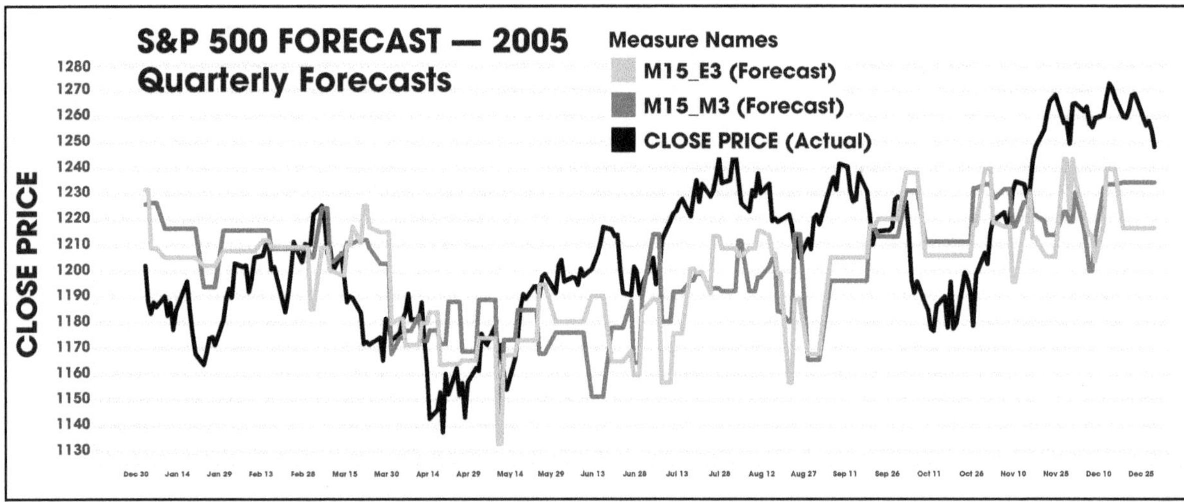

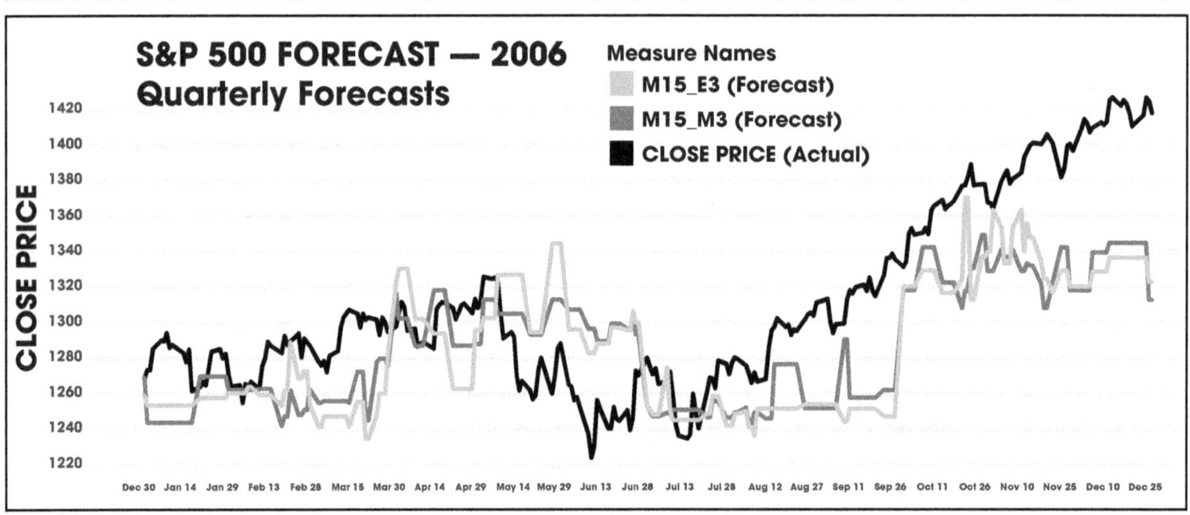

Figure 7: Comparison of M15_E3 and M15_M3 Quarterly Forecasts for the S&P 500, 2004-2006

5.2.2. Methodology to Combine the Seasonal Forecast Signal with Non-Seasonal Forecasts

All forecasts were generated in R using the traditional forecast options. These include

1. **ARIMA** (auto.arima)
2. **ESM** (ETS model zzz with multiplicative trend)
3. **MEAN**
4. **NAÏVE**
5. **HOLT** (two-parameter, additive, damped)

The seasonal influences were combined with each of these forecasts by combining the traditional forecast values with the **M15_E3** or **M15_M3** forecast values in either a 1:1 or a 3:1 combination. The 1:1 hybrid forecasts, designated as **(FORECAST)_E3** are the mean of the traditional forecast and the **M15_E3** forecast. The 3:1 hybrids, designated as **(FORECAST)3_E3**, are the average of 3x the traditional forecast and 1x the **M15_E3** forecast.

Figure 8 on page 80 compares the **ESM** forecast with the **ESM_E3** and the **ESM3_E3** forecast signals for the S&P 500 for 2004 through 2006. The most notable feature is how the integration of the **M15_E3** seasonal forecast signal transforms the **ESM** forecast from a virtual trend line to a forecast that contains peaks and troughs and noticeable adjustments within the forecast period.

The accuracy of each of the hybrid forecasts was compared to the accuracy of the traditional forecast to determine if any of the hybrid forecasts could be eliminated from consideration to limit the number of forecast methods included in the final study. In the case of the **ARIMA**, **HOLT**, **MEAN**, and **NAIVE** forecasts, one of the hybrid forecasts was significantly more accurate than the other (and consistently more accurate than the traditional forecast). In the case of the **ESM** forecasts, both hybrids performed equally well, and so both are included in the final study.

The final forecast methods used in the initial study of the **M15_E3** method include:

1. **ARIMA**
2. **ESM**
3. **HOLT**
4. **MEAN**
5. **NAIVE**
6. **M15_E3**
7. **ARIMA_E3**
8. **ESM_E3**
9. **ESM3_E3**
10. **HOLT3_E3**
11. **MEAN_E3**
12. **NAIVE3_E3**

The forecast methods used in the subsequent study of the **M15_M3** method include:

1. **ARIMA**
2. **ESM**
3. **HOLT**
4. **MEAN**
5. **NAIVE**
6. **M15_M3**
7. **ARIMA_M3**
8. **ESM_M3**
9. **ESM3_M3**
10. **HOLT3_M3**
11. **MEAN_M3**
12. **NAIVE3_M3**

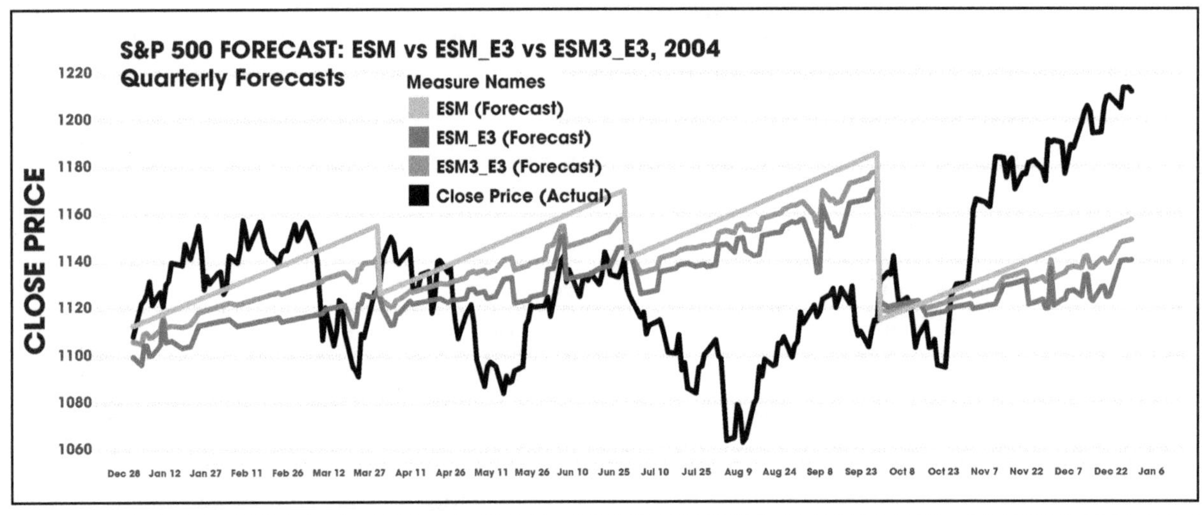

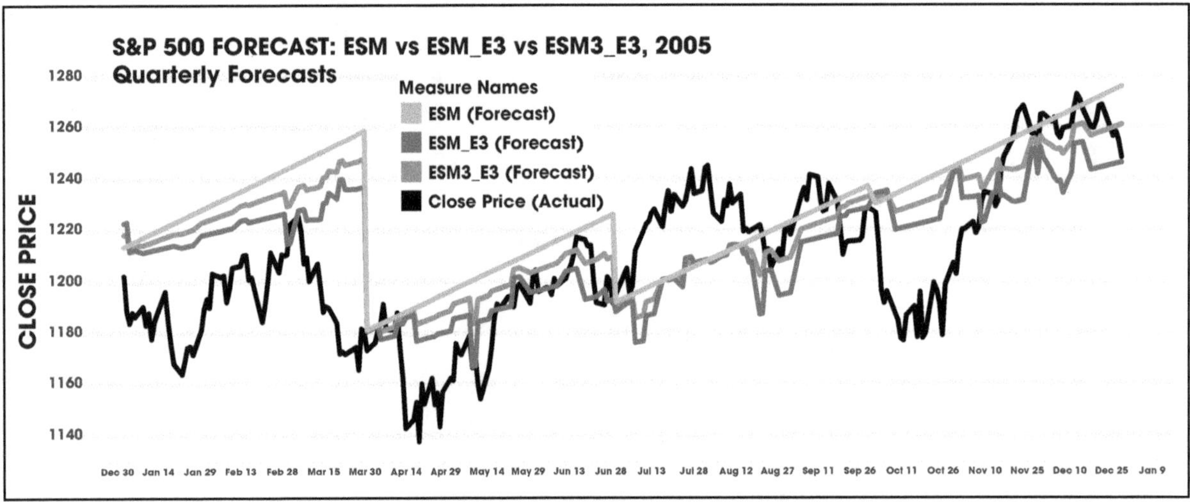

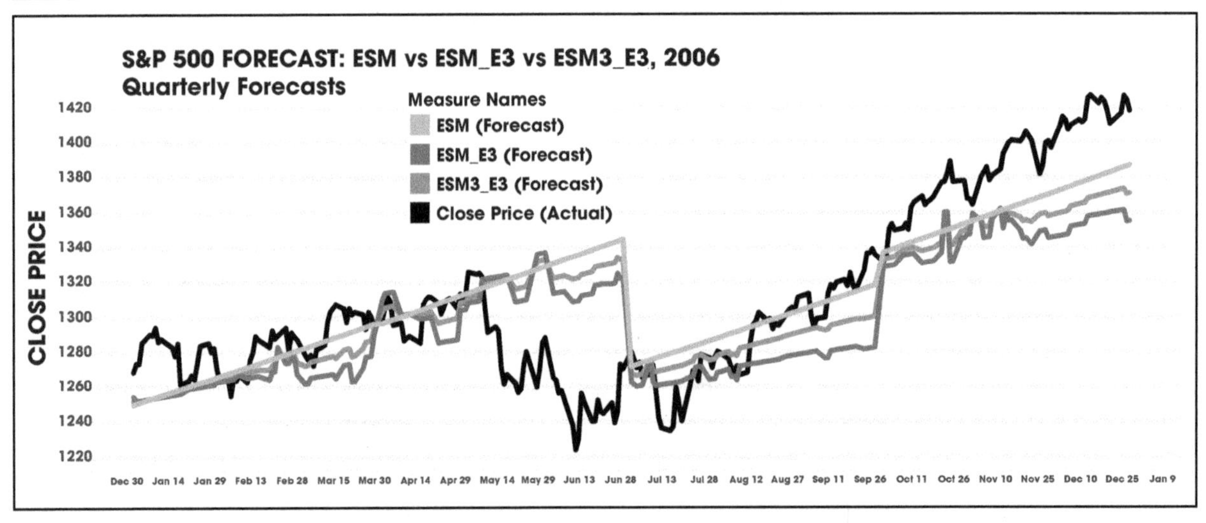

Figure 8: Comparison of ESM, ESM_E3, and ESM3_E3 Quarterly Forecasts for the S&P 500, 2000-2006

5.2.3. Methodology to Generate Forecast Accuracy

The aggregate accuracy of the entire 19-year forecast period for each forecast is calculated using both mean absolute percentage error (**MAPE**) and root mean square error (**RMSE**). It should be noted that **RMSE** penalizes larger errors more than smaller errors, and **RMSE** increases with the variance of the frequency distribution of error magnitudes. This is an important concern, especially when considering stock market indexes where the magnitude of the forecast errors is significantly greater than the magnitude of errors in other financial instruments, such as an individual stock or an interest rate.

When considering the accuracy of the Targeted Forecasts, the significant seasons are identified for each individual data set and then the forecast results are filtered so that only the dates that correspond to the significant seasons are included. The mean absolute percentage error (**MAPE**) and root mean square error (**RMSE**) accuracy for the significant seasons and dates for each data set are then calculated for each of the 12 forecast models.

Each individual set of time series data has its own seasonal patterns and its own subset of significant seasons that can be used for targeted forecasting. But the practical value of these significant seasons will change from year to year, because not every significant season will occur every year. Some illustrated examples will help clarify this very important point.

For the Dow Jones, 78 of the 202 **M15 Sign + Speed** seasons had significant variance, but only 22 of those seasons occur in 2020. Figure 9 on page 82 illustrates the 2020 dates of the significant **M15 Sign + Speed** seasons for the **Dow Jones Industrial Average**.

The S&P 500 had 68 out of 202 significant seasons, and 21 of those seasons occur in 2020. Figure 10 on page 83 illustrates the 2020 dates of the significant seasons for the **S&P 500.**

The NASDAQ Composite had 66 out of 202 significant seasons, with 18 of those seasons occurring in 2020. Figure 11 on page 84 illustrates the 2020 dates of the significant seasons for the **NASDAQ Composite**.

In each of these figures, the significant seasons are shown in color, and the rest of the **M15 Sign + Speed** seasons for 2020 are in black and white.

M15 2020 Forecast Dates for ^DJI

Month	Start date	Forecast
JAN	1	D1.9 ♑ CAPRICORN
JAN	17	D1.9 ♒ AQUARIUS
FEB	1	D1.9 ♒
FEB	3	D1.9 ♓
FEB	5	D1.8 ♓
FEB	8	D1.7 ♓
FEB	10	D1.6 ♓
FEB	12	D1.5 ♓
FEB	14	D1.2 ♓
FEB	15	D1.1 ♓
FEB	17	R1.1 ♓
FEB	18	R1.2 ♓
FEB	20	R1.5 ♓
FEB	22	R1.6 ♓
FEB	24	R1.7 ♓ PISCES
FEB	29	R1.6 ♓
MAR	1	R1.6 ♓
MAR	3	R1.5 ♓
MAR	4	R1.5 ♒
MAR	5	R1.4 ♒
MAR	8	R1.3 ♒
MAR	10	D1.3 ♒
MAR	13	D1.4 ♒
MAR	17	D1.5 ♓
MAR	19	D1.6 ♓
MAR	24	D1.7 ♓
MAR	31	D1.8 ♓
APR	1	D1.8 ♓ PISCES
APR	9	D1.9 ♓
APR	11	D1.9 ♈ ARIES
APR	18	D1.10 ♈ ARIES
APR	27	D1.11 ♈
APR	28	D1.11 ♉
MAY	1	D1.11 ♉ TAURUS
MAY	12	D1.11 ♊
MAY	15	D1.10 ♊ GEMINI
MAY	21	D1.9 ♊ GEMINI
MAY	26	D1.8 ♊
MAY	29	D1.8 ♋
MAY	30	D1.7 ♋
JUN	1	D1.7 ♋
JUN	4	D1.6 ♋
JUN	7	D1.5 ♋
JUN	11	D1.2 ♋
JUN	14	D1.1 ♋ CANCER
JUN	18	R1.1 ♋ CANCER
JUN	23	R1.2 ♋ CANCER
JUN	26	R1.5 ♋ CANCER
JUL	1	R1.5 ♋ CANCER
JUL	5	R1.4 ♋ CANCER
JUL	9	R1.3 ♋ CANCER
JUL	12	D1.3 ♋ CANCER
JUL	15	D1.4 ♋ CANCER
JUL	18	D1.5 ♋ CANCER
JUL	21	D1.6 ♋
JUL	23	D1.7 ♋ CANCER
JUL	26	D1.8 ♋ CANCER
JUL	30	D1.9 ♋
AUG	1	D1.9 ♋
AUG	2	D1.10 ♋ CANCER
AUG	5	D1.10 ♌ LEO
AUG	9	D1.11 ♌ LEO
AUG	17	D1.10 ♌ LEO
AUG	20	D1.10 ♍ VIRGO
AUG	29	D1.9 ♍ VIRGO
SEP	1	D1.9 ♍ VIRGO
SEP	6	D1.9 ♎ LIBRA
SEP	11	D1.8 ♎ LIBRA
SEP	23	D1.7 ♎ LIBRA
SEP	27	D1.7 ♏ SCORPIO
SEP	30	D1.6 ♏
OCT	1	D1.6 ♏ SCORPIO
OCT	5	D1.5 ♏ SCORPIO
OCT	9	D1.2 ♏ SCORPIO
OCT	12	D1.1 ♏
OCT	14	R1.1 ♏
OCT	16	R1.2 ♏
OCT	18	R1.5 ♏
OCT	19	R1.6 ♏
OCT	21	R1.7 ♏ SCORPIO
OCT	25	R1.8 ♏
OCT	26	R1.7 ♏
OCT	28	R1.7 ♎
OCT	29	R1.6 ♎
OCT	30	R1.5 ♎
NOV	1	R1.4 ♎
NOV	2	R1.3 ♎
NOV	4	D1.3 ♎
NOV	5	D1.4 ♎
NOV	7	D1.5 ♎
NOV	8	D1.6 ♎
NOV	11	D1.7 ♏
NOV	14	D1.8 ♏ SCORPIO
NOV	23	D1.9 ♏ SCORPIO
DEC	1	D1.9 ♏
DEC	2	D1.9 ♏ SAGITTARIUS
DEC	21	D1.9 ♑ CAPRICORN

Figure 9: 2020 Forecast Calendar for Significant Seasons of the Dow Jones Industrial Average Index

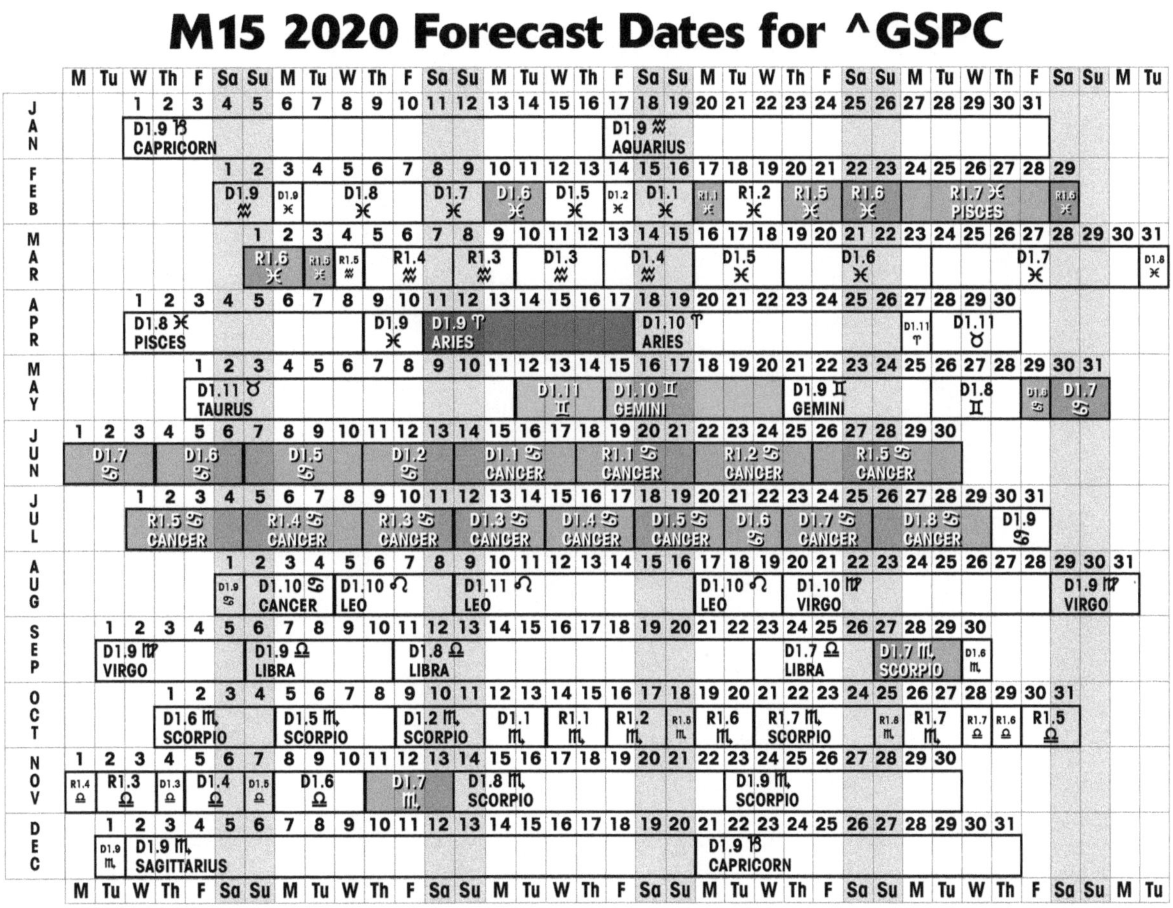
M15 2020 Forecast Dates for ^GSPC
M Tu W Th F Sa Su M Tu W Th F Sa Su M Tu W Th F Sa Su M Tu W Th F Sa Su M Tu W Th F Sa Su M Tu
JAN
D1.9 ♑ CAPRICORN
D1.9 ♒ AQUARIUS
FEB
D1.9 ♒
D1.9 ♓
D1.8 ♓
D1.7 ♓
D1.6 ♓
D1.5 ♓
D1.2 ♓
D1.1 ♓
R1.1 ♓
R1.2 ♓
R1.5 ♓
R1.6 ♓
R1.7 ♓ PISCES
R1.8 ♓
MAR
R1.6 ♓
R1.6 ♓
R1.5 ♒
R1.4 ♒
R1.3 ♒
D1.3 ♒
D1.4 ♒
D1.5 ♓
D1.6 ♓
D1.7 ♓
D1.8 ♓
APR
D1.8 ♓ PISCES
D1.9 ♓
D1.9 ♈ ARIES
D1.10 ♈ ARIES
D1.11 ♈
D1.11 ♉
MAY
D1.11 ♉ TAURUS
D1.11 ♊
D1.10 ♊ GEMINI
D1.9 ♊ GEMINI
D1.8 ♊
D1.8 ♋
D1.7 ♋
JUN
D1.7 ♋
D1.6 ♋
D1.5 ♋
D1.2 ♋
D1.1 ♋ CANCER
R1.1 ♋ CANCER
R1.2 ♋ CANCER
R1.5 ♋ CANCER
JUL
R1.5 ♋ CANCER
R1.4 ♋ CANCER
R1.3 ♋ CANCER
D1.3 ♋ CANCER
D1.4 ♋ CANCER
D1.5 ♋ CANCER
D1.6 ♋
D1.7 ♋ CANCER
D1.8 ♋ CANCER
D1.9 ♋
AUG
D1.9 ♋
D1.10 ♋ CANCER
D1.10 ♌ LEO
D1.11 ♌ LEO
D1.10 ♌ LEO
D1.10 ♍ VIRGO
D1.9 ♍ VIRGO
SEP
D1.9 ♍ VIRGO
D1.9 ♎ LIBRA
D1.8 ♎ LIBRA
D1.7 ♎ LIBRA
D1.7 ♏ SCORPIO
D1.6 ♏
OCT
D1.6 ♏ SCORPIO
D1.5 ♏ SCORPIO
D1.2 ♏ SCORPIO
D1.1 ♏
R1.1 ♏
R1.2 ♏
R1.5 ♏
R1.6 ♏
R1.7 ♏ SCORPIO
R1.8 ♏
R1.7 ♏
R1.7 ♎
R1.6 ♎
R1.5 ♎
NOV
R1.4 ♎
R1.3 ♎
D1.3 ♎
D1.4 ♎
D1.5 ♎
D1.6 ♎
D1.7 ♏
D1.8 ♏ SCORPIO
D1.9 ♏ SCORPIO
DEC
D1.9 ♏
D1.9 ♐ SAGITTARIUS
D1.9 ♑ CAPRICORN

Figure 10: 2020 Forecast Calendar for Significant Seasons of the S&P 500 Index

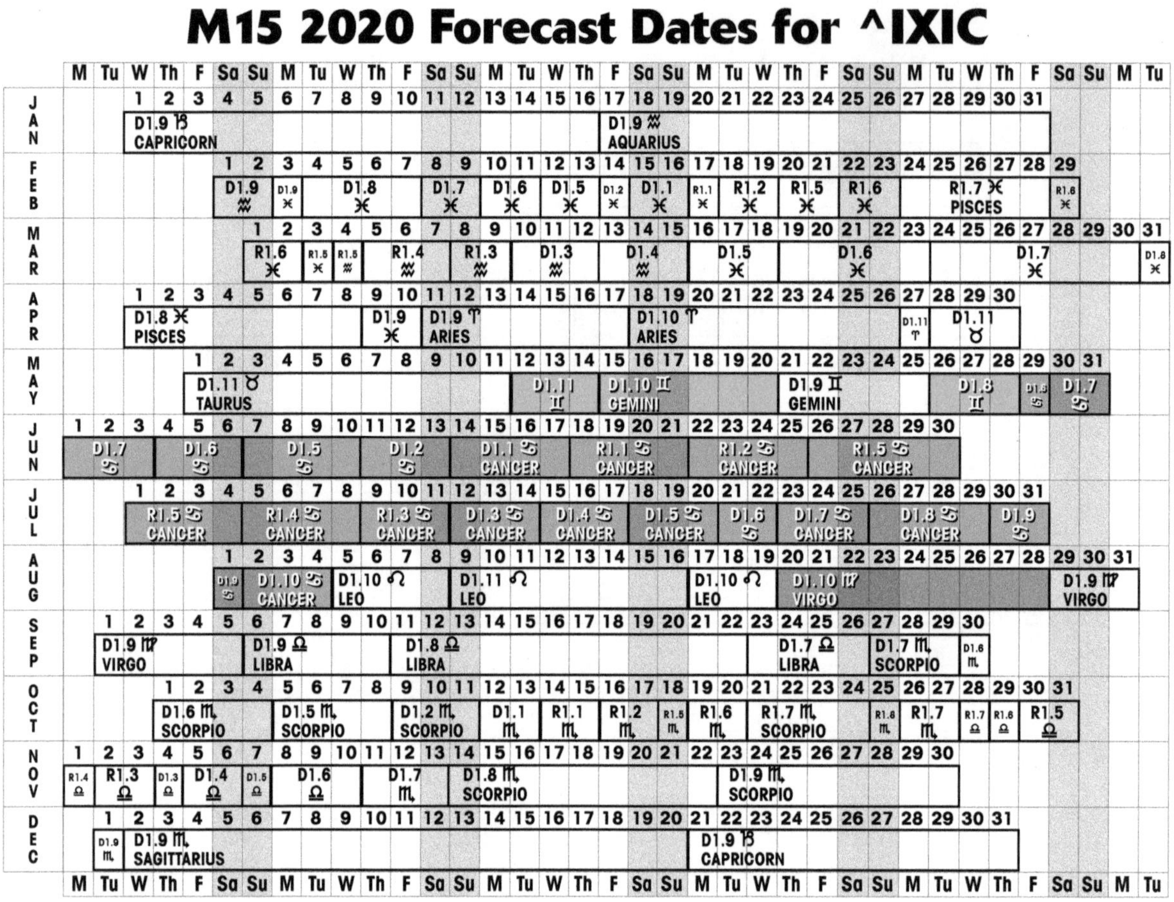

M15 2020 Forecast Dates for ^IXIC

Figure 11: 2020 Forecast Calendar for Significant Seasons of the NASDAQ Composite Index

5.2.4. Methodology to Evaluate and Compare the Accuracy of the Forecasts

The accuracy of the seasonal forecasts versus the non-seasonal forecasts can be evaluated in two different dimensions.

The first dimension ranks the accuracy of each of the 12 forecasts, and considers the number and percentage of the 430 data sets where one of the forecasts that included the seasonal influences is ranked #1, or is ranked in either the #1 or the #2 position, both by **MAPE** and by **RMSE**. This top two ranking may offer a more accurate picture because often the difference in accuracy score between the #1 and #2 ranked forecast is razor thin. The complete ranked **MAPE** and **RMSE** accuracy results for all 430 **M15_E3** forecasts are included in **Appendix B**, and the ranked **MAPE** and **RMSE** accuracy results for all 430 **M15_M3** forecasts are included in **Appendix C**.

The second dimension compares the accuracy of each of the traditional forecast methods to the accuracy of the hybrid versions of those forecast methods that include the seasonal influences, and considers the number and percentage of the 430 data sets where the hybrid seasonal forecast is more accurate than the traditional, non-seasonal counterpart.

For the purpose of this study, we will assume that any result that falls within 20 percentage points of 50% is not significant.

5.2.5. Methodology for Evaluating Accuracy of Targeted Forecasts

One of the questions under consideration is whether the variance of the historical effect sizes of a season can be used to anticipate the value of that season in forecasts. Because the seasonal influences have to be generated as a separate forecast, it becomes possible to include only the most significant seasons for the data set and to eliminate the influence of seasons that have a high historic variance and a low expectation of predictive value.

For the purposes of this study, any season with a variance of less than half of the mean variance of the entire data set is considered to be significant, and expected to have value when forecasting for that data set.

The targeted accuracy of the 430 data sets is ranked and compared in the same way as the full accuracy results. But the targeted accuracy results can also be compared to the full accuracy results. If the targeted seasonal forecasts are more accurate than the non-targeted seasonal forecasts, it supports the idea that the variance of the historic data correlates with the predictive value of the season.

5.3. E3 Forecast Accuracy for Financial Data Set

Table 56 contains the summary accuracy results for the set of 430 **M15_E3** forecasts and Table 57 contains the comparison between the **E3** hybrid seasonal forecasts and the traditional, non-seasonal counterparts. Significant results (less than 40% or greater than 60%) are highlighted.

Table 56: Accuracy Ranking of E3 Forecasts

	TOTAL FORECASTS	E3 HYBRID #1 RANK	% E3 HYBRID #1 RANK	E3 HYBRID #1/#2 RANK	% E3 HYBRID #1/#2 RANK
MAPE	430	239	53.49%	291	67.67%
RMSE	430	197	45.81%	270	62.79%

Table 57: Accuracy Comparison: E3 Hybrid Forecasts vs. Traditional Forecasts (Complete Set of 430 Forecasts)

	ARIMA_E3 beats ARIMA	ESM_E3 or ESM3_E3 beats ESM	NAIVE3_E3 beats NAIVE	MEAN_E3 beats MEAN	HOLT3_E3 beats HOLT	HYBRIDS beat TRADITIONAL
MAPE	66.28%	74.88%	60.70%	100%	59.07%	71.30%
RMSE	69.07%	69.53%	46.74%	99.30%	48.37%	66.93%

When considering the percentage of time that one of the seasonal forecasts ranked #1 in accuracy across the 12 forecast models, the results are not significant for either **MAPE** (53.49%) or for **RMSE** (45.81%). However, when considering the percentage of time that one of the seasonal forecasts ranked either #1 or #2 in accuracy, both the **MAPE** (67.67%) and **RMSE** (62.79%) results are significant.

When considering the comparison between the hybrid forecasts and their non-seasonal counterparts, the results are even more compelling. The results were significant by **MAPE** for the **ARIMA**, **ESM**, **NAIVE**, and **MEAN** models, where the hybrid forecasts were more accurate than the non-seasonal models more than 60% of the time, and the results for the hybrid **HOLT** forecasts only barely miss the threshold of significance. By **RMSE**, the hybrid forecasts were significantly more accurate than their non-seasonal counterparts for **ARIMA**, **ESM**, and **MEAN**. The results for **NAIVE** and **HOLT** are not significant, but they do suggest that including the **E3** seasonal data makes these forecasts slightly less accurate.

Overall, these results show compelling evidence that incorporating the irregular, astrology-based E3 seasonal data tends to improve the accuracy of the forecasts.

5.4. E3 Targeted Forecasts

Table 58 contains the summary accuracy results for the set of 430 targeted **M15_E3** forecasts and compares those results to the results of the untargeted forecasts, and Table 59 contains the comparison between the **E3** hybrid seasonal forecasts and the traditional, non-seasonal counterparts for both the targeted forecasts and the untargeted forecasts. Significant results are highlighted.

Table 58: Accuracy Ranking of E3 Forecasts vs E3 Targeted Forecasts

	TOTAL FORECASTS	E3 HYBRID #1 RANK	% E3 HYBRID #1 RANK	E3 HYBRID #1/#2 RANK	% E3 HYBRID #1/#2 RANK
MAPE (ALL)	430	239	53.49%	291	67.67%
MAPE (TARGET)	430	276	64.19%	340	79.07%
RMSE (ALL)	430	197	45.81%	270	62.79%
RMSE (TARGET)	430	280	65.12%	341	79.30%

Table 59: Accuracy Comparison: E3 Hybrid Forecasts vs. Traditional Forecasts (Complete Set of 430 Forecasts, All vs. Targeted Seasonality)

	ARIMA_E3 beats ARIMA	ESM_E3 or ESM3_E3 beats ESM	NAIVE3_E3 beats NAIVE	MEAN_E3 beats MEAN	HOLT3_E3 beats HOLT	HYBRIDS beat TRADITIONAL
MAPE (ALL)	66.28%	74.88%	60.70%	100%	59.07%	71.30%
MAPE (TARGET)	73.26%	78.84%	66.74%	100%	67.44%	77.26%
RMSE (ALL)	69.07%	69.53%	46.74%	99.30%	48.37%	66.93%
RMSE (TARGET)	75.35%	77.21%	65.81%	99.77%	67.91%	77.21%

When considering the accuracy of the targeted forecasts, every single result is now significant. One of the **E3** hybrid targeted forecasts ranked #1 in accuracy 64.19% of the time (**MAPE**) and 65.12% of the time (**RMSE**), and ranked either #1 or #2 79.07% of the time (**MAPE**) and 79.3% of the time (**RMSE**).

The **E3** hybrid targeted forecasts were significantly more accurate than every one of the traditional forecasts (targeted dates) both by **MAPE** and by **RMSE**, and were more accurate overall 77.26% of the time (**MAPE**) and 77.21% of the time (**RMSE**).

The accuracy results for the hybrid **E3** forecasts for only the targeted seasons were consistently higher than the accuracy results for the hybrid **E3** forecasts that included the full range of dates (both significant and insignificant seasons). **This suggests that the historical variance of the effect size of a season can be used to anticipate the relative value of that season in improving forecast accuracy.**

5.5. M3 Forecast Accuracy for Financial Data Set

Table 60 contains the summary accuracy results for the set of 430 **M15_M3** forecasts and Table 61 contains the comparison between the **M3** hybrid seasonal forecasts and the traditional, non-seasonal counterparts. Significant results are highlighted

Table 60: Accuracy Ranking of M3 Forecasts

	TOTAL FORECASTS	M3 HYBRID #1 RANK	% M3 HYBRID #1 RANK	M3 HYBRID #1/#2 RANK	% M3 HYBRID #1/#2 RANK
MAPE	430	259	60.23%	321	74.65%
RMSE	430	242	56.28%	301	70.00%

Table 61: Accuracy Comparison: M3 Hybrid Forecasts vs. Traditional Forecasts (Complete Set of 430 Forecasts)

	ARIMA_M3 beats ARIMA	ESM_M3 or ESM3_M3 beats ESM	NAIVE3_M3 beats NAIVE	MEAN_M3 beats MEAN	HOLT3_M3 beats HOLT	HYBRIDS beat TRADITIONAL
MAPE	66.98%	77.67%	62.79%	100.00%	64.42%	74.37%
RMSE	72.09%	74.65%	56.74%	99.53%	58.60%	72.33%

The first thing to note is that the percentages for the **M3** results are consistently higher in every instance than the percentages for the **E3** results. When considering the percentage of time that one of the seasonal forecasts ranked #1 in accuracy across the 12 forecast models, the **M3 MAPE** results are significant, and even the **RMSE** results suggest that the seasonal data improves forecast accuracy. The results for the #1 or #2 ranking are even more significant with the **M3** (74.65%/70%)than with the **E3** (71.3%/66.93%).

When considering the comparison between the hybrid forecasts and their non-seasonal counterparts, every single result is significant for **MAPE**, and while the **RMSE** results for **NAIVE** and **HOLT** do not reach the level of significance, they still suggest that incorporating the **M3** seasonal forecast will improve the overall accuracy of the forecast.

These results further support the conclusion that incorporating irregular, astrology-based seasonal influences can significantly improve the accuracy of time series forecasts. They also illustrate how big of a role the methodology used to generate those seasonal forecasts plays. Neither the **E3** nor the **M3** forecast methodologies has been optimized, and yet the **M3** forecast is noticeably more accurate than the **E3** forecast.

5.6. M3 Targeted Forecasts

Table 62 contains the summary accuracy results for the set of 430 targeted **M15_M3** forecasts and compares those results to the results of the untargeted forecasts, and Table 63 contains the comparison between the **M3** hybrid seasonal forecasts and the traditional, non-seasonal counterparts for both the targeted forecasts and the untargeted forecasts. Significant results are highlighted.

Table 62: Accuracy Ranking of E3 Forecasts vs E3 Targeted Forecasts

	TOTAL FORECASTS	M3 HYBRID #1 RANK	% M3 HYBRID #1 RANK	M3 HYBRID #1/#2 RANK	% M3 HYBRID #1/#2 RANK
MAPE (ALL)	430	259	60.23%	321	74.65%
MAPE (TARGET)	430	280	65.12%	335	77.91%
RMSE (ALL)	430	242	56.28%	301	70.00%
RMSE (TARGET)	430	264	66.05%	344	80.00%

Table 63: Accuracy Comparison: M3 Hybrid Forecasts vs. Traditional Forecasts (Complete Set of 430 Forecasts, All vs. Targeted Seasonality)

	ARIMA_M3 beats ARIMA	ESM_M3 or ESM3_M3 beats ESM	NAIVE3_M3 beats NAIVE	MEAN_M3 beats MEAN	HOLT3_M3 beats HOLT	HYBRIDS beat TRADITIONAL
MAPE (ALL)	66.98%	77.67%	62.79%	100.00%	64.42%	74.37%
MAPE (TARGET)	72.56%	78.60%	68.60%	99.77%	68.14%	77.53%
RMSE (ALL)	72.09%	74.65%	56.74%	99.53%	58.60%	72.33%
RMSE (TARGET)	74.65%	76.98%	65.35%	100.00%	68.14%	77.02%

The comparison between the targeted and untargeted **M3** forecasts provides the same results as with the comparison between the targeted and untargeted **E3** forecasts. Across every metric, the targeted forecasts were more accurate than the untargeted forecasts, and once again, every one of the targeted forecasts results is significant.

This strongly suggests that the historical variance of the effect size of a season can be used to anticipate the relative value of that season in improving forecast accuracy.

5.7. Conclusions

The most important conclusion based on the results of this study is that **the Mercury-based irregular seasonal model has clear and consistent value in quantitative time series forecasting**. Combining the seasonal forecast signal with each of the five traditional forecast methods consistently resulted in more accurate forecasts. The **E3** seasonal forecast was more accurate than the non-seasonal forecast 71.30% of the time for the **MAPE** and 66.93% of the time for the **RMSE**, and the **M3** seasonal forecast was more accurate than the non-seasonal forecast 74.37% of the time for the **MAPE** and 72.33% of the time for the **RMSE**.

What is most impressive about these results is that they were achieved using a Mercury-based irregular seasonal forecast signal that was by no means optimized. The methodology used to generate the **M15_E3** forecast does not adequately compensate for the considerable difference between the mean historical values used to generate the forecast and the current mean values of the stock. The gap between the Mercury forecast signal and the actual close price is therefore consistently larger than the gap between traditional forecasts and the actual close price, and this has an adverse effect on the accuracy of the Mercury forecast signal itself. And while the **M15_M3** forecasts were an improvement over the **M15_E3** forecasts, the **M15_M3** forecast methodology has also not yet been optimized.

The biggest advantage of the irregular seasonal forecasts can only be appreciated when the forecasts are visualized and compared to the forecasts produced by traditional methods. Daily fluctuations in the forecast values of traditional forecast methods are so small that the forecast signal looks more like a trend line. But the **M15_E3** and **M15_M3** forecast signals have peaks and troughs that frequently reflect the actual changes in the close price with a remarkable degree of fidelity (see Figure 7 on page 78). When the seasonal forecast signal is combined with a traditional forecast signal, it transforms the traditional forecast signal from a trend line to a forecast that more closely resembles the real-world changes in the stock price (See Figure 8 on page 80).

While the limited scope of this study means that only the **M15 Sign + Speed** seasonal model has been shown to have value in forecasting, it's reasonable to expect that other irregular seasonal models would show similar improvements. These results are encouraging enough to justify further research to explore the potential of irregular seasonal models.

The results of this study also strongly support the hypothesis that the variance of the historical effects of a season is a reliable indicator of the forecast value of that season. The targeted seasonal forecasts were significantly more accurate than the targeted non-seasonal forecasts, and also more accurate than the non-targeted seasonal forecasts. This also demonstrates the value of **Targeted Seasonality**, where only the seasonal influences of the significant seasons are used to adjust the forecast values.

6. Practical Applications and Further Study

This study has revisited how seasonality is applied in quantitative time series analysis and forecasting. It asserts that seasonality is a quality of time, not of data, and demonstrated how **Cohen's d** can be used to quantify and compare both the magnitude of the effect of any season, and the expected forecast value of any season.

This study also asserts that seasonal models are not limited to the regular units of calendar or clock time. We measure time based on the astrological cycles of the planets. The calendar and the clock are based on the cycles of the Sun and the Moon, but seasonal models can be based on the cycles of any planet. These new seasonal models constitute irregular seasonality. Examples of irregular seasonal models based on the cycles of the planet Mercury demonstrated how irregular seasonal models can not only reveal the presence of significant patterns in time series data, but can also be used to improve the accuracy of time series forecasts.

This study could represent an entirely new frontier in statistics. It certainly inspires further study and consideration. The Mercury-based seasonal models are ideally suited to daily-aggregated data, but are not useful for other types of data. Different astrology-based seasonal models can be created and evaluated to identify patterns of different frequencies.

The forecast methodologies used to generate the Mercury seasonal forecasts also merit further study and consideration.

6.1. Seasonal Models for Daily Aggregated Data

The Mercury-based seasonal models are ideal for data that is aggregated on a daily basis, including on-time transportation, financial and market data, sales and inventory data, or car crashes. Depending on the model used, the mean duration of the seasons can be kept to 4 or 5 days, allowing for multiple seasonal observations in each calendar month. These irregular seasonal models can be used both for analysis and for forecasting. Some of the more obvious applications are explored below.

6.1.1. On-Time Transportation Applications

The transportation industry routinely considers historical on-time performance statistics and has extensive need for the ability to identify and analyze any patterns in the historical performance data. Analysis of historical patterns and the ability to identify seasonal influences is far more important to this sector than the ability to forecast on-time performance. That the Mercury-based irregular seasonal models revealed the presence of seasonal influences that would otherwise be overlooked has significant potential. By isolating these seasonal influences and normalizing them to some degree, the resulting on-time performance data might offer a more accurate analysis of how the current schedules, infrastructure, and strategies are affecting the on-time performance.

6.1.2. Sales Projections and Financial Forecasts

While the Mercury-based seasonal forecasts do not make it possible to forecast the stock market, the fact that these models substantially improved the forecasts for the stock market shows that they have clear value when incorporated in financial forecasts, especially for less volatile businesses and industries. Incorporating these irregular seasonal models could improve the accuracy of such mundane data as quarterly sales forecasts or inventory projections.

The **M15 Sign + Speed** model, with its 202 seasons, requires approximately 20 years of historical data, but other models with fewer seasons would require less historical data (although they would also adjust for fewer seasons within each quarter). It would also be necessary to optimize the forecast methodology used to generate the seasonal forecasts. This is discussed below.

6.1.3. Auto Insurance

The results from the limited data available for car crashes strongly suggest that the Mercury-based irregular astrological seasonal influences may have significant practical value in the auto insurance industry. If there are historic irregular seasonal patterns of car crashes (that resulted in insurance claims), this information could be used to generate a seasonal forecast for the year, based on that historical data. This seasonal signal could then be integrated with the other actuarial forecasts and used to more accurately anticipate the number of crashes—and the resulting number of new claims—filed for each season.

6.2. Seasonal Models for Hourly Aggregated Data

The challenge with developing seasonal models that can identify cycles and patterns that occur within a 24-hour period is that the options are extensive. Because seasonality is a quality of time, any planet or combination of planets can be used to create an irregular daily seasonal model. But just because a seasonal model is valid doesn't mean that it will reveal any significant patterns in the data.

Two different approaches to creating seasonal models that can account for hourly aggregated data are **Lunar Aspects** and **Aspects by Primary Motion**.

6.2.1. Lunar Aspects

Lunar aspects are ideal lenses to view cycles that occur for specific windows of time on given days, but that do not necessarily occur on a daily basis. The Moon is the fastest moving "planet" in astrology, covering between 12° and 14° of longitude in a single day. During each 28-day cycle, the Moon will make at least 14 significant aspects (angular relationships) to every other planet. Each of these aspects could correspond to a 1- to 2-hour window of time. There are also regular periods when the Moon will make no more aspects until it changes signs (known as void-of-course).

Some examples of data sets where lunar aspect-based filters could provide interesting results are call centers (customer service, technical support), hospital admissions, and online sales for massive online merchants such as Amazon.com.

6.2.2. Aspects by Primary Motion

The most precise daily cycles involve primary motion, the rotation of the Earth. Each day, every planet appears to rise at the Ascendant (the eastern angle), culminate at the Midheaven (the southern angle), set at the Descendant (the western angle), and reach its nadir at the *Imum Coeli* (the northern angle). The timing of these events changes from day to day because of the secondary motion, the longitudinal movement of each of the planets measured along the ecliptic (each planet's orbit around the Sun as viewed from the Earth), and the duration of each season is less regular than with the Lunar aspects.

Primary motion-based models could be the most valuable when evaluating intra-day financial market fluctuations (including evaluating and forecasting volatility). These lenses could also reveal more specific and discrete patterns in data from call centers, hospitals, and online sales. They would probably need to be compared with other seasonal models and used to determine if outlying data is noise or if it correlates with a secondary irregular seasonal pattern.

6.3. Seasonal Models for Weekly or Monthly Aggregated Data

Exploring data that is aggregated on a weekly or monthly basis, such as mortgage rates or unemployment statistics requires an astrological filter that accounts for longer increments of time. Mercury and the Moon move far too swiftly for this type of data, but Mars, which has a 2.5-year cycle of the zodiac and biennial retrograde cycles could be ideally suited for these types of time series. As with Mercury, the Mars seasons would require a system that divides the movement of Mars into segments that last for at least the length of the smallest aggregated period (i.e., a week or a calendar month). The average daily motion of Mars may not be useful here, but other options are worth exploring. For example, each sign is divided into five irregular divisions known as Terms or Bounds, based on the degree of the sign. In general, Mars spends from one to three weeks in the same Term of a sign. Combining the sign, Term, and direction of Mars could produce a seasonality filter that could be used to identify irregular seasonality in weekly or monthly aggregated data.

6.4. Options to Improve the Forecast Accuracy of the Seasonal Forecasts

The **M15_E3** forecast methodology was "good enough" to demonstrate the practical applications of irregular seasonality, but it was by no means optimized to maximize the forecast accuracy. The **M15_M3** methodology was also not optimized. The fact that these prototypical seasonal forecast models resulted in such significant improvements in accuracy is extremely compelling.

6.4.2.1. Options to Improve the Accuracy of the M15_E3 Forecast Methodology

The **M15_E3** methodology generates a forecast using the actual mean historical stock prices for each season. These historical prices must be adjusted to compensate for the long-term market trends. The Dow Jones Industrial Average peaked at 12,500 in 2006, but the lowest point in 2014 was 15,373. Unadjusted references from seasons in these years would not produce an accurate forecast for the same season in 2017, when the low point of the Dow was over 20,000.

The adjustment of the historical seasonal means to bring them in line with the mean values immediately prior to the forecast period has the single greatest impact on the overall accuracy of the forecast signal.

This study considers the mean close value of a 30-day window centered on the historic reference period, and compares that to the mean close value of a 10-day period that immediately precedes the quarter being forecasted. The historic mean stock price is then multiplied by this ratio to generate the mean forecast value for that season. This approach was selected because it was accurate enough to demonstrate the predictive value of the irregular seasonal signal, and there were no resources available to further refine it or to explore alternative options.

A ratio-based adjustment is a sound principle. But the question of how to assemble that ratio has yet to be explored. Sampling a longer period of time arguably gives a more accurate reference mean, but it's unclear if a more broadly representative mean is beneficial. The fewer dates considered for each reference mean, the greater the variance of the adjustment ratio.

Other options that could affect the accuracy of the forecast signal involve adjustments to the moving average methodology. The current methodology gives equal weight to the three historic means, but a weighted moving average that prioritizes more recent results over more distant ones could be explored. In fact the nature of the irregular seasonality time scale means there are two different approaches to a weighted average: a fixed approach that consistently weights the historical references based on their positions relative to each other (most recent, next most recent, third most recent), and a variable approach that weighs the historical references based on their absolute positions on the timeline, adjusting the importance based on the amount of time that separates the historical period from the period being forecasted.

Additionally, the number of reference periods can be adjusted. The use of three periods for a moving average calculation is widely accepted, and more than three periods is impractical with the **M15 Sign + Speed** seasonal model because it would require even more than 20 years of historical data. In traditional moving average models, two historical data points are barely sufficient. But two historical seasons represent more than two historical data points because in most cases, the seasons include data from multiple consecutive days, and the two seasons are non-contiguous on the time scale. Two historical periods might be sufficient to create a reasonably accurate forecast. And of course, the question of whether to use an equal mean or a weighted mean to combine the historical periods applies here, too.

That being said, the applications of **Targeted Seasonality** may allow for less extensive reserves of historical data. The seasons that occur the most infrequently tend to have the least statistical significance (because they are also the shortest in duration, often lasting only a single day) and also the highest variance. This suggests that their value in forecasting is negligible. **Targeted Seasonality** might require only 10 years of historical data, rather than 20.

6.4.2.2. Options to Improve the Accuracy of the M15_M3 Forecast Methodology

The **M15_M3** forecast signal proved to be significantly more accurate than the **M15_E3** signal, and because it is percentage-based, it does not require any historical adjustments of the forecast values. For the forecast itself, the most critical variable is the quarterly historical reference period used as the baseline for each percentage-based forecast adjustment. The current model takes the mean close price of the 10 calendar days immediately prior to the quarter being forecasted. Adjusting this period to include either additional days or fewer days could significantly change the forecast values.

Additional adjustments include the use of weighted historical means rather than equal means, and the option to use two historical periods rather than three historical periods.

6.4.2.3. Options to Explore the Potential of the M15_D3 and M15_T3 Forecast Methodologies

When considering potential forecast methodologies, two additional methods were considered and then discarded, so they do not appear in this study. The **D3** model looks at the three most recent historical seasonal periods, and averages the percent change from each of these periods to create the forecast metric. The **T3** model is identical to the **D3** model, except it draws on a different set of historical periods for reference. The **T3** model considers only historical periods where both of the seasons of the transition are a precise match (e.g., the last time Mercury transitioned from **Aries D1.5** to **Aries D1.6**). The **T3** model requires the most extensive range of historical data, and even with a full 20 years of historical data to draw on, it is possible to fail to have even a single historical match for a given transition.

The forecast value for each season is the forecast value of the preceding season multiplied by the forecast metric (percent change). However, a starting value is required for the first forecast. The choice is between using the close price on the date immediately prior to the start of the forecast, or using some kind of mean close price as the starting value (such as the mean close value of the 10-day period immediately preceding the quarter being forecasted, as used in the **M15_E3** methodology). As with the **M15_E3** methodology, all of the options associated with calculating the moving average of the historic percentages apply.

The advantages of a percentage-based forecast metric are clear: because the signal is relative, it does not need to be normalized to conform with the more recent means. But the **D3** and **T3** signals do not necessarily measure the same thing that the **E3** or **M3** signals do. The **E3** and **M3** signals evaluate the actual historical data, but the **D3** and **T3** signals evaluate the relative change leading up to the data. It's possible that the **D3** or **T3** metrics could be combined with the **E3** or **M3** signal, adjusting the relative differences between the seasonal forecasts.

7. References

Cohen, Jacob. *Statistical Power Analysis for the Behavioral Sciences*, 2nd Edition. Lawrence Erlbaum Associates, 1988.

Cumming, G., Fidler, F., Kalinowski, P., and Lai, J. (2012), The statistical recommendations of the American Psychological Association Publication Manual: Effect sizes, confidence intervals, and meta-analysis. *Australian Journal of Psychology*, 64: 138-146. doi:10.1111/j.1742-9536.2011.00037.x

Sawilowsky, S. S. (2009). New effect size rules of thumb. *Journal of Modern Applied Statistical Methods*, 8(2), 597 – 599.

APPENDIX A: Financial Forecast Data List

This appendix includes a list of all of the financial data sets used in this study, grouped by sector or category. Each table includes the name of the security or financial data, the stock symbol (where appropriate), the name of the source file used for the forecast calculations, and the starting month of the historical data. A "fix" appended to a file name means that the data file was altered: missing dates and close prices were filled in with the previous close price to enable the R script to generate forecast data for the source file.

Source files, raw data, R scripts and Excel files used to generate the results of this study are available by request at **TheScienceofAstrology.com**.

Table A-1: Stock Market Indexes

SECURITY	SYMBOL	FILE NAME	START DATE
S&P 500	^GSPC	SP500_fix.csv	Jan-75
NASDAQ Composite	^IXIC	NASDAQ_fix.csv	Feb-75
Dow Jones Industrial Average	^DJI	DowJones_fix.csv	Feb-85
Dow Jones Transportation	^DJT	DJT_fix.csv	Jan-75
Dow Jones Utility	^DJU	DJU_fix.csv	Jan-75
Dow Jones Composite Average	^DJA	DJA.csv	Jan-81
NYSE Composite	^NYA	NYA.csv	Jan-75
Russell 2000	^RUT	Russel_fix.csv	Sep-87
Nikkei 225	^N225	N225.csv	Jan-75
CBOE Volatility Index	^VIX	VIX_fix.csv	Jan-90

Table A-2: Commodity Futures

SECURITY	SYMBOL	FILE NAME	START DATE
	METALS		
Gold Futures		gold_fix.csv	Jan-75
Platinum Prices		platinum_fix.csv	Jan-75
Copper Prices		copper_fix.csv	Jan-75
Silver Prices		silver_fix.csv	Jan-80
Palladium Prices		palladium_fix.csv	Jan-77
	GRAINS		
Corn Futures		corn_fix.csv	Jan-80
Oats Futures		oats_fix.csv	Jan-75
Soybean Prices		soybeans_fix.csv	Jan-75
Soybean Oil Futures		soybean_oil_fix.csv	Jan-80
Soybean Meal Futures		soybean_meal_fix.csv	Jan-80
Wheat Futures		wheat_fix.csv	Jan-75
	SOFTS		
U.S. Cocoa Futures		cocoa_fix.csv	Jan-80
Coffee Futures		coffee_fix.csv	Jan-80
Cotton Futures		cotton_fix.csv	Jan-75
Lumber Futures		lumber_fix.csv	Jan-80
Sugar Futures		sugar_fix.csv	Jan-75
	ENERGY		
Brent Oil Futures		Brent_oil_fix.csv	Jul-88
West Texas Oil Futures		WTI_oil_fix.csv	Apr-83
Heating Oil Futures		heat_oil_fix.csv	Jan-80
	MEAT		
Cattle Futures		cattle_fix.csv	Jan-80
Hog Futures		hog_fix.csv	Jan-80

Table A-3: Rates & Bonds

SECURITY	SYMBOL	FILE NAME	START DATE
BONDS			
United States 30-Year Bond		us30_fix.csv	Sep-87
United States 10-Year Bond		us10_fix.csv	Jan-80
United States 5-Year Bond		us05_fix.csv	Sep-87
United States 2-Year Bond		us02_fix.csv	Mar-88
FUTURES			
US 30 Year T-Bond Futures		us30T_fut_fix.csv	Jan-80
US 10 Year T-Bond Futures		us10T_fut_fix.csv	May-82
INTEREST RATES			
Federal Funds Rate		fedfunds_fix.csv	Jan-75
1 Year LIBOR Rate		libor_1yr_fix.csv	Jan-86
6 Month LIBOR Rate		libor_6mo_fix.csv	Jan-86
3 Month LIBOR Rate		libor_3mo_fix.csv	Jan-86

Table A-4: Aerospace Sector

SECURITY	SYMBOL	FILE NAME	START DATE
Aerospace and Defense General Stocks			
The Boeing Company	BA	BA.csv	Jan-75
Lockheed Martin Corporation	LMT	LMT.csv	Jan-77
General Dynamics Corporation	GD	GD.csv	Jan-77
Textron Inc.	TXT	TXT.csv	Mar-80
Northrop Grumman Corporation	NOC	NOC.csv	Dec-81
Aerospace and Defense Equipment Stocks			
Esterline Technologies Corporation	ESL	ESL.csv	Jan-75
HEICO Corporation	HEI	HEI.csv	Mar-80
Hexcel Corporation	HXL	HXL.csv	Mar-80
Curtiss-Wright Corporation	CW	CW.csv	Mar-80
Moog Inc.	MOG-A	MOG-A.csv	May-80
Raytheon Company	RTN	RTN.csv	Dec-81
Aerojet Rocketdyne Holdings, Inc.	AJRD	AJRD.csv	Feb-81

Table A-5: Auto/Tires/Trucks Sector

SECURITY	SYMBOL	FILE NAME	START DATE
Automotive Stocks			
Ford Motor Company	F	F.csv	Jan-75
Toyota Motor Corporation	TM	TM.csv	Aug-76
Honda Motor Co., Ltd.	HMC	HMC.csv	Mar-80
PACCAR Inc	PCAR	PCAR.csv	Mar-80
Harley-Davidson, Inc.	HOG	HOG.csv	Jul-86
Auto & Truck Original Equipment Manufacturers Stocks			
Gentex Corporation	GNTX	GNTX.csv	Dec-81
Tenneco Inc.	TEN	TEN.csv	Jan-82
Modine Manufacturing Company	MOD	MOD.csv	Sep-82
Magna international Inc.	MGA	MGA.csv	Oct-84
LCI Industries	LCII	LCII.csv	May-85
Oshkosh Corporation	OSK	OSK.csv	Mar-90
Wabash National Corporation	WNC	WNC.csv	Nov-91

Table A-6: Basic Materials Sector Stocks

SECURITY	SYMBOL	FILE NAME	START DATE
Chemicals Stocks			
Valhi, Inc.	VHI	VHI.csv	Mar-80
Air Products and Chemicals, Inc.	APD	APD.csv	Mar-80
FMC Corporation	FMC	FMC.csv	Mar-80
Olin Corporation	OLN	OLN.csv	Mar-80
Cabot Corporation	CBT	CBT.csv	Nov-80
PPG Industries, inc.	PPG	PPG.csv	Apr-83
Methanex Corporation	MEOH	MEOH.csv	May-92
Mining Stocks			
McEwen Mining Inc.	MUX	MUX.csv	May-80
BHP Group	BHP	BHP.csv	Mar-80
Cleveland-Cliffs Inc.	CLF	CLF.csv	Mar-80
Materion Corporation	MTRN	MTRN.csv	Mar-80
Rio Tinto plc	RIO	RIO.csv	Jun-90
Taseko Mines Limited	TGB	TGB.csv	Jun-94
Gold Miners Stocks			
U.S. Gold Corp.	USAU	USAU.csv	Jun-75
Newmont Mining Corporation	NEM	NEM.csv	Mar-80
Agnico Eagle Mines Limited	AEM	AEM.csv	Mar-80
Gold Fields Limited	GFI	GFI.csv	Mar-80
Royal Gold, Inc.	RGLD	RGLD.csv	Jun-81
Barrick Gold Corporation	GOLD	GOLD.csv	Feb-85
Vista Gold Corp.	VGZ	VGZ.csv	Apr-90
Steel Producers Stocks			
Nucor Corporation	NUE	NUE.csv	Mar-80
Commercial Metals Company	CMC	CMC.csv	Mar-80
L. B. Foster Company	FSTR	FSTR.csv	Jun-81
United States Steel Corporation	X	X.csv	Apr-91
Shiloh Industries, Inc.	SHLO	SHLO.csv	Jun-93
Schnitzer Steel Industries, Inc.	SCHN	SCHN.csv	Nov-93
Olympic Steel, Inc.	ZEUS	ZEUS.csv	Mar-94

Table A-7: Business Services Sector Stocks

SECURITY	SYMBOL	FILE NAME	START DATE
Financial Transaction Services Stocks			
Equifax Inc.	EFX	EFX.csv	Mar-80
Diebold Nixdorf, Incorporated	DBD	DBD.csv	Dec-81
Total System Services, Inc.	TSS	TSS.csv	Aug-83
Fiserv, Inc.	FISV	FISV.csv	Sep-86
Staffing Services Stocks			
Kelly Services, Inc.	KELYA	KELYA.csv	Mar-80
ManpowerGroup Inc.	MAN	MAN.csv	Oct-88
GEE Group, Inc.	JOB	JOB.csv	Mar-92
Robert Half International Inc.	RHI	RHI.csv	Mar-92
Outsourcing Services Stocks			
Automatic Data Processing, Inc.	ADP	ADP.csv	Mar-80
R.R. Donnelley & Sons Company	RRD	RRD.csv	Mar-80
Paychex, Inc.	PAYX	PAYX.csv	Aug-83
Barrett Business Services, Inc.	BBSI	BBSI.csv	Jun-93
Business Services Stocks			
Xerox Corporation	XRX	XRX.csv	Jan-77
Volt Information Sciences, Inc.	VISI	VISI.csv	Mar-80
Healthcare Services Group, Inc.	HCSG	HCSG.csv	Nov-83
Avis Budget Group, Inc.	CAR	CAR.csv	Sep-83
Spherix Incorporated	SPEX	SPEX.csv	Sep-84
Crawford & Company	CRD-A	CRD-A.csv	Jul-90
CorVel Corporation	CRVL	CRVL.csv	Jun-91
Advertising and Marketing Stocks			
Omnicom Group Inc.	OMC	OMC.csv	Mar-80
The Interpublic Group of Companies, Inc.	IPG	IPG.csv	Mar-80
WPP plc	WPP	WPP.csv	Dec-87
Insignia Systems, Inc.	ISIG	ISIG.csv	Mar-93
Harte Hanks, Inc.	HHS	HHS.csv	Nov-93

Table A-8: Computer & Technology Sector

SECURITY	SYMBOL	FILE NAME	START DATE
Computer Software Stocks			
Autodesk, Inc.	ADSK	ADKS.csv	Jun-85
Cadence Design Systems, Inc.	CDNS	CDNS.csv	Jun-87
Microsoft	MSFT	MSFT.csv	Mar-86
Oracle Corporation	ORCL	ORCL.csv	Mar-86
Adobe Inc.	ADBE	ADBE.csv	Aug-86
Symantec	SYMC	SYMC.csv	Jun-89
PTC Inc.	PTC	PTC.csv	Dec-89
Computer Stocks			
International Business Machine	IBM	IBM.csv	Jan-75
HP Inc.	HPQ	HPQ.csv	Jan-75
Agilysys, Inc.	AGYS	AGYS.csv	Mar-80
CSP Inc.	CSPI	CSPI.csv	Jan-82
Apple Inc.	AAPL	AAPL.csv	Dec-80
3D Systems Corporation	DDD	DDD.csv	Mar-88
PAR Technology Corporation	PAR	PAR.csv	Dec-88
Wireless Providers Stocks			
CenturyLink, Inc.	CTL	CTL.csv	Mar-80
Verizon Communications Inc.	VZ	VZ.csv	Nov-83
AT&T Inc.	T	ATT.csv	Nov-83
Sprint Corporation	S	Sprint.csv	Nov-84
Vodafone Group PLC	VOD	VOD.csv	Nov-88
United States Cellular Corp.	USM	USM.csv	May-88
ATN International, Inc.	ATNI	ATNI.csv	Nov-91
Semiconductor Stocks			
Texas Instruments Incorporated	TXN	TXN.csv	Jan-75
Intel Corporation	INTC	INTC.csv	Mar-80
Analog Devices, Inc.	ADI	ADI.csv	Mar-80
Semtech Corporation	SMTC	SMTC.csv	Mar-80
Amtech Systems, Inc.	ASYS	ASYS.csv	Aug-83
Maxim Integrated Products, Inc.	MXIM	MXIM.csv	Feb-88
Microchip Technology Incorporated	MCHP	MCHP.csv	Mar-93

SECURITY	SYMBOL	FILE NAME	START DATE
Communication Component Manufacturer Stocks			
Corning Incorporated	GLW	GLW.csv	Jan-82
Communications Systems, Inc.	JCS	JCS.csv	Mar-90
Technical Communications Corp	TCCO	TCCO.csv	Mar-90
Plantronics, Inc.	PLT	PLT.csv	Jan-94
Electrical Products Stocks			
Koninklijke Philips N.V.	PHG	PHG.csv	Dec-87
Cubic Corporation	CUB	CUB.csv	Mar-80
Bell Fuse Inc.	BELFA	BELFA.csv	Dec-83
Bonso Electronics International Inc	BNSO	BNSO.csv	Jun-89
Trimble Inc.	TRMB	TRMB.csv	Jul-90
Kopin Corporation	KOPN	KOPN.csv	Apr-92
Flex Ltd.	FLEX	FLEX.csv	Mar-94
Scientific Instrument Manufacturing Stocks			
MTS Systems Corporation	MTSC	MTSC.csv	Mar-80
Kewaunee Scientific Corporation	KEQU	KEQU.csv	Mar-80
PerkinElmer, Inc.	PKI	PKI.csv	Apr-83
Misonix, Inc.	MSON	MSON.csv	Jan-92

Table A-9: Construction Sector

SECURITY	SYMBOL	FILE NAME	START DATE
Building & Construction Products Stocks			
Masco Corporation	MAS	MAS.csv	Mar-80
USG Corporation	USG	USG.csv	Mar-80
Patrick Industries, Inc.	PATK	PATK.csv	Mar-80
Aegion Corporation	AEGN	AEGN.csv	Dec-81
CRH plc	CRH	CRH.csv	Jul-89
NCI Building Systems, Inc.	NCS	NCS.csv	Apr-92
Gibraltar Industries, Inc.	ROCK	ROCK.csv	Nov-93
Residential & Commercial Building Stocks			
Lennar Corporation	LEN	LEN.csv	Mar-80
M.D.C. Holdings, Inc.	MDC	MDC.csv	Mar-80
NVR, Inc.	NVR	NVR.csv	Jul-85
PulteGroup, Inc.	PHM	PHM.csv	Jul-85
Toll Brothers, Inc.	TOL	TOL.csv	Jul-86
KB Home	KBH	KBH.csv	Aug-86
D. R. Horton, Inc.	DHI	DHI.csv	Jun-92

Table A-10: Consumer Discretionary Sector

SECURITY	SYMBOL	FILE NAME	START DATE
Leisure & Recreation Stocks			
The Marcus Corporation	MCS	MCS.csv	Mar-80
Cedar Fair, L.P.	FUN	FUN.csv	Apr-87
Carnival Corporation	CCL	CCL.csv	Jul-87
Reading International, Inc.	RDI	RDI.csv	Mar-92
Royal Caribbean Cruises Ltd.	RCL	RCL.csv	Apr-93
Textile Stocks			
V.F. Corporation	VFC	VFC.csv	Mar-80
Oxford Industries, Inc.	OXM	OXM.csv	Mar-80
Interface, Inc.	TILE	TILE.csv	Apr-83
Culp, Inc.	CULP	CULP.csv	Aug-83
PVH Corp.	PVH	PVH.csv	Jul-87
G-III Apparel Group, Ltd.	GIII	GIII.csv	Dec-89
The Dixie Group, Inc.	DXYN	DXYN.csv	Mar-90
Schools Stocks			
Graham Holdings Company	GHC	GHC.csv	Mar-80
GP Strategies Corporation	GPX	GPX.csv	May-80
Adtalem Global Education Inc.	ATGE	ATGE.csv	Jun-91
Furniture Manufacturing Stocks			
Leggett & Platt, Incorporated	LEG	LEG.csv	Mar-80
La-Z-Boy Incorporated	LZB	LZB.csv	Mar-80
Kimball International, Inc.	KBAL	KBAL.csv	Mar-80
Bassett Furniture Industries, Incorporated	BSET	BSET.csv	Mar-80
Flexsteel Industries, Inc.	FLXS	FLXS	Mar-80
Virco Mfg. Corporation	VIRC	VIRC	Mar-80
American Woodmark Corporation	AMWD	AMWD.csv	Jul-86

Table A-11: Consumer Staples Sector

SECURITY	SYMBOL	FILE NAME	START DATE
Soft Drink Stocks			
The Coca-Cola Company	KO	KO.csv	Jan-75
PepsiCo., Inc.	PEP	PEP.csv	Jan-75
Coca-Cola European Partners	CCEP	CCEP.csv	Nov-86
National Beverage Corp.	FIZZ	FIZZ.csv	Sep-91
Cott Corporation	COT	COT.csv	Jun-92
Alcoholic Beverages Stocks			
Brown-Forman Corporation	BF-A	BF-A.csv	Jan-75
Molson Coors Brewing Company	TAP	TAP.csv	Mar-80
Constellation Brans, Inc.	STZ	STZ.csv	Mar-92
Compañia Cervecerias Unidas S.A.	CCU	CCU.csv	Sep-92
Willamette Valley Vineyards, Inc.	WVVI	WVVI.csv	Sep-94
Tobacco Products Stocks			
Altria Group, Inc.	MO	MO.csv	Jan-75
British American Tobacco p.l.c.	BTI	BTI.csv	Apr-80
Vector Group Ltd.	VGR	VGR.csv	Oct-87
Universal Corporation	UVV	UVV.csv	Mar-80
Schweitzer-Mauduit International, Inc.	SWM	SWM.csv	Nov-95
Food Item Stocks			
Sysco Corporation	SYY	SYY.csv	Jan-75
General Mills, Inc.	GIS	GIS.csv	Mar-80
Kellogg Company	K	K.csv	Mar-80
McCormick & Company, Incorporated	MKC	MKC.csv	Mar-80
Conagra Brands, Inc.	CAG	CAG.csv	Mar-80
Campbell Soup Company	CPG	CPG.csv	Mar-80
Seaboard Corporation	SEB	SEB.csv	Mar-80

Table A-12: Finance Sector (Part 1: Banks)

SECURITY	SYMBOL	FILE NAME	START DATE
Bank Stocks			
U.S. Bancorp	USB	USB.csv	Jan-75
Wells Fargo & Company	WFC	WFC.csv	Jan-75
The Bank of New York Mellon Corp.	BK	BK.csv	Jan-75
Citigroup Inc.	C	Citi.csv	Jan-77
JP Morgan Chase & Co	JPM	JPM.csv	Mar-80
Bank of America Corporation	BAC	BAC.csv	Mar-80
The PNC Financial Services Group	PNC	PNC.csv	Sep-88
Midwest Bank Stocks			
UMB Financial Corporation	UMBF	UMBF.csv	Mar-80
Huntington Bancshares Incorporated	HBAN	HBAN.csv	Mar-80
Commerce Bancshares, Inc.	CBSH	CBSH.csv	Mar-80
Associated Banc-Corp	ASB	ASB.csv	Mar-80
First Financial Bancorp.	FFBC	FFBC_fix.csv	Dec-83
Old National Bancorp	ONB	ONB_fix.csv	Jun-84
TCF Financial Corporation	TCF	TCV.csv	Jun-86
Investment Bank Stocks			
Siebert Financial Corp.	SIEB	SIEB.csv	Mar-80
Stifel Financial Corp.	SF	SF.csv	Jul-83
Raymond James Financial, Inc.	RJF	RJF.csv	Dec-87
The Charles Schwab Corporation	SCHW	SCHW.csv	Sep-87
WisdomTree Investments, Inc.	WETF	WETF.csv	Mar-93
Oppenheimer Holdings Inc.	OPY	OPY.csv	Aug-93
National Holdings Corporation	NHLD	NHLD.csv	Aug-95
Investment Management Stocks			
Eaton Vance Corp.	EV	EV.csv	Mar-80
Barings Corporate Investors	MCI	MCI.csv	Mar-80
SEI Investments Co.	SEIC	SEIC.csv	Mar-81
Franklin Resources, Inc.	BEN	BEN.csv	Sep-83
Legg Mason, Inc.	LM	LM.csv	Aug-83
T. Rowe Price Group Inc.	TROW	TROW.csv	Apr-86
AllianceBernstein Holding L.P.	AB	AB.csv	Apr-88

Table A-13: Finance Sector (Part 2: Insurance)

SECURITY	SYMBOL	FILE NAME	START DATE
Insurance Stocks (Property, Casualty, Title Insurance)			
W. R. Berkley Corporation	WRB	WRB.csv	Jan-75
Alleghany Corporation	Y	Y.csv	Jan-80
The Progressive Corporation	PGR	PGR.csv	Mar-80
The Travelers Companies, Inc.	TRV	TRV.csv	Mar-80
Cincinnati Financial Corporation	CINF	CINF.csv	Mar-80
CNA Financial Corporation	CNA	CNA.csv	Mar-80
Markel Corporation	MKL	MKL.csv	Dec-82
Insurance Brokers Stocks			
Marsh & McLennan Companies, Inc.	MMC	MMC.csv	Mar-80
Aon plc	AON	AON.csv	Jun-80
Brown & Brown, Inc.	BRO	BRO.csv	Feb-81
Arthur J. Gallagher & Co	AJG	AJG.csv	Jun-84

Table A-14: Finance Sector (Part 3: REIT & Real Estate)

SECURITY	SYMBOL	FILE NAME	START DATE
REIT – Other Equity Trusts Stocks			
Host Hotels & Resorts	HST	HST.csv	Mar-80
Public Storage	PSA	PSA.csv	Nov-80
Wellltower Inc.	WELL	WELL.csv	Mar-80
Vornado Realty Trust	VNO	VNO.csv	Mar-80
EastGroup Properties, Inc.	EGP	EGP.csv	Mar-80
HCP, Inc.	HCP	HCP.csv	May-85
Duke Realty Corporation	DRE	DRE.csv	Feb-86
REIT – Retail Equity Trusts Stocks			
Federal Realty Investment Trust	FRT	FRT.csv	Jan-75
Pennsylvania REIT	PEI	PEI.csv	Jan-75
Urstadt Biddle Properties, Inc.	UBP	UBP.csv	Mar-80
National Retail Properties, Inc.	NNN	NNN.csv	Oct-84
Weingarten Realty Investors	WRI	WRI.csv	Aug-85
Kimco Realty Corporation	KIM	KIM.csv	Nov-91
Taubman Centes, Inc.	TCO	TCO.csv	Nov-92
Real Estate Operations Stocks			
Texas Pacific Land Trust	TPL	TPL.csv	Mar-80
Tejon Ranch Co.	TRC	TRC.csv	Mar-80
American Realty Investors, Inc.	ARL	ARL.csv	Apr-82
Brookfield Asset Management Inc.	BAM	BAM.csv	Dec-83
FRP Holdings, Inc.	FRPH	FRPH.csv	Jul-86
Income Opportunity Realty Investors, Inc.	IOR	IOR.csv	Mar-92
J W Mays	MAYS	MAYS.csv	Feb-92

Table A-15: Industrial Products Sector Stocks

SECURITY	SYMBOL	FILE NAME	START DATE
General Industrial Machinery Stocks			
Ingersoll-Rand Plc	IR	IR.csv	Mar-80
Dover Corporation	DOV	DOV.csv	Mar-80
Nordson Corporation	NDSN	NDSN_fix.csv	Mar-80
Graco Inc.	GGG	GGG.csv	Mar-80
Illinois Tool Works Inc.	ITW	ITW.csv	Feb-81
Parker-Hannifin Corporation	PH	PH.csv	Jul-85
Roper Technologies, Inc.	ROP	ROP.csv	Feb-92
Electrical Machinery Stocks			
Emerson Electric Co.	EMR	EMR.csv	Jan-75
Eaton Corporation plc	ETN	ETN.csv	Jan-75
Franklin Electric Co., Inc.	FELE	FELE.csv	Mar-80
A. O. Smith Corporation	AOS	AOS.csv	Sep-83
Regal Beloit Corporation	RBC	RBC.csv	Dec-83
II-VI Incorporated	IIVI	IIVI.csv	Oct-87
ESCO Technologies Inc.	ESE	ESE.csv	Oct-90
Protection — Safety Equipment & Services Stocks			
MSA Safety Incorporated	MSA	MSA.csv	Mar-80
The Eastern Company	EML	EML.csv	Mar-80
Napco Security Technologies, Inc.	NSSC	NSSC.csv	Dec-81
Brady Corporation	BRC	BRC_fix.csv	Sep-84
Johnson Controls International plc	JCI	JCI.csv	Mar-85
Lakeland Industries, Inc.	LAKE	LAKE.csv	Sep-86
Magal Security Systems Ltd.	MAGS	MAGS.csv	Mar-93
Pollution Control Equipment & Services Stocks			
Donaldson Company, Inc.	DCI	DCI.csv	Mar-80
CECO Environmental Corp.	CECE	CECE.csv	Dec-80
Ecology & Environment, Inc.	EEI	EEI.csv	Mar-87
Tetra Tech, Inc.	TTEK	TTEK.csv	Dec-91
Appliance Recycling Centers of America	ARCI	ARCI.csv	Nov-91
Fuel Tech, Inc.	FTEK	FTEK.csv	Sep-93

Table A-16: Medical Sector Stocks

SECURITY	SYMBOL	FILE NAME	START DATE
Large Cap Pharmaceutical Stocks			
Johnson & Johnson	JNJ	JNJ.csv	Jan-75
Pfizer Inc.	PFE	PFE.csv	Jan-75
Merck & Co., Inc.	MRK	MRK.csv	Jan-75
Eli Lilly and Company	LLY	LLY.csv	Jan-75
Novo Nordisk A/S	NVO	NVO.csv	Jan-75
Bristol-Myers Squibb Company	BMY	BMY.csv	Jan-75
GlaxoSmithKline plc	GSK	GSK.csv	Mar-80
Biomedical and Genetics Stocks			
Amgen Inc.	AMGN	AMGN.csv	Jun-83
Celgene Corporation	CELG	CELG.csv	Jul-87
Bio-Techne Corporation	TECH	TECH.csv	Feb-89
Biogen Inc.	BIIB	BIIB.csv	Sep-91
Gilead Sciences, Inc.	GILD	GILD.csv	Jan-92
Vertex Pharmaceuticals Incorporated	VRTX	VRTX.csv	Jul-91
Regeneron Pharmaceuticals, Inc.	REGN	REGN.csv	Apr-91
Medical Products Manufacturing Stocks			
Abbott Laboratories	ABT	ABT.csv	Mar-80
Stryker Corporation	SYK	SYK.csv	Mar-80
Bio-Rad Laboratories, Inc.	BIO	BIO.csv	Feb-80
Hill-Rom Holdings, Inc	HRC	HRC.csv	Mar-80
Medtronic plc	MDT	MDT.csv	Dec-81
Baxter International Inc.	BAX	BAX.csv	Jan-82
Boston Scientific Corporation	BSX	BSX.csv	May-92
Medical Instruments Manufacturing Stocks			
Thermo Fisher Scientific Inc.	TMO	TMO.csv	Mar-80
Varian Medical Systems, Inc.	VAR	VAR.csv	Mar-80
ABIOMED, Inc.	ABMD	ABND.csv	Jul-87
Teleflex Incorporated	TFX	TFX.csv	Feb-88
Hologic, Inc.	HOLX	HOLX.csv	Mar-90
IDEXX Laboratories, Inc.	IDXX	IDXX.csv	Jun-91
STERIS plc	STE	STE.csv	Jun-92

Table A-17: Oil & Energy Sector

SECURITY	SYMBOL	FILE NAME	START DATE
Oil & Gas Stocks			
Exxon Mobile Corporation	XOM	XOM.csv	Jan-75
Chevron Corporation	CVX	CVX.csv	Jan-75
BP p.l.c.	BP	BP.csv	Jan-77
Royal Dutch Shell plc	RDS-B	RDS-B.csv	Mar-80
Sasol Limited	SSL	SSL.csv	Apr-82
TOTAL S.A.	TOT	TOT.csv	Oct-91
YPF Sociedad Anonima	YPF	YPF.csv	Jan-93
Oil & Gas US Exploration Stocks			
Apache Corporation	APA	APA.csv	May-79
Noble Energy, Inc.	NBL	NBL.csv	Mar-80
Murphy Oil Corporation	MUR	MUR.csv	Mar-80
Devon Energy Corporation	DVN	DVN.csv	Jul-85
Anadarko Petrolium Corporation	APC	APC.csv	Sep-86
EOG Resources, Inc.	EOG	EOG.csv	Oct-89
Cabot Oil & Gas Corporation	COG	COG.csv	Feb-90
Oil & Gas Production and Pipelines Stocks			
The Williams Companies, Inc.	WMB	WMB.csv	Dec-81
TransCanada Corporation	TRP	TRP.csv	Sep-82
Enbridge Inc.	ENB	ENB.csv	Mar-84
Oil Field Machinery & Equipment Stocks			
Weatherford International plc	WFT	WFT.csv	Jan-75
McDermott International, Inc.	MDR	MDR.csv	Dec-82
Matrix Service Company	MTRX	MTRX.csv	Sep-90
ION Geophysical Corporation	IO	IO.csv	Apr-91
Superior Energy Services, Inc.	SPN	SPN.csv	Jul-92
Oil & Gas Drilling Stocks			
Helmerich & Payne, Inc.	HP	HP.csv	Oct-80
Ensco plc	ESV	ESV.csv	Mar-80
Nabors Industries Ltd.	NBR	NBR.csv	Mar-80
Key Energy Services, Inc.	KEG	KEG.csv	Oct-80
Rowan Companies plc	RDC	RDC.csv	Apr-83
Noble Corporation plc	NE	NE.csv	Sep-85
Transocean Ltd.	RIG	RIG.csv	May-93

Table A-18: Retail/Wholesale Sector

SECURITY	SYMBOL	FILE NAME	START DATE
Food & Restaurant Stocks			
McDonald's Corporation	MCD	MCD.csv	Jan-75
The Wendy's Company	WEN	WEN_fix.csv	May-80
Cracker Barrel Old Country Store, Inc.	CBRL	CBRL.csv	Nov-81
Jack in the Box Inc.	JACK	JACK.csv	Mar-92
Dine Brands Global, Inc.	DIN	DIN.csv	Jul-91
Starbucks Corporation	SBUX	SBUX.csv	Jun-92
The Cheesecake Factory Incorporated	CAKE	CAKE.csv	Sep-92
Apparel & Shoes Stocks			
Foot Locker, Inc.	FL	FL.csv	Jan-75
Nordstrom, Inc.	JWN	JWN.csv	Mar-80
The Gap, Inc.	GPS	GPS.csv	Mar-80
Genesco Inc.	GCO	GCO.csv	Mar-80
L Brands, Inc.	LB	LB.csv	Mar-82
Ascena Retail Group, Inc.	ASNA	ASNA.csv	May-83
The Cato Corporation	CATO	CATO.csv	Apr-87
Discount & Variety Stocks			
Walmart Inc.	WMT	WMT.csv	Jan-75
Big Lots, Inc.	BIG	BIG.csv	Jun-85
Ross Stores, Inc.	ROST	ROST.csv	Aug-85
Costco Wholesale Corporation	COST	COST.csv	Jul-86
The TJX Companies, Inc.	TJX	TJX.csv	Jun-87
Fred's Inc.	FRED	FRED.csv	Mar-92
Kohl's Corporation	KSS	KSS.csv	May-92

Table A-19: Transportation Sector

SECURITY	SYMBOL	FILE NAME	START DATE
Airlines Stocks			
Southwest Airlines Co.	LUV	LUV.csv	Jan-80
Alaska Air Group, Inc.	ALK	ALK.csv	Mar-80
PHI, Inc.	PHII	PHII.csv	Sep-81
SkyWest, Inc.	SKYW	SKYW.csv	Jun-86
Hawaiian Holdings, Inc.	HA	HA.csv	Jun-95
Bristow Group Inc.	BRS	BRS.csv	Mar-90
Rail Stocks			
Union Pacific Corporation	UNP	UNP.csv	Jan-80
CSX Corporation	CSX	CSX.csv	Nov-80
Kansas City Southern	KSU	KSU.csv	Mar-80
Norfolk Southern Corporation	NSC	NSC.csv	Jun-82
Canadian Pacific Railway Limited	CP	CP.csv	Dec-83
Truck Transportation Stocks			
J. B. Hunt Transport Services, Inc.	JBHT	JBHT.csv	Nov-83
Werner Enterprises, Inc.	WERN	WERN.csv	Jun-86
Heartland Express, Inc.	HTLD	HTLD.csv	Nov-86
Marten Transport, Ltd.	MRTN	MRTN.csv	Sep-86
Old Dominion Freight Line, Inc.	ODFL	ODFL.csv	Oct-91
Landstar System, Inc.	LSTR	LSTR.csv	Mar-93
Forward Air Corporation	FWRD	FRWD.csv	Nov-93

Table A-20: Utilities Sector

SECURITY	SYMBOL	FILE NAME	START DATE
Electric Power Distribution Stocks			
American Electric Power Company, Inc.	AEP	AEP.csv	Jan-75
Exelon Corporation	EXC	EXC.csv	Jan-80
NextEra Energy, Inc.	NEE	NEE.csv	Mar-80
Duke Energy Corporation	DUK	DUK.csv	Mar-80
Dominion Energy, Inc.	D	D.csv	Mar-80
Public Service Enterprise Group Inc.	PEG	PEG.csv	Jan-80
The Southern Company	SO	SO.csv	Dec-81
Natural Gas Distribution Stocks			
National Fuel Gas Company	NFG	NFG.csv	Jan-75
ONEOK, Inc.	OKE	OKE.csv	Oct-80
UGI Corporation	UGI	UGI.csv	Mar-80
MDU Resources Group, Inc.	MDU	MDU.csv	Mar-80
New Jersey Resources Corporation	NJR	NJR.csv	Mar-80
Atmos Energy Corporation	ATO	ATO.csv	Dec-83
Southwest Gas Holdings, Inc.	SWX	SWX.csv	Dec-87
Water Supply Stocks			
SJW Group	SJW	SJW.csv	Jan-75
Aqua America, Inc.	WTR	WTR.csv	Mar-80
American States Water Company	AWR	AWR.csv	Mar-80
California Water Service Group	CWT	CWT.csv	Mar-80
Middlesex Water Company	MSEX	MSEX.csv	Mar-80
Conecticut Water Service, Inc.	CTWS	CTWS.csv	Mar-80

Table A-21: Currency Exchange Rates

SECURITY	SYMBOL	FILE NAME	START DATE
Euro US Dollar	EUR/USD	eur-usd_fix.csv	Jan-80
British Pound US Dollar	GBP/USD	gbp-usd_fix.csv	Jan-80
Australian Dollar US Dollar	AUD/USD	aud-usd_fix.csv	Jan-80
U.S. Dollar Currency Index	USDX	usdx_fix.csv	Jan-80
Euro Austraian Dollar	EUR/AUD	eur-aud_fix.csv	May-86
Euro British Pound	EUR/GBP	eur-gbp_fix.csv	May-86
British Pound Japanese Yen	GBP/JPY	gbp-jpy_fix.csv	Jan-80
U.S. Dollar Canadian Dollar	USD/CAD	usd-cad_fix.csv	Jan-82
U.S. Dollar Swiss Franc	USD/CHF	usd-chf.csv	Mar-88
U.S. Dollar Japanese Yen	USD/JPY	usd-jpy.csv	Mar-88

APPENDIX B: E3 Forecast Study Results

This appendix contains the full accuracy results for both **E3 MAPE** and **E3 RMSE** for the financial forecasts, organized by the sector. Each table includes the results of all 12 forecast models, ranked in descending order of accuracy. The summary results by sector and category are presented first. Significant results (60% or greater) are highlighted.

Source files, raw data, R scripts and Excel files used to generate the results of this study are available by request at **TheScienceofAstrology.com**.

B.1. Summary Results of E3 Forecast Study

Table B-1: E3 MAPE Accuracy Ranked by Sector

SECTOR	TOTAL	MAPE #1	% MAPE #1	SECTOR	TOTAL	MAPE #1/#2	% MAPE #1/#2
INDUSTRIAL PRODUCTS	27	22	81.48%	STOCK MARKET INDEXES	10	10	100.00%
STOCK MARKET INDEXES	10	8	80.00%	AEROSPACE	12	11	91.67%
TRANSPORTATION	18	14	77.78%	UTILITIES	20	18	90.00%
AEROSPACE	12	9	75.00%	TRANSPORTATION	18	16	88.89%
UTILITIES	20	15	75.00%	FINANCE (REIT/REAL ESTATE)	21	18	85.71%
FINANCE (INSURANCE)	11	8	72.73%	INDUSTRIAL PRODUCTS	27	23	85.19%
FINANCE (REIT/REAL ESTATE)	21	15	71.43%	FINANCE (INSURANCE)	11	8	72.73%
BUSINESS SERVICES	24	16	66.67%	BUSINESS SERVICES	24	17	70.83%
CONSUMER STAPLES	22	13	59.09%	COMPUTER & TECHNOLOGY	43	29	67.44%
AUTO/TIRES/TRUCKS	12	7	58.33%	AUTO/TIRES/TRUCKS	12	8	66.67%
COMPUTER & TECHNOLOGY	43	22	51.16%	RETAIL/WHOLESALE	21	14	66.67%
CONSUMER DISCRETIONARY	22	11	50.00%	OIL & ENERGY	29	19	65.52%
FINANCE (BANKS)	28	13	46.43%	FINANCE (BANKS)	28	18	64.29%
OIL & ENERGY	29	13	44.83%	MEDICAL SECTOR	28	17	60.71%
BASIC MATERIALS	27	12	44.44%	BASIC MATERIALS	27	16	59.26%
RATES & BONDS	10	4	40.00%	CONSUMER DISCRETIONARY	22	13	59.09%
MEDICAL SECTOR	28	10	35.71%	CONSUMER STAPLES	22	13	59.09%
RETAIL/WHOLESALE	21	7	33.33%	RATES & BONDS	10	5	50.00%
COMMODITIES	21	6	28.57%	CONSTRUCTION	14	7	50.00%
CONSTRUCTION	14	4	28.57%	COMMODITIES	21	8	38.10%
CURRENCY EXCHANGE RATES	10	1	10.00%	CURRENCY EXCHANGE RATES	10	3	30.00%
TOTAL STOCKS	430	230	53.49%	TOTAL STOCKS	430	291	67.67%

Table B-2: E3 RMSE Accuracy Ranked by Sector

SECTOR	TOTAL	RMSE #1	% RMSE #1
UTILITIES	20	17	85.00%
BUSINESS SERVICES	24	16	66.67%
FINANCE (REIT/REAL ESTATE)	21	13	61.90%
RATES & BONDS	10	6	60.00%
CONSUMER DISCRETIONARY	22	13	59.09%
TRANSPORTATION	18	10	55.56%
FINANCE (INSURANCE)	11	6	54.55%
INDUSTRIAL PRODUCTS	27	14	51.85%
MEDICAL SECTOR	28	14	50.00%
RETAIL/WHOLESALE	21	10	47.62%
CONSUMER STAPLES	22	10	45.45%
BASIC MATERIALS	27	12	44.44%
AEROSPACE	12	5	41.67%
AUTO/TIRES/TRUCKS	12	5	41.67%
STOCK MARKET INDEXES	10	4	40.00%
COMPUTER & TECHNOLOGY	43	17	39.53%
OIL & ENERGY	29	10	34.48%
COMMODITIES	21	6	28.57%
CONSTRUCTION	14	3	21.43%
FINANCE (BANKS)	28	6	21.43%
CURRENCY EXCHANGE RATES	10	0	0.00%
TOTAL STOCKS	430	197	45.81%

SECTOR	TOTAL	RMSE #1/#2	% RMSE #1/#2
UTILITIES	20	19	95.00%
TRANSPORTATION	18	16	88.89%
RATES & BONDS	10	8	80.00%
AEROSPACE	12	9	75.00%
BUSINESS SERVICES	24	18	75.00%
MEDICAL SECTOR	28	21	75.00%
STOCK MARKET INDEXES	10	7	70.00%
BASIC MATERIALS	27	18	66.67%
FINANCE (REIT/REAL ESTATE)	21	14	66.67%
COMPUTER & TECHNOLOGY	43	28	65.12%
CONSUMER DISCRETIONARY	22	14	63.64%
CONSUMER STAPLES	22	14	63.64%
FINANCE (INSURANCE)	11	7	63.64%
INDUSTRIAL PRODUCTS	27	17	62.96%
RETAIL/WHOLESALE	21	12	57.14%
OIL & ENERGY	29	15	51.72%
AUTO/TIRES/TRUCKS	12	6	50.00%
COMMODITIES	21	9	42.86%
CONSTRUCTION	14	6	42.86%
FINANCE (BANKS)	28	12	42.86%
CURRENCY EXCHANGE RATES	10	0	0.00%
TOTAL STOCKS	430	270	62.79%

Table B-3: #1 E3 MAPE Accuracy Ranked by Category

CATEGORY	#	MAPE #1	% MAPE #1	CATEGORY	#	MAPE #1	% MAPE #1
Meat (Commodity)	2	2	100.00%	Insurance Brokers Stocks	4	2	50.00%
Automotive Stocks	5	5	100.00%	Chemicals	7	3	42.86%
Staffing Services	4	4	100.00%	Computer Software	7	3	42.86%
Schools	3	3	100.00%	Computer Hardware	7	3	42.86%
General Industrial Machinery	7	7	100.00%	Semiconductor Stocks	7	3	42.86%
Protection — Safety Equipment	7	7	100.00%	Textile	7	3	42.86%
Water Supply	6	6	100.00%	Oil & Gas U.S. Exploration	7	3	42.86%
Aerospace Equipment	7	6	85.71%	Oil & Gas Drilling	7	3	42.86%
Steel Producers	7	6	85.71%	Apparel & Shoes	7	3	42.86%
Wireless Providers	7	6	85.71%	Rates & Bonds	10	4	40.00%
Insurance Stocks	7	6	85.71%	Oil Field Machinery & Equipment	5	2	40.00%
REIT—Other Equity Trust Stocks	7	6	85.71%	Energy (Commodity)	3	1	33.33%
Electrical Machinery	7	6	85.71%	Mining	6	2	33.33%
Airlines	6	5	83.33%	Pollution Control Equipment	6	2	33.33%
Stock Market Indexes	10	8	80.00%	Oil & Gas Production & Pipeline	3	1	33.33%
Alcoholic Beverages	5	4	80.00%	OEM Manufacturer Stocks	7	2	28.57%
Rail	5	4	80.00%	Building & Construction	7	2	28.57%
Financial Transaction Services	4	3	75.00%	Residential & Commercial	7	2	28.57%
Electrical Products	7	5	71.43%	Furniture Manufacturing	7	2	28.57%
Food Item Stocks	7	5	71.43%	Bank Stocks	7	2	28.57%
Midwest Bank Stocks	7	5	71.43%	Investment Management Stocks	7	2	28.57%
Real Estate Operations Stocks	7	5	71.43%	Biomedical and Genetics	7	2	28.57%
Truck Transportation	7	5	71.43%	Medical Products Manufacturing	7	2	28.57%
Natural Gas Distribution	7	5	71.43%	Medical Instruments Manufacturing	7	2	28.57%
Aerospace General	5	3	60.00%	Food & Restaurant	7	2	28.57%
Advertising and Marketing	5	3	60.00%	Discount & Variety	7	2	28.57%
Leisure & Recreation	5	3	60.00%	Communication Components	4	1	25.00%
Soft Drinks	5	3	60.00%	Scientific Instruments	4	1	25.00%
Business Services	7	4	57.14%	Metals (Commodity)	5	1	20.00%
Investment Bank Stocks	7	4	57.14%	Softs (Commodity)	5	1	20.00%
REIT—Retail Equity Trusts Stocks	7	4	57.14%	Tobacco Products	5	1	20.00%
Large Cap Pharmaceutical	7	4	57.14%	Grains (Commodity)	6	1	16.67%
Oil & Gas Stocks	7	4	57.14%	Gold Mining Stocks	7	1	14.29%
Electric Power Distribution	7	4	57.14%	Currency Exchange Rates	10	1	10.00%
Outsourcing Services	4	2	50.00%				

Table B-4: #1/#2 E3 MAPE Accuracy Ranked by Category

CATEGORY	#	MAPE #1/#2	% MAPE #1/#2
Stock Market Indexes	10	10	100.00%
Meat (Commodity)	2	2	100.00%
Aerospace Equipment	7	7	100.00%
Automotive Stocks	5	5	100.00%
Staffing Services	4	4	100.00%
Schools	3	3	100.00%
REIT—Other Equity Trust Stocks	7	7	100.00%
General Industrial Machinery	7	7	100.00%
Electrical Machinery	7	7	100.00%
Protection — Safety Equipment	7	7	100.00%
Truck Transportation	7	7	100.00%
Water Supply	6	6	100.00%
Steel Producers	7	6	85.71%
Computer Software	7	6	85.71%
Wireless Providers	7	6	85.71%
Insurance Stocks	7	6	85.71%
REIT—Retail Equity Trusts Stocks	7	6	85.71%
Medical Instruments Manufacturing	7	6	85.71%
Oil & Gas Stocks	7	6	85.71%
Electric Power Distribution	7	6	85.71%
Natural Gas Distribution	7	6	85.71%
Airlines	6	5	83.33%
Aerospace General	5	4	80.00%
Advertising and Marketing	5	4	80.00%
Leisure & Recreation	5	4	80.00%
Alcoholic Beverages	5	4	80.00%
Rail	5	4	80.00%
Financial Transaction Services	4	3	75.00%
Communication Components	4	3	75.00%
Chemicals	7	5	71.43%
Electrical Products	7	5	71.43%
Food Item Stocks	7	5	71.43%
Midwest Bank Stocks	7	5	71.43%
Investment Bank Stocks	7	5	71.43%
Real Estate Operations Stocks	7	5	71.43%
Medical Products Manufacturing	7	5	71.43%
Oil & Gas U.S. Exploration	7	5	71.43%
Apparel & Shoes	7	5	71.43%
Discount & Variety	7	5	71.43%
Energy (Commodity)	3	2	66.67%
Oil & Gas Production & Pipeline	3	2	66.67%
Soft Drinks	5	3	60.00%
Business Services	7	4	57.14%
Computer Hardware	7	4	57.14%
Building & Construction	7	4	57.14%
Textile	7	4	57.14%
Bank Stocks	7	4	57.14%
Investment Management Stocks	7	4	57.14%
Large Cap Pharmaceutical	7	4	57.14%
Oil & Gas Drilling	7	4	57.14%
Food & Restaurant	7	4	57.14%
Rates & Bonds	10	5	50.00%
Mining	6	3	50.00%
Outsourcing Services	4	2	50.00%
Scientific Instruments	4	2	50.00%
Insurance Brokers Stocks	4	2	50.00%
OEM Manufacturer Stocks	7	3	42.86%
Semiconductor Stocks	7	3	42.86%
Residential & Commercial	7	3	42.86%
Metals (Commodity)	5	2	40.00%
Oil Field Machinery & Equipment	5	2	40.00%
Pollution Control Equipment	6	2	33.33%
Currency Exchange Rates	10	3	30.00%
Gold Mining Stocks	7	2	28.57%
Furniture Manufacturing	7	2	28.57%
Biomedical and Genetics	7	2	28.57%
Softs (Commodity)	5	1	20.00%
Tobacco Products	5	1	20.00%
Grains (Commodity)	6	1	16.67%

Table B-5: #1 E3 RMSE Accuracy Ranked by Category

CATEGORY	#	RMSE #1	% RMSE #1	CATEGORY	#	RMSE #1	% RMSE #1
Meat (Commodity)	2	2	100.00%	Mining	6	3	50.00%
Water Supply	6	6	100.00%	Gold Mining Stocks	7	3	42.86%
Natural Gas Distribution	7	6	85.71%	Furniture Manufacturing	7	3	42.86%
Automotive Stocks	5	4	80.00%	REIT—Retail Equity Trusts Stocks	7	3	42.86%
Leisure & Recreation	5	4	80.00%	Stock Market Indexes	10	4	40.00%
Alcoholic Beverages	5	4	80.00%	Softs (Commodity)	5	2	40.00%
Financial Transaction Services	4	3	75.00%	Soft Drinks	5	2	40.00%
Staffing Services	4	3	75.00%	Rail	5	2	40.00%
Outsourcing Services	4	3	75.00%	Pollution Control Equipment	6	2	33.33%
Electrical Products	7	5	71.43%	Oil & Gas Production & Pipeline	3	1	33.33%
Insurance Stocks	7	5	71.43%	Aerospace Equipment	7	2	28.57%
REIT—Other Equity Trust Stocks	7	5	71.43%	Chemicals	7	2	28.57%
Real Estate Operations Stocks	7	5	71.43%	Computer Software	7	2	28.57%
Biomedical and Genetics	7	5	71.43%	Computer Hardware	7	2	28.57%
Electric Power Distribution	7	5	71.43%	Semiconductor Stocks	7	2	28.57%
Schools	3	2	66.67%	Building & Construction	7	2	28.57%
Airlines	6	4	66.67%	Investment Bank Stocks	7	2	28.57%
Rates & Bonds	10	6	60.00%	Oil & Gas U.S. Exploration	7	2	28.57%
Aerospace General	5	3	60.00%	Oil & Gas Drilling	7	2	28.57%
Advertising and Marketing	5	3	60.00%	Apparel & Shoes	7	2	28.57%
Steel Producers	7	4	57.14%	Communication Components	4	1	25.00%
Business Services	7	4	57.14%	Scientific Instruments	4	1	25.00%
Wireless Providers	7	4	57.14%	Insurance Brokers Stocks	4	1	25.00%
Textile	7	4	57.14%	Metals (Commodity)	5	1	20.00%
Food Item Stocks	7	4	57.14%	Oil Field Machinery & Equipment	5	1	20.00%
Midwest Bank Stocks	7	4	57.14%	Grains (Commodity)	6	1	16.67%
General Industrial Machinery	7	4	57.14%	OEM Manufacturer Stocks	7	1	14.29%
Electrical Machinery	7	4	57.14%	Residential & Commercial	7	1	14.29%
Protection — Safety Equipment	7	4	57.14%	Medical Instruments Manufacturing	7	1	14.29%
Large Cap Pharmaceutical	7	4	57.14%	Energy (Commodity)	3	0	0.00%
Medical Products Manufacturing	7	4	57.14%	Tobacco Products	5	0	0.00%
Oil & Gas Stocks	7	4	57.14%	Bank Stocks	7	0	0.00%
Food & Restaurant	7	4	57.14%	Investment Management Stocks	7	0	0.00%
Discount & Variety	7	4	57.14%	Currency Exchange Rates	10	0	0.00%
Truck Transportation	7	4	57.14%				

Table B-6: #1/#2 E3 RMSE Accuracy Ranked by Category

CATEGORY	#	RMSE #1/#2	% RMSE #1/#2
Meat (Commodity)	2	2	100.00%
Truck Transportation	7	7	100.00%
Electric Power Distribution	7	7	100.00%
Water Supply	6	6	100.00%
Computer Software	7	6	85.71%
Electrical Products	7	6	85.71%
Electrical Machinery	7	6	85.71%
Biomedical and Genetics	7	6	85.71%
Oil & Gas Stocks	7	6	85.71%
Natural Gas Distribution	7	6	85.71%
Airlines	6	5	83.33%
Rates & Bonds	10	8	80.00%
Aerospace General	5	4	80.00%
Automotive Stocks	5	4	80.00%
Advertising and Marketing	5	4	80.00%
Leisure & Recreation	5	4	80.00%
Alcoholic Beverages	5	4	80.00%
Rail	5	4	80.00%
Financial Transaction Services	4	3	75.00%
Staffing Services	4	3	75.00%
Outsourcing Services	4	3	75.00%
Scientific Instruments	4	3	75.00%
Aerospace Equipment	7	5	71.43%
Chemicals	7	5	71.43%
Steel Producers	7	5	71.43%
Business Services	7	5	71.43%
Textile	7	5	71.43%
Food Item Stocks	7	5	71.43%
Midwest Bank Stocks	7	5	71.43%
Insurance Stocks	7	5	71.43%
REIT—Other Equity Trust Stocks	7	5	71.43%
Real Estate Operations Stocks	7	5	71.43%
General Industrial Machinery	7	5	71.43%
Large Cap Pharmaceutical	7	5	71.43%
Medical Products Manufacturing	7	5	71.43%
Medical Instruments Manufacturing	7	5	71.43%
Oil & Gas U.S. Exploration	7	5	71.43%
Stock Market Indexes	10	7	70.00%
Mining	6	4	66.67%
Schools	3	2	66.67%
Softs (Commodity)	5	3	60.00%
Soft Drinks	5	3	60.00%
Gold Mining Stocks	7	4	57.14%
Wireless Providers	7	4	57.14%
Semiconductor Stocks	7	4	57.14%
Building & Construction	7	4	57.14%
REIT—Retail Equity Trusts Stocks	7	4	57.14%
Protection — Safety Equipment	7	4	57.14%
Food & Restaurant	7	4	57.14%
Apparel & Shoes	7	4	57.14%
Discount & Variety	7	4	57.14%
Communication Components	4	2	50.00%
Insurance Brokers Stocks	4	2	50.00%
Computer Hardware	7	3	42.86%
Furniture Manufacturing	7	3	42.86%
Investment Bank Stocks	7	3	42.86%
Tobacco Products	5	2	40.00%
Grains (Commodity)	6	2	33.33%
Energy (Commodity)	3	1	33.33%
Pollution Control Equipment	6	2	33.33%
Oil & Gas Production & Pipeline	3	1	33.33%
OEM Manufacturer Stocks	7	2	28.57%
Residential & Commercial	7	2	28.57%
Bank Stocks	7	2	28.57%
Investment Management Stocks	7	2	28.57%
Oil & Gas Drilling	7	2	28.57%
Metals (Commodity)	5	1	20.00%
Oil Field Machinery & Equipment	5	1	20.00%
Currency Exchange Rates	10	0	0.00%

B.2. Stock Market Indexes

Table B-7: E3 MAPE Accuracy Results: Stock Market Indexes

SECURITY	MAPE #1	MAPE #2	MAPE #3	MAPE #4	MAPE #5	MAPE #6	MAPE #7	MAPE #8	MAPE #9	MAPE #10	MAPE #11	MAPE #12
STOCK MARKET INDEXES												
S&P 500 (^GSPC)	ESM3_E3 4.4216	ESM_E3 4.4321	ESM 4.4770	HOLT3_E3 4.5430	NAIVE3_E3 4.5469	HOLT 4.5473	NAÏVE 4.5530	ARIMA_E3 4.5904	M15_E3 4.6337	ARIMA 4.8700	MEAN_E3 4.9520	MEAN 6.0296
NASDAQ Composite (^IXIC)	ESM3_E3 6.4796	ESM 6.4818	ESM_E3 6.5293	ARIMA_E3 6.5668	HOLT3_E3 6.6824	NAIVE3_E3 6.7068	HOLT 6.7195	NAÏVE 6.7380	M15_E3 6.7714	MEAN_E3 7.3798	ARIMA 7.7890	MEAN 8.9988
Dow Jones Industrial (^DJI)	ESM_E3 4.3041	HOLT3_E3 4.3048	NAIVE3_E3 4.3063	ESM3_E3 4.3069	HOLT 4.3194	NAÏVE 4.3220	ESM 4.3345	ARIMA_E3 4.3561	M15_E3 4.3701	ARIMA 4.4405	MEAN_E3 4.5815	MEAN 5.4157
Dow Jones Transportation (^DJT)	ESM_E3 6.1309	ARIMA_E3 6.1337	ARIMA 6.1699	ESM3_E3 6.1814	NAIVE3_E3 6.1999	HOLT3_E3 6.2060	M15_E3 6.2380	NAÏVE 6.2505	HOLT 6.2646	ESM 6.3006	MEAN_E3 6.5893	MEAN 7.8809
Dow Jones Utility (^DJU)	ESM3_E3 4.4824	ESM_E3 4.4857	HOLT3_E3 4.5061	NAIVE3_E3 4.5075	ESM 4.5102	ARIMA_E3 4.5224	HOLT 4.5396	NAÏVE 4.5409	ARIMA 4.5711	M15_E3 4.5750	MEAN_E3 5.1876	MEAN 6.5040
Dow Jones Composite (^DJA)	ESM_E3 4.0647	ESM3_E3 4.0828	ESM 4.1708	HOLT3_E3 4.1954	NAIVE3_E3 4.1957	ARIMA_E3 4.2073	HOLT 4.2189	NAÏVE 4.2199	M15_E3 4.2427	ARIMA 4.3157	MEAN_E3 4.6062	MEAN 5.6715
NYSE Composite (^NYA)	ESM_E3 4.4129	ESM3_E3 4.4181	NAÏVE 4.5042	ESM 4.5047	HOLT 4.5048	NAIVE3_E3 4.5051	HOLT3_E3 4.5056	M15_E3 4.6255	ARIMA_E3 4.8097	MEAN_E3 4.9685	ARIMA 5.3086	MEAN 6.0735
Russell 2000 (^RUT)	HOLT3_E3 6.0728	NAIVE3_E3 6.0808	ARIMA_E3 6.0839	ARIMA 6.0857	ESM_E3 6.0878	ESM3_E3 6.0916	HOLT 6.0938	NAÏVE 6.1000	ESM 6.1293	M15_E3 6.1897	MEAN_E3 6.5665	MEAN 7.8769
Nikkei 225 (^N225)	HOLT 6.0260	HOLT3_E3 6.0263	NAIVE3_E3 6.0449	NAÏVE 6.0480	ARIMA 6.0504	ESM3_E3 6.0704	ESM 6.0814	ESM_E3 6.0941	ARIMA_E3 6.1069	M15_E3 6.2342	MEAN_E3 6.9983	MEAN 8.6752
CBOE Volatility Index (^VIX)	ARIMA 18.1236	MEAN_E3 18.1524	ARIMA_E3 18.4723	NAIVE3_E3 18.5371	HOLT3_E3 18.5493	NAÏVE 18.6753	HOLT 18.6762	M15_E3 19.0003	MEAN 19.7358	ESM_E3 19.9532	ESM3_E3 20.8881	ESM 22.0285

Table B-8: E3 RMSE Accuracy Results: Stock Market Indexes

SECURITY	RMSE #1	RMSE #2	RMSE #3	RMSE #4	RMSE #5	RMSE #6	RMSE #7	RMSE #8	RMSE #9	RMSE #10	RMSE #11	RMSE #12
STOCK MARKET INDEXES												
S&P 500 (^GSPC)	HOLT 1.1883	NAÏVE 1.1897	HOLT3_E3 1.1921	ESM_E3 1.1925	NAIVE3_E3 1.1931	ESM3_E3 1.1992	ARIMA_E3 1.2129	ESM 1.2225	M15_E3 1.2292	MEAN_E3 1.2974	ARIMA 1.3422	MEAN 1.5163
NASDAQ Composite (^IXIC)	ESM 3.8871	ESM3_E3 3.9063	HOLT 3.9442	ESM_E3 3.9633	HOLT3_E3 3.9730	NAÏVE 3.9792	NAIVE3_E3 4.0122	M15_E3 4.1840	ARIMA_E3 4.1978	MEAN_E3 4.4173	MEAN 5.1930	ARIMA 7.1975
Dow Jones Industrial (^DJI)	ESM3_E3 10.4189	HOLT 10.4205	NAÏVE 10.4258	HOLT3_E3 10.4368	NAIVE3_E3 10.4398	ESM 10.4464	ESM_E3 10.4537	ARIMA_E3 10.6264	M15_E3 10.7071	ARIMA 10.9845	MEAN_E3 11.1304	MEAN 12.9940
Dow Jones Transportation (^DJT)	ARIMA 5.4146	ARIMA_E3 5.5666	ESM_E3 5.6344	ESM3_E3 5.6727	NAIVE3_E3 5.6823	HOLT3_E3 5.6862	NAÏVE 5.6996	HOLT 5.7058	M15_E3 5.7857	ESM 5.7859	MEAN_E3 6.1469	MEAN 7.1776
Dow Jones Utility (^DJU)	ESM_E3 0.3582	ARIMA_E3 0.3592	ESM3_E3 0.3593	HOLT3_E3 0.3599	NAIVE3_E3 0.3599	ESM 0.3628	HOLT 0.3632	NAÏVE 0.3633	M15_E3 0.3636	ARIMA 0.3645	MEAN_E3 0.4092	MEAN 0.5025
Dow Jones Composite (^DJA)	ESM_E3 3.2912	ESM3_E3 3.3063	ARIMA_E3 3.3365	HOLT3_E3 3.3528	NAIVE3_E3 3.3548	HOLT 3.3555	NAÏVE 3.3586	ESM 3.3756	ARIMA 3.3978	M15_E3 3.4238	MEAN_E3 3.7071	MEAN 4.4031
NYSE Composite (^NYA)	HOLT 6.7368	NAÏVE 6.7406	HOLT3_E3 6.7625	NAIVE3_E3 6.7657	ESM_E3 6.8450	ESM3_E3 6.9130	M15_E3 7.0017	ESM 7.0760	ARIMA_E3 7.2613	MEAN_E3 7.6325	ARIMA 8.7339	MEAN 8.9923
Russell 2000 (^RUT)	HOLT 0.8656	NAÏVE 0.8662	ARIMA 0.8663	ESM 0.8670	HOLT3_E3 0.8684	ESM3_E3 0.8689	NAIVE3_E3 0.8692	ESM_E3 0.8754	ARIMA_E3 0.8803	M15_E3 0.9019	MEAN_E3 0.9509	MEAN 1.0872
Nikkei 225 (^N225)	NAIVE3_E3 15.9072	HOLT3_E3 15.9078	NAÏVE 15.9467	HOLT 15.9596	ARIMA 15.9673	ARIMA_E3 15.9782	ESM_E3 15.9991	ESM3_E3 16.0118	ESM 16.0908	M15_E3 16.1729	MEAN_E3 17.4106	MEAN 20.8469
CBOE Volatility Index (^VIX)	ARIMA 0.0858	ARIMA_E3 0.0862	HOLT3_E3 0.0863	NAIVE3_E3 0.0867	HOLT 0.0868	ESM_E3 0.0873	NAÏVE 0.0873	M15_E3 0.0877	ESM3_E3 0.0883	MEAN_E3 0.0893	ESM 0.0902	MEAN 0.1013

B.3. Commodities

Table B-9: E3 MAPE Accuracy Results: Commodities

SECURITY	MAPE #1	MAPE #2	MAPE #3	MAPE #4	MAPE #5	MAPE #6	MAPE #7	MAPE #8	MAPE #9	MAPE #10	MAPE #11	MAPE #12
METALS												
Gold Futures	ESM_E3 4.3309	ESM3_E3 4.3464	NAÏVE 4.3573	HOLT 4.3601	NAIVE3_E3 4.3706	HOLT3_E3 4.3736	ESM 4.4388	M15_E3 4.5520	MEAN_E3 4.7007	ARIMA_E3 4.7840	MEAN 5.5928	ARIMA 5.8827
Platinum Futures	HOLT 5.8462	NAÏVE 5.9000	ESM 5.9062	HOLT3_E3 5.9353	ESM3_E3 5.9999	NAIVE3_E3 6.0039	ESM_E3 6.1378	ARIMA_E3 6.4611	M15_E3 6.5086	ARIMA 6.6274	MEAN_E3 7.6880	MEAN 9.5002
Silver Futures	ESM 8.7322	ESM3_E3 8.7822	NAÏVE 8.8066	HOLT 8.8295	NAIVE3_E3 8.8887	HOLT3_E3 8.9074	ESM_E3 8.9358	ARIMA 8.9386	ARIMA_E3 9.1785	M15_E3 9.5109	MEAN_E3 9.8066	MEAN 11.5908
Copper Futures	NAÏVE 7.5696	ESM 7.5717	NAIVE3_E3 7.6060	ESM3_E3 7.6083	ESM_E3 7.6957	HOLT3_E3 7.7092	HOLT 7.7242	ARIMA_E3 7.9293	M15_E3 7.9994	ARIMA 8.2911	MEAN_E3 8.8093	MEAN 10.9439
Palladium Futures	ESM 8.3682	HOLT 8.4074	NAÏVE 8.4262	HOLT3_E3 8.4870	ESM3_E3 8.5098	NAIVE3_E3 8.5969	ARIMA 8.7083	ESM_E3 8.7625	ARIMA_E3 9.0636	M15_E3 9.5105	MEAN_E3 11.3855	MEAN 14.5746

APPENDIX B: E3 Forecast Study Results

SECURITY	MAPE #1	MAPE #2	MAPE #3	MAPE #4	MAPE #5	MAPE #6	MAPE #7	MAPE #8	MAPE #9	MAPE #10	MAPE #11	MAPE #12
SOFTS												
Coffee Futures	ARIMA	ESM	HOLT	NAÏVE	ESM3_E3	HOLT3_E3	NAIVE3_E3	ESM_E3	ARIMA_E3	M15_E3	MEAN_E3	MEAN
	7.2553	7.2640	7.2641	7.2642	7.2677	7.2678	7.2678	7.3290	7.3770	7.6062	8.6138	10.9282
Lumber Futures	HOLT	ESM	NAÏVE	HOLT3_E3	ARIMA	ESM3_E3	NAIVE3_E3	ESM_E3	ARIMA_E3	M15_E3	MEAN_E3	MEAN
	8.7711	8.7825	8.7825	8.7842	8.7972	8.7993	8.7993	8.8714	8.9289	9.1707	10.2886	13.0247
Cocoa Futures	ESM3_E3	NAIVE3_E3	HOLT3_E3	ARIMA	ESM_E3	ARIMA_E3	ESM	NAÏVE	HOLT	M15_E3	MEAN_E3	MEAN
	7.3293	7.3301	7.3302	7.3340	7.3341	7.3400	7.4019	7.4031	7.4035	7.5366	8.0381	9.5613
Sugar Futures	NAÏVE	ESM	NAIVE3_E3	ESM3_E3	HOLT3_E3	HOLT	ESM_E3	M15_E3	MEAN_E3	ARIMA_E3	MEAN	ARIMA
	8.9223	8.9409	8.9832	8.9870	9.0280	9.1022	9.2007	10.0292	11.0533	12.1776	13.9239	16.6941
Cotton Futures	HOLT	ARIMA	NAÏVE	ESM	HOLT3_E3	ESM3_E3	NAIVE3_E3	ESM_E3	ARIMA_E3	M15_E3	MEAN_E3	MEAN
	7.0340	7.0695	7.0972	7.0998	7.1612	7.2128	7.2483	7.4264	7.5897	8.0974	9.5203	11.9867
GRAINS												
Corn Futures	ARIMA	HOLT	NAÏVE	ESM	HOLT3_E3	NAIVE3_E3	ESM3_E3	ESM_E3	ARIMA_E3	M15_E3	MEAN_E3	MEAN
	7.0312	7.2033	7.2816	7.2816	7.3839	7.4653	7.4654	7.7268	7.7736	8.4089	9.4177	11.2777
Oats Futures	ESM	ARIMA	ESM3_E3	NAÏVE	HOLT3_E3	NAIVE3_E3	HOLT	ESM_E3	ARIMA_E3	M15_E3	MEAN_E3	MEAN
	8.2940	8.3176	8.3223	8.3284	8.3562	8.3691	8.3845	8.4388	8.5364	8.8955	9.6455	11.6271
Soybean Futures	ARIMA	HOLT	NAÏVE	ESM	HOLT3_E3	ESM3_E3	NAIVE3_E3	ESM_E3	ARIMA_E3	M15_E3	MEAN_E3	MEAN
	6.9217	6.9496	6.9664	6.9694	7.0019	7.0130	7.0427	7.1458	7.2576	7.6749	8.5064	10.3551
Soybean Oil Futures	ESM	NAÏVE	HOLT	HOLT3_E3	NAIVE3_E3	ESM3_E3	ESM_E3	ARIMA_E3	M15_E3	ARIMA	MEAN_E3	MEAN
	6.3800	6.3800	6.3887	6.4257	6.4314	6.4314	6.5456	6.8542	6.9506	6.9886	7.9386	9.5350
Soybean Meal Futures	HOLT3_E3	ESM3_E3	NAIVE3_E3	ESM	NAÏVE	HOLT	ESM_E3	ARIMA	ARIMA_E3	M15_E3	MEAN_E3	MEAN
	8.2038	8.2046	8.2046	8.2371	8.2371	8.2431	8.2437	8.2493	8.2960	8.5353	9.3691	11.5485
Wheat Futures	NAÏVE	ESM	HOLT	ESM3_E3	NAIVE3_E3	HOLT3_E3	ESM_E3	ARIMA_E3	M15_E3	ARIMA	MEAN_E3	MEAN
	7.5016	7.5026	7.5107	7.5156	7.5158	7.5214	7.5966	7.8913	7.9670	8.1350	8.3371	9.7449
ENERGY												
Brent Oil Futures	NAÏVE	NAIVE3_E3	ESM	ESM3_E3	ESM_E3	HOLT3_E3	HOLT	M15_E3	ARIMA_E3	ARIMA	MEAN_E3	MEAN
	9.3512	9.3708	9.4067	9.4123	9.4857	9.5415	9.5927	9.7658	9.8225	10.1397	11.0123	13.8578
Crude Oil WTI Futures	NAÏVE	ESM	NAIVE3_E3	HOLT3_E3	ESM3_E3	HOLT	ESM_E3	M15_E3	ARIMA_E3	ARIMA	MEAN_E3	MEAN
	9.7596	9.7759	9.7771	9.7845	9.7906	9.7943	9.8562	10.1286	10.3746	11.1985	11.3836	14.2917
Heating Oil Futures	HOLT3_E3	HOLT	NAIVE3_E3	ESM3_E3	NAÏVE	ESM	ESM_E3	ARIMA_E3	M15_E3	ARIMA	MEAN_E3	MEAN
	8.4756	8.4761	8.4794	8.4801	8.4836	8.4838	8.5449	8.7448	8.8450	8.8972	10.2920	13.1985
MEATS												
Cattle Futures	HOLT3_E3	ESM3_E3	NAIVE3_E3	NAÏVE	ESM	ARIMA	HOLT	ESM_E3	ARIMA_E3	M15_E3	MEAN_E3	MEAN
	4.3388	4.3416	4.3416	4.3446	4.3485	4.3515	4.3521	4.3807	4.4212	4.6185	5.1922	6.4061
Hog Futures	NAIVE3_E3	HOLT3_E3	ESM3_E3	NAÏVE	ARIMA	ESM_E3	HOLT	ESM	ARIMA_E3	M15_E3	MEAN_E3	MEAN
	10.5201	10.5304	10.5347	10.5444	10.5584	10.5596	10.5622	10.5639	10.5877	10.7545	11.1529	13.1042

Table B-10: E3 RMSE Accuracy Results: Commodities

SECURITY	RMSE #1	RMSE #2	RMSE #3	RMSE #4	RMSE #5	RMSE #6	RMSE #7	RMSE #8	RMSE #9	RMSE #10	RMSE #11	RMSE #12
METALS												
Gold Futures	ESM_E3	NAIVE3_E3	HOLT3_E3	M15_E3	NAÏVE	HOLT	ESM3_E3	MEAN_E3	ESM	ARIMA_E3	MEAN	ARIMA
	0.9408	0.9433	0.9441	0.9476	0.9505	0.9517	0.9677	0.9820	1.0131	1.1351	1.1414	2.0937
Platinum Futures	HOLT	NAÏVE	HOLT3_E3	NAIVE3_E3	ESM3_E3	ESM_E3	ESM	M15_E3	ARIMA_E3	MEAN_E3	MEAN	ARIMA
	1.6349	1.6411	1.6426	1.6511	1.6866	1.6897	1.6912	1.7188	1.8895	1.9875	2.4434	2.5574
Silver Futures	NAÏVE	ARIMA	HOLT	NAIVE3_E3	ESM3_E3	HOLT3_E3	ESM	ESM_E3	ARIMA_E3	MEAN_E3	M15_E3	MEAN
	0.0359	0.0360	0.0360	0.0360	0.0360	0.0360	0.0362	0.0362	0.0365	0.0374	0.0376	0.0457
Copper Futures	NAÏVE	ESM	NAIVE3_E3	ESM3_E3	ESM_E3	HOLT3_E3	HOLT	M15_E3	ARIMA_E3	MEAN_E3	ARIMA	MEAN
	0.0040	0.0040	0.0041	0.0041	0.0041	0.0041	0.0041	0.0043	0.0043	0.0048	0.0053	0.0057
Palladium Futures	ESM	HOLT	NAÏVE	HOLT3_E3	ESM3_E3	NAIVE3_E3	ESM_E3	ARIMA_E3	M15_E3	ARIMA	MEAN_E3	MEAN
	0.8666	0.8673	0.8685	0.8716	0.8723	0.8761	0.8851	0.9264	0.9303	0.9826	1.0996	1.3746
SOFTS												
Coffee Futures	HOLT3_E3	ESM3_E3	NAIVE3_E3	ARIMA	HOLT	ESM	NAÏVE	ESM_E3	ARIMA_E3	M15_E3	MEAN_E3	MEAN
	0.1864	0.1864	0.1864	0.1868	0.1869	0.1869	0.1870	0.1871	0.1880	0.1922	0.2273	0.2891
Lumber Futures	HOLT	ESM	NAÏVE	HOLT3_E3	ARIMA	ESM3_E3	NAIVE3_E3	ESM_E3	ARIMA_E3	M15_E3	MEAN_E3	MEAN
	0.5182	0.5182	0.5182	0.5188	0.5190	0.5192	0.5192	0.5222	0.5248	0.5345	0.5817	0.7097
Cocoa Futures	ARIMA	ESM3_E3	NAIVE3_E3	HOLT3_E3	ESM_E3	ARIMA_E3	ESM	NAÏVE	HOLT	M15_E3	MEAN_E3	MEAN
	3.1025	3.1056	3.1058	3.1059	3.1123	3.1184	3.1214	3.1217	3.1218	3.1917	3.3559	3.9238
Sugar Futures	NAIVE3_E3	ESM3_E3	HOLT3_E3	NAÏVE	ESM_E3	ESM	HOLT	M15_E3	MEAN_E3	MEAN	ARIMA_E3	ARIMA
	0.0003	0.0003	0.0003	0.0003	0.0003	0.0003	0.0003	0.0003	0.0003	0.0004	0.0008	0.0018
Cotton Futures	HOLT	ESM	ARIMA	NAÏVE	HOLT3_E3	ESM3_E3	NAIVE3_E3	ESM_E3	ARIMA_E3	M15_E3	MEAN_E3	MEAN
	0.0014	0.0014	0.0014	0.0014	0.0014	0.0014	0.0014	0.0015	0.0015	0.0016	0.0018	0.0022
GRAINS												
Corn Futures	ARIMA	HOLT	NAÏVE	ESM	HOLT3_E3	NAIVE3_E3	ESM3_E3	ARIMA_E3	ESM_E3	M15_E3	MEAN_E3	MEAN
	0.6390	0.6789	0.6835	0.6835	0.6891	0.6943	0.6943	0.6976	0.7104	0.7574	0.8297	0.9765
Oats Futures	ESM	ESM3_E3	NAÏVE	ARIMA	HOLT3_E3	NAIVE3_E3	ESM_E3	HOLT	ARIMA_E3	M15_E3	MEAN_E3	MEAN
	0.0042	0.0042	0.0042	0.0042	0.0042	0.0042	0.0043	0.0043	0.0043	0.0044	0.0049	0.0058
Soybean Futures	ARIMA	NAÏVE	HOLT	ESM	HOLT3_E3	ESM3_E3	NAIVE3_E3	ESM_E3	ARIMA_E3	M15_E3	MEAN_E3	MEAN
	0.0141	0.0142	0.0142	0.0142	0.0142	0.0143	0.0143	0.0144	0.0146	0.0152	0.0166	0.0199
Soybean Oil Futures	ESM	NAÏVE	ESM3_E3	NAIVE3_E3	HOLT3_E3	HOLT	ESM_E3	M15_E3	ARIMA_E3	MEAN_E3	MEAN	ARIMA
	0.0506	0.0506	0.0507	0.0507	0.0508	0.0509	0.0510	0.0528	0.0602	0.0607	0.0729	0.0808
Soybean Meal Futures	ARIMA_E3	ESM_E3	M15_E3	ESM3_E3	NAIVE3_E3	HOLT3_E3	ESM	NAÏVE	ARIMA	HOLT	MEAN_E3	MEAN
	0.5117	0.5123	0.5147	0.5156	0.5156	0.5160	0.5219	0.5219	0.5222	0.5228	0.5374	0.6415
Wheat Futures	ESM	NAÏVE	HOLT	ESM3_E3	NAIVE3_E3	HOLT3_E3	ESM_E3	M15_E3	ARIMA_E3	MEAN_E3	MEAN	ARIMA
	0.0090	0.0090	0.0090	0.0090	0.0090	0.0090	0.0091	0.0095	0.0099	0.0102	0.0121	0.0126

SECURITY	RMSE #1	RMSE #2	RMSE #3	RMSE #4	RMSE #5	RMSE #6	RMSE #7	RMSE #8	RMSE #9	RMSE #10	RMSE #11	RMSE #12
ENERGY												
Brent Oil Futures	NAÏVE 0.1367	NAIVE3_E3 0.1375	HOLT3_E3 0.1390	HOLT 0.1390	ESM3_E3 0.1399	ESM 0.1399	ESM_E3 0.1404	M15_E3 0.1428	MEAN_E3 0.1553	ARIMA_E3 0.1567	MEAN 0.1858	ARIMA 0.1935
Crude Oil WTI Futures	NAÏVE 0.1361	ESM 0.1365	HOLT 0.1371	NAIVE3_E3 0.1373	ESM3_E3 0.1376	HOLT3_E3 0.1379	ESM_E3 0.1391	M15_E3 0.1437	MEAN_E3 0.1562	ARIMA_E3 0.1583	MEAN 0.1861	ARIMA 0.2028
Heating Oil Futures	HOLT 0.0034	ESM 0.0034	NAÏVE 0.0034	HOLT3_E3 0.0034	ESM3_E3 0.0034	NAIVE3_E3 0.0034	ESM_E3 0.0035	M15_E3 0.0036	ARIMA_E3 0.0038	MEAN_E3 0.0040	ARIMA 0.0046	MEAN 0.0049
MEATS												
Cattle Futures	NAIVE3_E3 0.0830	ESM3_E3 0.0830	HOLT3_E3 0.0830	NAÏVE 0.0834	ESM_E3 0.0835	ESM 0.0835	ARIMA 0.0836	HOLT 0.0836	ARIMA_E3 0.0841	M15_E3 0.0873	MEAN_E3 0.0964	MEAN 0.1182
Hog Futures	M15_E3 0.1416	ARIMA_E3 0.1417	MEAN_E3 0.1421	ESM_E3 0.1423	NAIVE3_E3 0.1430	HOLT3_E3 0.1433	ESM3_E3 0.1434	NAÏVE 0.1445	ARIMA 0.1447	HOLT 0.1450	ESM 0.1452	MEAN 0.1643

B.4. Rates & Bonds

Table B-11: Rates & Bonds E3 MAPE Accuracy Results

SECURITY	MAPE #1	MAPE #2	MAPE #3	MAPE #4	MAPE #5	MAPE #6	MAPE #7	MAPE #8	MAPE #9	MAPE #10	MAPE #11	MAPE #12
RATES & BONDS												
U.S. 30-Year Bond	ARIMA 2.8115	ESM 2.8134	NAÏVE 2.8134	HOLT 2.8184	NAIVE3_E3 2.8185	ESM3_E3 2.8185	HOLT3_E3 2.8221	ESM_E3 2.8462	ARIMA_E3 2.8664	M15_E3 2.9594	MEAN_E3 2.9706	MEAN 3.6398
U.S. 10-Year Bond	NAIVE3_E3 8.0667	NAÏVE 8.0904	ARIMA 8.0981	ARIMA_E3 8.1116	HOLT3_E3 8.1174	ESM_E3 8.1175	ESM3_E3 8.1247	HOLT 8.1670	ESM 8.1742	M15_E3 8.2541	MEAN_E3 8.4523	MEAN 10.3773
U.S. 5-Year Bond	ESM_E3 11.6116	HOLT3_E3 11.6240	ESM3_E3 11.6258	NAIVE3_E3 11.6360	HOLT 11.7008	ESM 11.7028	ARIMA_E3 11.7161	NAÏVE 11.7185	M15_E3 11.7520	ARIMA 12.1120	MEAN_E3 12.3581	MEAN 15.1648
U.S. 2-Year Bond	ARIMA_E3 14.8026	NAIVE3_E3 14.9236	M15_E3 14.9277	ESM_E3 14.9578	NAÏVE 15.1278	ESM3_E3 15.1424	HOLT3_E3 15.2024	ESM 15.4402	HOLT 15.5404	ARIMA 15.8344	MEAN_E3 16.3891	MEAN 20.7118
U.S. 30-Year T-Bond Futures	ARIMA 2.8115	ESM 2.8134	NAÏVE 2.8134	HOLT 2.8184	NAIVE3_E3 2.8185	ESM3_E3 2.8185	HOLT3_E3 2.8221	ESM_E3 2.8462	ARIMA_E3 2.8664	M15_E3 2.9594	MEAN_E3 2.9706	MEAN 3.6398
U.S. 10-Year T-Bond Futures	ESM 1.6661	NAÏVE 1.6661	HOLT 1.6699	ESM3_E3 1.6702	NAIVE3_E3 1.6702	HOLT3_E3 1.6725	ARIMA 1.6765	ESM_E3 1.6863	ARIMA_E3 1.7021	M15_E3 1.7520	MEAN_E3 1.7881	MEAN 2.1844
Federal Funds Rate	M15_E3 18.4682	ARIMA_E3 18.7624	ARIMA 20.5169	ESM_E3 21.3695	HOLT3_E3 22.7098	ESM3_E3 23.9418	HOLT 24.9032	MEAN_E3 25.4969	NAIVE3_E3 26.0699	ESM 26.9142	NAÏVE 29.6013	MEAN 34.8371
1 Year LIBOR Rate	ESM 8.0831	HOLT 8.1746	ESM3_E3 8.1835	HOLT3_E3 8.2745	NAÏVE 8.3597	ESM_E3 8.3747	NAIVE3_E3 8.4511	ARIMA 8.6312	ARIMA_E3 8.6634	M15_E3 8.9374	MEAN_E3 11.6240	MEAN 15.5246
6 Month LIBOR Rate	HOLT 8.6441	ESM 8.7244	ESM3_E3 8.7614	HOLT3_E3 8.8010	ARIMA 8.9739	ESM_E3 8.9880	NAÏVE 9.1381	NAIVE3_E3 9.2740	ARIMA_E3 9.3195	M15_E3 9.8749	MEAN_E3 13.8390	MEAN 18.9925
3 Month LIBOR Rate	ESM 9.6205	ESM3_E3 9.7514	NAÏVE 9.9104	ESM_E3 9.9659	NAIVE3_E3 10.0342	HOLT 10.0407	HOLT3_E3 10.1156	ARIMA 10.1436	ARIMA_E3 10.3083	M15_E3 10.6326	MEAN_E3 15.5283	MEAN 21.4382

Table B-12: Rates & Bonds E3 RMSE Accuracy Results

SECURITY	RMSE #1	RMSE #2	RMSE #3	RMSE #4	RMSE #5	RMSE #6	RMSE #7	RMSE #8	RMSE #9	RMSE #10	RMSE #11	RMSE #12
RATES & BONDS												
U.S. 30-Year Bond	ARIMA 0.0727	NAÏVE 0.0727	ESM 0.0727	NAIVE3_E3 0.0727	ESM3_E3 0.0727	HOLT 0.0728	HOLT3_E3 0.0728	ESM_E3 0.0731	ARIMA_E3 0.0734	MEAN_E3 0.0738	M15_E3 0.0748	MEAN 0.0892
U.S. 10-Year Bond	NAIVE3_E3 0.0051	NAÏVE 0.0051	ARIMA 0.0051	HOLT3_E3 0.0051	ESM3_E3 0.0051	ESM_E3 0.0051	ARIMA_E3 0.0051	HOLT 0.0051	ESM 0.0051	MEAN_E3 0.0051	M15_E3 0.0052	MEAN 0.0060
U.S. 5-Year Bond	HOLT3_E3 0.0054	ESM3_E3 0.0054	HOLT 0.0054	NAIVE3_E3 0.0054	ESM 0.0054	NAÏVE 0.0054	ESM_E3 0.0054	ARIMA_E3 0.0055	ARIMA 0.0055	M15_E3 0.0055	MEAN_E3 0.0057	MEAN 0.0067
U.S. 2-Year Bond	ESM 0.0048	ESM3_E3 0.0048	HOLT 0.0048	HOLT3_E3 0.0048	NAÏVE 0.0049	NAIVE3_E3 0.0049	ESM_E3 0.0049	ARIMA 0.0049	ARIMA_E3 0.0049	M15_E3 0.0050	MEAN_E3 0.0057	MEAN 0.0070
U.S. 30-Year T-Bond Futures	ARIMA 0.0727	NAÏVE 0.0727	ESM 0.0727	NAIVE3_E3 0.0727	ESM3_E3 0.0727	HOLT 0.0728	HOLT3_E3 0.0728	ESM_E3 0.0731	ARIMA_E3 0.0734	MEAN_E3 0.0738	M15_E3 0.0748	MEAN 0.0892
U.S. 10-Year T-Bond Futures	ESM3_E3 0.0393	NAIVE3_E3 0.0393	ESM 0.0393	NAÏVE 0.0393	HOLT3_E3 0.0393	HOLT 0.0394	ESM_E3 0.0394	ARIMA 0.0395	ARIMA_E3 0.0396	M15_E3 0.0403	MEAN_E3 0.0410	MEAN 0.0501
Federal Funds Rate	ARIMA_E3 0.0044	ARIMA 0.0044	M15_E3 0.0045	ESM_E3 0.0045	ESM3_E3 0.0046	HOLT3_E3 0.0047	ESM 0.0048	HOLT 0.0049	NAIVE3_E3 0.0050	NAÏVE 0.0055	MEAN_E3 0.0059	MEAN 0.0077
1 Year LIBOR Rate	ESM 0.0047	ESM3_E3 0.0047	ESM_E3 0.0047	HOLT3_E3 0.0048	NAÏVE 0.0048	HOLT 0.0048	NAIVE3_E3 0.0048	ARIMA 0.0048	ARIMA_E3 0.0048	M15_E3 0.0049	MEAN_E3 0.0058	MEAN 0.0075
6 Month LIBOR Rate	ESM_E3 0.0048	HOLT3_E3 0.0048	NAÏVE 0.0048	NAIVE3_E3 0.0048	ESM3_E3 0.0048	HOLT 0.0049	M15_E3 0.0049	ARIMA_E3 0.0049	ESM 0.0050	ARIMA 0.0051	MEAN_E3 0.0060	MEAN 0.0077
3 Month LIBOR Rate	ESM_E3 0.0050	M15_E3 0.0050	ESM3_E3 0.0050	NAIVE3_E3 0.0050	NAÏVE 0.0051	ESM 0.0051	ARIMA_E3 0.0052	HOLT3_E3 0.0053	HOLT 0.0054	ARIMA 0.0056	MEAN_E3 0.0062	MEAN 0.0080

B.5. Aerospace Sector

Table B-13: Aerospace Sector E3 MAPE Accuracy Results

SECURITY	MAPE #1	MAPE #2	MAPE #3	MAPE #4	MAPE #5	MAPE #6	MAPE #7	MAPE #8	MAPE #9	MAPE #10	MAPE #11	MAPE #12
AEROSPACE AND DEFENSE GENERAL STOCKS												
The Boeing Company (BA)	ESM3_E3 7.5936	ESM 7.6103	ESM_E3 7.6571	NAIVE3_E3 7.7399	HOLT3_E3 7.7515	NAÏVE 7.7613	HOLT 7.7765	ARIMA_E3 7.8955	M15_E3 7.9593	ARIMA 8.1593	MEAN_E3 9.3702	MEAN 11.8609
Lockheed Martin Corporation (LMT)	ESM 6.4349	ESM3_E3 6.4520	ESM_E3 6.5552	NAÏVE 6.5948	HOLT 6.6052	NAIVE3_E3 6.6311	HOLT3_E3 6.6396	M15_E3 6.9682	MEAN_E3 7.6125	ARIMA_E3 7.7190	MEAN 8.9634	ARIMA 10.1463
General Dynamics Corporation (GD)	ESM3_E3 7.3194	ESM 7.3361	ESM_E3 7.3670	NAIVE3_E3 7.4159	NAÏVE 7.4198	HOLT3_E3 7.4218	HOLT 7.4363	M15_E3 7.6421	MEAN_E3 8.1106	ARIMA_E3 8.5076	MEAN 9.5658	ARIMA 10.7653
Textron Inc. (TXT)	HOLT3_E3 12.1305	HOLT 12.1364	NAIVE3_E3 12.2430	NAÏVE 12.2533	ARIMA 12.2562	ARIMA_E3 12.3330	ESM3_E3 12.3398	ESM_E3 12.3445	ESM 12.4038	M15_E3 12.5440	MEAN_E3 14.1441	MEAN 17.2741
Northrop Grumman Corporation (NOC)	HOLT 6.4275	NAÏVE 6.4307	ESM 6.4309	ESM3_E3 6.4330	HOLT3_E3 6.4341	NAIVE3_E3 6.4415	ESM_E3 6.4840	M15_E3 6.7249	ARIMA_E3 7.2336	MEAN_E3 7.5079	ARIMA 8.8857	MEAN 9.2746
AEROSPACE AND DEFENSE EQUIPMENT STOCKS												
Esterline Technologies Corporation (ESL)	ESM_E3 8.7583	ESM3_E3 8.7658	HOLT3_E3 8.7888	NAIVE3_E3 8.7929	HOLT 8.8864	ARIMA_E3 8.8906	NAÏVE 8.8909	ESM 8.9000	M15_E3 9.0684	ARIMA 9.1009	MEAN_E3 10.2961	MEAN 13.5833
HEICO Corporation (HEI)	NAIVE3_E3 8.5090	ESM3_E3 8.5219	ESM_E3 8.5403	NAÏVE 8.5578	HOLT3_E3 8.5629	ESM 8.6466	HOLT 8.7373	M15_E3 8.9184	ARIMA_E3 9.5634	MEAN_E3 9.8758	MEAN 12.6701	ARIMA 12.8070
Hexcel Corporation (HXL)	HOLT3_E3 11.4894	HOLT 11.5224	NAIVE3_E3 11.6082	NAÏVE 11.6697	ARIMA_E3 11.6698	ESM_E3 11.7117	ARIMA 11.7171	ESM3_E3 11.8253	M15_E3 11.8630	ESM 12.0478	MEAN_E3 15.3405	MEAN 20.0097
Curtiss-Wright Corporation (CW)	NAÏVE 7.8402	NAIVE3_E3 7.8666	ESM3_E3 7.8893	HOLT3_E3 7.8955	HOLT 7.8993	ESM 7.9070	ESM_E3 7.9398	ARIMA_E3 8.2627	M15_E3 8.2644	MEAN_E3 8.8069	ARIMA 9.9104	MEAN 10.3230
Moog Inc. (MOG-A)	ESM_E3 8.7577	NAIVE3_E3 8.8287	M15_E3 8.8331	ESM3_E3 8.8754	HOLT3_E3 8.8915	NAÏVE 8.9584	MEAN_E3 9.0085	HOLT 9.0529	ESM 9.0713	ARIMA_E3 9.6284	MEAN 10.6506	ARIMA 12.5834
Raytheon Company (RTN)	ESM3_E3 6.4727	ESM 6.4737	NAIVE3_E3 6.4781	NAÏVE 6.4809	HOLT3_E3 6.4879	HOLT 6.5180	ESM_E3 6.5284	M15_E3 6.7838	ARIMA_E3 6.7882	ARIMA 7.1079	MEAN_E3 7.7256	MEAN 9.8807
Aerojet Rocketdyne Holdings, Inc. (AJRD)	NAIVE3_E3 12.0545	NAÏVE 12.0784	ARIMA 12.0986	HOLT3_E3 12.1088	ESM3_E3 12.1098	ESM 12.1494	ESM_E3 12.1651	HOLT 12.1931	ARIMA_E3 12.2071	M15_E3 12.5632	MEAN_E3 13.6970	MEAN 16.9416

Table B-14: Aerospace Sector E3 RMSE Accuracy Results

SECURITY	RMSE #1	RMSE #2	RMSE #3	RMSE #4	RMSE #5	RMSE #6	RMSE #7	RMSE #8	RMSE #9	RMSE #10	RMSE #11	RMSE #12
AEROSPACE AND DEFENSE GENERAL STOCKS												
The Boeing Company (BA)	ESM3_E3 0.1651	ESM 0.1656	ESM_E3 0.1658	NAIVE3_E3 0.1688	NAÏVE 0.1690	HOLT3_E3 0.1704	M15_E3 0.1705	HOLT 0.1712	ARIMA_E3 0.1900	MEAN_E3 0.1987	MEAN 0.2414	ARIMA 0.2474
Lockheed Martin Corporation (LMT)	MEAN_E3 0.1607	NAIVE3_E3 0.1610	HOLT3_E3 0.1612	M15_E3 0.1615	NAÏVE 0.1618	ESM_E3 0.1620	HOLT 0.1621	ESM3_E3 0.1649	ESM 0.1695	MEAN 0.1768	ARIMA_E3 0.2406	ARIMA 0.4585
General Dynamics Corporation (GD)	ESM 0.1125	ESM3_E3 0.1129	ESM_E3 0.1138	HOLT 0.1138	NAÏVE 0.1139	HOLT3_E3 0.1141	NAIVE3_E3 0.1142	M15_E3 0.1171	MEAN_E3 0.1223	ARIMA_E3 0.1393	MEAN 0.1400	ARIMA 0.2191
Textron Inc. (TXT)	HOLT 0.0694	NAÏVE 0.0694	ARIMA 0.0696	HOLT3_E3 0.0697	NAIVE3_E3 0.0699	ARIMA_E3 0.0710	ESM_E3 0.0720	ESM3_E3 0.0722	M15_E3 0.0725	ESM 0.0728	MEAN_E3 0.0786	MEAN 0.0927
Northrop Grumman Corporation (NOC)	HOLT3_E3 0.1441	HOLT 0.1445	ESM_E3 0.1453	NAIVE3_E3 0.1454	M15_E3 0.1457	ESM3_E3 0.1458	NAÏVE 0.1461	ESM 0.1467	MEAN_E3 0.1611	MEAN 0.1884	ARIMA_E3 0.2184	ARIMA 0.4143
AEROSPACE AND DEFENSE EQUIPMENT STOCKS												
Esterline Technologies Corporation (ESL)	HOLT 0.0980	NAÏVE 0.0980	HOLT3_E3 0.0980	NAIVE3_E3 0.0980	ESM_E3 0.0989	ESM3_E3 0.0990	ESM 0.0998	ARIMA_E3 0.1004	M15_E3 0.1005	MEAN_E3 0.1111	ARIMA 0.1112	MEAN 0.1356
HEICO Corporation (HEI)	ESM3_E3 0.0354	ESM 0.0354	ESM_E3 0.0357	M15_E3 0.0371	NAIVE3_E3 0.0372	HOLT3_E3 0.0372	NAÏVE 0.0374	HOLT 0.0375	MEAN_E3 0.0392	MEAN 0.0476	ARIMA_E3 0.0521	ARIMA 0.0952
Hexcel Corporation (HXL)	ESM 0.0376	ESM3_E3 0.0377	HOLT 0.0377	NAÏVE 0.0378	ARIMA 0.0379	HOLT3_E3 0.0379	NAIVE3_E3 0.0380	ESM_E3 0.0382	ARIMA_E3 0.0389	M15_E3 0.0403	MEAN_E3 0.0451	MEAN 0.0542
Curtiss-Wright Corporation (CW)	ESM 0.0808	NAÏVE 0.0810	HOLT 0.0811	ESM3_E3 0.0815	NAIVE3_E3 0.0817	HOLT3_E3 0.0818	ESM_E3 0.0825	M15_E3 0.0852	ARIMA_E3 0.0861	MEAN_E3 0.0869	MEAN 0.0977	ARIMA 0.1168
Moog Inc. (MOG-A)	NAÏVE 0.0637	NAIVE3_E3 0.0638	HOLT 0.0638	HOLT3_E3 0.0638	ESM3_E3 0.0641	ESM_E3 0.0643	ESM 0.0644	MEAN_E3 0.0653	M15_E3 0.0662	ARIMA_E3 0.0696	MEAN 0.0736	ARIMA 0.0992
Raytheon Company (RTN)	ESM_E3 0.0965	ESM3_E3 0.0966	NAIVE3_E3 0.0968	HOLT3_E3 0.0968	ESM 0.0970	NAÏVE 0.0972	M15_E3 0.0973	HOLT 0.0973	MEAN_E3 0.1039	ARIMA_E3 0.1167	MEAN 0.1219	ARIMA 0.1672
Aerojet Rocketdyne Holdings, Inc. (AJRD)	NAÏVE 0.0272	NAIVE3_E3 0.0272	ARIMA 0.0272	ESM 0.0273	ESM3_E3 0.0273	ESM_E3 0.0274	HOLT3_E3 0.0274	ARIMA_E3 0.0275	HOLT 0.0275	M15_E3 0.0281	MEAN_E3 0.0293	MEAN 0.0355

B.6. Auto/Tires/Trucks Sector

Table B-15: Auto/Tires/Trucks Sector E3 MAPE Accuracy Results

SECURITY	MAPE #1	MAPE #2	MAPE #3	MAPE #4	MAPE #5	MAPE #6	MAPE #7	MAPE #8	MAPE #9	MAPE #10	MAPE #11	MAPE #12
AUTOMOTIVE STOCKS												
Ford Motor Company (F)	HOLT3_E3 11.5624	HOLT 11.6043	NAIVE3_E3 11.6163	ARIMA 11.6398	ARIMA_E3 11.6438	ESM_E3 11.6578	NAÏVE 11.6728	ESM3_E3 11.6854	ESM 11.7898	M15_E3 11.8545	MEAN_E3 13.1775	MEAN 15.9354
Toyota Motor Corporation (TM)	NAIVE3_E3 6.1169	HOLT3_E3 6.1178	NAÏVE 6.1579	HOLT 6.1586	ESM_E3 6.3438	M15_E3 6.3615	ESM3_E3 6.4752	ESM 6.7189	ARIMA_E3 6.8957	MEAN_E3 7.1728	ARIMA 8.1450	MEAN 9.4176
Honda Motor Co., Ltd (HMC)	ESM3_E3 6.7194	ESM 6.7336	HOLT3_E3 6.7557	NAIVE3_E3 6.7573	HOLT 6.7581	NAÏVE 6.7607	ESM_E3 6.7721	ARIMA_E3 6.9237	ARIMA 6.9322	M15_E3 7.0491	MEAN_E3 7.3338	MEAN 8.8451
PACCAR Inc. (PCAR)	HOLT3_E3 8.4107	NAIVE3_E3 8.4294	ESM_E3 8.4626	HOLT 8.4788	NAÏVE 8.5059	M15_E3 8.5067	ESM3_E3 8.5370	ESM 8.6653	ARIMA_E3 8.8448	MEAN_E3 8.8459	MEAN 10.4019	ARIMA 11.6289
Harley-Davidson, Inc. (HOG)	ARIMA_E3 10.1217	HOLT3_E3 10.1542	NAIVE3_E3 10.1547	HOLT 10.1562	NAÏVE 10.1570	M15_E3 10.4295	ESM_E3 10.4898	ESM3_E3 10.7001	ESM 11.0321	MEAN_E3 11.1378	ARIMA 11.2199	MEAN 13.1090
AUTO & TRUCK ORIGINAL EQUIPMENT MANUFACTURERS STOCKS												
Gentex Corporation (GNTX)	ARIMA 9.6025	HOLT 9.8632	NAÏVE 9.8638	ARIMA_E3 9.8778	HOLT3_E3 9.8993	NAIVE3_E3 9.8998	ESM_E3 10.1148	ESM3_E3 10.2587	M15_E3 10.3138	ESM 10.5666	MEAN_E3 10.9382	MEAN 12.8868
Tenneco Inc. (TEN)	HOLT 17.5557	ESM 17.5558	NAÏVE 17.5559	HOLT3_E3 17.6190	ESM3_E3 17.6193	NAIVE3_E3 17.6193	ARIMA 17.6941	ESM_E3 17.8085	ARIMA_E3 18.0521	M15_E3 18.5566	MEAN_E3 23.4198	MEAN 30.9881
Modine Manufacturing Company (MOD)	ARIMA 15.9458	HOLT 15.9564	HOLT3_E3 15.9721	NAIVE3_E3 16.0120	NAÏVE 16.0173	ARIMA_E3 16.1288	ESM3_E3 16.1727	ESM_E3 16.1799	ESM 16.2456	M15_E3 16.4678	MEAN_E3 19.3366	MEAN 24.1208
Magna International, Inc. (MGA)	HOLT 8.7945	ESM 8.8026	NAÏVE 8.8031	ARIMA 8.8058	ESM3_E3 8.8644	HOLT3_E3 8.8788	NAIVE3_E3 8.8945	ESM_E3 8.9928	ARIMA_E3 9.1122	M15_E3 9.4180	MEAN_E3 10.6859	MEAN 13.1190
LCI Industries (LC II)	ESM3_E3 10.3161	NAIVE3_E3 10.3172	NAÏVE 10.3253	ESM 10.3353	HOLT3_E3 10.3392	HOLT 10.3416	ESM_E3 10.3867	M15_E3 10.7223	ARIMA_E3 12.4046	MEAN_E3 12.5125	MEAN 15.3174	ARIMA 17.7164
Oshkosh Corporation (OSK)	NAÏVE 12.0411	NAIVE3_E3 12.0996	HOLT 12.1222	HOLT3_E3 12.1430	ESM3_E3 12.1996	ESM_E3 12.2389	ESM 12.2902	M15_E3 12.7000	ARIMA_E3 13.5822	MEAN_E3 15.1213	ARIMA 15.8696	MEAN 19.1566
Wabash National Corporation (WNC)	ESM_E3 17.4015	ESM3_E3 17.4228	NAIVE3_E3 17.4441	ARIMA_E3 17.4448	HOLT3_E3 17.4543	ESM 17.5654	NAÏVE 17.5922	ARIMA 17.5932	M15_E3 17.6157	HOLT 17.6163	MEAN_E3 20.0677	MEAN 25.1145

Table B-16: Auto/Tires/Trucks Sector E3 RMSE Accuracy Results

SECURITY	RMSE #1	RMSE #2	RMSE #3	RMSE #4	RMSE #5	RMSE #6	RMSE #7	RMSE #8	RMSE #9	RMSE #10	RMSE #11	RMSE #12
AUTOMOTIVE STOCKS												
Ford Motor Company (F)	HOLT3_E3 0.0250	HOLT 0.0250	NAIVE3_E3 0.0250	NAÏVE 0.0251	ARIMA 0.0251	ESM_E3 0.0252	ARIMA_E3 0.0252	ESM3_E3 0.0252	ESM 0.0254	M15_E3 0.0256	MEAN_E3 0.0266	MEAN 0.0317
Toyota Motor Corporation (TM)	HOLT3_E3 0.1053	NAIVE3_E3 0.1053	HOLT 0.1058	NAÏVE 0.1058	ESM_E3 0.1068	ESM3_E3 0.1084	M15_E3 0.1085	ESM 0.1113	MEAN_E3 0.1210	ARIMA_E3 0.1349	MEAN 0.1508	ARIMA 0.2361
Honda Motor Co., Ltd (HMC)	HOLT3_E3 0.0382	NAIVE3_E3 0.0382	HOLT 0.0382	NAÏVE 0.0382	ESM3_E3 0.0383	ESM_E3 0.0383	ESM 0.0385	ARIMA_E3 0.0388	ARIMA 0.0391	M15_E3 0.0391	MEAN_E3 0.0404	MEAN 0.0469
PACCAR Inc. (PCAR)	HOLT3_E3 0.0625	HOLT 0.0626	NAIVE3_E3 0.0626	NAÏVE 0.0627	ESM_E3 0.0636	ESM3_E3 0.0639	M15_E3 0.0642	ESM 0.0646	ARIMA_E3 0.0657	MEAN_E3 0.0657	MEAN 0.0753	ARIMA 0.0954
Harley-Davidson, Inc. (HOG)	HOLT 0.0752	NAÏVE 0.0752	NAIVE3_E3 0.0760	HOLT3_E3 0.0760	ARIMA_E3 0.0772	ESM_E3 0.0793	M15_E3 0.0808	ESM3_E3 0.0810	MEAN_E3 0.0819	ESM 0.0844	ARIMA 0.0926	MEAN 0.0927
AUTO & TRUCK ORIGINAL EQUIPMENT MANUFACTURERS STOCKS												
Gentex Corporation (GNTX)	ARIMA 0.0183	NAÏVE 0.0186	NAIVE3_E3 0.0186	HOLT 0.0186	HOLT3_E3 0.0186	ARIMA_E3 0.0186	ESM_E3 0.0189	M15_E3 0.0192	ESM3_E3 0.0193	ESM 0.0203	MEAN_E3 0.0203	MEAN 0.0239
Tenneco Inc. (TEN)	ESM 0.0602	NAÏVE 0.0602	HOLT 0.0602	ARIMA 0.0603	ESM3_E3 0.0604	NAIVE3_E3 0.0604	HOLT3_E3 0.0604	ESM_E3 0.0612	ARIMA_E3 0.0620	M15_E3 0.0647	MEAN_E3 0.0686	MEAN 0.0821
Modine Manufacturing Company (MOD)	HOLT 0.0383	ARIMA 0.0383	NAÏVE 0.0383	HOLT3_E3 0.0384	NAIVE3_E3 0.0384	ARIMA_E3 0.0389	ESM3_E3 0.0389	ESM_E3 0.0390	ESM 0.0391	M15_E3 0.0398	MEAN_E3 0.0426	MEAN 0.0506
Magna International, Inc. (MGA)	ARIMA 0.0402	NAÏVE 0.0402	ESM 0.0404	HOLT 0.0404	ESM3_E3 0.0407	NAIVE3_E3 0.0408	HOLT3_E3 0.0408	ESM_E3 0.0414	ARIMA_E3 0.0420	M15_E3 0.0439	MEAN_E3 0.0484	MEAN 0.0582
LCI Industries (LC II)	NAÏVE 0.0705	HOLT 0.0706	NAIVE3_E3 0.0713	HOLT3_E3 0.0714	ESM3_E3 0.0738	ESM_E3 0.0740	ESM 0.0743	M15_E3 0.0760	MEAN_E3 0.0886	ARIMA_E3 0.0969	MEAN 0.1099	ARIMA 0.2100
Oshkosh Corporation (OSK)	NAÏVE 0.0726	NAIVE3_E3 0.0734	HOLT 0.0736	HOLT3_E3 0.0739	ESM3_E3 0.0751	ESM_E3 0.0752	ESM 0.0762	M15_E3 0.0786	MEAN_E3 0.0887	ARIMA_E3 0.0982	MEAN 0.1079	ARIMA 0.1655
Wabash National Corporation (WNC)	ESM_E3 0.0325	NAIVE3_E3 0.0326	ARIMA_E3 0.0326	ESM3_E3 0.0326	HOLT3_E3 0.0326	ARIMA 0.0328	NAÏVE 0.0328	ESM 0.0329	HOLT 0.0329	M15_E3 0.0331	MEAN_E3 0.0355	MEAN 0.0423

B.7. Basic Materials Sector

Table B-17: Basic Materials Sector E3 MAPE Accuracy Results

SECURITY	MAPE #1	MAPE #2	MAPE #3	MAPE #4	MAPE #5	MAPE #6	MAPE #7	MAPE #8	MAPE #9	MAPE #10	MAPE #11	MAPE #12
CHEMICALS												
Valhi, Inc. (VHI)	HOLT	NAÏVE	ESM	HOLT3_E3	NAIVE3_E3	ESM3_E3	ESM_E3	ARIMA_E3	M15_E3	ARIMA	MEAN_E3	MEAN
	15.1211	15.1815	15.2130	15.2317	15.2928	15.3159	15.5674	15.5991	16.3940	17.4846	19.2978	24.8044
Air Products and Chemicals, Inc. (APD)	ARIMA	ARIMA_E3	NAIVE3_E3	HOLT3_E3	ESM_E3	ESM3_E3	NAÏVE	HOLT	ESM	M15_E3	MEAN_E3	MEAN
	6.1771	6.2681	6.2717	6.2722	6.2785	6.2969	6.2982	6.2983	6.3766	6.4186	6.7375	8.2110
FMC Corporation (FMC)	NAÏVE	NAIVE3_E3	HOLT3_E3	HOLT	ESM_E3	ARIMA_E3	ESM3_E3	ESM	ARIMA	M15_E3	MEAN_E3	MEAN
	8.8509	8.8553	8.8845	8.9070	8.9810	8.9927	9.0096	9.1136	9.1564	9.1945	10.6980	13.5957
Olin Corporation (OLN)	NAIVE3_E3	HOLT3_E3	ARIMA	NAÏVE	ESM3_E3	HOLT	ESM_E3	ARIMA_E3	ESM	M15_E3	MEAN_E3	MEAN
	8.8451	8.8578	8.8878	8.8919	8.8995	8.9115	8.9139	8.9311	8.9704	9.1994	9.5408	11.8832
Cabot Corporation (CBT)	ARIMA_E3	M15_E3	ESM_E3	NAIVE3_E3	ESM3_E3	HOLT3_E3	NAÏVE	ARIMA	ESM	HOLT	MEAN_E3	MEAN
	10.9684	10.9871	10.9980	11.0688	11.0892	11.1085	11.2124	11.2239	11.2405	11.2681	11.8283	14.2242
PPG Industries, Inc. (PPG)	ESM_E3	ESM3_E3	ESM	HOLT3_E3	NAIVE3_E3	ARIMA_E3	M15_E3	HOLT	NAÏVE	MEAN_E3	ARIMA	MEAN
	6.9424	6.9515	7.0085	7.0181	7.0222	7.0465	7.0486	7.0814	7.0878	7.4752	8.1473	8.9133
Methanex Corporation (MEOH)	ESM	NAÏVE	ESM3_E3	NAIVE3_E3	HOLT	HOLT3_E3	ESM_E3	ARIMA_E3	M15_E3	ARIMA	MEAN_E3	MEAN
	11.5683	11.5973	11.6491	11.6742	11.6885	11.7317	11.8049	11.8798	12.3537	13.0546	13.6357	16.8716
MINING STOCKS												
McEwen Mining, Inc. (MUX)	HOLT3_E3	NAIVE3_E3	HOLT	NAÏVE	M15_E3	ARIMA_E3	ESM_E3	ARIMA	MEAN_E3	ESM3_E3	ESM	MEAN
	19.6462	19.6538	19.6592	19.6630	20.5418	20.8956	22.2185	22.4446	23.7636	23.9326	25.9059	29.7466
BHP Group (BHP)	NAÏVE	HOLT	NAIVE3_E3	HOLT3_E3	ESM3_E3	ESM	ESM_E3	M15_E3	ARIMA_E3	MEAN_E3	ARIMA	MEAN
	8.7807	8.7867	8.8267	8.8310	8.9032	8.9304	8.9579	9.2897	9.7757	10.4275	11.9643	12.8216
Cleveland-Cliffs, Inc. (CLF)	NAÏVE	NAIVE3_E3	HOLT	HOLT3_E3	ESM_E3	M15_E3	ESM3_E3	ESM	ARIMA_E3	MEAN_E3	ARIMA	MEAN
	17.8959	17.9589	18.1207	18.1212	18.5109	18.5559	18.6350	18.8852	19.2516	22.2013	22.3218	28.6194
Materion Corporation (MTRN)	ARIMA	NAÏVE	HOLT	HOLT3_E3	NAIVE3_E3	ESM	ESM3_E3	ESM_E3	ARIMA_E3	M15_E3	MEAN_E3	MEAN
	12.7652	12.7869	12.7887	12.8611	12.8691	12.9304	12.9629	13.0688	13.0994	13.4758	14.9257	18.1814
Rio Tinto plc (RIO)	ESM3_E3	HOLT3_E3	ESM	NAIVE3_E3	HOLT	NAÏVE	ESM_E3	M15_E3	ARIMA_E3	MEAN_E3	ARIMA	MEAN
	10.3000	10.3103	10.3186	10.3219	10.3315	10.3479	10.3491	10.6576	11.2276	12.3161	13.6841	15.5787
Taseko Mines Limited (TGB)	NAÏVE	HOLT	NAIVE3_E3	HOLT3_E3	ESM3_E3	ESM_E3	ESM	M15_E3	MEAN_E3	ARIMA_E3	MEAN	ARIMA
	20.1960	20.2222	20.3415	20.3658	21.0577	21.0943	21.1855	21.5723	25.8213	27.5603	33.0341	47.4164

SECURITY	MAPE #1	MAPE #2	MAPE #3	MAPE #4	MAPE #5	MAPE #6	MAPE #7	MAPE #8	MAPE #9	MAPE #10	MAPE #11	MAPE #12
GOLD MINER STOCKS												
U.S. Gold Corp (USAU)	NAÏVE 16.9962	NAIVE3_E3 17.1079	HOLT3_E3 17.2424	HOLT 17.2989	M15_E3 18.3819	ARIMA_E3 18.4959	ESM_E3 19.4228	ARIMA 20.1032	ESM3_E3 20.5710	ESM 21.8515	MEAN_E3 22.9694	MEAN 30.5156
Newmont Mining Corporation (NEM)	ESM 9.3630	NAÏVE 9.3630	HOLT 9.3717	ARIMA 9.3821	NAIVE3_E3 9.3920	ESM3_E3 9.3920	HOLT3_E3 9.3961	ESM_E3 9.4984	ARIMA_E3 9.5793	M15_E3 9.8943	MEAN_E3 10.5019	MEAN 12.7347
Agnico Eagle Mines Limited (AEM)	NAÏVE 11.7903	HOLT 11.8059	ESM 11.8088	NAIVE3_E3 11.8731	HOLT3_E3 11.8755	ESM3_E3 11.8870	ESM_E3 12.0539	M15_E3 12.6288	ARIMA_E3 12.9339	MEAN_E3 13.3290	ARIMA 14.0037	MEAN 15.5416
Gold Fields Limited (GFI)	HOLT 10.9949	NAÏVE 11.0046	HOLT3_E3 11.0548	NAIVE3_E3 11.0604	ARIMA 11.0902	ESM3_E3 11.1659	ESM 11.1873	ESM_E3 11.2270	ARIMA_E3 11.3216	M15_E3 11.6470	MEAN_E3 13.0323	MEAN 16.5002
Royal Gold, Inc. (RGLD)	NAIVE3_E3 11.5998	HOLT3_E3 11.6175	NAÏVE 11.6292	HOLT 11.6655	ESM_E3 11.7266	ESM3_E3 11.7347	ESM 11.8463	ARIMA_E3 12.0216	M15_E3 12.0249	MEAN_E3 12.0452	ARIMA 12.4504	MEAN 14.6906
Barrick Gold Corporation (GOLD)	NAÏVE 10.9457	HOLT 10.9521	NAIVE3_E3 11.0041	HOLT3_E3 11.0088	ARIMA 11.0636	ESM 11.1693	ESM3_E3 11.1720	ESM_E3 11.2356	ARIMA_E3 11.2533	M15_E3 11.5099	MEAN_E3 11.5810	MEAN 13.3653
Vista Gold Corp. (VGZ)	HOLT 18.6508	NAÏVE 18.6794	ESM 18.6964	HOLT3_E3 18.7405	NAIVE3_E3 18.7449	ESM3_E3 18.7464	ARIMA 18.9057	ESM_E3 18.9183	ARIMA_E3 19.2082	M15_E3 19.6489	MEAN_E3 21.8128	MEAN 26.4962
STEEL PRODUCERS												
Nucor Corporation (NUE)	HOLT3_E3 8.8797	NAIVE3_E3 8.8863	HOLT 8.9049	ESM_E3 8.9073	NAÏVE 8.9163	ESM3_E3 8.9226	ESM 9.0010	M15_E3 9.0733	ARIMA_E3 9.2565	MEAN_E3 9.7950	ARIMA 10.3274	MEAN 12.0849
Commercial Metals Company (CMC)	HOLT3_E3 10.7529	HOLT 10.7569	NAIVE3_E3 10.7578	NAÏVE 10.7619	ESM3_E3 10.8565	ESM_E3 10.8881	ESM 10.9056	M15_E3 11.1644	MEAN_E3 12.3951	ARIMA_E3 12.4206	MEAN 14.9830	ARIMA 15.5348
L. B. Foster Company (FSTR)	HOLT3_E3 11.5373	ESM_E3 11.5544	ESM3_E3 11.5615	NAIVE3_E3 11.5883	HOLT 11.6395	ESM 11.6616	NAÏVE 11.7247	M15_E3 11.7755	ARIMA_E3 12.3570	MEAN_E3 13.2865	ARIMA 15.6943	MEAN 16.6893
United States Steel Corporation (X)	ARIMA_E3 17.1463	HOLT3_E3 17.3037	HOLT 17.3596	ARIMA 17.3977	NAIVE3_E3 17.4374	ESM3_E3 17.4435	ESM_E3 17.4576	NAÏVE 17.4769	ESM 17.4871	M15_E3 17.6474	MEAN_E3 21.0044	MEAN 27.2162
Shiloh Industries Inc. (SHLO)	NAÏVE 16.9777	ARIMA 17.0064	HOLT 17.0068	NAIVE3_E3 17.1621	HOLT3_E3 17.2026	ESM 17.5326	ESM3_E3 17.5655	ESM_E3 17.7484	ARIMA_E3 17.7776	M15_E3 18.5995	MEAN_E3 21.8066	MEAN 27.4445
Schnitzer Steel Industries, Inc. (SCHN)	HOLT3_E3 13.5767	NAIVE3_E3 13.6019	NAÏVE 13.6212	HOLT 13.6643	ESM_E3 13.9840	M15_E3 14.0169	ESM3_E3 14.0827	ESM 14.2555	MEAN_E3 14.5327	ARIMA_E3 15.8319	MEAN 17.7771	ARIMA 19.3458
Olympic Steel, Inc. (ZEUS)	NAIVE3_E3 15.0448	HOLT3_E3 15.1143	ESM_E3 15.1152	NAÏVE 15.1801	ESM3_E3 15.2046	ARIMA_E3 15.2894	HOLT 15.2953	M15_E3 15.3453	ESM 15.4122	ARIMA 15.7650	MEAN_E3 18.0045	MEAN 23.4836

Table B-18: Basic Materials Sector E3 RMSE Accuracy Results

SECURITY	RMSE #1	RMSE #2	RMSE #3	RMSE #4	RMSE #5	RMSE #6	RMSE #7	RMSE #8	RMSE #9	RMSE #10	RMSE #11	RMSE #12
CHEMICALS												
Valhi, Inc. (VHI)	NAÏVE	NAIVE3_E3	HOLT3_E3	HOLT	ESM3_E3	ESM	ESM_E3	M15_E3	ARIMA_E3	MEAN_E3	MEAN	ARIMA
	0.0221	0.0221	0.0221	0.0221	0.0222	0.0222	0.0223	0.0228	0.0245	0.0254	0.0309	0.0421
Air Products and Chemicals, Inc. (APD)	ARIMA	HOLT3_E3	NAIVE3_E3	HOLT	NAÏVE	ESM3_E3	ARIMA_E3	ESM_E3	ESM	M15_E3	MEAN_E3	MEAN
	0.0849	0.0862	0.0862	0.0863	0.0863	0.0865	0.0865	0.0866	0.0872	0.0889	0.0973	0.1192
FMC Corporation (FMC)	NAÏVE	HOLT	NAIVE3_E3	ESM	HOLT3_E3	ESM3_E3	ESM_E3	ARIMA_E3	M15_E3	ARIMA	MEAN_E3	MEAN
	0.0500	0.0502	0.0505	0.0506	0.0507	0.0507	0.0514	0.0516	0.0542	0.0556	0.0597	0.0725
Olin Corporation (OLN)	ARIMA	NAÏVE	NAIVE3_E3	HOLT	HOLT3_E3	ESM3_E3	ESM	ESM_E3	ARIMA_E3	M15_E3	MEAN_E3	MEAN
	0.0352	0.0352	0.0352	0.0352	0.0352	0.0353	0.0354	0.0355	0.0357	0.0368	0.0388	0.0470
Cabot Corporation (CBT)	ESM_E3	ARIMA_E3	NAIVE3_E3	ESM3_E3	HOLT3_E3	NAÏVE	M15_E3	ARIMA	ESM	HOLT	MEAN_E3	MEAN
	0.0690	0.0690	0.0691	0.0692	0.0692	0.0695	0.0696	0.0696	0.0697	0.0698	0.0726	0.0841
PPG Industries, Inc. (PPG)	ESM	ESM3_E3	ESM_E3	HOLT	HOLT3_E3	NAÏVE	NAIVE3_E3	M15_E3	MEAN_E3	ARIMA_E3	MEAN	ARIMA
	0.0673	0.0674	0.0678	0.0679	0.0679	0.0679	0.0679	0.0698	0.0698	0.0728	0.0806	0.1250
Methanex Corporation (MEOH)	ARIMA_E3	NAÏVE	HOLT	ESM	NAIVE3_E3	HOLT3_E3	ESM3_E3	ESM_E3	M15_E3	MEAN_E3	ARIMA	MEAN
	0.0664	0.0676	0.0678	0.0678	0.0685	0.0685	0.0686	0.0695	0.0723	0.0735	0.0788	0.0831
MINING STOCKS												
McEwen Mining, Inc. (MUX)	ESM_E3	NAIVE3_E3	HOLT3_E3	ESM3_E3	NAÏVE	M15_E3	HOLT	ESM	MEAN_E3	ARIMA_E3	MEAN	ARIMA
	0.0111	0.0112	0.0112	0.0113	0.0113	0.0114	0.0114	0.0115	0.0125	0.0150	0.0154	0.0265
BHP Group (BHP)	NAÏVE	HOLT	HOLT3_E3	NAIVE3_E3	ESM_E3	ESM3_E3	ESM	M15_E3	ARIMA_E3	MEAN_E3	MEAN	ARIMA
	0.0846	0.0847	0.0850	0.0850	0.0871	0.0874	0.0886	0.0890	0.0997	0.0998	0.1209	0.1597
Cleveland-Cliffs, Inc. (CLF)	M15_E3	NAIVE3_E3	NAÏVE	ARIMA_E3	HOLT3_E3	ESM_E3	HOLT	MEAN_E3	ESM3_E3	ESM	ARIMA	MEAN
	0.1216	0.1219	0.1231	0.1299	0.1305	0.1327	0.1353	0.1409	0.1418	0.1528	0.1605	0.1766
Materion Corporation (MTRN)	ARIMA	NAÏVE	HOLT	ESM	ESM3_E3	HOLT3_E3	NAIVE3_E3	ESM_E3	ARIMA_E3	M15_E3	MEAN_E3	MEAN
	0.0600	0.0600	0.0600	0.0600	0.0604	0.0604	0.0604	0.0610	0.0615	0.0633	0.0696	0.0839
Rio Tinto plc (RIO)	ESM	ESM3_E3	HOLT3_E3	HOLT	ESM_E3	NAIVE3_E3	NAÏVE	M15_E3	MEAN_E3	ARIMA_E3	MEAN	ARIMA
	0.1136	0.1136	0.1139	0.1139	0.1140	0.1144	0.1147	0.1164	0.1355	0.1571	0.1659	0.2817
Taseko Mines Limited (TGB)	NAIVE3_E3	HOLT3_E3	HOLT	NAÏVE	ESM3_E3	ESM_E3	ESM	M15_E3	MEAN_E3	MEAN	ARIMA_E3	ARIMA
	0.0076	0.0076	0.0076	0.0076	0.0077	0.0077	0.0077	0.0079	0.0090	0.0110	0.0166	0.0392

SECURITY	RMSE #1	RMSE #2	RMSE #3	RMSE #4	RMSE #5	RMSE #6	RMSE #7	RMSE #8	RMSE #9	RMSE #10	RMSE #11	RMSE #12
GOLD MINER STOCKS												
U.S. Gold Corp (USAU)	NAÏVE 1.4374	NAIVE3_E3 1.4429	ESM 1.4470	ESM3_E3 1.4503	ESM_E3 1.4716	M15_E3 1.5644	HOLT3_E3 1.5856	HOLT 1.7433	MEAN_E3 1.9330	ARIMA_E3 2.0664	MEAN 2.6727	ARIMA 4.4900
Newmont Mining Corporation (NEM)	NAÏVE 0.0608	ESM 0.0608	HOLT 0.0608	ARIMA 0.0609	NAIVE3_E3 0.0610	ESM3_E3 0.0610	HOLT3_E3 0.0610	ESM_E3 0.0615	ARIMA_E3 0.0619	M15_E3 0.0637	MEAN_E3 0.0691	MEAN 0.0832
Agnico Eagle Mines Limited (AEM)	HOLT 0.0859	NAÏVE 0.0862	ESM 0.0865	HOLT3_E3 0.0869	NAIVE3_E3 0.0872	ESM3_E3 0.0874	ESM_E3 0.0889	M15_E3 0.0932	MEAN_E3 0.0989	MEAN 0.1139	ARIMA_E3 0.1193	ARIMA 0.2029
Gold Fields Limited (GFI)	ESM3_E3 0.0174	HOLT3_E3 0.0174	NAIVE3_E3 0.0174	ESM_E3 0.0174	HOLT 0.0175	NAÏVE 0.0175	ESM 0.0175	ARIMA 0.0175	ARIMA_E3 0.0176	M15_E3 0.0180	MEAN_E3 0.0207	MEAN 0.0257
Royal Gold, Inc. (RGLD)	NAIVE3_E3 0.0936	NAÏVE 0.0937	HOLT3_E3 0.0938	HOLT 0.0940	ESM_E3 0.0946	ESM3_E3 0.0947	ESM 0.0955	M15_E3 0.0964	MEAN_E3 0.0997	ARIMA_E3 0.1092	MEAN 0.1181	ARIMA 0.1535
Barrick Gold Corporation (GOLD)	NAIVE3_E3 0.0506	HOLT3_E3 0.0506	NAÏVE 0.0506	HOLT 0.0507	ARIMA_E3 0.0511	ARIMA 0.0512	ESM_E3 0.0513	ESM3_E3 0.0514	ESM 0.0517	M15_E3 0.0518	MEAN_E3 0.0541	MEAN 0.0615
Vista Gold Corp. (VGZ)	NAÏVE 0.0110	HOLT 0.0110	NAIVE3_E3 0.0110	HOLT3_E3 0.0111	ESM 0.0111	ESM3_E3 0.0112	ARIMA 0.0112	ESM_E3 0.0113	ARIMA_E3 0.0114	M15_E3 0.0117	MEAN_E3 0.0126	MEAN 0.0146
STEEL PRODUCERS												
Nucor Corporation (NUE)	HOLT 0.0697	HOLT3_E3 0.0698	ESM3_E3 0.0698	NAÏVE 0.0698	NAIVE3_E3 0.0699	ESM_E3 0.0700	ESM 0.0700	M15_E3 0.0717	ARIMA_E3 0.0773	MEAN_E3 0.0778	MEAN 0.0930	ARIMA 0.1209
Commercial Metals Company (CMC)	NAÏVE 0.0383	HOLT 0.0383	NAIVE3_E3 0.0384	HOLT3_E3 0.0384	ESM_E3 0.0396	M15_E3 0.0396	ESM3_E3 0.0398	ESM 0.0402	MEAN_E3 0.0425	ARIMA_E3 0.0485	MEAN 0.0501	ARIMA 0.0890
L. B. Foster Company (FSTR)	HOLT3_E3 0.0585	NAIVE3_E3 0.0586	ESM_E3 0.0587	ESM3_E3 0.0587	HOLT 0.0587	NAÏVE 0.0589	ESM 0.0590	M15_E3 0.0594	MEAN_E3 0.0651	ARIMA_E3 0.0683	MEAN 0.0789	ARIMA 0.1205
United States Steel Corporation (X)	ARIMA_E3 0.1514	NAÏVE 0.1729	ESM 0.1729	HOLT 0.1748	NAIVE3_E3 0.1753	ESM3_E3 0.1753	HOLT3_E3 0.1763	ESM_E3 0.1780	M15_E3 0.1842	ARIMA 0.1851	MEAN_E3 0.2078	MEAN 0.2543
Shiloh Industries Inc. (SHLO)	NAÏVE 0.0287	HOLT 0.0287	ARIMA 0.0287	NAIVE3_E3 0.0288	HOLT3_E3 0.0289	ESM3_E3 0.0293	ESM 0.0293	ESM_E3 0.0295	ARIMA_E3 0.0295	M15_E3 0.0305	MEAN_E3 0.0318	MEAN 0.0366
Schnitzer Steel Industries, Inc. (SCHN)	M15_E3 0.1164	NAIVE3_E3 0.1187	ESM_E3 0.1197	HOLT3_E3 0.1200	NAÏVE 0.1202	MEAN_E3 0.1208	HOLT 0.1224	ESM3_E3 0.1225	ESM 0.1261	MEAN 0.1437	ARIMA_E3 0.1676	ARIMA 0.2800
Olympic Steel, Inc. (ZEUS)	M15_E3 0.0749	NAIVE3_E3 0.0752	ESM_E3 0.0753	NAÏVE 0.0758	ESM3_E3 0.0758	ESM 0.0767	HOLT3_E3 0.0794	MEAN_E3 0.0800	HOLT 0.0816	ARIMA_E3 0.0840	MEAN 0.0985	ARIMA 0.1033

B.8. Business Services Sector

Table B-19: Business Services E3 MAPE Accuracy Results

SECURITY	MAPE #1	MAPE #2	MAPE #3	MAPE #4	MAPE #5	MAPE #6	MAPE #7	MAPE #8	MAPE #9	MAPE #10	MAPE #11	MAPE #12
FINANCIAL TRANSACTIONS												
Equifax Inc. (EFX)	HOLT 7.7677	NAÏVE 7.7776	HOLT3_E3 7.7839	NAIVE3_E3 7.7907	ESM3_E3 7.7910	ESM_E3 7.8005	ESM 7.8679	ARIMA_E3 7.9821	M15_E3 8.0283	MEAN_E3 8.4095	ARIMA 8.7821	MEAN 9.9536
Diebold Nixdorf, Incorporated (DBD)	ARIMA_E3 10.1157	ESM_E3 10.1316	HOLT3_E3 10.1488	NAIVE3_E3 10.1560	ESM3_E3 10.1797	M15_E3 10.2115	ARIMA 10.2170	HOLT 10.2354	NAÏVE 10.2456	ESM 10.2784	MEAN_E3 11.8072	MEAN 14.6414
Total System Services, Inc (TSS)	ESM3_E3 8.2202	ESM_E3 8.2517	HOLT3_E3 8.2559	HOLT 8.2624	ESM 8.2675	NAIVE3_E3 8.2921	NAÏVE 8.2933	ARIMA_E3 8.4913	M15_E3 8.5778	ARIMA 8.6410	MEAN_E3 9.8961	MEAN 12.0264
Fiserv, Inc. (FISV)	ESM_E3 6.5990	ESM3_E3 6.6795	HOLT3_E3 6.7881	NAIVE3_E3 6.8139	HOLT 6.8419	M15_E3 6.8672	NAÏVE 6.8831	ESM 6.8960	MEAN_E3 7.0976	ARIMA_E3 7.9880	MEAN 8.4563	ARIMA 10.4831
STAFFING												
Kelly Services, Inc. (KELYA)	ARIMA_E3 9.2093	M15_E3 9.2509	HOLT3_E3 9.2863	ESM_E3 9.2872	NAIVE3_E3 9.3346	ESM3_E3 9.4031	ARIMA 9.4039	HOLT 9.4118	NAÏVE 9.4818	ESM 9.5977	MEAN_E3 9.8820	MEAN 12.0018
ManpowerGroup Inc. (MAN)	ESM_E3 10.1615	NAIVE3_E3 10.1631	ESM3_E3 10.1635	HOLT3_E3 10.1689	ARIMA_E3 10.1803	NAÏVE 10.2024	ESM 10.2028	ARIMA 10.2099	HOLT 10.2113	M15_E3 10.2815	MEAN_E3 11.0480	MEAN 13.4925
GEE Group., Inc. (JOB)	HOLT3_E3 18.0153	ARIMA_E3 18.0211	NAIVE3_E3 18.0567	ARIMA 18.1214	HOLT 18.1756	NAÏVE 18.2294	M15_E3 18.2975	ESM_E3 20.7480	MEAN_E3 22.5539	ESM3_E3 22.7274	ESM 24.9935	MEAN 29.9891
Robert Half International, Inc. (RHI)	NAIVE3_E3 9.9817	HOLT3_E3 9.9818	HOLT 9.9912	NAÏVE 9.9917	ARIMA 10.0746	ARIMA_E3 10.0892	M15_E3 10.2449	ESM_E3 10.2602	ESM3_E3 10.4266	ESM 10.7081	MEAN_E3 11.0145	MEAN 12.8875
OUTSOURCING												
Automatic Data Processing, Inc. (ADP)	NAIVE3_E3 5.7379	HOLT3_E3 5.7404	NAÏVE 5.7519	HOLT 5.7534	ESM_E3 5.8523	M15_E3 5.8700	ESM3_E3 5.9248	MEAN_E3 6.0494	ESM 6.0563	ARIMA_E3 6.5638	MEAN 7.0391	ARIMA 8.2441
R.R. Donnelley & Sons Company (RRD)	HOLT 10.3291	ARIMA 10.3300	NAÏVE 10.3319	ESM 10.3559	HOLT3_E3 10.3829	NAIVE3_E3 10.3858	ESM3_E3 10.3945	ESM_E3 10.4929	ARIMA_E3 10.5584	M15_E3 10.8224	MEAN_E3 12.2433	MEAN 14.8914
Paychex, Inc. (PAYX)	NAÏVE 6.7251	HOLT 6.7291	NAIVE3_E3 6.7663	HOLT3_E3 6.7744	ESM_E3 7.0161	ESM3_E3 7.1174	M15_E3 7.1868	ESM 7.3982	ARIMA_E3 7.4501	MEAN_E3 7.7803	ARIMA 8.6891	MEAN 9.1645
Barrett Business Services, Inc. (BBSI)	NAIVE3_E3 11.5782	HOLT3_E3 11.5871	NAÏVE 11.5892	HOLT 11.5928	ESM3_E3 11.6244	ESM 11.6679	ESM_E3 11.6986	M15_E3 12.1580	ARIMA_E3 12.5140	MEAN_E3 13.7895	ARIMA 14.7379	MEAN 17.1869

SECURITY	MAPE #1	MAPE #2	MAPE #3	MAPE #4	MAPE #5	MAPE #6	MAPE #7	MAPE #8	MAPE #9	MAPE #10	MAPE #11	MAPE #12
BUSINESS SERVICES												
Xerox Corporation (XRX)	ESM_E3 11.3823	ESM3_E3 11.4078	NAIVE3_E3 11.4120	ARIMA_E3 11.4149	HOLT3_E3 11.5053	ARIMA 11.5362	ESM 11.5776	NAÏVE 11.5861	M15_E3 11.6804	HOLT 11.7364	MEAN_E3 12.5382	MEAN 15.4212
Volt Information Services, Inc. (VISI)	ARIMA 12.7116	NAÏVE 12.7175	NAIVE3_E3 12.7805	HOLT 12.7924	HOLT3_E3 12.8017	ESM3_E3 13.0839	ESM 13.1114	ARIMA_E3 13.1710	ESM_E3 13.2007	M15_E3 13.8022	MEAN_E3 15.6112	MEAN 19.9618
Healthcare Services Group, Inc. (HCSG)	HOLT3_E3 7.6844	NAIVE3_E3 7.6928	HOLT 7.7064	ESM3_E3 7.7386	NAÏVE 7.7444	ESM_E3 7.7591	ESM 7.8664	M15_E3 8.1869	MEAN_E3 8.6875	ARIMA_E3 10.2385	MEAN 10.3358	ARIMA 18.6814
Avis Budget Group, Inc. (CAR)	HOLT 20.3083	NAÏVE 20.3086	HOLT3_E3 20.3498	NAIVE3_E3 20.3501	ARIMA 20.3831	ESM3_E3 20.6722	ARIMA_E3 20.6882	ESM_E3 20.6969	ESM 20.8142	M15_E3 21.1781	MEAN_E3 27.4225	MEAN 36.1545
Spherix Incorporated (SPEX)	HOLT 25.3934	NAÏVE 25.3944	NAIVE3_E3 25.3955	HOLT3_E3 25.4401	ARIMA 25.8764	ESM_E3 26.1665	ESM3_E3 26.2403	ARIMA_E3 26.3022	ESM 26.5787	M15_E3 26.9474	MEAN_E3 32.8449	MEAN 42.2901
Crawford & Company (CRD-A)	HOLT3_E3 10.0740	ESM3_E3 10.0764	NAIVE3_E3 10.0766	ESM_E3 10.1189	ESM 10.1372	NAÏVE 10.1374	ARIMA 10.1447	HOLT 10.1449	ARIMA_E3 10.1821	M15_E3 10.5221	MEAN_E3 12.3742	MEAN 15.8346
CorVel Corporation (CRVL	NAIVE3_E3 9.1053	NAÏVE 9.1390	HOLT3_E3 9.2007	ESM_E3 9.2137	ESM3_E3 9.2241	HOLT 9.2585	ESM 9.3196	M15_E3 9.4487	ARIMA_E3 9.4566	ARIMA 10.1512	MEAN_E3 10.7068	MEAN 13.2917
ADVERTISING & MARKETING												
Omnicom Group, Inc (OMC)	HOLT3_E3 7.0976	NAIVE3_E3 7.1033	HOLT 7.1209	NAÏVE 7.1284	ARIMA_E3 7.2062	ESM_E3 7.2875	M15_E3 7.3221	MEAN_E3 7.3671	ESM3_E3 7.4279	ESM 7.6732	ARIMA 8.2965	MEAN 8.7662
The Interpublic Group of Co., Inc. (IPG)	NAIVE3_E3 10.4880	HOLT3_E3 10.5067	NAÏVE 10.5601	HOLT 10.5891	ESM_E3 10.5919	M15_E3 10.5933	ESM3_E3 10.6987	ESM 10.8657	ARIMA_E3 11.3389	MEAN_E3 11.7074	ARIMA 12.8952	MEAN 14.6869
WPP plc (WPP)	NAIVE3_E3 8.3713	NAÏVE 8.3742	HOLT3_E3 8.3873	HOLT 8.4005	ESM3_E3 8.4209	ESM_E3 8.4458	ESM 8.4741	M15_E3 8.7080	MEAN_E3 9.5710	ARIMA_E3 9.8426	MEAN 11.6831	ARIMA 12.4003
Insignia Systems, Inc. (ISIG)	ARIMA 18.2666	HOLT3_E3 18.2915	NAIVE3_E3 18.3244	HOLT 18.3349	NAÏVE 18.3861	ARIMA_E3 18.5507	ESM_E3 18.5607	ESM3_E3 18.6734	ESM 19.1015	M15_E3 19.1491	MEAN_E3 21.4388	MEAN 27.6159
Harte Hanks, Inc. (HHS)	HOLT 11.1890	NAÏVE 11.1904	NAIVE3_E3 11.2215	HOLT3_E3 11.2255	ESM_E3 11.4418	ESM3_E3 11.4598	ESM 11.5889	ARIMA_E3 11.6812	M15_E3 11.7865	ARIMA 13.7468	MEAN_E3 14.2546	MEAN 17.8338

Table B-20: Business Services E3 RMSE Accuracy Results

SECURITY	RMSE #1	RMSE #2	RMSE #3	RMSE #4	RMSE #5	RMSE #6	RMSE #7	RMSE #8	RMSE #9	RMSE #10	RMSE #11	RMSE #12
FINANCIAL TRANSACTIONS												
Equifax Inc. (EFX)	HOLT 0.0860	NAÏVE 0.0861	HOLT3_E3 0.0866	NAIVE3_E3 0.0867	MEAN_E3 0.0902	M15_E3 0.0902	ESM_E3 0.0906	ESM3_E3 0.0918	ESM 0.0936	ARIMA_E3 0.0976	MEAN 0.1031	ARIMA 0.1638
Diebold Nixdorf, Incorporated (DBD)	ARIMA_E3 0.0520	ESM_E3 0.0521	HOLT3_E3 0.0523	NAIVE3_E3 0.0523	M15_E3 0.0524	ESM3_E3 0.0524	ARIMA 0.0528	HOLT 0.0529	NAÏVE 0.0529	ESM 0.0530	MEAN_E3 0.0565	MEAN 0.0690
Total System Services, Inc (TSS)	ESM_E3 0.0517	ESM3_E3 0.0519	HOLT3_E3 0.0526	NAIVE3_E3 0.0526	ESM 0.0527	NAÏVE 0.0528	HOLT 0.0528	M15_E3 0.0531	MEAN_E3 0.0571	ARIMA_E3 0.0579	MEAN 0.0660	ARIMA 0.0696
Fiserv, Inc. (FISV)	ESM_E3 0.0239	ESM3_E3 0.0245	HOLT3_E3 0.0263	ESM 0.0263	NAIVE3_E3 0.0263	M15_E3 0.0263	HOLT 0.0266	NAÏVE 0.0267	MEAN_E3 0.0289	MEAN 0.0345	ARIMA_E3 0.0481	ARIMA 0.1017
STAFFING												
Kelly Services, Inc. (KELYA)	ARIMA_E3 0.0337	M15_E3 0.0337	ESM_E3 0.0340	HOLT3_E3 0.0340	NAIVE3_E3 0.0342	HOLT 0.0344	ARIMA 0.0344	ESM3_E3 0.0345	NAÏVE 0.0347	ESM 0.0352	MEAN_E3 0.0353	MEAN 0.0418
ManpowerGroup Inc. (MAN)	ESM3_E3 0.1064	NAIVE3_E3 0.1064	HOLT3_E3 0.1065	ESM 0.1066	NAÏVE 0.1066	HOLT 0.1066	ESM_E3 0.1067	ARIMA 0.1070	ARIMA_E3 0.1071	M15_E3 0.1082	MEAN_E3 0.1186	MEAN 0.1430
GEE Group., Inc. (JOB)	ARIMA 0.0359	HOLT 0.0360	HOLT3_E3 0.0360	NAIVE3_E3 0.0360	NAÏVE 0.0360	ARIMA_E3 0.0365	ESM_E3 0.0373	M15_E3 0.0375	ESM3_E3 0.0383	ESM 0.0398	MEAN_E3 0.0413	MEAN 0.0524
Robert Half International, Inc. (RHI)	NAIVE3_E3 0.0580	HOLT3_E3 0.0580	NAÏVE 0.0581	HOLT 0.0581	ARIMA 0.0584	ARIMA_E3 0.0585	ESM_E3 0.0591	M15_E3 0.0594	ESM3_E3 0.0603	ESM 0.0624	MEAN_E3 0.0635	MEAN 0.0736
OUTSOURCING												
Automatic Data Processing, Inc. (ADP)	NAIVE3_E3 0.0615	HOLT3_E3 0.0616	NAÏVE 0.0616	HOLT 0.0617	ESM_E3 0.0620	ESM3_E3 0.0624	M15_E3 0.0626	ESM 0.0633	MEAN_E3 0.0646	MEAN 0.0755	ARIMA_E3 0.0764	ARIMA 0.1237
R.R. Donnelley & Sons Company (RRD)	HOLT3_E3 0.0388	NAIVE3_E3 0.0388	ARIMA_E3 0.0388	ESM_E3 0.0389	HOLT 0.0389	NAÏVE 0.0389	ARIMA 0.0389	ESM3_E3 0.0390	M15_E3 0.0392	ESM 0.0393	MEAN_E3 0.0427	MEAN 0.0504
Paychex, Inc. (PAYX)	NAÏVE 0.0469	HOLT 0.0470	NAIVE3_E3 0.0472	HOLT3_E3 0.0473	M15_E3 0.0500	ESM_E3 0.0500	ESM3_E3 0.0522	MEAN_E3 0.0543	ESM 0.0556	ARIMA_E3 0.0579	MEAN 0.0649	ARIMA 0.0971
Barrett Business Services, Inc. (BBSI)	ESM_E3 0.0855	M15_E3 0.0858	ESM3_E3 0.0859	NAIVE3_E3 0.0861	ESM 0.0867	HOLT3_E3 0.0868	NAÏVE 0.0869	HOLT 0.0879	MEAN_E3 0.0942	ARIMA_E3 0.1059	MEAN 0.1168	ARIMA 0.1740

SECURITY	RMSE #1	RMSE #2	RMSE #3	RMSE #4	RMSE #5	RMSE #6	RMSE #7	RMSE #8	RMSE #9	RMSE #10	RMSE #11	RMSE #12
BUSINESS SERVICES												
Xerox Corporation (XRX)	ARIMA 0.0634	ESM 0.0636	ESM3_E3 0.0636	NAIVE3_E3 0.0636	NAÏVE 0.0636	HOLT3_E3 0.0636	HOLT 0.0637	ESM_E3 0.0639	ARIMA_E3 0.0642	M15_E3 0.0657	MEAN_E3 0.0757	MEAN 0.0976
Volt Information Services, Inc. (VISI)	ARIMA 0.0299	NAÏVE 0.0300	NAIVE3_E3 0.0300	ESM3_E3 0.0302	ESM 0.0303	HOLT3_E3 0.0303	ARIMA_E3 0.0303	ESM_E3 0.0303	HOLT 0.0306	M15_E3 0.0312	MEAN_E3 0.0349	MEAN 0.0461
Healthcare Services Group, Inc. (HCSG)	NAIVE3_E3 0.0247	NAÏVE 0.0247	ESM3_E3 0.0247	HOLT3_E3 0.0247	HOLT 0.0247	ESM_E3 0.0248	ESM 0.0248	M15_E3 0.0257	MEAN_E3 0.0280	MEAN 0.0333	ARIMA_E3 0.0375	ARIMA 0.0870
Avis Budget Group, Inc. (CAR)	NAIVE3_E3 0.0610	HOLT3_E3 0.0610	NAÏVE 0.0610	HOLT 0.0610	ARIMA 0.0610	ESM3_E3 0.0623	ARIMA_E3 0.0625	ESM_E3 0.0625	ESM 0.0632	M15_E3 0.0656	MEAN_E3 0.0722	MEAN 0.0894
Spherix Incorporated (SPEX)	NAÏVE 51.4268	NAIVE3_E3 51.4797	HOLT 51.6388	HOLT3_E3 51.6838	ARIMA 51.8677	MEAN_E3 52.2265	ARIMA_E3 52.3664	M15_E3 53.0637	ESM_E3 55.4538	ESM3_E3 57.6925	MEAN 57.8041	ESM 60.5321
Crawford & Company (CRD-A)	ESM3_E3 0.0143	NAIVE3_E3 0.0143	ARIMA 0.0143	ESM 0.0143	NAÏVE 0.0143	HOLT3_E3 0.0143	ESM_E3 0.0144	HOLT 0.0144	ARIMA_E3 0.0144	M15_E3 0.0147	MEAN_E3 0.0152	MEAN 0.0183
CorVel Corporation (CRVL	ESM3_E3 0.0387	ESM_E3 0.0388	ESM 0.0389	NAÏVE 0.0391	NAIVE3_E3 0.0392	HOLT3_E3 0.0398	HOLT 0.0401	M15_E3 0.0404	ARIMA_E3 0.0417	MEAN_E3 0.0465	ARIMA 0.0518	MEAN 0.0584
ADVERTISING & MARKETING												
Omnicom Group, Inc (OMC)	HOLT3_E3 0.0619	NAIVE3_E3 0.0619	HOLT 0.0621	NAÏVE 0.0621	MEAN_E3 0.0627	ESM_E3 0.0629	ARIMA_E3 0.0634	M15_E3 0.0634	ESM3_E3 0.0639	ESM 0.0656	MEAN 0.0740	ARIMA 0.0844
The Interpublic Group of Co.s, Inc. (IPG)	MEAN_E3 0.0362	M15_E3 0.0368	NAIVE3_E3 0.0381	HOLT3_E3 0.0385	ESM_E3 0.0386	NAÏVE 0.0389	HOLT 0.0394	ESM3_E3 0.0400	ESM 0.0417	MEAN 0.0454	ARIMA_E3 0.0685	ARIMA 0.1385
WPP plc (WPP)	NAÏVE 0.0961	NAIVE3_E3 0.0961	HOLT3_E3 0.0963	HOLT 0.0964	ESM3_E3 0.0966	ESM_E3 0.0967	ESM 0.0971	M15_E3 0.0994	MEAN_E3 0.1073	MEAN 0.1293	ARIMA_E3 0.1673	ARIMA 0.3569
Insignia Systems, Inc. (ISIG)	MEAN_E3 0.0139	ESM3_E3 0.0144	ESM_E3 0.0144	NAIVE3_E3 0.0144	HOLT3_E3 0.0145	ARIMA 0.0145	HOLT 0.0145	NAÏVE 0.0145	ARIMA_E3 0.0145	ESM 0.0145	M15_E3 0.0148	MEAN 0.0163
Harte Hanks, Inc. (HHS)	HOLT 0.1993	NAÏVE 0.1996	HOLT3_E3 0.2008	NAIVE3_E3 0.2010	ARIMA_E3 0.2024	ESM3_E3 0.2049	ESM_E3 0.2052	ESM 0.2064	M15_E3 0.2114	MEAN_E3 0.2409	ARIMA 0.2657	MEAN 0.2936

B.9. Computer & Technology Sector

Table B-21: Computer & Technology Sector E3 MAPE Accuracy Results

SECURITY	MAPE #1	MAPE #2	MAPE #3	MAPE #4	MAPE #5	MAPE #6	MAPE #7	MAPE #8	MAPE #9	MAPE #10	MAPE #11	MAPE #12
COMPUTER SOFTWARE												
Autodesk, Inc. (ADSK)	ESM_E3 12.3572	ESM3_E3 12.3815	NAIVE3_E3 12.3933	HOLT3_E3 12.3992	NAÏVE 12.4471	HOLT 12.4573	ESM 12.5145	M15_E3 12.6060	ARIMA_E3 13.2278	MEAN_E3 13.3008	ARIMA 15.2136	MEAN 15.5680
Cadence Design Systems, Inc. (CDNS)	ARIMA 10.3507	NAIVE3_E3 10.3691	HOLT3_E3 10.3692	NAÏVE 10.3752	HOLT 10.3754	ARIMA_E3 10.4859	ESM_E3 10.4962	ESM3_E3 10.4966	ESM 10.5687	M15_E3 10.7317	MEAN_E3 11.7831	MEAN 14.5574
Microsoft (MSFT)	ARIMA 8.0481	MEAN_E3 8.0940	ARIMA_E3 8.2908	NAIVE3_E3 8.4523	HOLT3_E3 8.4632	NAÏVE 8.4780	HOLT 8.4927	M15_E3 8.6145	ESM_E3 8.9365	MEAN 9.1435	ESM3_E3 9.1823	ESM 9.4858
Oracle Corporation (ORCL)	HOLT3_E3 7.6369	HOLT 7.6820	NAIVE3_E3 7.6972	NAÏVE 7.7307	ARIMA_E3 7.8941	ESM_E3 8.0077	M15_E3 8.0770	ESM3_E3 8.3863	ARIMA 8.5315	MEAN_E3 8.6192	ESM 9.0369	MEAN 11.2493
Adobe, Inc. (ADBE)	NAIVE3_E3 10.5620	ESM3_E3 10.5628	HOLT3_E3 10.5638	ESM_E3 10.5926	HOLT 10.5998	NAÏVE 10.6040	ESM 10.6314	M15_E3 10.8728	ARIMA_E3 11.1752	MEAN_E3 12.0337	ARIMA 12.7309	MEAN 15.1365
Symantec (SYMC)	HOLT 10.2761	HOLT3_E3 10.2969	NAÏVE 10.3026	NAIVE3_E3 10.3152	ARIMA_E3 10.6713	ESM_E3 10.6725	M15_E3 10.7140	ESM3_E3 10.8384	ESM 11.1301	MEAN_E3 11.3208	ARIMA 11.7582	MEAN 13.7008
PTC, Inc. (PTC)	ARIMA 14.9423	HOLT 14.9643	NAÏVE 14.9650	ESM3_E3 14.9661	ESM 14.9689	NAIVE3_E3 15.0053	HOLT3_E3 15.0055	ESM_E3 15.0800	ARIMA_E3 15.2097	M15_E3 15.6158	MEAN_E3 15.6277	MEAN 18.0233
COMPUTERS												
International Business Machines (IBM)	HOLT 7.2533	NAÏVE 7.2548	HOLT3_E3 7.2586	NAIVE3_E3 7.2588	ARIMA_E3 7.3441	ESM_E3 7.3498	ESM3_E3 7.3578	ESM 7.4140	M15_E3 7.4861	ARIMA 7.5073	MEAN_E3 7.5074	MEAN 8.3999
HP Inc. (HPQ)	HOLT3_E3 9.2788	NAIVE3_E3 9.2861	HOLT 9.3163	NAÏVE 9.3280	ESM_E3 9.4423	ESM3_E3 9.5197	M15_E3 9.5242	ESM 9.6954	ARIMA_E3 9.7871	MEAN_E3 10.7378	ARIMA 11.2630	MEAN 14.0009
Agilsys, Inc. (AGYS)	ARIMA 12.6197	HOLT 12.6271	NAÏVE 12.6366	NAIVE3_E3 12.6690	HOLT3_E3 12.6763	ESM 12.7136	ESM3_E3 12.7413	ESM_E3 12.8821	ARIMA_E3 12.9703	M15_E3 13.4650	MEAN_E3 14.3930	MEAN 16.9250
CSP Inc. (CSPI)	ARIMA 10.8455	NAIVE3_E3 10.8506	HOLT3_E3 10.8629	HOLT 10.8735	NAÏVE 10.8776	ESM3_E3 10.9015	ESM 10.9187	ESM_E3 10.9694	ARIMA_E3 11.0145	M15_E3 11.3339	MEAN_E3 12.0448	MEAN 14.5891
Apple Inc. (AAPL)	HOLT 11.2023	NAÏVE 11.3290	ESM 11.5225	HOLT3_E3 11.7350	NAIVE3_E3 11.8483	ESM3_E3 11.9617	ESM_E3 12.4565	M15_E3 13.6341	ARIMA_E3 14.0840	MEAN_E3 15.0218	MEAN 18.5417	ARIMA 19.8208
3D Systems Corporation (DDD)	NAIVE3_E3 18.4987	ESM_E3 18.5768	HOLT3_E3 18.5819	M15_E3 18.6580	NAÏVE 18.6628	HOLT 18.7773	ESM3_E3 18.8031	ESM 19.1868	ARIMA_E3 19.3016	MEAN_E3 20.3162	ARIMA 21.3402	MEAN 24.6898
PAR Technology Corporation (PAR)	NAIVE3_E3 12.9907	NAÏVE 13.0044	HOLT3_E3 13.0124	HOLT 13.0554	ESM3_E3 13.0735	ESM 13.1144	ESM_E3 13.1447	ARIMA 13.2653	ARIMA_E3 13.2844	M15_E3 13.5737	MEAN_E3 14.9682	MEAN 18.3670

SECURITY	MAPE #1	MAPE #2	MAPE #3	MAPE #4	MAPE #5	MAPE #6	MAPE #7	MAPE #8	MAPE #9	MAPE #10	MAPE #11	MAPE #12
WIRELESS PROVIDERS												
CenturyLink, Inc (CTL)	NAIVE3_E3 7.8204	HOLT3_E3 7.8225	NAÏVE 7.8257	HOLT 7.8303	M15_E3 8.0630	ESM_E3 8.0810	ARIMA_E3 8.1191	ESM3_E3 8.2067	ESM 8.4023	ARIMA 8.4198	MEAN_E3 8.4385	MEAN 9.9794
Verizon Communications Inc. (VZ)	HOLT 5.9636	NAÏVE 5.9691	HOLT3_E3 5.9778	NAIVE3_E3 5.9806	ARIMA 6.0227	ESM3_E3 6.0410	ESM 6.0556	ESM_E3 6.0640	ARIMA_E3 6.0906	M15_E3 6.2195	MEAN_E3 6.2582	MEAN 7.0969
AT&T Inc. (T)	HOLT3_E3 5.9568	NAIVE3_E3 5.9589	HOLT 5.9811	NAÏVE 5.9849	ARIMA 5.9976	ESM_E3 6.0331	ESM3_E3 6.0339	ARIMA_E3 6.0399	ESM 6.0875	M15_E3 6.1688	MEAN_E3 6.1903	MEAN 7.3076
Sprint Corporation (S)	HOLT3_E3 14.7704	HOLT 14.7902	NAIVE3_E3 14.8180	NAÏVE 14.8362	ARIMA_E3 14.8504	ARIMA 15.0612	ESM_E3 15.1531	M15_E3 15.1591	ESM3_E3 15.3707	ESM 15.7333	MEAN_E3 17.3896	MEAN 22.3875
Vodafone Group PLC (VOD)	ARIMA_E3 7.2586	HOLT3_E3 7.3583	NAIVE3_E3 7.3639	HOLT 7.3776	NAÏVE 7.3869	ESM_E3 7.5213	M15_E3 7.5589	ESM3_E3 7.5886	ESM 7.7197	ARIMA 7.7625	MEAN_E3 8.7225	MEAN 11.1029
United States Cellular Corp. (USM)	ESM_E3 8.1886	HOLT3_E3 8.2056	NAIVE3_E3 8.2087	ESM3_E3 8.2356	ARIMA_E3 8.2510	NAÏVE 8.2914	M15_E3 8.2920	HOLT 8.3076	ESM 8.3538	ARIMA 8.4476	MEAN_E3 8.9087	MEAN 10.7071
ATN International, Inc. (ATNI)	ESM3_E3 9.4442	ESM 9.4536	HOLT3_E3 9.4574	HOLT 9.4718	NAIVE3_E3 9.4742	NAÏVE 9.4923	ESM_E3 9.5053	MEAN_E3 9.7997	M15_E3 9.9538	ARIMA_E3 11.0054	MEAN 11.3348	ARIMA 14.6015
SEMICONDUCTORS												
Texas Instruments Inc. (TXN)	NAÏVE 9.0909	HOLT 9.0912	NAIVE3_E3 9.0955	HOLT3_E3 9.0957	ESM_E3 9.2361	ESM3_E3 9.2503	ESM 9.3438	M15_E3 9.4747	ARIMA_E3 10.0864	MEAN_E3 10.5854	ARIMA 12.8284	MEAN 13.1634
Intel Corporation (INTC)	NAÏVE 8.5304	HOLT 8.5528	NAIVE3_E3 8.5601	HOLT3_E3 8.5676	ESM3_E3 8.6263	ESM_E3 8.6313	ESM 8.7036	ARIMA_E3 8.8506	M15_E3 8.8960	ARIMA 9.2934	MEAN_E3 10.2782	MEAN 13.2636
Analog Devices, Inc. (ADI)	HOLT3_E3 9.2180	NAIVE3_E3 9.2720	HOLT 9.2812	NAÏVE 9.3388	M15_E3 9.4334	ESM_E3 9.4455	ESM3_E3 9.7002	ARIMA_E3 9.8140	MEAN_E3 10.0903	ESM 10.1187	MEAN 12.1147	ARIMA 12.4356
Semtech Corporation (SMTC)	HOLT 11.9062	NAÏVE 11.9065	HOLT3_E3 12.0252	NAIVE3_E3 12.0255	ESM 12.1620	ESM3_E3 12.1703	ARIMA_E3 12.1723	ESM_E3 12.2885	M15_E3 12.7941	MEAN_E3 13.2779	MEAN 16.1003	ARIMA 16.6362
Amtech Systems, Inc. (ASYS)	NAIVE3_E3 18.2031	HOLT3_E3 18.2139	ESM3_E3 18.2141	ESM 18.2627	HOLT 18.2636	NAÏVE 18.2648	ARIMA 18.3206	ESM_E3 18.3845	ARIMA_E3 18.6480	M15_E3 19.2252	MEAN_E3 21.6119	MEAN 27.7400
Maxim Integrated Products, Inc. (MXIM)	HOLT3_E3 9.8342	NAIVE3_E3 9.8649	M15_E3 9.9041	HOLT 9.9112	NAÏVE 9.9598	MEAN_E3 10.0889	ESM_E3 10.0928	ARIMA_E3 10.1067	ESM3_E3 10.3472	ESM 10.6964	ARIMA 10.6988	MEAN 11.7529
Microchip Technology Inc. (MCHP)	HOLT 8.6922	NAÏVE 8.7122	HOLT3_E3 8.7530	ESM3_E3 8.7568	NAIVE3_E3 8.7626	ARIMA 8.7712	ESM 8.7929	ESM_E3 8.8201	ARIMA_E3 9.0132	M15_E3 9.2906	MEAN_E3 9.5966	MEAN 11.2663

APPENDIX B: E3 Forecast Study Results

SECURITY	MAPE #1	MAPE #2	MAPE #3	MAPE #4	MAPE #5	MAPE #6	MAPE #7	MAPE #8	MAPE #9	MAPE #10	MAPE #11	MAPE #12
ELECTRICAL PRODUCTS												
Koninkijke Philips N.V. (PHG)	ESM_E3 9.3914	HOLT3_E3 9.3972	NAIVE3_E3 9.3994	M15_E3 9.4304	ESM3_E3 9.4645	HOLT 9.4993	NAÏVE 9.5032	ARIMA_E3 9.5834	ESM 9.6001	MEAN_E3 10.0230	ARIMA 10.5737	MEAN 12.4082
Cubic Corporation (CUB)	NAIVE3_E3 9.5303	HOLT3_E3 9.5382	NAÏVE 9.5932	HOLT 9.6059	ARIMA_E3 9.6095	ESM_E3 9.6525	M15_E3 9.6705	ESM3_E3 9.7706	ARIMA 9.7847	ESM 9.9540	MEAN_E3 10.4259	MEAN 12.3719
Bell Fuse Inc. (BELFA)	NAIVE3_E3 11.2180	HOLT3_E3 11.2520	ESM3_E3 11.2640	ESM_E3 11.2667	NAÏVE 11.2838	HOLT 11.3307	ESM 11.3461	M15_E3 11.5511	MEAN_E3 11.8627	ARIMA_E3 12.4205	ARIMA 14.3479	MEAN 14.4172
Bonso Electronics International, Inc. (BNSO)	ARIMA 14.8268	HOLT 14.8475	HOLT3_E3 14.8611	NAIVE3_E3 14.8977	NAÏVE 14.9080	ARIMA_E3 15.1357	ESM_E3 15.3764	ESM3_E3 15.4310	ESM 15.6357	M15_E3 15.6899	MEAN_E3 17.3922	MEAN 21.8215
Trimble Inc. (TRMB)	ESM_E3 11.1860	ESM3_E3 11.1908	NAIVE3_E3 11.2954	ESM 11.2999	HOLT3_E3 11.3065	NAÏVE 11.3865	HOLT 11.4012	ARIMA_E3 11.4307	M15_E3 11.4812	ARIMA 12.0019	MEAN_E3 13.1264	MEAN 16.5463
Kopin Corporation (KOPN)	HOLT 14.7506	NAÏVE 14.8216	HOLT3_E3 14.9291	NAIVE3_E3 14.9862	ARIMA 15.3765	ESM3_E3 15.5408	ESM 15.6004	ESM_E3 15.6196	ARIMA_E3 15.6315	M15_E3 16.2608	MEAN_E3 18.2119	MEAN 22.6991
Flex Ltd. (FLEX)	ARIMA_E3 12.8433	NAIVE3_E3 13.0686	HOLT3_E3 13.0686	NAÏVE 13.0778	HOLT 13.0779	ESM_E3 13.3659	ESM3_E3 13.4150	ESM 13.5936	M15_E3 13.6391	ARIMA 13.7295	MEAN_E3 15.1826	MEAN 18.8526
COMMUNICATIONS COMPONENT												
Corning Incorporated (GLW)	ARIMA_E3 12.6180	HOLT 12.7791	HOLT3_E3 12.8117	NAÏVE 12.9103	NAIVE3_E3 12.9118	ARIMA 13.2602	M15_E3 13.3432	ESM_E3 13.4051	ESM3_E3 13.7050	ESM 14.1851	MEAN_E3 17.4055	MEAN 23.3375
Communications Systems, Inc. (JCS)	HOLT 8.2018	NAÏVE 8.2208	HOLT3_E3 8.2461	NAIVE3_E3 8.2545	ARIMA 8.2842	ARIMA_E3 8.5439	ESM_E3 8.6073	ESM3_E3 8.6387	ESM 8.7893	M15_E3 8.8587	MEAN_E3 9.7562	MEAN 12.1522
Plantronics (PLT)	NAÏVE 12.1236	NAIVE3_E3 12.1544	ARIMA 12.1899	HOLT3_E3 12.1917	HOLT 12.1999	ARIMA_E3 12.3617	ESM_E3 12.5072	ESM3_E3 12.5566	ESM 12.7053	M15_E3 12.7246	MEAN_E3 14.2848	MEAN 17.3219
Technical Communications Corp. (TCCO)	ARIMA 17.3917	ARIMA_E3 17.5999	NAIVE3_E3 17.6381	HOLT3_E3 17.7104	NAÏVE 17.7342	HOLT 17.8148	M15_E3 18.1043	MEAN_E3 20.3544	ESM_E3 22.5430	MEAN 25.3780	ESM3_E3 26.4018	ESM 30.7622
SCIENTIFIC INSTRUMENTS												
MTS Systems Corporation (MTSC)	ESM3_E3 9.5615	ESM_E3 9.5762	HOLT3_E3 9.5973	NAIVE3_E3 9.6285	ESM 9.6447	HOLT 9.6742	NAÏVE 9.7487	ARIMA_E3 9.8058	M15_E3 9.8415	MEAN_E3 10.0009	ARIMA 10.2474	MEAN 11.9715
Kewaunee Scientific Corp. (KEQU)	NAÏVE 6.6090	HOLT 6.6147	ARIMA 6.6395	ESM 6.6630	NAIVE3_E3 6.6725	HOLT3_E3 6.6883	ESM3_E3 6.7075	ESM_E3 6.8711	ARIMA_E3 7.0230	M15_E3 7.5224	MEAN_E3 8.6541	MEAN 11.4645
PerkinElmer, Inc. (PKI)	NAÏVE 11.0089	NAIVE3_E3 11.0114	HOLT3_E3 11.0409	HOLT 11.0582	ESM3_E3 11.0732	ESM_E3 11.0807	ESM 11.1013	M15_E3 11.2507	ARIMA_E3 11.6036	ARIMA 12.4711	MEAN_E3 12.9126	MEAN 16.4499
Mixonix, Inc. (MSON)	ARIMA 15.0436	HOLT 15.0684	NAÏVE 15.0942	NAIVE3_E3 15.2522	HOLT3_E3 15.2554	ARIMA_E3 15.7773	ESM_E3 16.0529	ESM3_E3 16.1402	ESM 16.4063	M15_E3 16.5124	MEAN_E3 17.6247	MEAN 21.6729

Table B-22: Computer & Technology Sector E3 RMSE Accuracy Results

SECURITY	RMSE #1	RMSE #2	RMSE #3	RMSE #4	RMSE #5	RMSE #6	RMSE #7	RMSE #8	RMSE #9	RMSE #10	RMSE #11	RMSE #12
COMPUTER SOFTWARE												
Autodesk, Inc. (ADSK)	MEAN_E3 0.0925	ESM_E3 0.0936	ESM3_E3 0.0936	ESM 0.0942	NAIVE3_E3 0.0951	HOLT3_E3 0.0952	M15_E3 0.0953	NAÏVE 0.0957	HOLT 0.0958	MEAN 0.1018	ARIMA_E3 0.1218	ARIMA 0.2066
Cadence Design Systems, Inc. (CDNS)	ARIMA 0.0321	NAIVE3_E3 0.0321	HOLT3_E3 0.0321	ARIMA_E3 0.0322	NAÏVE 0.0322	HOLT 0.0322	ESM_E3 0.0324	ESM3_E3 0.0327	M15_E3 0.0327	ESM 0.0332	MEAN_E3 0.0344	MEAN 0.0415
Microsoft (MSFT)	ARIMA 0.0586	ARIMA_E3 0.0613	MEAN_E3 0.0620	NAIVE3_E3 0.0642	NAÏVE 0.0643	HOLT3_E3 0.0643	HOLT 0.0644	M15_E3 0.0649	MEAN 0.0674	ESM_E3 0.0677	ESM3_E3 0.0702	ESM 0.0733
Oracle Corporation (ORCL)	HOLT 0.0377	HOLT3_E3 0.0379	NAÏVE 0.0391	NAIVE3_E3 0.0395	ARIMA_E3 0.0404	M15_E3 0.0423	ESM_E3 0.0437	MEAN_E3 0.0458	ESM3_E3 0.0462	ESM 0.0495	ARIMA 0.0558	MEAN 0.0588
Adobe, Inc. (ADBE)	NAIVE3_E3 0.1189	NAÏVE 0.1190	HOLT3_E3 0.1192	HOLT 0.1193	M15_E3 0.1208	ESM_E3 0.1231	ESM3_E3 0.1261	ESM 0.1302	MEAN_E3 0.1344	ARIMA_E3 0.1480	MEAN 0.1577	ARIMA 0.2370
Symantec (SYMC)	HOLT 0.0319	HOLT3_E3 0.0319	NAÏVE 0.0319	NAIVE3_E3 0.0319	M15_E3 0.0328	ESM_E3 0.0333	ARIMA_E3 0.0337	ESM3_E3 0.0342	MEAN_E3 0.0342	ESM 0.0356	MEAN 0.0406	ARIMA 0.0490
PTC, Inc. (PTC)	ARIMA 0.0849	NAÏVE 0.0850	HOLT 0.0851	ESM 0.0869	NAIVE3_E3 0.0869	HOLT3_E3 0.0870	ESM3_E3 0.0879	ESM_E3 0.0895	ARIMA_E3 0.0904	MEAN_E3 0.0934	M15_E3 0.0947	MEAN 0.1018
COMPUTERS												
International Business Machines (IBM)	HOLT 0.1645	NAÏVE 0.1645	HOLT3_E3 0.1651	NAIVE3_E3 0.1652	ARIMA_E3 0.1679	ESM_E3 0.1685	MEAN_E3 0.1685	ESM3_E3 0.1691	M15_E3 0.1710	ESM 0.1710	ARIMA 0.1732	MEAN 0.1851
HP Inc. (HPQ)	HOLT 0.0262	NAÏVE 0.0262	HOLT3_E3 0.0263	NAIVE3_E3 0.0263	ESM_E3 0.0273	M15_E3 0.0276	ESM3_E3 0.0276	ESM 0.0281	MEAN_E3 0.0307	ARIMA_E3 0.0329	MEAN 0.0380	ARIMA 0.0638
Agilsys, Inc. (AGYS)	ARIMA 0.0260	HOLT 0.0260	NAÏVE 0.0260	NAIVE3_E3 0.0261	HOLT3_E3 0.0261	ESM3_E3 0.0264	ARIMA_E3 0.0264	ESM_E3 0.0265	ESM 0.0265	M15_E3 0.0270	MEAN_E3 0.0280	MEAN 0.0323
CSP Inc. (CSPI)	NAÏVE 0.0166	ARIMA 0.0166	NAIVE3_E3 0.0166	HOLT 0.0167	HOLT3_E3 0.0167	ESM 0.0168	ESM3_E3 0.0168	ESM_E3 0.0169	ARIMA_E3 0.0170	M15_E3 0.0175	MEAN_E3 0.0178	MEAN 0.0201
Apple Inc. (AAPL)	NAIVE3_E3 0.1238	HOLT3_E3 0.1238	NAÏVE 0.1243	HOLT 0.1244	M15_E3 0.1244	MEAN_E3 0.1261	ESM_E3 0.1284	ESM3_E3 0.1325	ESM 0.1378	MEAN 0.1399	ARIMA_E3 0.1699	ARIMA 0.3118
3D Systems Corporation (DDD)	NAIVE3_E3 0.0697	NAÏVE 0.0698	ESM_E3 0.0700	ESM3_E3 0.0700	ESM 0.0703	M15_E3 0.0707	MEAN_E3 0.0748	HOLT3_E3 0.0750	HOLT 0.0773	MEAN 0.0903	ARIMA_E3 0.0989	ARIMA 0.1704
PAR Technology Corporation (PAR)	NAÏVE 0.0227	NAIVE3_E3 0.0227	ESM 0.0228	ESM3_E3 0.0228	HOLT3_E3 0.0228	HOLT 0.0228	ESM_E3 0.0229	M15_E3 0.0235	ARIMA_E3 0.0238	ARIMA 0.0255	MEAN_E3 0.0261	MEAN 0.0319

APPENDIX B: E3 Forecast Study Results

SECURITY	RMSE #1	RMSE #2	RMSE #3	RMSE #4	RMSE #5	RMSE #6	RMSE #7	RMSE #8	RMSE #9	RMSE #10	RMSE #11	RMSE #12
WIRELESS PROVIDERS												
CenturyLink, Inc (CTL)	NAIVE3_E3 0.0487	HOLT3_E3 0.0487	NAÏVE 0.0488	HOLT 0.0488	M15_E3 0.0495	ESM_E3 0.0503	MEAN_E3 0.0504	ESM3_E3 0.0514	ESM 0.0529	ARIMA_E3 0.0533	MEAN 0.0570	ARIMA 0.0643
Verizon Communications Inc. (VZ)	HOLT 0.0448	NAÏVE 0.0448	NAIVE3_E3 0.0451	HOLT3_E3 0.0451	ARIMA 0.0455	ESM 0.0456	ESM3_E3 0.0456	ESM_E3 0.0460	ARIMA_E3 0.0464	MEAN_E3 0.0468	M15_E3 0.0476	MEAN 0.0517
AT&T Inc. (T)	HOLT3_E3 0.0388	NAIVE3_E3 0.0388	HOLT 0.0390	NAÏVE 0.0390	ARIMA_E3 0.0393	ESM_E3 0.0394	ARIMA 0.0397	M15_E3 0.0397	ESM3_E3 0.0397	ESM 0.0403	MEAN_E3 0.0404	MEAN 0.0476
Sprint Corporation (S)	NAÏVE 0.0405	HOLT 0.0406	NAIVE3_E3 0.0417	HOLT3_E3 0.0418	ARIMA_E3 0.0435	ESM3_E3 0.0452	ESM 0.0453	ESM_E3 0.0454	M15_E3 0.0467	ARIMA 0.0497	MEAN_E3 0.0518	MEAN 0.0630
Vodafone Group PLC (VOD)	ARIMA_E3 0.0413	HOLT 0.0446	NAÏVE 0.0447	HOLT3_E3 0.0451	NAIVE3_E3 0.0451	M15_E3 0.0478	ESM_E3 0.0487	ESM3_E3 0.0497	ESM 0.0511	MEAN_E3 0.0517	MEAN 0.0619	ARIMA 0.0680
United States Cellular Corp. (USM)	HOLT3_E3 0.0885	NAIVE3_E3 0.0888	HOLT 0.0889	ESM_E3 0.0889	ESM3_E3 0.0891	NAÏVE 0.0892	M15_E3 0.0897	ESM 0.0897	ARIMA_E3 0.0919	MEAN_E3 0.0921	ARIMA 0.0976	MEAN 0.1075
ATN International, Inc. (ATNI)	ESM 0.0721	HOLT 0.0724	NAÏVE 0.0724	ESM3_E3 0.0726	HOLT3_E3 0.0728	NAIVE3_E3 0.0728	ESM_E3 0.0733	MEAN_E3 0.0738	M15_E3 0.0756	MEAN 0.0844	ARIMA_E3 0.0850	ARIMA 0.1415
SEMICONDUCTORS												
Texas Instruments Incorporated (TXN)	HOLT 0.0751	NAÏVE 0.0751	HOLT3_E3 0.0754	NAIVE3_E3 0.0754	ESM_E3 0.0769	ESM3_E3 0.0769	ESM 0.0778	M15_E3 0.0790	MEAN_E3 0.0879	MEAN 0.1065	ARIMA_E3 0.1252	ARIMA 0.2795
Intel Corporation (INTC)	NAÏVE 0.0519	ESM3_E3 0.0521	ESM 0.0522	NAIVE3_E3 0.0523	HOLT 0.0523	HOLT3_E3 0.0525	ESM_E3 0.0525	M15_E3 0.0549	ARIMA_E3 0.0592	MEAN_E3 0.0636	MEAN 0.0836	ARIMA 0.0855
Analog Devices, Inc. (ADI)	HOLT 0.0875	HOLT3_E3 0.0885	NAÏVE 0.0905	NAIVE3_E3 0.0910	M15_E3 0.0959	ESM_E3 0.0970	ESM3_E3 0.0994	MEAN_E3 0.1025	ESM 0.1031	MEAN 0.1225	ARIMA_E3 0.1418	ARIMA 0.3283
Semtech Corporation (SMTC)	NAÏVE 0.0607	HOLT 0.0607	NAIVE3_E3 0.0624	HOLT3_E3 0.0624	ESM3_E3 0.0662	ESM 0.0662	ESM_E3 0.0668	ARIMA_E3 0.0668	MEAN_E3 0.0681	M15_E3 0.0697	MEAN 0.0765	ARIMA 0.1765
Amtech Systems, Inc. (ASYS)	ARIMA_E3 0.0344	ESM_E3 0.0345	NAIVE3_E3 0.0346	M15_E3 0.0346	ESM3_E3 0.0346	HOLT3_E3 0.0346	ARIMA 0.0347	NAÏVE 0.0349	ESM 0.0349	HOLT 0.0349	MEAN_E3 0.0362	MEAN 0.0435
Maxim Integrated Products, Inc. (MXIM)	MEAN_E3 0.0772	M15_E3 0.0792	NAIVE3_E3 0.0797	HOLT3_E3 0.0798	NAÏVE 0.0804	HOLT 0.0805	ESM_E3 0.0821	ESM3_E3 0.0857	MEAN 0.0878	ESM 0.0905	ARIMA_E3 0.0919	ARIMA 0.1241
Microchip Technology Inc. (MCHP)	ARIMA 0.0572	ESM 0.0574	ESM3_E3 0.0575	HOLT 0.0575	NAÏVE 0.0576	HOLT3_E3 0.0578	NAIVE3_E3 0.0579	ESM_E3 0.0581	ARIMA_E3 0.0590	M15_E3 0.0609	MEAN_E3 0.0661	MEAN 0.0768

SECURITY	RMSE #1	RMSE #2	RMSE #3	RMSE #4	RMSE #5	RMSE #6	RMSE #7	RMSE #8	RMSE #9	RMSE #10	RMSE #11	RMSE #12
ELECTRICAL PRODUCTS												
Koninkijke Philips N.V. (PHG)	HOLT3_E3 0.0504	NAIVE3_E3 0.0504	HOLT 0.0505	NAÏVE 0.0505	ESM_E3 0.0508	ESM3_E3 0.0508	ESM 0.0512	M15_E3 0.0517	ARIMA_E3 0.0559	MEAN_E3 0.0559	MEAN 0.0674	ARIMA 0.0877
Cubic Corporation (CUB)	M15_E3 0.0607	HOLT3_E3 0.0609	NAIVE3_E3 0.0609	ARIMA_E3 0.0609	HOLT 0.0614	NAÏVE 0.0614	ESM_E3 0.0615	ARIMA 0.0625	ESM3_E3 0.0627	ESM 0.0643	MEAN_E3 0.0652	MEAN 0.0762
Bell Fuse Inc. (BELFA)	ESM_E3 0.0465	ESM3_E3 0.0467	NAIVE3_E3 0.0469	M15_E3 0.0472	HOLT3_E3 0.0474	ESM 0.0474	NAÏVE 0.0477	HOLT 0.0484	MEAN_E3 0.0489	MEAN 0.0594	ARIMA_E3 0.0824	ARIMA 0.1818
Bonso Electronics International, Inc. (BNSO)	NAÏVE 0.0159	HOLT 0.0159	ARIMA 0.0160	NAIVE3_E3 0.0162	HOLT3_E3 0.0162	ESM 0.0165	ESM3_E3 0.0167	ESM_E3 0.0171	ARIMA_E3 0.0171	M15_E3 0.0183	MEAN_E3 0.0195	MEAN 0.0222
Trimble Inc. (TRMB)	ESM_E3 0.0344	NAIVE3_E3 0.0345	HOLT3_E3 0.0345	ESM3_E3 0.0346	NAÏVE 0.0347	HOLT 0.0347	M15_E3 0.0348	ARIMA_E3 0.0349	ESM 0.0350	MEAN_E3 0.0370	ARIMA 0.0390	MEAN 0.0446
Kopin Corporation (KOPN)	HOLT 0.0362	HOLT3_E3 0.0376	NAÏVE 0.0388	NAIVE3_E3 0.0396	ARIMA_E3 0.0425	M15_E3 0.0428	ESM_E3 0.0449	ESM3_E3 0.0467	ESM 0.0491	MEAN_E3 0.0500	ARIMA 0.0586	MEAN 0.0604
Flex Ltd. (FLEX)	ARIMA_E3 0.0379	MEAN_E3 0.0405	NAIVE3_E3 0.0423	HOLT3_E3 0.0423	NAÏVE 0.0425	HOLT 0.0425	ESM_E3 0.0431	M15_E3 0.0431	ESM3_E3 0.0437	ESM 0.0448	MEAN 0.0468	ARIMA 0.0532
COMMUNICATIONS COMPONENT												
Corning Incorporated (GLW)	ARIMA_E3 0.0623	HOLT 0.0729	HOLT3_E3 0.0753	NAÏVE 0.0771	NAIVE3_E3 0.0787	M15_E3 0.0848	ESM_E3 0.0950	MEAN_E3 0.1004	ESM3_E3 0.1022	ARIMA 0.1057	ESM 0.1105	MEAN 0.1300
Communications Systems, Inc. (JCS)	HOLT 0.0184	NAÏVE 0.0185	HOLT3_E3 0.0186	NAIVE3_E3 0.0186	ARIMA 0.0187	ESM3_E3 0.0188	ESM 0.0188	ESM_E3 0.0189	ARIMA_E3 0.0191	M15_E3 0.0195	MEAN_E3 0.0202	MEAN 0.0235
Plantronics (PLT)	NAÏVE 0.0748	HOLT 0.0752	ARIMA 0.0753	NAIVE3_E3 0.0755	HOLT3_E3 0.0756	ARIMA_E3 0.0774	ESM_E3 0.0784	ESM3_E3 0.0785	ESM 0.0793	M15_E3 0.0799	MEAN_E3 0.0856	MEAN 0.1011
Technical Communications Corp. (TCCO)	ARIMA 0.0173	ARIMA_E3 0.0175	M15_E3 0.0180	NAIVE3_E3 0.0181	HOLT3_E3 0.0182	NAÏVE 0.0184	MEAN_E3 0.0184	HOLT 0.0184	ESM_E3 0.0189	ESM3_E3 0.0201	ESM 0.0217	MEAN 0.0218
SCIENTIFIC INSTRUMENTS												
MTS Systems Corporation (MTSC)	MEAN_E3 0.0682	HOLT3_E3 0.0702	NAIVE3_E3 0.0702	HOLT 0.0703	NAÏVE 0.0704	ESM3_E3 0.0705	ESM_E3 0.0705	ESM 0.0708	M15_E3 0.0716	ARIMA_E3 0.0723	MEAN 0.0744	ARIMA 0.0814
Kewaunee Scientific Corp. (KEQU)	HOLT 0.0200	NAÏVE 0.0200	ARIMA 0.0200	ESM 0.0201	HOLT3_E3 0.0201	NAIVE3_E3 0.0201	ESM3_E3 0.0202	ESM_E3 0.0205	ARIMA_E3 0.0208	M15_E3 0.0218	MEAN_E3 0.0246	MEAN 0.0312
PerkinElmer, Inc. (PKI)	NAÏVE 0.0673	NAIVE3_E3 0.0673	ESM_E3 0.0676	ESM3_E3 0.0678	HOLT3_E3 0.0679	MEAN_E3 0.0680	ESM 0.0682	HOLT 0.0682	M15_E3 0.0683	ARIMA_E3 0.0781	MEAN 0.0807	ARIMA 0.1035
Mixonix, Inc. (MSON)	ARIMA 0.0186	HOLT3_E3 0.0187	NAIVE3_E3 0.0187	HOLT 0.0187	NAÏVE 0.0187	ESM3_E3 0.0188	ESM_E3 0.0188	ARIMA_E3 0.0189	ESM 0.0189	M15_E3 0.0195	MEAN_E3 0.0203	MEAN 0.0242

B.10. Construction Sector

Table B-23: Construction Sector E3 MAPE Accuracy Results

SECURITY	MAPE #1	MAPE #2	MAPE #3	MAPE #4	MAPE #5	MAPE #6	MAPE #7	MAPE #8	MAPE #9	MAPE #10	MAPE #11	MAPE #12
BUILDING AND CONSTRUCTION PRODUCTS												
Masco Corporation (MAS)	ARIMA 9.5372	HOLT3_E3 9.5433	NAIVE3_E3 9.5435	HOLT 9.5612	NAÏVE 9.5613	ESM3_E3 9.6157	ESM_E3 9.6521	ESM 9.6653	ARIMA_E3 9.6716	M15_E3 9.9343	MEAN_E3 10.7959	MEAN 13.3594
USG Corporation (USG)	ESM3_E3 17.7796	NAIVE3_E3 17.7879	HOLT3_E3 17.7971	ESM 17.8015	NAÏVE 17.8151	ARIMA 17.8349	HOLT 17.8427	ESM_E3 17.8809	ARIMA_E3 17.9797	M15_E3 18.3856	MEAN_E3 20.8857	MEAN 26.9389
Patrick Industries, Inc. (PATK)	NAÏVE 17.8209	HOLT 17.9053	NAIVE3_E3 17.9673	ESM 18.0234	HOLT3_E3 18.0430	ESM3_E3 18.1348	ESM_E3 18.3760	M15_E3 19.2191	ARIMA_E3 20.5821	MEAN_E3 24.5358	ARIMA 25.3927	MEAN 32.1665
Aegion Corporation (AEGN)	ARIMA 11.2127	NAÏVE 11.2349	HOLT 11.2357	NAIVE3_E3 11.2639	HOLT3_E3 11.2648	ARIMA_E3 11.4444	ESM3_E3 11.5005	ESM_E3 11.5010	ESM 11.5926	M15_E3 11.7419	MEAN_E3 11.7913	MEAN 14.3282
CRH plc (CRH)	NAÏVE 7.1895	HOLT 7.1944	ARIMA 7.1945	NAIVE3_E3 7.2021	HOLT3_E3 7.2137	ARIMA_E3 7.3586	ESM3_E3 7.3855	ESM_E3 7.3899	ESM 7.4588	M15_E3 7.5858	MEAN_E3 8.2008	MEAN 10.0378
NCI Building Systems (NCS)	ARIMA 15.2338	HOLT3_E3 15.2947	NAIVE3_E3 15.3038	NAÏVE 15.3435	HOLT 15.3466	ARIMA_E3 15.4212	M15_E3 15.7993	ESM_E3 15.9087	ESM3_E3 16.4612	ESM 17.2938	MEAN_E3 17.8989	MEAN 23.5704
Gibraltar Industries (ROCK)	NAIVE3_E3 13.4153	NAÏVE 13.4336	HOLT3_E3 13.4429	HOLT 13.4579	ARIMA 13.5437	ARIMA_E3 13.5952	ESM_E3 13.6644	ESM3_E3 13.7176	M15_E3 13.8053	ESM 13.8513	MEAN_E3 14.2762	MEAN 17.0860
RESIDENTIAL AND COMMERCIAL BUILDING												
Lennar Corporation (LEN)	NAIVE3_E3 9.9962	HOLT3_E3 9.9964	NAÏVE 10.0325	HOLT 10.0337	ESM_E3 10.1131	ESM3_E3 10.1515	ESM 10.3508	ARIMA_E3 10.4847	M15_E3 10.5499	ARIMA 11.1496	MEAN_E3 12.0054	MEAN 15.1575
M.D.C. Holdings, Inc. (MDC)	NAÏVE 8.9443	HOLT 8.9469	NAIVE3_E3 9.0372	HOLT3_E3 9.0398	ESM_E3 9.2805	ESM3_E3 9.3107	ESM 9.5244	M15_E3 9.6951	ARIMA_E3 10.4203	MEAN_E3 10.5949	MEAN 12.8455	ARIMA 13.6869
NVR, Inc. (NVR)	ESM3_E3 8.3044	HOLT3_E3 8.3157	ESM_E3 8.3222	HOLT 8.3548	ESM 8.3804	NAIVE3_E3 8.3985	NAÏVE 8.4041	M15_E3 8.7474	MEAN_E3 10.3497	ARIMA_E3 10.6014	MEAN 13.0967	ARIMA 18.8004
PulteGroup, Inc. (PHM)	NAÏVE 11.3518	HOLT 11.3590	ESM 11.3648	NAIVE3_E3 11.4406	HOLT3_E3 11.4459	ESM3_E3 11.4493	ESM_E3 11.6079	ARIMA_E3 11.8530	M15_E3 12.0815	MEAN_E3 12.6676	ARIMA 13.1188	MEAN 15.1063
Toll Brothers, Inc. (TOL)	HOLT 9.2300	HOLT3_E3 9.2438	NAÏVE 9.2624	NAIVE3_E3 9.2689	ESM_E3 9.4744	ESM3_E3 9.4935	ESM 9.6018	ARIMA_E3 9.6434	M15_E3 9.6496	MEAN_E3 10.5213	ARIMA 10.5737	MEAN 13.0062
KB Home (KBH)	NAÏVE 11.5299	HOLT 11.5391	ESM 11.5512	NAIVE3_E3 11.7063	HOLT3_E3 11.7126	ESM3_E3 11.7270	ESM_E3 11.9800	ARIMA_E3 12.2359	ARIMA 12.4757	M15_E3 12.6911	MEAN_E3 14.0257	MEAN 17.3160
D.R. Horton, Inc. (DHI)	NAÏVE 10.7533	HOLT 10.7538	HOLT3_E3 10.7683	NAIVE3_E3 10.7695	ESM3_E3 10.8543	ESM 10.8787	ESM_E3 10.9217	M15_E3 11.2358	ARIMA_E3 11.5123	MEAN_E3 12.3052	ARIMA 12.5433	MEAN 15.0436

Table B-24: Construction Sector E3 RMSE Accuracy Results

SECURITY	RMSE #1	RMSE #2	RMSE #3	RMSE #4	RMSE #5	RMSE #6	RMSE #7	RMSE #8	RMSE #9	RMSE #10	RMSE #11	RMSE #12
BUILDING AND CONSTRUCTION PRODUCTS												
Masco Corporation (MAS)	HOLT 0.0318	NAÏVE 0.0318	HOLT3_E3 0.0319	NAIVE3_E3 0.0319	ARIMA 0.0319	ESM3_E3 0.0320	ESM 0.0320	ESM_E3 0.0322	ARIMA_E3 0.0323	M15_E3 0.0331	MEAN_E3 0.0354	MEAN 0.0420
USG Corporation (USG)	ESM3_E3 0.0906	NAIVE3_E3 0.0906	ESM_E3 0.0907	HOLT3_E3 0.0907	ARIMA_E3 0.0909	ESM 0.0909	NAÏVE 0.0909	ARIMA 0.0910	HOLT 0.0911	M15_E3 0.0920	MEAN_E3 0.1046	MEAN 0.1271
Patrick Industries, Inc. (PATK)	HOLT 0.0432	NAÏVE 0.0433	HOLT3_E3 0.0435	NAIVE3_E3 0.0436	ESM 0.0437	ESM3_E3 0.0438	ESM_E3 0.0442	M15_E3 0.0459	MEAN_E3 0.0491	MEAN 0.0575	ARIMA_E3 0.0609	ARIMA 0.1398
Aegion Corporation (AEGN)	ARIMA 0.0436	NAIVE3_E3 0.0438	HOLT3_E3 0.0438	NAÏVE 0.0440	HOLT 0.0440	ARIMA_E3 0.0440	ESM_E3 0.0443	ESM3_E3 0.0445	M15_E3 0.0448	MEAN_E3 0.0451	ESM 0.0452	MEAN 0.0549
CRH plc (CRH)	ARIMA 0.0337	NAIVE3_E3 0.0342	NAÏVE 0.0342	HOLT3_E3 0.0342	HOLT 0.0343	ARIMA_E3 0.0344	ESM_E3 0.0348	ESM3_E3 0.0349	ESM 0.0353	M15_E3 0.0353	MEAN_E3 0.0388	MEAN 0.0480
NCI Building Systems (NCS)	ARIMA 0.2530	HOLT 0.2542	NAÏVE 0.2547	HOLT3_E3 0.2560	NAIVE3_E3 0.2564	ARIMA_E3 0.2599	ESM_E3 0.2609	ESM3_E3 0.2617	ESM 0.2650	M15_E3 0.2665	MEAN_E3 0.2930	MEAN 0.3585
Gibraltar Industries (ROCK)	NAIVE3_E3 0.0448	NAÏVE 0.0448	HOLT3_E3 0.0448	HOLT 0.0448	ARIMA 0.0450	ARIMA_E3 0.0453	ESM_E3 0.0455	ESM3_E3 0.0456	M15_E3 0.0460	ESM 0.0460	MEAN_E3 0.0474	MEAN 0.0538
RESIDENTIAL AND COMMERCIAL BUILDING												
Lennar Corporation (LEN)	NAÏVE 0.0539	HOLT 0.0539	HOLT3_E3 0.0542	NAIVE3_E3 0.0542	ESM_E3 0.0554	ESM3_E3 0.0557	ESM 0.0567	M15_E3 0.0574	ARIMA_E3 0.0602	MEAN_E3 0.0642	ARIMA 0.0792	MEAN 0.0797
M.D.C. Holdings, Inc. (MDC)	NAÏVE 0.0550	HOLT 0.0551	NAIVE3_E3 0.0553	HOLT3_E3 0.0553	ESM_E3 0.0562	ESM3_E3 0.0565	ESM 0.0574	M15_E3 0.0580	MEAN_E3 0.0624	ARIMA_E3 0.0720	MEAN 0.0742	ARIMA 0.1311
NVR, Inc. (NVR)	M15_E3 1.6933	NAIVE3_E3 1.6972	HOLT3_E3 1.7159	NAÏVE 1.7185	ESM_E3 1.7215	HOLT 1.7500	ESM3_E3 1.7653	ESM 1.8272	MEAN_E3 1.9789	MEAN 2.4519	ARIMA_E3 3.2029	ARIMA 7.5166
PulteGroup, Inc. (PHM)	NAÏVE 0.0342	HOLT 0.0342	ESM 0.0343	NAIVE3_E3 0.0346	HOLT3_E3 0.0346	ESM3_E3 0.0346	ESM_E3 0.0352	ARIMA_E3 0.0367	M15_E3 0.0367	MEAN_E3 0.0396	MEAN 0.0483	ARIMA 0.0584
Toll Brothers, Inc. (TOL)	HOLT 0.0424	HOLT3_E3 0.0425	NAÏVE 0.0427	NAIVE3_E3 0.0429	ESM_E3 0.0433	ESM3_E3 0.0433	ESM 0.0438	M15_E3 0.0447	ARIMA_E3 0.0472	MEAN_E3 0.0510	MEAN 0.0631	ARIMA 0.0669
KB Home (KBH)	HOLT 0.0517	NAÏVE 0.0517	ESM 0.0519	HOLT3_E3 0.0522	NAIVE3_E3 0.0522	ESM3_E3 0.0523	ESM_E3 0.0531	M15_E3 0.0555	ARIMA_E3 0.0584	MEAN_E3 0.0648	MEAN 0.0823	ARIMA 0.0887
D.R. Horton, Inc. (DHI)	HOLT 0.0391	NAÏVE 0.0391	HOLT3_E3 0.0394	NAIVE3_E3 0.0394	ESM3_E3 0.0398	ESM 0.0398	ESM_E3 0.0400	M15_E3 0.0414	MEAN_E3 0.0443	ARIMA_E3 0.0445	MEAN 0.0527	ARIMA 0.0584

B.11. Consumer Discretionary Sector

Table B-25: Consumer Discretionary E3 MAPE Accuracy Results

SECURITY	MAPE #1	MAPE #2	MAPE #3	MAPE #4	MAPE #5	MAPE #6	MAPE #7	MAPE #8	MAPE #9	MAPE #10	MAPE #11	MAPE #12
LEISURE												
The Marcus Corporation (MCS)	ARIMA_E3 8.7920	ESM_E3 8.8247	HOLT3_E3 8.8738	ARIMA 8.8790	NAIVE3_E3 8.9037	ESM3_E3 8.9344	M15_E3 8.9664	HOLT 9.0077	NAÏVE 9.0577	ESM 9.1531	MEAN_E3 9.2213	MEAN 10.9104
Cedar Fair, L.P. (FUN)	HOLT 6.5667	HOLT3_E3 6.5821	NAÏVE 6.5884	NAIVE3_E3 6.5966	ESM3_E3 6.6158	ESM 6.6398	ESM_E3 6.6440	ARIMA 6.6736	ARIMA_E3 6.7067	M15_E3 6.8539	MEAN_E3 7.7170	MEAN 9.4845
Carnival Corporation (CCL)	HOLT3_E3 7.7764	NAIVE3_E3 7.7766	ESM3_E3 7.7865	ESM_E3 7.7906	NAÏVE 7.8178	HOLT 7.8210	ARIMA_E3 7.8380	ESM 7.8503	ARIMA 7.8586	M15_E3 7.9783	MEAN_E3 9.0235	MEAN 11.6124
Reading International, Inc. (RDI)	ESM 7.6651	HOLT 7.6659	NAÏVE 7.6660	HOLT3_E3 7.7390	NAIVE3_E3 7.7396	ESM3_E3 7.7522	ESM_E3 7.9073	M15_E3 8.3718	MEAN_E3 9.2042	ARIMA_E3 10.7479	MEAN 11.5887	ARIMA 19.6823
Royal Caribbean Cruises, Ltd. (RCL)	ESM_E3 13.3091	NAIVE3_E3 13.3391	ESM3_E3 13.3399	HOLT3_E3 13.3553	ARIMA_E3 13.3829	NAÏVE 13.4333	ARIMA 13.4550	HOLT 13.4631	ESM 13.4952	M15_E3 13.5851	MEAN_E3 15.1447	MEAN 19.1067
TEXTILES												
V.F. Corporation (VFC)	ESM3_E3 6.4095	ESM 6.4203	ESM_E3 6.4580	HOLT3_E3 6.5241	NAIVE3_E3 6.5266	HOLT 6.5320	NAÏVE 6.5360	M15_E3 6.7119	ARIMA_E3 7.0001	MEAN_E3 7.4309	ARIMA 8.4428	MEAN 9.1405
Oxford Industries, Inc. (OXM)	HOLT3_E3 13.7215	HOLT 13.7225	NAÏVE 13.7404	ESM3_E3 13.7734	NAIVE3_E3 13.7778	ESM 13.8588	ESM_E3 13.8912	ARIMA_E3 14.3001	M15_E3 14.5173	ARIMA 14.6712	MEAN_E3 15.7034	MEAN 18.2599
Interface, Inc. (TILE)	HOLT 13.3819	ARIMA 13.3833	NAÏVE 13.3848	ESM 13.3932	ESM3_E3 13.4027	NAIVE3_E3 13.4108	HOLT3_E3 13.4115	ESM_E3 13.5260	ARIMA_E3 13.6383	M15_E3 14.0371	MEAN_E3 16.5343	MEAN 20.5301
Culp, Inc. (CULP)	ESM 13.1837	ESM3_E3 13.2486	ARIMA 13.2488	NAÏVE 13.2918	NAIVE3_E3 13.3254	HOLT 13.3978	HOLT3_E3 13.4131	ESM_E3 13.4222	ARIMA_E3 13.5771	M15_E3 14.0412	MEAN_E3 16.0768	MEAN 20.6370
PVH Corp (PVH)	ESM3_E3 9.4644	ESM_E3 9.5020	NAIVE3_E3 9.5239	HOLT3_E3 9.5343	ESM 9.5356	NAÏVE 9.5922	HOLT 9.6133	M15_E3 9.8234	ARIMA_E3 9.9889	ARIMA 10.9674	MEAN_E3 11.6066	MEAN 14.7761
G-III Apparel Group, Ltd. (GIII)	NAÏVE 12.9340	HOLT 12.9633	ESM 13.0078	NAIVE3_E3 13.0280	HOLT3_E3 13.1085	ESM3_E3 13.1363	ESM_E3 13.4183	ARIMA_E3 14.3179	M15_E3 14.3548	ARIMA 16.6041	MEAN_E3 16.7540	MEAN 21.5521
The Dixie Group (DXYN)	NAÏVE 15.4309	ARIMA 15.4410	HOLT 15.4411	NAIVE3_E3 15.6110	HOLT3_E3 15.6326	ARIMA_E3 16.2350	ESM3_E3 16.2957	ESM 16.3247	ESM_E3 16.4349	M15_E3 17.0662	MEAN_E3 20.9882	MEAN 26.9382
SCHOOLS												
Graham Holdings Company (GHC)	NAIVE3_E3 5.7956	HOLT3_E3 5.8058	NAÏVE 5.8296	HOLT 5.8446	ARIMA_E3 5.8536	ESM_E3 5.8548	ARIMA 5.9084	ESM3_E3 5.9164	M15_E3 6.0267	ESM 6.0450	MEAN_E3 6.8698	MEAN 8.5837
GP Strategies Corporation (GPX)	NAIVE3_E3 10.1468	HOLT3_E3 10.1469	NAÏVE 10.1732	HOLT 10.1736	ARIMA 10.1752	ARIMA_E3 10.3355	ESM_E3 10.4515	ESM3_E3 10.7013	M15_E3 10.7261	ESM 11.2499	MEAN_E3 11.7502	MEAN 14.5517
Adtalem Global Education (ATGE)	NAIVE3_E3 10.9833	HOLT3_E3 11.0002	NAÏVE 11.0202	HOLT 11.0320	ARIMA 11.1826	ARIMA_E3 11.2201	ESM_E3 11.4447	M15_E3 11.4823	ESM3_E3 11.6302	ESM 11.9069	MEAN_E3 12.9045	MEAN 15.4043

SECURITY	MAPE #1	MAPE #2	MAPE #3	MAPE #4	MAPE #5	MAPE #6	MAPE #7	MAPE #8	MAPE #9	MAPE #10	MAPE #11	MAPE #12
FURNITURE MANUFACTURERS												
Leggett & Platt, Incorporated (LEG)	NAÏVE 7.1270	ARIMA 7.1358	HOLT 7.1358	NAIVE3_E3 7.1471	HOLT3_E3 7.1505	ESM3_E3 7.2005	ESM 7.2365	ESM_E3 7.2467	ARIMA_E3 7.2876	M15_E3 7.5327	MEAN_E3 7.8553	MEAN 9.4163
La-Z-Boy Incorporated (LZB)	NAÏVE 13.2283	HOLT 13.2574	ARIMA 13.2837	NAIVE3_E3 13.3708	ESM 13.3833	HOLT3_E3 13.3965	ESM3_E3 13.4252	ESM_E3 13.6180	ARIMA_E3 13.8188	M15_E3 14.3716	MEAN_E3 16.8229	MEAN 21.0107
Kimball International, Inc. (KBAL)	ARIMA_E3 9.9766	M15_E3 10.0264	ESM_E3 10.0622	HOLT3_E3 10.0983	NAIVE3_E3 10.1649	ESM3_E3 10.2016	ARIMA 10.2465	HOLT 10.2684	NAÏVE 10.3797	ESM 10.4096	MEAN_E3 10.7500	MEAN 12.9975
Bassett Furniture Industries Inc. (BSET)	NAÏVE 13.1292	ESM 13.1546	HOLT 13.1627	ARIMA 13.1888	ESM3_E3 13.2469	NAIVE3_E3 13.2495	HOLT3_E3 13.2899	ESM_E3 13.4935	ARIMA_E3 13.7080	M15_E3 14.3878	MEAN_E3 15.8383	MEAN 19.0029
Flexsteel Industies, Inc. (FLXS)	HOLT 8.8737	ARIMA 8.8878	NAÏVE 8.9047	NAIVE3_E3 8.9251	HOLT3_E3 8.9290	ESM 8.9466	ESM3_E3 8.9753	ESM_E3 9.0802	ARIMA_E3 9.1439	M15_E3 9.4519	MEAN_E3 9.6869	MEAN 11.2930
Virco Mfg. Corporation (VIRC)	NAIVE3_E3 9.1700	HOLT3_E3 9.2017	NAÏVE 9.2358	ESM3_E3 9.2567	ESM_E3 9.2610	HOLT 9.3061	ARIMA_E3 9.3457	ESM 9.3528	ARIMA 9.4241	M15_E3 9.6143	MEAN_E3 11.4033	MEAN 15.0312
American Woodmark Corp. (AMWD)	NAÏVE 11.4845	HOLT 11.5248	ESM 11.5305	NAIVE3_E3 11.5481	HOLT3_E3 11.5836	ESM3_E3 11.5965	ESM_E3 11.7299	M15_E3 12.1626	ARIMA_E3 12.4413	MEAN_E3 13.2789	ARIMA 13.4789	MEAN 15.9270

Table B-26: Consumer Discretionary E3 RMSE Accuracy Results

SECURITY	RMSE #1	RMSE #2	RMSE #3	RMSE #4	RMSE #5	RMSE #6	RMSE #7	RMSE #8	RMSE #9	RMSE #10	RMSE #11	RMSE #12
LEISURE												
The Marcus Corporation (MCS)	ESM_E3 0.0294	ARIMA_E3 0.0294	HOLT3_E3 0.0296	NAIVE3_E3 0.0296	ARIMA 0.0297	ESM3_E3 0.0297	M15_E3 0.0298	HOLT 0.0300	NAÏVE 0.0301	ESM 0.0304	MEAN_E3 0.0316	MEAN 0.0377
Cedar Fair, L.P. (FUN)	ESM3_E3 0.0374	ESM 0.0374	HOLT 0.0375	NAÏVE 0.0376	HOLT3_E3 0.0376	NAIVE3_E3 0.0376	ARIMA 0.0377	ESM_E3 0.0377	ARIMA_E3 0.0381	M15_E3 0.0392	MEAN_E3 0.0426	MEAN 0.0509
Carnival Corporation (CCL)	NAIVE3_E3 0.0579	HOLT3_E3 0.0579	NAÏVE 0.0581	HOLT 0.0581	ESM3_E3 0.0582	ESM_E3 0.0582	ESM 0.0586	ARIMA_E3 0.0586	ARIMA 0.0588	M15_E3 0.0594	MEAN_E3 0.0633	MEAN 0.0772
Reading International, Inc. (RDI)	ESM 0.0113	HOLT 0.0113	NAÏVE 0.0114	HOLT3_E3 0.0114	NAIVE3_E3 0.0114	ESM3_E3 0.0114	ESM_E3 0.0115	M15_E3 0.0119	MEAN_E3 0.0130	MEAN 0.0155	ARIMA_E3 0.0199	ARIMA 0.0482
Royal Caribbean Cruises, Ltd. (RCL)	NAIVE3_E3 0.0998	HOLT3_E3 0.0999	ARIMA_E3 0.0999	M15_E3 0.1003	ESM_E3 0.1005	NAÏVE 0.1007	HOLT 0.1008	ARIMA 0.1020	ESM3_E3 0.1023	MEAN_E3 0.1036	ESM 0.1052	MEAN 0.1204
TEXTILES												
V.F. Corporation (VFC)	ESM 0.0460	ESM3_E3 0.0461	ESM_E3 0.0464	NAÏVE 0.0465	HOLT 0.0465	NAIVE3_E3 0.0466	HOLT3_E3 0.0466	M15_E3 0.0478	MEAN_E3 0.0509	ARIMA_E3 0.0589	MEAN 0.0590	ARIMA 0.0942
Oxford Industries, Inc. (OXM)	NAÏVE 0.0808	HOLT 0.0809	ESM3_E3 0.0810	HOLT3_E3 0.0810	NAIVE3_E3 0.0811	ESM 0.0813	ESM_E3 0.0820	ARIMA_E3 0.0863	M15_E3 0.0878	MEAN_E3 0.0926	ARIMA 0.0953	MEAN 0.1085
Interface, Inc. (TILE)	HOLT3_E3 0.0270	NAIVE3_E3 0.0271	HOLT 0.0271	ESM3_E3 0.0271	NAÏVE 0.0271	ARIMA 0.0272	ESM_E3 0.0272	ESM 0.0272	ARIMA_E3 0.0273	M15_E3 0.0279	MEAN_E3 0.0304	MEAN 0.0362
Culp, Inc. (CULP)	NAIVE3_E3 0.0261	ESM_E3 0.0261	ESM3_E3 0.0261	ARIMA_E3 0.0262	ARIMA 0.0262	NAÏVE 0.0263	ESM 0.0264	HOLT3_E3 0.0266	M15_E3 0.0268	HOLT 0.0270	MEAN_E3 0.0297	MEAN 0.0373
PVH Corp (PVH)	NAIVE3_E3 0.1143	HOLT3_E3 0.1143	ESM3_E3 0.1144	NAÏVE 0.1145	HOLT 0.1146	ESM_E3 0.1149	ESM 0.1149	M15_E3 0.1187	ARIMA_E3 0.1234	MEAN_E3 0.1296	MEAN 0.1544	ARIMA 0.1577

APPENDIX B: E3 Forecast Study Results

SECURITY	RMSE #1	RMSE #2	RMSE #3	RMSE #4	RMSE #5	RMSE #6	RMSE #7	RMSE #8	RMSE #9	RMSE #10	RMSE #11	RMSE #12
G-III Apparel Group, Ltd. (GIII)	ARIMA_E3 0.0462	NAIVE3_E3 0.0476	HOLT3_E3 0.0476	NAÏVE 0.0477	HOLT 0.0477	ESM3_E3 0.0480	ESM_E3 0.0480	ESM 0.0483	M15_E3 0.0491	MEAN_E3 0.0546	MEAN 0.0691	ARIMA 0.0784
The Dixie Group (DXYN)	ARIMA 0.0190	NAÏVE 0.0190	HOLT 0.0191	ESM 0.0191	NAIVE3_E3 0.0192	HOLT3_E3 0.0193	ESM3_E3 0.0193	ESM_E3 0.0196	ARIMA_E3 0.0198	M15_E3 0.0207	MEAN_E3 0.0225	MEAN 0.0272
SCHOOLS												
Graham Holdings Company (GHC)	NAIVE3_E3 0.4601	HOLT3_E3 0.4605	NAÏVE 0.4612	HOLT 0.4618	ESM_E3 0.4630	ARIMA_E3 0.4630	ESM3_E3 0.4645	ARIMA 0.4645	ESM 0.4693	M15_E3 0.4704	MEAN_E3 0.5331	MEAN 0.6505
GP Strategies Corporation (GPX)	NAÏVE 0.0271	HOLT 0.0271	ARIMA 0.0272	NAIVE3_E3 0.0273	HOLT3_E3 0.0273	ARIMA_E3 0.0279	ESM_E3 0.0281	ESM3_E3 0.0282	ESM 0.0286	M15_E3 0.0288	MEAN_E3 0.0291	MEAN 0.0339
Adtalem Global Education (ATGE)	NAIVE3_E3 0.0726	HOLT3_E3 0.0727	NAÏVE 0.0729	HOLT 0.0730	ARIMA_E3 0.0734	ARIMA 0.0736	M15_E3 0.0745	ESM_E3 0.0754	ESM3_E3 0.0770	MEAN_E3 0.0787	ESM 0.0793	MEAN 0.0914
FURNITURE MANUFACTURERS												
Leggett & Platt, Incorporated (LEG)	NAÏVE 0.0345	HOLT 0.0345	ARIMA 0.0346	NAIVE3_E3 0.0346	HOLT3_E3 0.0346	ESM3_E3 0.0349	ESM_E3 0.0351	ESM 0.0352	ARIMA_E3 0.0352	M15_E3 0.0364	MEAN_E3 0.0372	MEAN 0.0433
La-Z-Boy Incorporated (LZB)	ARIMA 0.0333	HOLT 0.0333	NAÏVE 0.0333	HOLT3_E3 0.0334	NAIVE3_E3 0.0334	ESM3_E3 0.0339	ESM_E3 0.0340	ARIMA_E3 0.0340	ESM 0.0341	MEAN_E3 0.0349	M15_E3 0.0352	MEAN 0.0413
Kimball International, Inc. (KBAL)	ARIMA_E3 0.0198	ESM_E3 0.0199	HOLT3_E3 0.0199	M15_E3 0.0200	NAIVE3_E3 0.0200	ESM3_E3 0.0201	ARIMA 0.0202	HOLT 0.0202	NAÏVE 0.0203	ESM 0.0204	MEAN_E3 0.0213	MEAN 0.0254
Bassett Furniture Industries Inc. (BSET)	NAÏVE 0.0333	ESM 0.0333	HOLT 0.0334	ESM3_E3 0.0334	NAIVE3_E3 0.0334	ARIMA 0.0334	HOLT3_E3 0.0335	ESM_E3 0.0340	ARIMA_E3 0.0345	M15_E3 0.0365	MEAN_E3 0.0378	MEAN 0.0434
Flexsteel Industies, Inc. (FLXS)	MEAN_E3 0.0458	NAÏVE 0.0479	NAIVE3_E3 0.0481	HOLT 0.0483	HOLT3_E3 0.0484	ARIMA 0.0485	ESM3_E3 0.0488	ESM 0.0488	ESM_E3 0.0490	ARIMA_E3 0.0491	M15_E3 0.0500	MEAN 0.0507
Virco Mfg. Corporation (VIRC)	NAIVE3_E3 0.0104	HOLT3_E3 0.0105	NAÏVE 0.0105	ESM3_E3 0.0106	ESM_E3 0.0106	ARIMA_E3 0.0106	HOLT 0.0106	ARIMA 0.0107	ESM 0.0107	M15_E3 0.0110	MEAN_E3 0.0120	MEAN 0.0150
American Woodmark Corp. (AMWD)	NAÏVE 0.0825	HOLT 0.0826	ESM 0.0831	NAIVE3_E3 0.0836	HOLT3_E3 0.0837	ESM3_E3 0.0841	ESM_E3 0.0855	M15_E3 0.0893	ARIMA_E3 0.0947	MEAN_E3 0.0980	MEAN 0.1218	ARIMA 0.1272

B.12. Consumer Staples Sector

Table B-27: Consumer Staples Sector E3 MAPE Accuracy Results

SECURITY	MAPE #1	MAPE #2	MAPE #3	MAPE #4	MAPE #5	MAPE #6	MAPE #7	MAPE #8	MAPE #9	MAPE #10	MAPE #11	MAPE #12
SOFT DRINKS												
The Coca-Cola Company (KO)	HOLT3_E3 4.4860	NAIVE3_E3 4.4872	HOLT 4.4990	NAÏVE 4.5004	ARIMA 4.5230	ARIMA_E3 4.5287	ESM_E3 4.5697	M15_E3 4.6245	ESM3_E3 4.6770	ESM 4.8797	MEAN_E3 5.0086	MEAN 6.2218
PepsiCo, Inc. (PEP)	ARIMA 4.4485	HOLT 4.5005	NAÏVE 4.5089	HOLT3_E3 4.5563	NAIVE3_E3 4.5591	ESM3_E3 4.5713	ESM_E3 4.5843	ESM 4.6773	ARIMA_E3 4.6825	M15_E3 4.8968	MEAN_E3 4.9694	MEAN 5.5439
Coca-Cola European Partners (CCEP)	HOLT 7.9007	ESM 7.9092	NAÏVE 7.9103	ARIMA 7.9170	HOLT3_E3 7.9743	ESM3_E3 7.9915	NAIVE3_E3 7.9918	ESM_E3 8.1354	ARIMA_E3 8.2260	M15_E3 8.5490	MEAN_E3 8.8478	MEAN 10.5138
National Beverage Corp. (FIZZ)	ESM3_E3 8.3052	ESM_E3 8.3364	NAIVE3_E3 8.3537	ESM 8.3687	NAÏVE 8.4196	HOLT3_E3 8.4311	HOLT 8.5748	M15_E3 8.6473	MEAN_E3 9.2120	ARIMA_E3 9.6565	MEAN 11.4322	ARIMA 12.8036
Cott Corporation (COT)	HOLT3_E3 12.2550	HOLT 12.2732	ARIMA 12.3270	NAIVE3_E3 12.3349	NAÏVE 12.3686	ARIMA_E3 12.4495	ESM_E3 12.5735	M15_E3 12.7872	ESM3_E3 12.8603	ESM 13.3788	MEAN_E3 16.0269	MEAN 20.6629
ALCOHOLIC BEVERAGES												
Brown-Forman Corporation (BF-A)	ESM3_E3 5.3746	HOLT3_E3 5.3824	ESM_E3 5.3911	NAIVE3_E3 5.4007	HOLT 5.4079	ESM 5.4125	NAÏVE 5.4271	M15_E3 5.5517	MEAN_E3 6.0300	ARIMA_E3 7.1064	MEAN 7.3936	ARIMA 11.8273
Molson Coors Brewing Company (TAP)	HOLT3_E3 6.7656	NAIVE3_E3 6.7698	ESM_E3 6.7917	ESM3_E3 6.8004	HOLT 6.8055	NAÏVE 6.8127	ESM 6.8634	M15_E3 6.9095	ARIMA_E3 7.1579	MEAN_E3 7.3116	ARIMA 7.7671	MEAN 8.8696
Constellation Brans, Inc. (STZ)	ESM 7.8615	NAÏVE 7.8679	HOLT 7.8783	ESM3_E3 7.9288	NAIVE3_E3 7.9650	HOLT3_E3 7.9746	ESM_E3 8.0748	M15_E3 8.6443	MEAN_E3 9.3871	ARIMA_E3 9.6415	MEAN 11.0387	ARIMA 13.1527
Compañia Cervecerias Unidas S.A. (CCU)	ARIMA_E3 6.9874	NAIVE3_E3 6.9925	ESM_E3 6.9999	HOLT3_E3 7.0371	NAÏVE 7.0762	ESM3_E3 7.0775	M15_E3 7.0950	ARIMA 7.1120	HOLT 7.1758	ESM 7.2437	MEAN_E3 7.6235	MEAN 9.3280
Willamette Valley Vineyards, Inc. (WVVI)	HOLT3_E3 7.6817	HOLT 7.7168	ESM_E3 7.7959	ESM3_E3 7.8013	NAIVE3_E3 7.8174	ESM 7.8954	NAÏVE 7.9455	M15_E3 8.0161	ARIMA_E3 8.0430	ARIMA 8.5869	MEAN_E3 8.6081	MEAN 10.6927
TOBACCO PRODUCTS												
Altria Group, Inc. (MO)	ESM 6.1163	NAÏVE 6.2874	HOLT 6.4090	ARIMA 6.6549	ESM3_E3 6.7228	NAIVE3_E3 6.8927	HOLT3_E3 6.9887	ESM_E3 7.4186	ARIMA_E3 8.0512	M15_E3 9.0543	MEAN_E3 10.7667	MEAN 13.5035
British American Tobacco p.l.c. (BTI)	ESM 5.7321	NAÏVE 5.7653	HOLT 5.7660	ESM3_E3 5.8304	NAIVE3_E3 5.8562	HOLT3_E3 5.8589	ESM_E3 5.9771	M15_E3 6.3825	MEAN_E3 7.0810	ARIMA_E3 7.1285	MEAN 8.7161	ARIMA 11.0103
Vector Group Ltd. (VGR)	NAÏVE 7.5813	HOLT 7.6200	NAIVE3_E3 7.6237	HOLT3_E3 7.6412	ESM3_E3 7.6479	ESM 7.6524	ESM_E3 7.7492	M15_E3 8.1499	ARIMA_E3 8.2750	MEAN_E3 8.8648	ARIMA 8.8841	MEAN 10.8477
Universal Corporation (UVV)	NAIVE3_E3 7.2623	HOLT3_E3 7.2720	ESM3_E3 7.2807	NAÏVE 7.3117	ESM_E3 7.3117	ARIMA 7.3198	HOLT 7.3240	ESM 7.3392	ARIMA_E3 7.3599	M15_E3 7.6013	MEAN_E3 8.6112	MEAN 11.0832
Schweitzer-Mauduit Intl., Inc. (SWM)	ARIMA 9.7176	NAÏVE 9.7388	HOLT 9.7599	ESM 9.7704	NAIVE3_E3 9.8765	HOLT3_E3 9.8808	ESM3_E3 9.8999	ARIMA_E3 10.0359	ESM_E3 10.1171	MEAN_E3 10.5591	M15_E3 10.7850	MEAN 12.1291

SECURITY	MAPE #1	MAPE #2	MAPE #3	MAPE #4	MAPE #5	MAPE #6	MAPE #7	MAPE #8	MAPE #9	MAPE #10	MAPE #11	MAPE #12
FOOD ITEMS												
Sysco Corporation (SYY)	ESM_E3 5.6098	HOLT3_E3 5.6105	NAIVE3_E3 5.6153	ESM3_E3 5.6266	HOLT 5.6639	NAÏVE 5.6742	ESM 5.7004	M15_E3 5.7410	MEAN_E3 5.7812	MEAN 6.9035	ARIMA_E3 7.5810	ARIMA 13.4759
General Mills, Inc. (GIS)	HOLT 3.7031	NAÏVE 3.7046	NAIVE3_E3 3.7095	HOLT3_E3 3.7097	ESM3_E3 3.7363	ESM 3.7760	ESM_E3 3.7903	M15_E3 4.0857	ARIMA_E3 4.2292	MEAN_E3 4.6571	ARIMA 4.9050	MEAN 5.8997
Kellogg Company (K)	ESM3_E3 4.5749	ESM 4.5905	HOLT 4.6078	NAÏVE 4.6145	ESM_E3 4.6158	HOLT3_E3 4.6186	NAIVE3_E3 4.6187	ARIMA 4.6187	ARIMA_E3 4.7061	M15_E3 4.8558	MEAN_E3 5.2081	MEAN 6.1757
McCormick & Company, Inc. (MKC)	ESM_E3 4.4437	ESM3_E3 4.4476	NAIVE3_E3 4.4576	HOLT3_E3 4.4582	HOLT 4.4838	NAÏVE 4.4857	ESM 4.5443	M15_E3 4.7045	MEAN_E3 5.3872	ARIMA_E3 6.6069	MEAN 6.7833	ARIMA 11.6031
Conagra Brands, Inc. (CAG)	HOLT 5.6467	NAÏVE 5.6468	ESM3_E3 5.6485	ARIMA 5.6499	NAIVE3_E3 5.6604	HOLT3_E3 5.6607	ESM 5.6638	ESM_E3 5.7139	ARIMA_E3 5.7930	M15_E3 6.0341	MEAN_E3 6.1731	MEAN 7.6344
Campbell Soup Company (CPG)	HOLT3_E3 5.0636	NAIVE3_E3 5.0691	ARIMA_E3 5.0888	HOLT 5.1079	NAÏVE 5.1202	ARIMA 5.1229	ESM_E3 5.1640	M15_E3 5.1810	ESM3_E3 5.2296	ESM 5.3430	MEAN_E3 5.6277	MEAN 7.1908
Seaboard Corporation (SEB)	NAIVE3_E3 8.5520	HOLT3_E3 8.5712	NAÏVE 8.5863	HOLT 8.6764	M15_E3 8.9238	ESM_E3 8.9341	ESM3_E3 9.1441	ESM 9.4858	ARIMA_E3 9.8195	MEAN_E3 10.1415	MEAN 12.8411	ARIMA 13.4104

Table B-28: Consumer Staples Sector E3 RMSE Accuracy Results

SECURITY	RMSE #1	RMSE #2	RMSE #3	RMSE #4	RMSE #5	RMSE #6	RMSE #7	RMSE #8	RMSE #9	RMSE #10	RMSE #11	RMSE #12
SOFT DRINKS												
The Coca-Cola Company (KO)	HOLT3_E3 0.0251	NAIVE3_E3 0.0251	HOLT 0.0252	NAÏVE 0.0252	ARIMA_E3 0.0253	ARIMA 0.0254	ESM_E3 0.0257	M15_E3 0.0257	ESM3_E3 0.0266	MEAN_E3 0.0271	ESM 0.0280	MEAN 0.0330
PepsiCo, Inc. (PEP)	ARIMA 0.0576	HOLT 0.0577	NAÏVE 0.0578	HOLT3_E3 0.0579	NAIVE3_E3 0.0580	ARIMA_E3 0.0588	ESM_E3 0.0592	M15_E3 0.0605	ESM3_E3 0.0606	MEAN_E3 0.0612	ESM 0.0632	MEAN 0.0684
Coca-Cola European Partners (CCEP)	ARIMA 0.0404	ESM 0.0404	HOLT 0.0405	NAÏVE 0.0405	HOLT3_E3 0.0405	ESM3_E3 0.0405	NAIVE3_E3 0.0405	ESM_E3 0.0407	ARIMA_E3 0.0409	MEAN_E3 0.0418	M15_E3 0.0418	MEAN 0.0489
National Beverage Corp. (FIZZ)	M15_E3 0.0780	ESM_E3 0.0795	ESM3_E3 0.0808	NAIVE3_E3 0.0815	ESM 0.0825	NAÏVE 0.0833	HOLT3_E3 0.0834	MEAN_E3 0.0842	HOLT 0.0870	MEAN 0.1029	ARIMA_E3 0.1375	ARIMA 0.2791
Cott Corporation (COT)	HOLT 0.0238	HOLT3_E3 0.0239	ARIMA 0.0239	NAÏVE 0.0239	NAIVE3_E3 0.0240	ESM_E3 0.0240	ESM3_E3 0.0241	ARIMA_E3 0.0243	ESM 0.0245	M15_E3 0.0250	MEAN_E3 0.0284	MEAN 0.0353
ALCOHOLIC BEVERAGES												
Brown-Forman Corporation (BF-A)	HOLT3_E3 0.0251	ESM_E3 0.0251	ESM3_E3 0.0251	NAIVE3_E3 0.0251	HOLT 0.0251	NAÏVE 0.0252	ESM 0.0252	M15_E3 0.0253	MEAN_E3 0.0292	ARIMA_E3 0.0350	MEAN 0.0362	ARIMA 0.0716
Molson Coors Brewing Company (TAP)	ESM_E3 0.0593	M15_E3 0.0594	HOLT3_E3 0.0595	NAIVE3_E3 0.0595	ESM3_E3 0.0598	HOLT 0.0601	NAÏVE 0.0601	ESM 0.0607	MEAN_E3 0.0647	ARIMA_E3 0.0708	MEAN 0.0796	ARIMA 0.1052
Constellation Brans, Inc. (STZ)	ESM 0.0869	HOLT 0.0871	NAÏVE 0.0872	ESM3_E3 0.0874	NAIVE3_E3 0.0878	HOLT3_E3 0.0878	ESM_E3 0.0889	M15_E3 0.0948	MEAN_E3 0.1064	MEAN 0.1245	ARIMA_E3 0.1910	ARIMA 0.5012
Compañia Cervecerias Unidas S.A. (CCU)	M15_E3 0.0219	ARIMA_E3 0.0219	ESM_E3 0.0221	NAIVE3_E3 0.0223	HOLT3_E3 0.0224	ESM3_E3 0.0226	NAÏVE 0.0227	ARIMA 0.0229	HOLT 0.0230	MEAN_E3 0.0231	ESM 0.0233	MEAN 0.0282
Willamette Valley Vineyards, Inc. (WVVI)	MEAN_E3 0.0077	HOLT3_E3 0.0077	M15_E3 0.0077	HOLT 0.0077	NAIVE3_E3 0.0078	ESM_E3 0.0078	NAÏVE 0.0079	ESM3_E3 0.0079	ESM 0.0081	ARIMA_E3 0.0087	MEAN 0.0090	ARIMA 0.0114

SECURITY	RMSE #1	RMSE #2	RMSE #3	RMSE #4	RMSE #5	RMSE #6	RMSE #7	RMSE #8	RMSE #9	RMSE #10	RMSE #11	RMSE #12
TOBACCO PRODUCTS												
Altria Group, Inc. (MO)	ESM 0.0625	NAÏVE 0.0632	HOLT 0.0633	ARIMA 0.0642	ESM3_E3 0.0643	NAIVE3_E3 0.0654	HOLT3_E3 0.0662	ESM_E3 0.0714	ARIMA_E3 0.0795	M15_E3 0.0957	MEAN_E3 0.1054	MEAN 0.1218
British American Tobacco p.l.c. (BTI)	HOLT 0.0396	NAÏVE 0.0397	ESM 0.0397	ESM3_E3 0.0400	HOLT3_E3 0.0400	NAIVE3_E3 0.0400	ESM_E3 0.0406	M15_E3 0.0426	MEAN_E3 0.0460	MEAN 0.0539	ARIMA_E3 0.0557	ARIMA 0.1222
Vector Group Ltd. (VGR)	NAÏVE 0.0180	HOLT 0.0180	ESM 0.0180	ESM3_E3 0.0181	NAIVE3_E3 0.0181	HOLT3_E3 0.0181	ESM_E3 0.0182	M15_E3 0.0190	MEAN_E3 0.0212	ARIMA_E3 0.0226	MEAN 0.0257	ARIMA 0.0382
Universal Corporation (UVV)	NAÏVE 0.0672	NAIVE3_E3 0.0673	HOLT 0.0673	HOLT3_E3 0.0673	ARIMA 0.0675	ESM3_E3 0.0676	ESM 0.0678	ESM_E3 0.0679	ARIMA_E3 0.0681	M15_E3 0.0694	MEAN_E3 0.0736	MEAN 0.0904
Schweitzer-Mauduit Intl., Inc. (SWM)	ARIMA 0.0490	ARIMA_E3 0.0491	NAÏVE 0.0492	HOLT 0.0495	ESM 0.0496	NAIVE3_E3 0.0498	HOLT3_E3 0.0500	ESM3_E3 0.0501	ESM_E3 0.0508	M15_E3 0.0527	MEAN_E3 0.0547	MEAN 0.0639
FOOD ITEMS												
Sysco Corporation (SYY)	HOLT3_E3 0.0370	NAIVE3_E3 0.0370	HOLT 0.0372	ESM_E3 0.0372	NAÏVE 0.0372	ESM3_E3 0.0374	M15_E3 0.0375	ESM 0.0379	MEAN_E3 0.0379	MEAN 0.0442	ARIMA_E3 0.0506	ARIMA 0.1028
General Mills, Inc. (GIS)	HOLT3_E3 0.0266	NAIVE3_E3 0.0266	HOLT 0.0266	NAÏVE 0.0266	ESM3_E3 0.0267	ESM_E3 0.0268	ESM 0.0269	M15_E3 0.0285	MEAN_E3 0.0328	MEAN 0.0430	ARIMA_E3 0.0526	ARIMA 0.1259
Kellogg Company (K)	HOLT 0.0433	HOLT3_E3 0.0434	NAIVE3_E3 0.0434	NAÏVE 0.0434	ESM3_E3 0.0436	ESM_E3 0.0436	ARIMA 0.0437	ESM 0.0441	ARIMA_E3 0.0441	M15_E3 0.0452	MEAN_E3 0.0480	MEAN 0.0562
McCormick & Company, Inc. (MKC)	HOLT 0.0473	NAÏVE 0.0474	HOLT3_E3 0.0477	NAIVE3_E3 0.0477	ESM3_E3 0.0488	ESM_E3 0.0489	ESM 0.0498	M15_E3 0.0526	MEAN_E3 0.0617	MEAN 0.0756	ARIMA_E3 0.1188	ARIMA 0.3284
Conagra Brands, Inc. (CAG)	HOLT 0.0236	NAÏVE 0.0236	ARIMA 0.0236	HOLT3_E3 0.0237	NAIVE3_E3 0.0237	ESM3_E3 0.0239	ESM 0.0240	ESM_E3 0.0241	ARIMA_E3 0.0244	M15_E3 0.0255	MEAN_E3 0.0260	MEAN 0.0308
Campbell Soup Company (CPG)	M15_E3 0.0380	ARIMA_E3 0.0381	HOLT3_E3 0.0383	NAIVE3_E3 0.0384	ESM_E3 0.0386	HOLT 0.0388	NAÏVE 0.0389	ARIMA 0.0392	ESM3_E3 0.0394	ESM 0.0404	MEAN_E3 0.0420	MEAN 0.0521
Seaboard Corporation (SEB)	NAIVE3_E3 3.1534	NAÏVE 3.1605	HOLT3_E3 3.1732	HOLT 3.1901	M15_E3 3.2845	ESM_E3 3.3049	ESM3_E3 3.4365	MEAN_E3 3.4722	ESM 3.6392	ARIMA_E3 3.9885	MEAN 4.1647	ARIMA 7.3415

B.13. Financial Sector (Banks)

Table B-29: Financial Sector (Banks) E3 MAPE Accuracy Results

SECURITY	MAPE #1	MAPE #2	MAPE #3	MAPE #4	MAPE #5	MAPE #6	MAPE #7	MAPE #8	MAPE #9	MAPE #10	MAPE #11	MAPE #12
BANKS												
US Bancorp (USB)	ARIMA_E3 6.5727	M15_E3 6.6062	ESM_E3 6.6451	NAIVE3_E3 6.6703	HOLT3_E3 6.6863	ARIMA 6.7689	ESM3_E3 6.7728	NAÏVE 6.7837	HOLT 6.8074	MEAN_E3 6.8244	ESM 6.9591	MEAN 8.1143
Wells Fargo & Company (WFC)	HOLT3_E3 6.8792	NAIVE3_E3 6.8908	ESM_E3 6.9174	M15_E3 6.9351	HOLT 6.9650	ARIMA_E3 6.9816	NAÏVE 6.9833	MEAN_E3 7.0426	ESM3_E3 7.0688	ARIMA 7.3146	ESM 7.3636	MEAN 8.3135
The Bank of New York Mellon Corp. (BK)	HOLT 6.6889	NAÏVE 6.7037	HOLT3_E3 6.7162	NAIVE3_E3 6.7239	ESM3_E3 6.8072	ESM_E3 6.8218	ESM 6.8565	M15_E3 7.0636	ARIMA_E3 7.4051	MEAN_E3 7.7133	MEAN 9.5313	ARIMA 9.9136
Citigroup Inc. (C)	NAIVE3_E3 12.2586	HOLT3_E3 12.2588	NAÏVE 12.2970	HOLT 12.2974	ESM_E3 12.4275	M15_E3 12.4569	ESM3_E3 12.5491	ESM 12.7755	ARIMA_E3 12.9476	MEAN_E3 14.0018	ARIMA 14.2046	MEAN 17.6316
JP Morgan Chase & Co (JPM)	HOLT3_E3 8.2798	NAIVE3_E3 8.2860	ESM3_E3 8.2921	HOLT 8.3038	NAÏVE 8.3126	ESM_E3 8.3211	ESM 8.3284	ARIMA 8.3801	ARIMA_E3 8.3865	M15_E3 8.5368	MEAN_E3 8.5606	MEAN 10.1802
Bank of America Corporation (BAC)	ESM 11.3572	HOLT 11.3603	ESM3_E3 11.3704	NAÏVE 11.3742	HOLT3_E3 11.3796	NAIVE3_E3 11.3898	ESM_E3 11.4322	ARIMA 11.4568	ARIMA_E3 11.5256	M15_E3 11.7124	MEAN_E3 12.9977	MEAN 15.5536
The PNC Financial Services Group (PNC)	ESM_E3 7.0237	HOLT3_E3 7.0243	NAIVE3_E3 7.0280	ESM3_E3 7.0342	ARIMA_E3 7.0446	HOLT 7.0899	NAÏVE 7.0964	ESM 7.1031	ARIMA 7.1153	M15_E3 7.1561	MEAN_E3 7.8865	MEAN 9.5629
MIDWEST BANKS												
UMB Financial Corporation (UMBF)	HOLT3_E3 6.0470	ESM_E3 6.0649	ESM3_E3 6.1113	HOLT 6.1139	NAIVE3_E3 6.1300	M15_E3 6.1335	ESM 6.2212	NAÏVE 6.2469	ARIMA_E3 6.3216	MEAN_E3 6.3384	ARIMA 6.9048	MEAN 7.4640
Huntington Bancshares Inc. (HBAN)	HOLT3_E3 11.5138	ESM_E3 11.5312	NAIVE3_E3 11.5534	ARIMA_E3 11.5684	HOLT 11.6074	ESM3_E3 11.6093	M15_E3 11.6258	NAÏVE 11.6762	ARIMA 11.7654	ESM 11.7734	MEAN_E3 12.0743	MEAN 14.2711
Commerce Bancshares, Inc. (CBSH)	ARIMA_E3 4.9985	ARIMA 5.0245	HOLT3_E3 5.0404	NAIVE3_E3 5.0577	M15_E3 5.0735	HOLT 5.0790	ESM_E3 5.1031	NAÏVE 5.1038	MEAN_E3 5.1983	ESM3_E3 5.2042	ESM 5.3609	MEAN 6.0447
Associated Banc-Corp (ASB)	ARIMA 6.7592	ESM 6.8115	NAÏVE 6.8305	HOLT 6.8311	ESM3_E3 6.8380	NAIVE3_E3 6.8524	HOLT3_E3 6.8535	ESM_E3 6.9232	ARIMA_E3 6.9730	M15_E3 7.2589	MEAN_E3 7.4344	MEAN 8.8603
First Financial Bancorp (FFBC)	HOLT3_E3 7.2307	ARIMA 7.2499	MEAN_E3 7.2611	ARIMA_E3 7.2669	HOLT 7.2788	NAIVE3_E3 7.2980	ESM_E3 7.3049	ESM3_E3 7.3343	M15_E3 7.4105	NAÏVE 7.4145	ESM 7.4175	MEAN 8.2879
Old National Bancorp (ONB)	MEAN_E3 6.4258	ARIMA_E3 6.6598	ESM_E3 6.6717	M15_E3 6.6952	ESM3_E3 6.7301	HOLT3_E3 6.7402	NAIVE3_E3 6.7517	ESM 6.8304	ARIMA 6.8410	HOLT 6.8476	NAÏVE 6.8714	MEAN 7.2023
TCF Financial Corporation (TCF)	HOLT 7.8163	NAÏVE 7.8182	ESM 7.8252	ARIMA_E3 7.8289	HOLT3_E3 7.8352	NAIVE3_E3 7.8385	ESM3_E3 7.8417	ESM_E3 7.9111	ARIMA 8.1215	M15_E3 8.2165	MEAN_E3 8.6007	MEAN 10.4540

SECURITY	MAPE #1	MAPE #2	MAPE #3	MAPE #4	MAPE #5	MAPE #6	MAPE #7	MAPE #8	MAPE #9	MAPE #10	MAPE #11	MAPE #12
INVESTMENT BANKS												
Siebert Financial Corp. (SIEB)	NAÏVE 12.7973	HOLT 12.7975	NAIVE3_E3 12.8725	HOLT3_E3 12.8726	ARIMA 13.0250	ARIMA_E3 13.5886	M15_E3 14.1581	MEAN_E3 14.2696	ESM_E3 15.1502	ESM3_E3 16.3542	MEAN 17.3765	ESM 17.8691
Stifel Financial Corp. (SF)	ESM_E3 8.8449	NAIVE3_E3 8.8661	HOLT3_E3 8.8714	ESM3_E3 8.8719	M15_E3 8.9447	NAÏVE 8.9544	ESM 8.9609	HOLT 8.9653	MEAN_E3 9.7500	ARIMA_E3 10.9031	MEAN 12.1075	ARIMA 15.1268
Raymond James Financial, Inc. (RJF)	ESM_E3 9.1805	HOLT3_E3 9.1922	ARIMA_E3 9.2228	HOLT 9.2394	ARIMA 9.2675	NAIVE3_E3 9.2721	ESM3_E3 9.3006	NAÏVE 9.3511	M15_E3 9.3713	ESM 9.5397	MEAN_E3 9.6999	MEAN 11.0160
The Charles Schwab Corp. (SCHW)	ARIMA 10.4831	HOLT3_E3 10.5171	NAIVE3_E3 10.5215	HOLT 10.5306	NAÏVE 10.5398	ARIMA_E3 10.6047	M15_E3 10.8123	MEAN_E3 10.9832	ESM_E3 11.1091	ESM3_E3 11.5216	ESM 12.1091	MEAN 13.1911
WisdomTree Investments, Inc. (WETF)	NAÏVE 22.6224	HOLT 22.6478	ARIMA 22.6939	ESM 22.8337	NAIVE3_E3 23.0580	HOLT3_E3 23.0756	ESM3_E3 23.1455	ESM_E3 23.6651	ARIMA_E3 23.9666	M15_E3 25.0843	MEAN_E3 29.5196	MEAN 37.6617
Oppenheimer Holdings, Inc. (OPY)	NAIVE3_E3 10.6317	HOLT3_E3 10.6622	ARIMA_E3 10.6768	NAÏVE 10.7039	ESM_E3 10.7163	HOLT 10.7454	ESM3_E3 10.7656	ARIMA 10.7706	M15_E3 10.8531	ESM 10.8866	MEAN_E3 12.1020	MEAN 14.8277
National Holdings Corp. (NHLD)	ARIMA_E3 21.3472	ARIMA 21.6422	M15_E3 21.9985	HOLT3_E3 22.0573	NAIVE3_E3 22.3906	HOLT 23.0283	NAÏVE 23.5797	MEAN_E3 24.2178	MEAN 30.0433	ESM_E3 36.9209	ESM3_E3 48.1758	ESM 59.9233
INVESTMENT MANAGEMENT												
Eaton Vance Corp. (EV)	ESM_E3 8.4176	HOLT3_E3 8.5210	HOLT 8.5222	NAIVE3_E3 8.5555	NAÏVE 8.5696	ESM3_E3 8.5734	M15_E3 8.8035	ESM 8.9990	MEAN_E3 9.5768	ARIMA_E3 9.9670	MEAN 11.4185	ARIMA 15.1013
Barings Corporate Investors (MCI)	ESM 4.0111	ESM3_E3 4.0299	ESM_E3 4.1041	NAÏVE 4.1127	HOLT 4.1145	NAIVE3_E3 4.1156	HOLT3_E3 4.1176	ARIMA 4.1391	ARIMA_E3 4.2248	M15_E3 4.3619	MEAN_E3 4.6626	MEAN 5.7397
SEI Investments Co. (SEIC)	HOLT 10.3462	NAÏVE 10.3674	HOLT3_E3 10.4158	NAIVE3_E3 10.4306	ESM_E3 10.7907	ESM3_E3 10.8561	M15_E3 10.8724	MEAN_E3 10.9842	ESM 11.0066	ARIMA_E3 11.3344	MEAN 12.5418	ARIMA 13.5138
Franklin Resources, Inc. (BEN)	NAÏVE 8.2508	HOLT 8.2552	NAIVE3_E3 8.2663	HOLT3_E3 8.2709	ARIMA 8.2825	ESM3_E3 8.3000	ESM 8.3003	ESM_E3 8.3428	ARIMA_E3 8.3532	M15_E3 8.5607	MEAN_E3 9.3226	MEAN 11.3217
Legg Mason, Inc. (LM)	HOLT 10.1744	HOLT3_E3 10.2150	NAÏVE 10.2333	NAIVE3_E3 10.2584	ESM_E3 10.4716	ESM3_E3 10.4824	ESM 10.5563	M15_E3 10.6585	ARIMA_E3 11.1401	MEAN_E3 12.0300	ARIMA 13.2399	MEAN 14.5445
T. Rowe Price Group, Inc. (TROW)	HOLT3_E3 8.4392	HOLT 8.4410	NAIVE3_E3 8.4606	NAÏVE 8.4713	ESM3_E3 8.4900	ARIMA_E3 8.4920	ESM_E3 8.5083	ESM 8.5381	ARIMA 8.6791	M15_E3 8.7616	MEAN_E3 9.1886	MEAN 11.0461
Alliance-Bernstein Holding L.P. (AB)	HOLT 9.0092	NAÏVE 9.0097	HOLT3_E3 9.0876	NAIVE3_E3 9.0880	ESM3_E3 9.3880	ESM_E3 9.4016	ESM 9.4808	ARIMA_E3 9.5830	M15_E3 9.7096	ARIMA 10.0130	MEAN_E3 11.7289	MEAN 14.5893

Table B-30: Financial Sector (Banks) E3 RMSE Accuracy Results

SECURITY	RMSE #1	RMSE #2	RMSE #3	RMSE #4	RMSE #5	RMSE #6	RMSE #7	RMSE #8	RMSE #9	RMSE #10	RMSE #11	RMSE #12
BANKS												
US Bancorp (USB)	ARIMA_E3 0.0369	M15_E3 0.0369	ESM_E3 0.0372	NAIVE3_E3 0.0373	HOLT3_E3 0.0373	NAÏVE 0.0379	HOLT 0.0380	ESM3_E3 0.0380	MEAN_E3 0.0381	ARIMA 0.0383	ESM 0.0391	MEAN 0.0446
Wells Fargo & Company (WFC)	MEAN_E3 0.0450	HOLT3_E3 0.0456	M15_E3 0.0457	NAIVE3_E3 0.0457	ARIMA_E3 0.0460	ESM_E3 0.0461	HOLT 0.0462	NAÏVE 0.0464	ESM3_E3 0.0474	ESM 0.0494	ARIMA 0.0502	MEAN 0.0516
The Bank of New York Mellon Corp. (BK)	HOLT 0.0452	NAÏVE 0.0452	HOLT3_E3 0.0456	NAIVE3_E3 0.0456	ESM3_E3 0.0473	ESM_E3 0.0474	ESM 0.0477	M15_E3 0.0489	MEAN_E3 0.0511	MEAN 0.0614	ARIMA_E3 0.0626	ARIMA 0.1416
Citigroup Inc. (C)	NAÏVE 0.4894	HOLT 0.4894	NAIVE3_E3 0.4896	HOLT3_E3 0.4896	M15_E3 0.5003	ESM_E3 0.5088	MEAN_E3 0.5101	ESM3_E3 0.5239	ESM 0.5455	MEAN 0.5978	ARIMA_E3 0.6201	ARIMA 0.9745
JP Morgan Chase & Co (JPM)	ESM 0.0715	ESM3_E3 0.0716	HOLT 0.0722	HOLT3_E3 0.0723	ESM_E3 0.0723	NAÏVE 0.0723	NAIVE3_E3 0.0723	ARIMA 0.0727	ARIMA_E3 0.0733	MEAN_E3 0.0742	M15_E3 0.0750	MEAN 0.0851
Bank of America Corporation (BAC)	HOLT3_E3 0.0466	ESM_E3 0.0466	ESM3_E3 0.0466	NAIVE3_E3 0.0467	ARIMA_E3 0.0467	HOLT 0.0468	ARIMA 0.0469	ESM 0.0469	NAÏVE 0.0470	M15_E3 0.0473	MEAN_E3 0.0485	MEAN 0.0561
The PNC Financial Services Group (PNC)	HOLT3_E3 0.0923	NAIVE3_E3 0.0923	ESM_E3 0.0924	ESM3_E3 0.0924	ARIMA_E3 0.0925	ARIMA 0.0927	HOLT 0.0928	NAÏVE 0.0929	ESM 0.0930	M15_E3 0.0941	MEAN_E3 0.1020	MEAN 0.1230
MIDWEST BANKS												
UMB Financial Corporation (UMBF)	HOLT3_E3 0.0545	ESM_E3 0.0548	HOLT 0.0549	NAIVE3_E3 0.0550	ESM3_E3 0.0552	M15_E3 0.0553	NAÏVE 0.0555	ESM 0.0559	MEAN_E3 0.0576	ARIMA_E3 0.0606	MEAN 0.0666	ARIMA 0.0875
Huntington Bancshares Inc. (HBAN)	ARIMA_E3 0.0204	HOLT3_E3 0.0204	M15_E3 0.0205	NAIVE3_E3 0.0205	ESM_E3 0.0206	HOLT 0.0207	ARIMA 0.0208	NAÏVE 0.0208	ESM3_E3 0.0210	ESM 0.0215	MEAN_E3 0.0223	MEAN 0.0266
Commerce Bancshares, Inc. (CBSH)	ARIMA 0.0311	ARIMA_E3 0.0312	ESM_E3 0.0312	HOLT3_E3 0.0313	NAIVE3_E3 0.0314	HOLT 0.0314	NAÏVE 0.0316	ESM3_E3 0.0317	M15_E3 0.0318	ESM 0.0325	MEAN_E3 0.0327	MEAN 0.0370
Associated Banc-Corp (ASB)	ESM 0.0257	ARIMA 0.0257	ESM3_E3 0.0257	HOLT3_E3 0.0258	HOLT 0.0258	NAIVE3_E3 0.0258	NAÏVE 0.0258	ESM_E3 0.0259	ARIMA_E3 0.0261	M15_E3 0.0268	MEAN_E3 0.0286	MEAN 0.0337
First Financial Bancorp (FFBC)	HOLT3_E3 0.0231	ARIMA 0.0231	HOLT 0.0231	NAIVE3_E3 0.0233	ARIMA_E3 0.0233	ESM_E3 0.0233	ESM3_E3 0.0233	NAÏVE 0.0235	ESM 0.0235	M15_E3 0.0239	MEAN_E3 0.0241	MEAN 0.0277
Old National Bancorp (ONB)	MEAN_E3 0.0202	M15_E3 0.0211	ARIMA_E3 0.0213	ESM_E3 0.0215	ESM3_E3 0.0219	HOLT3_E3 0.0220	NAIVE3_E3 0.0220	MEAN 0.0222	ESM 0.0225	ARIMA 0.0225	HOLT 0.0225	NAÏVE 0.0227
TCF Financial Corporation (TCF)	NAÏVE 0.0274	HOLT 0.0274	ESM 0.0274	ESM3_E3 0.0275	HOLT3_E3 0.0275	ARIMA_E3 0.0275	NAIVE3_E3 0.0275	ESM_E3 0.0277	M15_E3 0.0287	MEAN_E3 0.0299	ARIMA 0.0313	MEAN 0.0354

SECURITY	RMSE #1	RMSE #2	RMSE #3	RMSE #4	RMSE #5	RMSE #6	RMSE #7	RMSE #8	RMSE #9	RMSE #10	RMSE #11	RMSE #12
INVESTMENT BANKS												
Siebert Financial Corp. (SIEB)	NAÏVE 0.0178	HOLT 0.0178	NAIVE3_E3 0.0185	HOLT3_E3 0.0185	ARIMA 0.0187	MEAN_E3 0.0194	ARIMA_E3 0.0200	M15_E3 0.0210	MEAN 0.0220	ESM_E3 0.0247	ESM3_E3 0.0280	ESM 0.0319
Stifel Financial Corp. (SF)	HOLT3_E3 0.0530	NAIVE3_E3 0.0530	HOLT 0.0531	ESM3_E3 0.0531	NAÏVE 0.0532	ESM_E3 0.0532	ESM 0.0533	M15_E3 0.0542	MEAN_E3 0.0586	ARIMA_E3 0.0657	MEAN 0.0711	ARIMA 0.1035
Raymond James Financial, Inc. (RJF)	HOLT 0.0628	HOLT3_E3 0.0629	ARIMA 0.0629	ESM_E3 0.0632	NAIVE3_E3 0.0633	NAÏVE 0.0634	ARIMA_E3 0.0635	ESM3_E3 0.0637	M15_E3 0.0646	ESM 0.0648	MEAN_E3 0.0650	MEAN 0.0711
The Charles Schwab Corp. (SCHW)	HOLT 0.0433	NAÏVE 0.0433	ARIMA 0.0433	HOLT3_E3 0.0434	NAIVE3_E3 0.0434	ARIMA_E3 0.0440	MEAN_E3 0.0446	M15_E3 0.0449	ESM_E3 0.0472	ESM3_E3 0.0496	MEAN 0.0522	ESM 0.0528
WisdomTree Investments, Inc. (WETF)	HOLT 0.0204	NAÏVE 0.0204	ESM 0.0204	ARIMA 0.0204	HOLT3_E3 0.0206	NAIVE3_E3 0.0206	ESM3_E3 0.0206	ESM_E3 0.0210	ARIMA_E3 0.0212	M15_E3 0.0221	MEAN_E3 0.0226	MEAN 0.0271
Oppenheimer Holdings, Inc. (OPY)	NAIVE3_E3 0.0529	ARIMA_E3 0.0529	HOLT3_E3 0.0531	NAÏVE 0.0532	ESM_E3 0.0532	M15_E3 0.0532	HOLT 0.0534	ESM3_E3 0.0535	ARIMA 0.0536	ESM 0.0540	MEAN_E3 0.0576	MEAN 0.0691
National Holdings Corp. (NHLD)	NAÏVE 0.0923	HOLT 0.0928	ARIMA 0.0948	NAIVE3_E3 0.0968	HOLT3_E3 0.0974	ESM 0.1006	ESM3_E3 0.1016	MEAN_E3 0.1026	ESM_E3 0.1063	MEAN 0.1087	ARIMA_E3 0.1103	M15_E3 0.1247
INVESTMENT MANAGEMENT												
Eaton Vance Corp. (EV)	HOLT 0.0541	HOLT3_E3 0.0542	NAIVE3_E3 0.0545	NAÏVE 0.0545	ARIMA_E3 0.0546	M15_E3 0.0554	ESM_E3 0.0564	MEAN_E3 0.0570	ESM3_E3 0.0582	ESM 0.0608	MEAN 0.0637	ARIMA 0.0777
Barings Corporate Investors (MCI)	ESM 0.0112	ESM3_E3 0.0113	NAÏVE 0.0114	HOLT 0.0114	NAIVE3_E3 0.0114	ESM_E3 0.0114	HOLT3_E3 0.0114	ARIMA 0.0115	ARIMA_E3 0.0117	M15_E3 0.0119	MEAN_E3 0.0127	MEAN 0.0152
SEI Investments Co. (SEIC)	HOLT 0.0522	NAÏVE 0.0523	HOLT3_E3 0.0524	NAIVE3_E3 0.0525	M15_E3 0.0538	ESM_E3 0.0549	MEAN_E3 0.0553	ESM3_E3 0.0564	ARIMA_E3 0.0583	ESM 0.0585	MEAN 0.0629	ARIMA 0.0742
Franklin Resources, Inc. (BEN)	HOLT 0.0455	NAÏVE 0.0456	HOLT3_E3 0.0458	NAIVE3_E3 0.0459	ESM 0.0461	ESM3_E3 0.0463	ARIMA 0.0466	ESM_E3 0.0467	ARIMA_E3 0.0470	M15_E3 0.0483	MEAN_E3 0.0519	MEAN 0.0622
Legg Mason, Inc. (LM)	HOLT 0.0864	NAÏVE 0.0868	HOLT3_E3 0.0874	NAIVE3_E3 0.0877	ESM 0.0884	ESM3_E3 0.0888	ESM_E3 0.0898	M15_E3 0.0933	MEAN_E3 0.1076	ARIMA_E3 0.1119	MEAN 0.1303	ARIMA 0.2335
T. Rowe Price Group, Inc. (TROW)	HOLT 0.0783	NAÏVE 0.0785	ESM 0.0786	HOLT3_E3 0.0790	NAIVE3_E3 0.0791	ESM3_E3 0.0792	ESM_E3 0.0801	ARIMA_E3 0.0806	ARIMA 0.0827	M15_E3 0.0832	MEAN_E3 0.0891	MEAN 0.1033
Alliance-Bernstein Holding L.P. (AB)	HOLT 0.0630	NAÏVE 0.0630	HOLT3_E3 0.0641	NAIVE3_E3 0.0642	ESM_E3 0.0668	ESM3_E3 0.0669	ESM 0.0679	M15_E3 0.0695	ARIMA_E3 0.0727	MEAN_E3 0.0819	ARIMA 0.0994	MEAN 0.1001

B.14. Financial Sector (Insurance)

Table B-31: Financial Sector (Insurance) E3 MAPE Accuracy Results

SECURITY	MAPE #1	MAPE #2	MAPE #3	MAPE #4	MAPE #5	MAPE #6	MAPE #7	MAPE #8	MAPE #9	MAPE #10	MAPE #11	MAPE #12
INSURANCE STOCKS												
W. R. Berkley Corporation (WRB)	HOLT3_E3 5.8525	NAIVE3_E3 5.8639	HOLT 5.8820	ESM3_E3 5.8832	ESM_E3 5.8908	NAÏVE 5.8943	ESM 5.9738	M15_E3 6.4370	MEAN_E3 6.8780	ARIMA_E3 7.1836	MEAN 8.2270	ARIMA 10.8211
Alleghany Corporation (Y)	HOLT3_E3 4.4024	NAIVE3_E3 4.4230	ARIMA_E3 4.4481	HOLT 4.4528	ARIMA 4.4651	ESM_E3 4.4793	NAÏVE 4.4832	M15_E3 4.5887	ESM3_E3 4.6105	ESM 4.8606	MEAN_E3 4.9304	MEAN 5.9173
The Progressive Corporation (PGR)	ESM3_E3 6.6477	HOLT3_E3 6.6628	HOLT 6.6691	ESM 6.6808	ESM_E3 6.6978	NAÏVE 6.6978	NAIVE3_E3 6.7024	ARIMA_E3 6.8687	M15_E3 7.0083	MEAN_E3 7.4534	ARIMA 7.4831	MEAN 9.0234
The Travelers Companies, Inc. (TRV)	ESM3_E3 6.7275	HOLT3_E3 6.7318	HOLT 6.7420	ESM 6.7513	ARIMA 6.7538	NAIVE3_E3 6.7540	NAÏVE 6.7633	ESM_E3 6.7850	ARIMA_E3 6.8533	MEAN_E3 6.8935	M15_E3 7.0773	MEAN 7.6028
Cincinnati Financial Corp. (CINF)	HOLT3_E3 5.6188	HOLT 5.6340	NAIVE3_E3 5.6359	ESM_E3 5.6407	ARIMA 5.6614	NAÏVE 5.6644	ESM3_E3 5.6650	ARIMA_E3 5.6768	ESM 5.7869	M15_E3 5.8115	MEAN_E3 6.0586	MEAN 7.0352
CNA Financial Corporation (CNA)	NAÏVE 8.6338	HOLT 8.6391	HOLT3_E3 8.6500	ARIMA 8.6511	NAIVE3_E3 8.6527	ESM 8.6825	ESM3_E3 8.6849	ESM_E3 8.7405	ARIMA_E3 8.7728	M15_E3 8.9944	MEAN_E3 9.5928	MEAN 11.4471
Markel Corporation (MKL)	HOLT3_E3 4.6382	HOLT 4.6442	NAIVE3_E3 4.6498	NAÏVE 4.6600	ESM_E3 4.6798	ESM3_E3 4.7202	ESM 4.8709	M15_E3 5.0342	ARIMA_E3 5.6691	MEAN_E3 5.7270	MEAN 7.0191	ARIMA 8.3080
INSURANCE BROKERS												
March & McLennan Co, Inc. (MMC)	NAIVE3_E3 6.3503	HOLT3_E3 6.3611	ESM_E3 6.4111	NAÏVE 6.4222	HOLT 6.4440	ESM3_E3 6.4514	ESM 6.5789	M15_E3 6.6203	MEAN_E3 6.8058	ARIMA_E3 7.7748	MEAN 7.9841	ARIMA 10.2731
Aon plc (AON)	ESM3_E3 7.4601	ESM 7.4646	NAÏVE 7.4688	HOLT 7.4731	NAIVE3_E3 7.4756	HOLT3_E3 7.4782	ESM_E3 7.5073	ARIMA_E3 7.6264	ARIMA 7.7007	M15_E3 7.7568	MEAN_E3 8.0862	MEAN 9.2922
Brown & Brown, Inc. (BRO)	HOLT 6.0230	NAÏVE 6.0761	HOLT3_E3 6.0861	NAIVE3_E3 6.1296	ESM_E3 6.2837	ESM3_E3 6.3572	ESM 6.5559	M15_E3 6.5837	MEAN_E3 7.0169	ARIMA_E3 7.2872	MEAN 8.4400	ARIMA 13.1718
Arthur J. Gallagher & Co (AJG)	HOLT 6.2133	NAÏVE 6.2143	HOLT3_E3 6.2181	NAIVE3_E3 6.2204	ESM_E3 6.3627	ESM3_E3 6.4349	M15_E3 6.5032	ARIMA_E3 6.5318	ESM 6.5840	MEAN_E3 6.6265	MEAN 7.7122	ARIMA 7.7993

Table B-32: Financial Sector (Insurance) E3 RMSE Accuracy Results

SECURITY	RMSE #1	RMSE #2	RMSE #3	RMSE #4	RMSE #5	RMSE #6	RMSE #7	RMSE #8	RMSE #9	RMSE #10	RMSE #11	RMSE #12
INSURANCE STOCKS												
W. R. Berkley Corporation (WRB)	ESM3_E3 0.0342	HOLT 0.0343	NAÏVE 0.0343	NAIVE3_E3 0.0343	HOLT3_E3 0.0343	ESM 0.0344	ESM_E3 0.0345	M15_E3 0.0362	MEAN_E3 0.0368	MEAN 0.0423	ARIMA_E3 0.0571	ARIMA 0.1284
Alleghany Corporation (Y)	ARIMA_E3 0.3170	HOLT3_E3 0.3180	ESM_E3 0.3189	ARIMA 0.3192	HOLT 0.3203	NAIVE3_E3 0.3205	M15_E3 0.3216	NAÏVE 0.3238	MEAN_E3 0.3250	ESM3_E3 0.3274	ESM 0.3419	MEAN 0.3724
The Progressive Corporation (PGR)	ESM_E3 0.0296	NAIVE3_E3 0.0296	NAÏVE 0.0298	HOLT3_E3 0.0298	ESM3_E3 0.0298	M15_E3 0.0299	HOLT 0.0301	ESM 0.0304	MEAN_E3 0.0320	ARIMA_E3 0.0334	MEAN 0.0387	ARIMA 0.0468
The Travelers Companies, Inc. (TRV)	ARIMA 0.0736	ESM 0.0737	HOLT 0.0737	NAÏVE 0.0738	ESM3_E3 0.0738	HOLT3_E3 0.0739	MEAN_E3 0.0740	NAIVE3_E3 0.0740	ESM_E3 0.0745	ARIMA_E3 0.0750	M15_E3 0.0773	MEAN 0.0809
Cincinnati Financial Co. (CINF)	HOLT3_E3 0.0434	ARIMA 0.0434	HOLT 0.0435	NAIVE3_E3 0.0435	ESM_E3 0.0436	NAÏVE 0.0437	ARIMA_E3 0.0437	ESM3_E3 0.0439	ESM 0.0449	M15_E3 0.0450	MEAN_E3 0.0457	MEAN 0.0522
CNA Financial Corporation (CNA)	HOLT 0.0477	NAÏVE 0.0478	ARIMA 0.0479	HOLT3_E3 0.0481	NAIVE3_E3 0.0482	ESM 0.0487	ESM3_E3 0.0488	ESM_E3 0.0491	ARIMA_E3 0.0491	M15_E3 0.0505	MEAN_E3 0.0530	MEAN 0.0615
Markel Corporation (MKL)	ESM_E3 0.4698	HOLT 0.4701	NAÏVE 0.4710	HOLT3_E3 0.4741	NAIVE3_E3 0.4747	ESM3_E3 0.4760	ESM 0.4974	M15_E3 0.5040	MEAN_E3 0.5132	MEAN 0.5819	ARIMA_E3 0.6551	ARIMA 1.3620
INSURANCE BROKERS												
March & McLennan Co., Inc. (MMC)	NAIVE3_E3 0.0529	HOLT3_E3 0.0530	NAÏVE 0.0531	HOLT 0.0533	ESM_E3 0.0536	ESM3_E3 0.0540	M15_E3 0.0546	MEAN_E3 0.0548	ESM 0.0549	MEAN 0.0627	ARIMA_E3 0.0913	ARIMA 0.2018
Aon plc (AON)	NAÏVE 0.0666	HOLT 0.0667	NAIVE3_E3 0.0669	HOLT3_E3 0.0669	ESM 0.0673	ESM3_E3 0.0673	ESM_E3 0.0677	M15_E3 0.0698	ARIMA_E3 0.0704	MEAN_E3 0.0721	ARIMA 0.0822	MEAN 0.0849
Brown & Brown, Inc. (BRO)	HOLT 0.0141	HOLT3_E3 0.0142	NAÏVE 0.0142	NAIVE3_E3 0.0142	ESM_E3 0.0143	ESM3_E3 0.0146	M15_E3 0.0148	ESM 0.0151	MEAN_E3 0.0155	ARIMA_E3 0.0172	MEAN 0.0183	ARIMA 0.0330
Arthur J. Gallagher & Co (AJG)	HOLT 0.0381	NAÏVE 0.0381	HOLT3_E3 0.0382	NAIVE3_E3 0.0382	ESM_E3 0.0387	ESM3_E3 0.0391	M15_E3 0.0396	ESM 0.0399	MEAN_E3 0.0412	ARIMA_E3 0.0423	MEAN 0.0484	ARIMA 0.0710

B.15. Financial Sector (REIT/Real Estate)

Table B-33: Financial Sector (REIT/Real Estate) E3 MAPE Accuracy Results

SECURITY	MAPE #1	MAPE #2	MAPE #3	MAPE #4	MAPE #5	MAPE #6	MAPE #7	MAPE #8	MAPE #9	MAPE #10	MAPE #11	MAPE #12
REIT—OTHER EQUITY TRUSTS												
Host Hotels & Resorts (HST)	ESM3_E3 9.1338	ESM 9.1728	ESM_E3 9.1737	HOLT3_E3 9.2118	NAIVE3_E3 9.2119	HOLT 9.2213	NAÏVE 9.2214	ARIMA 9.2300	ARIMA_E3 9.2923	M15_E3 9.4754	MEAN_E3 10.6720	MEAN 13.4545
Public Storage (PSA)	M15_E3 6.2779	ESM_E3 6.2957	HOLT3_E3 6.3182	NAIVE3_E3 6.3337	ESM3_E3 6.3513	HOLT 6.3907	NAÏVE 6.4132	ESM 6.4363	MEAN_E3 6.8513	ARIMA_E3 7.8929	MEAN 8.1151	ARIMA 13.2784
Welltower Inc. (WELL)	NAIVE3_E3 6.5902	HOLT3_E3 6.5924	ESM_E3 6.5993	ARIMA 6.6011	ESM3_E3 6.6055	ARIMA_E3 6.6130	HOLT 6.6204	NAÏVE 6.6219	ESM 6.6737	M15_E3 6.7140	MEAN_E3 7.1660	MEAN 8.4427
Vornado Realty Trust (VNO)	ESM3_E3 7.1060	ESM 7.1124	ESM_E3 7.1500	HOLT3_E3 7.2196	HOLT 7.2296	NAIVE3_E3 7.2330	NAÏVE 7.2506	M15_E3 7.3694	ARIMA_E3 7.5527	MEAN_E3 8.0626	ARIMA 8.6769	MEAN 9.5240
EastGroup Properties, Inc. (EGP)	ESM 5.5654	ESM3_E3 5.5766	HOLT 5.5810	HOLT3_E3 5.5902	NAÏVE 5.5903	NAIVE3_E3 5.5916	ARIMA 5.6152	ESM_E3 5.6358	ARIMA_E3 5.6985	M15_E3 5.8735	MEAN_E3 6.3259	MEAN 7.3692
HCP, Inc. (HCP)	ESM3_E3 7.3028	ESM_E3 7.3220	NAIVE3_E3 7.3225	HOLT3_E3 7.3270	ESM 7.3328	NAÏVE 7.3436	HOLT 7.3482	ARIMA 7.3654	ARIMA_E3 7.3713	M15_E3 7.4887	MEAN_E3 7.7884	MEAN 9.0297
Duke Realty Corporation (DRE)	ESM3_E3 8.6575	HOLT3_E3 8.6578	NAIVE3_E3 8.6648	ESM 8.6680	HOLT 8.6684	ARIMA 8.6735	NAÏVE 8.6830	ESM_E3 8.6973	ARIMA_E3 8.7404	M15_E3 8.9088	MEAN_E3 9.6619	MEAN 11.1878
REIT—RETAIL EQUITY TRUSTS												
Federal Realty Investment Trust (FRT)	ESM 6.1443	ESM3_E3 6.1632	ESM_E3 6.2245	HOLT3_E3 6.2924	NAIVE3_E3 6.2957	HOLT 6.3178	NAÏVE 6.3262	M15_E3 6.4380	MEAN_E3 6.5400	ARIMA_E3 6.9957	MEAN 7.4641	ARIMA 9.0607
Pennsylvania REIT (PEI)	ESM 12.1037	NAÏVE 12.1166	ARIMA 12.1186	HOLT 12.1188	ESM3_E3 12.1919	NAIVE3_E3 12.2021	HOLT3_E3 12.2038	ESM_E3 12.3445	ARIMA_E3 12.4481	M15_E3 12.8300	MEAN_E3 14.1862	MEAN 17.0285
Urstadt Biddle Properties, Inc. (UBP)	ESM_E3 4.4651	ARIMA_E3 4.4740	ESM3_E3 4.5206	HOLT3_E3 4.5211	NAIVE3_E3 4.5559	ARIMA 4.5906	M15_E3 4.6021	ESM 4.6393	HOLT 4.6401	NAÏVE 4.7017	MEAN_E3 4.8832	MEAN 5.7340
National Retail Properties, Inc. (NNN)	NAIVE3_E3 6.1515	ESM3_E3 6.1516	HOLT3_E3 6.1519	ESM 6.1552	HOLT 6.1563	NAÏVE 6.1750	ARIMA 6.1796	ESM_E3 6.1852	ARIMA_E3 6.2238	M15_E3 6.3400	MEAN_E3 6.4341	MEAN 7.3536
Weingarten Realty Investors (WRI)	ARIMA 7.8138	HOLT3_E3 7.8493	ESM3_E3 7.8521	NAIVE3_E3 7.8577	HOLT 7.8662	ESM 7.8691	NAÏVE 7.8788	ESM_E3 7.8897	ARIMA_E3 7.9004	M15_E3 8.0793	MEAN_E3 8.2288	MEAN 9.3366
Kimco Realty Corporation (KIM)	NAIVE3_E3 8.6508	NAÏVE 8.6531	HOLT3_E3 8.6541	HOLT 8.6558	ESM3_E3 8.6912	ESM_E3 8.7164	ESM 8.7330	ARIMA_E3 8.9192	M15_E3 8.9332	MEAN_E3 9.2360	ARIMA 9.6053	MEAN 11.3338
Taubman Centes, Inc. (TCO)	ESM3_E3 7.7873	ESM 7.8001	NAIVE3_E3 7.8014	HOLT3_E3 7.8035	ESM_E3 7.8204	NAÏVE 7.8269	HOLT 7.8275	M15_E3 7.9719	ARIMA_E3 8.5389	MEAN_E3 8.7086	MEAN 10.4381	ARIMA 10.6035

SECURITY	MAPE #1	MAPE #2	MAPE #3	MAPE #4	MAPE #5	MAPE #6	MAPE #7	MAPE #8	MAPE #9	MAPE #10	MAPE #11	MAPE #12
REAL ESTATE OPERATIONS STOCKS												
Texas Pacific Land Trust (TPL)	ESM3_E3 9.5020	ESM_E3 9.5385	ESM 9.5519	NAIVE3_E3 9.5769	NAÏVE 9.6007	HOLT3_E3 9.6164	HOLT 9.6953	M15_E3 9.9094	MEAN_E3 11.5902	ARIMA_E3 12.2715	MEAN 14.5043	ARIMA 21.0625
Tejon Ranch Co. (TRC)	NAIVE3_E3 8.7847	NAÏVE 8.7923	HOLT3_E3 8.7969	ARIMA 8.7995	HOLT 8.8112	ESM3_E3 8.8344	ESM_E3 8.8412	ARIMA_E3 8.8427	ESM 8.8637	M15_E3 8.9686	MEAN_E3 9.4737	MEAN 11.4011
American Realty Investors, Inc. (ARL)	ARIMA 12.7004	HOLT 12.7318	NAÏVE 12.7348	HOLT3_E3 12.7844	NAIVE3_E3 12.7908	ESM 12.8677	ESM3_E3 12.8839	ESM_E3 13.0521	ARIMA_E3 13.1433	M15_E3 13.8887	MEAN_E3 14.1701	MEAN 17.5852
Brookfield Asset Management, Inc. (BAM)	NAIVE3_E3 7.2911	NAÏVE 7.3299	HOLT3_E3 7.3419	ESM_E3 7.3546	ESM3_E3 7.3906	HOLT 7.4101	ESM 7.4926	M15_E3 7.4992	ARIMA_E3 8.5207	MEAN_E3 8.5680	MEAN 10.5997	ARIMA 13.1632
FRP Holdings, Inc. (FRPH)	ESM_E3 8.7033	ESM3_E3 8.7132	HOLT3_E3 8.7198	NAIVE3_E3 8.7575	ESM 8.7914	HOLT 8.8157	MEAN_E3 8.8668	M15_E3 8.8827	NAÏVE 8.8914	ARIMA_E3 9.0871	ARIMA 9.9085	MEAN 10.3390
Income Opportunity Realty Investors, Inc. (IOR)	NAIVE3_E3 11.1104	NAÏVE 11.1446	ESM3_E3 11.1815	HOLT3_E3 11.1910	ESM_E3 11.2377	ESM 11.2489	HOLT 11.2630	ARIMA_E3 11.4573	M15_E3 11.6238	MEAN_E3 11.7674	ARIMA 11.9588	MEAN 14.1795
J W Mays (MAYS)	NAÏVE 9.0293	HOLT 9.0469	ESM 9.0822	NAIVE3_E3 9.1059	HOLT3_E3 9.1251	ESM3_E3 9.1413	ESM_E3 9.2877	ARIMA_E3 9.4391	M15_E3 9.7856	MEAN_E3 10.3338	ARIMA 10.4396	MEAN 12.7140

Table B-34: Financial Sector (REIT/Real Estate) E3 RMSE Accuracy Results

SECURITY	RMSE #1	RMSE #2	RMSE #3	RMSE #4	RMSE #5	RMSE #6	RMSE #7	RMSE #8	RMSE #9	RMSE #10	RMSE #11	RMSE #12
REIT—OTHER EQUITY TRUSTS												
Host Hotels & Resorts (HST)	HOLT 0.0233	NAÏVE 0.0233	HOLT3_E3 0.0233	NAIVE3_E3 0.0233	ARIMA 0.0233	ESM3_E3 0.0234	ESM_E3 0.0234	ARIMA_E3 0.0236	ESM 0.0236	M15_E3 0.0241	MEAN_E3 0.0259	MEAN 0.0314
Public Storage (PSA)	M15_E3 0.1322	ESM_E3 0.1332	HOLT3_E3 0.1340	NAIVE3_E3 0.1345	ESM3_E3 0.1349	HOLT 0.1361	NAÏVE 0.1368	ESM 0.1374	MEAN_E3 0.1375	MEAN 0.1616	ARIMA_E3 0.2288	ARIMA 0.4924
Welltower Inc. (WELL)	HOLT3_E3 0.0581	NAIVE3_E3 0.0581	ARIMA_E3 0.0582	ESM_E3 0.0583	HOLT 0.0585	NAÏVE 0.0586	M15_E3 0.0587	ARIMA 0.0588	ESM3_E3 0.0588	ESM 0.0597	MEAN_E3 0.0626	MEAN 0.0726
Vornado Realty Trust (VNO)	ESM3_E3 0.0781	ESM_E3 0.0781	HOLT3_E3 0.0784	ESM 0.0785	HOLT 0.0787	NAIVE3_E3 0.0787	NAÏVE 0.0790	M15_E3 0.0795	ARIMA_E3 0.0840	MEAN_E3 0.0858	MEAN 0.0999	ARIMA 0.1089
EastGroup Properties, Inc. (EGP)	HOLT 0.0523	ESM 0.0523	ESM3_E3 0.0523	HOLT3_E3 0.0523	NAIVE3_E3 0.0524	NAÏVE 0.0524	ESM_E3 0.0526	ARIMA 0.0528	ARIMA_E3 0.0530	M15_E3 0.0541	MEAN_E3 0.0570	MEAN 0.0655
HCP, Inc. (HCP)	M15_E3 0.0403	ARIMA_E3 0.0403	ESM_E3 0.0406	HOLT3_E3 0.0408	NAIVE3_E3 0.0409	ESM3_E3 0.0410	MEAN_E3 0.0411	ARIMA 0.0412	HOLT 0.0413	NAÏVE 0.0414	ESM 0.0416	MEAN 0.0463
Duke Realty Corporation (DRE)	ESM3_E3 0.0358	HOLT3_E3 0.0358	ESM 0.0359	HOLT 0.0359	ESM_E3 0.0359	NAIVE3_E3 0.0359	ARIMA 0.0359	NAÏVE 0.0359	ARIMA_E3 0.0360	M15_E3 0.0363	MEAN_E3 0.0387	MEAN 0.0442

APPENDIX B: E3 Forecast Study Results

SECURITY	RMSE #1	RMSE #2	RMSE #3	RMSE #4	RMSE #5	RMSE #6	RMSE #7	RMSE #8	RMSE #9	RMSE #10	RMSE #11	RMSE #12
REIT—RETAIL EQUITY TRUSTS												
Federal Realty Investment Trust (FRT)	ESM_E3 0.0961	ESM3_E3 0.0961	HOLT3_E3 0.0962	NAIVE3_E3 0.0963	ESM 0.0967	HOLT 0.0968	NAÏVE 0.0969	MEAN_E3 0.0971	M15_E3 0.0977	MEAN 0.1073	ARIMA_E3 0.1234	ARIMA 0.2330
Pennsylvania REIT (PEI)	NAÏVE 0.0395	HOLT 0.0395	ESM 0.0395	ARIMA 0.0396	NAIVE3_E3 0.0396	HOLT3_E3 0.0396	ESM3_E3 0.0397	ESM_E3 0.0400	ARIMA_E3 0.0402	M15_E3 0.0410	MEAN_E3 0.0446	MEAN 0.0523
Urstadt Biddle Properties, Inc. (UBP)	ARIMA 0.0138	ESM_E3 0.0138	ESM3_E3 0.0138	HOLT3_E3 0.0138	ARIMA_E3 0.0139	NAIVE3_E3 0.0139	ESM 0.0140	HOLT 0.0140	NAÏVE 0.0142	M15_E3 0.0142	MEAN_E3 0.0144	MEAN 0.0165
National Retail Properties, Inc. (NNN)	ESM 0.0322	HOLT 0.0322	ESM3_E3 0.0323	HOLT3_E3 0.0323	NAIVE3_E3 0.0324	NAÏVE 0.0324	ARIMA 0.0324	ESM_E3 0.0325	ARIMA_E3 0.0328	M15_E3 0.0334	MEAN_E3 0.0338	MEAN 0.0381
Weingarten Realty Investors (WRI)	ARIMA_E3 0.0445	ESM_E3 0.0446	HOLT3_E3 0.0447	M15_E3 0.0448	NAIVE3_E3 0.0448	ESM3_E3 0.0448	ARIMA 0.0449	MEAN_E3 0.0450	HOLT 0.0450	NAÏVE 0.0451	ESM 0.0452	MEAN 0.0499
Kimco Realty Corporation (KIM)	HOLT 0.0442	NAÏVE 0.0442	HOLT3_E3 0.0442	NAIVE3_E3 0.0442	M15_E3 0.0450	ESM_E3 0.0451	ESM3_E3 0.0454	ESM 0.0458	MEAN_E3 0.0464	ARIMA_E3 0.0473	MEAN 0.0527	ARIMA 0.0635
Taubman Centes, Inc. (TCO)	ESM_E3 0.0666	ESM3_E3 0.0666	HOLT3_E3 0.0668	NAIVE3_E3 0.0668	ESM 0.0669	HOLT 0.0672	NAÏVE 0.0673	M15_E3 0.0675	MEAN_E3 0.0721	ARIMA_E3 0.0731	MEAN 0.0845	ARIMA 0.1117
REAL ESTATE OPERATIONS STOCKS												
Texas Pacific Land Trust (TPL)	M15_E3 0.5629	NAIVE3_E3 0.5649	NAÏVE 0.5663	HOLT3_E3 0.5779	ESM_E3 0.5822	HOLT 0.5845	MEAN_E3 0.5857	ESM3_E3 0.5979	ESM 0.6172	MEAN 0.6477	ARIMA_E3 0.6557	ARIMA 0.9608
Tejon Ranch Co. (TRC)	ARIMA_E3 0.0486	NAIVE3_E3 0.0488	ESM_E3 0.0488	M15_E3 0.0489	HOLT3_E3 0.0489	ESM3_E3 0.0490	NAÏVE 0.0491	ARIMA 0.0492	HOLT 0.0493	ESM 0.0495	MEAN_E3 0.0504	MEAN 0.0591
American Realty Investors, Inc. (ARL)	HOLT3_E3 0.0212	NAIVE3_E3 0.0212	ESM3_E3 0.0212	ESM_E3 0.0213	ARIMA_E3 0.0213	ARIMA 0.0214	HOLT 0.0214	NAÏVE 0.0214	ESM 0.0214	MEAN_E3 0.0215	M15_E3 0.0220	MEAN 0.0263
Brookfield Asset Management, Inc. (BAM)	NAIVE3_E3 0.0245	NAÏVE 0.0245	HOLT3_E3 0.0246	HOLT 0.0246	ESM_E3 0.0248	ESM3_E3 0.0249	M15_E3 0.0252	ESM 0.0253	MEAN_E3 0.0267	ARIMA_E3 0.0290	MEAN 0.0310	ARIMA 0.0545
FRP Holdings, Inc. (FRPH)	NAÏVE 0.0456	HOLT 0.0457	NAIVE3_E3 0.0459	ESM 0.0459	HOLT3_E3 0.0460	ESM3_E3 0.0461	ESM_E3 0.0465	MEAN_E3 0.0473	M15_E3 0.0479	ARIMA_E3 0.0493	MEAN 0.0521	ARIMA 0.0616
Income Opportunity Realty Investors, Inc. (IOR)	MEAN_E3 0.0123	ESM3_E3 0.0128	ESM_E3 0.0128	NAIVE3_E3 0.0129	ESM 0.0129	NAÏVE 0.0130	ARIMA_E3 0.0131	M15_E3 0.0132	HOLT3_E3 0.0133	HOLT 0.0135	ARIMA 0.0135	MEAN 0.0138
J W Mays (MAYS)	NAÏVE 0.0436	HOLT 0.0445	NAIVE3_E3 0.0446	ESM 0.0447	HOLT3_E3 0.0453	ESM3_E3 0.0454	ESM_E3 0.0463	ARIMA_E3 0.0481	M15_E3 0.0490	MEAN_E3 0.0498	MEAN 0.0576	ARIMA 0.0984

B.16. Industrial Products Sector

Table B-35: Industrial Products Sector E3 MAPE Accuracy Results

SECURITY	MAPE #1	MAPE #2	MAPE #3	MAPE #4	MAPE #5	MAPE #6	MAPE #7	MAPE #8	MAPE #9	MAPE #10	MAPE #11	MAPE #12
GENERAL INDUSTRIAL MACHINERY												
Ingersoll-Rand Plc (IR)	ARIMA_E3	ESM_E3	HOLT3_E3	NAIVE3_E3	ESM3_E3	M15_E3	HOLT	ARIMA	NAÏVE	ESM	MEAN_E3	MEAN
	9.3026	9.3357	9.3641	9.3688	9.3730	9.4409	9.4563	9.4569	9.4627	9.4718	10.3451	12.5942
Dover Corporation (DOV)	ESM_E3	HOLT3_E3	ESM3_E3	NAIVE3_E3	HOLT	M15_E3	NAÏVE	ARIMA_E3	ESM	ARIMA	MEAN_E3	MEAN
	8.0026	8.0491	8.0498	8.0527	8.0964	8.0986	8.1016	8.1254	8.1587	8.3839	8.4239	9.8603
Nordson Corporation (NDSN)	HOLT3_E3	NAIVE3_E3	ESM_E3	HOLT	ESM3_E3	NAÏVE	ARIMA_E3	ESM	M15_E3	MEAN_E3	ARIMA	MEAN
	8.7620	8.7841	8.7950	8.8010	8.8246	8.8379	8.8710	8.9002	8.9056	9.5143	9.6813	11.7960
Graco Inc. (GGG)	ESM3_E3	ESM_E3	NAIVE3_E3	HOLT3_E3	NAÏVE	HOLT	ESM	M15_E3	MEAN_E3	ARIMA_E3	MEAN	ARIMA
	7.6433	7.6536	7.7794	7.7805	7.7869	7.7962	7.8000	8.1200	8.6265	9.5098	10.0694	14.3069
Illinois Tools Works, Inc. (ITW)	ESM_E3	ARIMA_E3	ESM3_E3	HOLT3_E3	NAIVE3_E3	HOLT	NAÏVE	ESM	M15_E3	ARIMA	MEAN_E3	MEAN
	6.4431	6.4468	6.4828	6.5408	6.5441	6.5789	6.5834	6.6015	6.6162	6.7037	6.9224	8.3587
Parker-Hannifn Corporation (PH)	ARIMA_E3	M15_E3	NAIVE3_E3	HOLT3_E3	ESM_E3	NAÏVE	HOLT	ESM3_E3	ARIMA	ESM	MEAN_E3	MEAN
	8.8707	8.8999	8.9054	8.9204	8.9284	8.9811	9.0017	9.0248	9.0279	9.1693	9.1983	10.7132
Roper Technologies, Inc. (ROP)	HOLT3_E3	NAIVE3_E3	ESM_E3	HOLT	NAÏVE	ESM3_E3	M15_E3	ESM	ARIMA_E3	MEAN_E3	MEAN	ARIMA
	6.9724	6.9733	7.0182	7.0313	7.0336	7.0760	7.2562	7.2596	7.4747	7.7414	9.0835	9.4383
ELECTRICAL MACHINERY												
Emerson Electric Co (EMR)	ESM_E3	ESM3_E3	HOLT3_E3	NAIVE3_E3	ARIMA_E3	ESM	HOLT	NAÏVE	ARIMA	M15_E3	MEAN_E3	MEAN
	6.8670	6.8699	6.8750	6.8847	6.9084	6.9176	6.9189	6.9328	6.9583	6.9804	7.1488	8.4490
Eaton Corporation plc (ETN)	ESM3_E3	ESM	ESM_E3	NAIVE3_E3	HOLT3_E3	ARIMA_E3	NAÏVE	HOLT	M15_E3	ARIMA	MEAN_E3	MEAN
	8.3174	8.3388	8.3408	8.3940	8.4055	8.4103	8.4149	8.4316	8.5060	8.5398	8.9468	10.5685
Franlin Electric Co. Inc. (FELE)	ESM_E3	HOLT3_E3	NAIVE3_E3	ESM3_E3	HOLT	NAÏVE	ESM	M15_E3	ARIMA_E3	MEAN_E3	ARIMA	MEAN
	7.3144	7.3177	7.3469	7.3482	7.3680	7.4170	7.4618	7.4788	7.8392	7.9541	9.0918	9.3392
A. O. Smith Corporation (AOS)	ESM_E3	ESM3_E3	NAIVE3_E3	M15_E3	HOLT3_E3	ESM	NAÏVE	HOLT	ARIMA_E3	MEAN_E3	MEAN	ARIMA
	9.5826	9.6272	9.6779	9.7090	9.7334	9.7432	9.7755	9.8628	9.9463	10.1527	11.7995	13.0693
Regal Beloit Corporation (RBC)	ESM_E3	ARIMA_E3	HOLT3_E3	NAIVE3_E3	ESM3_E3	M15_E3	HOLT	NAÏVE	ARIMA	ESM	MEAN_E3	MEAN
	8.4170	8.4205	8.4522	8.4555	8.4746	8.4988	8.5469	8.5514	8.5886	8.5986	8.9651	10.6831
II-VI Incorporated (IIVI)	HOLT3_E3	ARIMA	NAIVE3_E3	HOLT	NAÏVE	ARIMA_E3	M15_E3	ESM_E3	MEAN_E3	ESM3_E3	ESM	MEAN
	11.7190	11.7349	11.7452	11.7946	11.7995	11.8756	12.2336	12.3281	12.5865	12.6015	12.9866	14.7872
ESCO Technologies, Inc. (ESE)	ESM	ESM3_E3	NAIVE3_E3	HOLT3_E3	NAÏVE	ESM_E3	HOLT	M15_E3	ARIMA_E3	MEAN_E3	ARIMA	MEAN
	8.0023	8.0063	8.0594	8.0670	8.0704	8.0779	8.0805	8.4023	8.7314	8.8776	10.4028	10.6552

APPENDIX B: E3 Forecast Study Results

SECURITY	MAPE #1	MAPE #2	MAPE #3	MAPE #4	MAPE #5	MAPE #6	MAPE #7	MAPE #8	MAPE #9	MAPE #10	MAPE #11	MAPE #12
PROTECTION—SAFETY EQUIPMENT & SERVICES												
MSA Safety Incorporated (MSA)	NAIVE3_E3 8.4135	HOLT 8.4161	HOLT3_E3 8.4235	NAÏVE 8.4374	ESM_E3 8.6873	M15_E3 8.7867	ESM3_E3 8.8534	ARIMA_E3 8.8621	ESM 9.1860	MEAN_E3 9.5400	ARIMA 10.8797	MEAN 11.5215
The Eastern Company (EML)	NAIVE3_E3 8.0946	HOLT3_E3 8.0967	NAÏVE 8.1125	ESM3_E3 8.1149	HOLT 8.1398	ESM 8.1436	ESM_E3 8.1941	ARIMA_E3 8.3638	ARIMA 8.4203	M15_E3 8.5861	MEAN_E3 9.0763	MEAN 10.7177
Napco Security Technologies, Inc. (NSSC)	ESM3_E3 10.4459	ESM 10.4491	NAIVE3_E3 10.5049	HOLT3_E3 10.5110	NAÏVE 10.5203	ESM_E3 10.5261	HOLT 10.5330	M15_E3 10.9615	ARIMA_E3 11.6492	MEAN_E3 13.4732	ARIMA 13.7007	MEAN 17.6654
Brady Corporation (BRC)	ESM_E3 7.8048	HOLT3_E3 7.8270	ARIMA_E3 7.8349	NAIVE3_E3 7.8456	ESM3_E3 7.8889	M15_E3 7.8999	HOLT 7.9483	NAÏVE 7.9826	ESM 8.0560	ARIMA 8.0788	MEAN_E3 8.3050	MEAN 9.8766
Johnson Controls Intl. plc (JCI)	NAIVE3_E3 11.5976	HOLT3_E3 11.6020	NAÏVE 11.6207	HOLT 11.6289	ESM_E3 11.6395	ARIMA_E3 11.6490	ARIMA 11.6515	ESM3_E3 11.6602	ESM 11.7650	M15_E3 11.8845	MEAN_E3 13.3855	MEAN 16.3377
Lakeland Industries, Inc. (LAKE)	NAIVE3_E3 10.8390	HOLT3_E3 10.8450	ESM_E3 10.8868	NAÏVE 10.9092	HOLT 10.9199	ESM3_E3 10.9249	ESM 11.0945	M15_E3 11.2644	MEAN_E3 12.1794	ARIMA_E3 13.2386	MEAN 15.0239	ARIMA 17.1903
Magal Security Systems Ltd. (MAGS)	M15_E3 12.4079	NAIVE3_E3 12.0324	HOLT3_E3 12.1031	NAÏVE 12.1621	HOLT 12.2657	ESM_E3 12.3846	ESM3_E3 12.6305	MEAN_E3 12.9947	ESM 13.0164	ARIMA_E3 15.0930	MEAN 15.4071	ARIMA 20.6161
POLLUTION CONTROL EQUIPMENT & SERVICES												
Donaldson Company, Inc. (DCI)	NAIVE3_E3 6.4847	HOLT3_E3 6.4901	NAÏVE 6.5017	HOLT 6.5077	ESM_E3 6.5604	ESM3_E3 6.5679	ESM 6.6417	M15_E3 6.7364	MEAN_E3 6.9361	ARIMA_E3 7.3117	MEAN 8.2179	ARIMA 9.0341
CECO Environmental Corp. (CECE)	NAÏVE 18.7880	HOLT 18.9017	NAIVE3_E3 19.0741	HOLT3_E3 19.2022	ARIMA 19.2445	ARIMA_E3 19.7395	M15_E3 20.7850	ESM_E3 21.0697	ESM3_E3 21.7646	ESM 22.7908	MEAN_E3 22.9068	MEAN 28.0183
Ecology & Environment, Inc. (EEI)	ESM3_E3 6.6079	HOLT3_E3 6.6151	NAIVE3_E3 6.6179	ESM 6.6201	HOLT 6.6308	NAÏVE 6.6346	ARIMA 6.6408	ESM_E3 6.6638	ARIMA_E3 6.7203	M15_E3 6.9315	MEAN_E3 7.5565	MEAN 9.6600
Tetra Tech, Inc. (TTEK)	ARIMA 10.3585	NAÏVE 10.3670	HOLT 10.3684	NAIVE3_E3 10.3926	HOLT3_E3 10.3949	ESM 10.4110	ESM3_E3 10.4224	ESM_E3 10.5136	ARIMA_E3 10.5832	M15_E3 10.8710	MEAN_E3 11.0421	MEAN 12.7417
Appliance Recycling Centers of America (ARCI)	NAÏVE 18.7880	HOLT 18.9017	NAIVE3_E3 19.0741	HOLT3_E3 19.2022	ARIMA 19.2445	ARIMA_E3 19.7395	M15_E3 20.7850	ESM_E3 21.0697	ESM3_E3 21.7646	ESM 22.7908	MEAN_E3 22.9068	MEAN 28.0183
Fuel Tech, Inc. (FTEK)	HOLT 13.6975	NAÏVE 13.7109	HOLT3_E3 13.7521	NAIVE3_E3 13.7603	ARIMA_E3 14.2521	M15_E3 14.5687	ESM_E3 14.8362	ARIMA 15.0240	ESM3_E3 15.2351	MEAN_E3 15.6604	ESM 15.8090	MEAN 19.7447

Table B-36: Industrial Products Sector E3 RMSE Accuracy Results

SECURITY	RMSE #1	RMSE #2	RMSE #3	RMSE #4	RMSE #5	RMSE #6	RMSE #7	RMSE #8	RMSE #9	RMSE #10	RMSE #11	RMSE #12
GENERAL INDUSTRIAL MACHINERY												
Ingersoll-Rand Plc (IR)	ARIMA_E3 0.0583	ARIMA 0.0583	HOLT3_E3 0.0584	NAIVE3_E3 0.0584	HOLT 0.0585	NAÏVE 0.0585	ESM3_E3 0.0586	ESM_E3 0.0588	ESM 0.0588	M15_E3 0.0602	MEAN_E3 0.0642	MEAN 0.0758
Dover Corporation (DOV)	HOLT3_E3 0.0545	NAIVE3_E3 0.0545	HOLT 0.0545	NAÏVE 0.0546	ESM_E3 0.0546	ESM3_E3 0.0547	ARIMA_E3 0.0552	ESM 0.0552	M15_E3 0.0557	ARIMA 0.0564	MEAN_E3 0.0575	MEAN 0.0670
Nordson Corporation (NDSN)	ESM 0.0764	HOLT 0.0766	NAÏVE 0.0766	ESM3_E3 0.0767	HOLT3_E3 0.0770	NAIVE3_E3 0.0770	ESM_E3 0.0774	M15_E3 0.0801	ARIMA_E3 0.0810	MEAN_E3 0.0823	MEAN 0.0978	ARIMA 0.1323
Graco Inc. (GGG)	HOLT 0.0233	NAÏVE 0.0234	ESM_E3 0.0235	HOLT3_E3 0.0235	NAIVE3_E3 0.0235	ESM3_E3 0.0235	ESM 0.0240	M15_E3 0.0244	MEAN_E3 0.0262	ARIMA_E3 0.0268	MEAN 0.0304	ARIMA 0.0445
Illinois Tools Works, Inc. (ITW)	ESM3_E3 0.0728	ARIMA_E3 0.0730	ESM_E3 0.0733	ESM 0.0735	HOLT 0.0740	NAÏVE 0.0741	HOLT3_E3 0.0745	NAIVE3_E3 0.0746	M15_E3 0.0779	MEAN_E3 0.0840	ARIMA 0.0941	MEAN 0.0989
Parker-Hannifn Corporation (PH)	NAÏVE 0.1204	NAIVE3_E3 0.1207	HOLT 0.1208	HOLT3_E3 0.1209	ARIMA_E3 0.1226	ARIMA 0.1230	M15_E3 0.1237	ESM_E3 0.1238	MEAN_E3 0.1246	ESM3_E3 0.1253	ESM 0.1278	MEAN 0.1448
Roper Technologies, Inc. (ROP)	ESM_E3 0.1179	NAÏVE 0.1186	HOLT 0.1187	ESM3_E3 0.1193	NAIVE3_E3 0.1193	HOLT3_E3 0.1194	ESM 0.1234	M15_E3 0.1237	MEAN_E3 0.1249	ARIMA_E3 0.1309	MEAN 0.1392	ARIMA 0.2358
ELECTRICAL MACHINERY												
Emerson Electric Co (EMR)	HOLT3_E3 0.0563	NAIVE3_E3 0.0564	ESM_E3 0.0565	ARIMA_E3 0.0565	HOLT 0.0566	NAÏVE 0.0568	ESM3_E3 0.0568	M15_E3 0.0569	ARIMA 0.0571	ESM 0.0576	MEAN_E3 0.0589	MEAN 0.0687
Eaton Corporation plc (ETN)	NAIVE3_E3 0.0644	HOLT3_E3 0.0644	NAÏVE 0.0644	HOLT 0.0644	ESM_E3 0.0648	ARIMA_E3 0.0649	ESM3_E3 0.0649	ESM 0.0654	M15_E3 0.0658	MEAN_E3 0.0671	ARIMA 0.0675	MEAN 0.0776
Franlin Electric Co. Inc. (FELE)	ESM_E3 0.0341	HOLT3_E3 0.0342	NAIVE3_E3 0.0342	ESM3_E3 0.0343	HOLT 0.0344	NAÏVE 0.0345	ESM 0.0348	M15_E3 0.0349	MEAN_E3 0.0358	ARIMA_E3 0.0367	MEAN 0.0415	ARIMA 0.0515
A. O. Smith Corporation (AOS)	ESM 0.0268	ESM3_E3 0.0273	HOLT 0.0275	NAÏVE 0.0275	HOLT3_E3 0.0281	NAIVE3_E3 0.0281	ESM_E3 0.0282	M15_E3 0.0308	MEAN_E3 0.0346	ARIMA_E3 0.0381	MEAN 0.0409	ARIMA 0.1059
Regal Beloit Corporation (RBC)	HOLT3_E3 0.0754	NAIVE3_E3 0.0755	HOLT 0.0756	NAÏVE 0.0756	ESM_E3 0.0760	ARIMA_E3 0.0762	ESM3_E3 0.0762	ARIMA 0.0763	ESM 0.0771	M15_E3 0.0773	MEAN_E3 0.0781	MEAN 0.0893
II-VI Incorporated (IIVI)	HOLT 0.0406	NAÏVE 0.0406	ARIMA 0.0407	HOLT3_E3 0.0407	NAIVE3_E3 0.0407	ARIMA_E3 0.0414	M15_E3 0.0422	MEAN_E3 0.0425	ESM_E3 0.0435	ESM3_E3 0.0446	ESM 0.0461	MEAN 0.0471
ESCO Technologies, Inc. (ESE)	ESM 0.0551	ESM3_E3 0.0553	NAÏVE 0.0555	HOLT 0.0555	NAIVE3_E3 0.0556	HOLT3_E3 0.0556	ESM_E3 0.0559	M15_E3 0.0577	MEAN_E3 0.0595	ARIMA_E3 0.0655	MEAN 0.0687	ARIMA 0.1081

APPENDIX B: E3 Forecast Study Results

SECURITY	RMSE #1	RMSE #2	RMSE #3	RMSE #4	RMSE #5	RMSE #6	RMSE #7	RMSE #8	RMSE #9	RMSE #10	RMSE #11	RMSE #12
PROTECTION—SAFETY EQUIPMENT & SERVICES												
MSA Safety Incorporated (MSA)	NAÏVE 0.0605	HOLT 0.0605	NAIVE3_E3 0.0610	HOLT3_E3 0.0610	ESM_E3 0.0626	ESM3_E3 0.0638	M15_E3 0.0643	ARIMA_E3 0.0644	ESM 0.0663	MEAN_E3 0.0670	MEAN 0.0782	ARIMA 0.1023
The Eastern Company (EML)	NAÏVE 0.0300	HOLT 0.0300	HOLT3_E3 0.0301	ESM 0.0301	NAIVE3_E3 0.0301	ESM3_E3 0.0302	ESM_E3 0.0304	M15_E3 0.0312	ARIMA_E3 0.0320	MEAN_E3 0.0327	ARIMA 0.0348	MEAN 0.0372
Napco Security Technologies, Inc. (NSSC)	NAÏVE 0.0096	HOLT 0.0097	NAIVE3_E3 0.0097	ESM 0.0097	HOLT3_E3 0.0097	ESM3_E3 0.0097	ESM_E3 0.0098	M15_E3 0.0103	MEAN_E3 0.0124	ARIMA_E3 0.0141	MEAN 0.0163	ARIMA 0.0253
Brady Corporation (BRC)	HOLT3_E3 0.0409	NAIVE3_E3 0.0409	ARIMA_E3 0.0410	ESM_E3 0.0411	M15_E3 0.0411	HOLT 0.0412	NAÏVE 0.0413	ESM3_E3 0.0415	ARIMA 0.0420	ESM 0.0421	MEAN_E3 0.0438	MEAN 0.0518
Johnson Controls Intl. plc (JCI)	MEAN_E3 0.1433	ESM 0.1433	ESM3_E3 0.1436	NAÏVE 0.1436	HOLT 0.1436	HOLT3_E3 0.1442	NAIVE3_E3 0.1442	ESM_E3 0.1445	ARIMA 0.1449	ARIMA_E3 0.1458	M15_E3 0.1480	MEAN 0.1567
Lakeland Industries, Inc. (LAKE)	ESM_E3 0.0232	NAIVE3_E3 0.0233	HOLT3_E3 0.0233	ESM3_E3 0.0233	NAÏVE 0.0235	M15_E3 0.0235	HOLT 0.0235	ESM 0.0236	MEAN_E3 0.0243	MEAN 0.0281	ARIMA_E3 0.0464	ARIMA 0.1051
Magal Security Systems Ltd. (MAGS)	M15_E3 0.0209	NAIVE3_E3 0.0203	ESM_E3 0.0206	HOLT3_E3 0.0207	NAÏVE 0.0207	ESM3_E3 0.0209	HOLT 0.0214	ESM 0.0215	MEAN_E3 0.0224	MEAN 0.0273	ARIMA_E3 0.0821	ARIMA 0.2167
POLLUTION CONTROL EQUIPMENT & SERVICES												
Donaldson Company, Inc. (DCI)	NAÏVE 0.0331	HOLT 0.0331	NAIVE3_E3 0.0332	HOLT3_E3 0.0332	MEAN_E3 0.0333	ESM3_E3 0.0336	ESM_E3 0.0336	ESM 0.0338	M15_E3 0.0343	ARIMA_E3 0.0343	MEAN 0.0380	ARIMA 0.0446
CECO Environmental Corp. (CECE)	HOLT3_E3 0.0122	ARIMA_E3 0.0122	NAIVE3_E3 0.0122	HOLT 0.0122	ARIMA 0.0122	NAÏVE 0.0123	M15_E3 0.0124	ESM_E3 0.0125	ESM3_E3 0.0127	MEAN_E3 0.0130	ESM 0.0131	MEAN 0.0153
Ecology & Environment, Inc. (EEI)	ESM 0.0140	HOLT 0.0140	NAÏVE 0.0140	HOLT3_E3 0.0140	ESM3_E3 0.0140	NAIVE3_E3 0.0140	ARIMA 0.0140	ESM_E3 0.0142	ARIMA_E3 0.0143	M15_E3 0.0148	MEAN_E3 0.0165	MEAN 0.0211
Tetra Tech, Inc. (TTEK)	NAÏVE 0.0468	HOLT 0.0469	ESM 0.0470	NAIVE3_E3 0.0470	HOLT3_E3 0.0471	ESM3_E3 0.0472	ARIMA 0.0474	ESM_E3 0.0475	ARIMA_E3 0.0480	M15_E3 0.0489	MEAN_E3 0.0492	MEAN 0.0549
Appliance Recycling Centers of America (ARCI)	HOLT3_E3 0.0122	ARIMA_E3 0.0122	NAIVE3_E3 0.0122	HOLT 0.0122	ARIMA 0.0122	NAÏVE 0.0123	M15_E3 0.0124	ESM_E3 0.0125	ESM3_E3 0.0127	MEAN_E3 0.0130	ESM 0.0131	MEAN 0.0153
Fuel Tech, Inc. (FTEK)	NAÏVE 0.0292	HOLT 0.0292	NAIVE3_E3 0.0295	HOLT3_E3 0.0295	ESM 0.0297	ESM3_E3 0.0298	ESM_E3 0.0301	M15_E3 0.0312	MEAN_E3 0.0314	ARIMA_E3 0.0318	MEAN 0.0365	ARIMA 0.0429

B.17. Medical Sector

Table B-37: Medical Sector E3 MAPE Accuracy Results

SECURITY	MAPE #1	MAPE #2	MAPE #3	MAPE #4	MAPE #5	MAPE #6	MAPE #7	MAPE #8	MAPE #9	MAPE #10	MAPE #11	MAPE #12
LARGE CAP PHARMACEUTICAL STOCKS												
Johnson & Johnson (JNJ)	ESM_E3 4.4489	HOLT3_E3 4.4598	ESM3_E3 4.4599	NAIVE3_E3 4.4602	HOLT 4.4926	NAÏVE 4.4932	ESM 4.5077	M15_E3 4.5670	MEAN_E3 4.9258	ARIMA_E3 5.7532	MEAN 5.9657	ARIMA 8.6005
Pfizer Inc. (PFE)	NAÏVE 5.8151	HOLT 5.8152	NAIVE3_E3 5.8158	HOLT3_E3 5.8159	ESM_E3 5.9471	ESM3_E3 6.0256	M15_E3 6.0949	ESM 6.1831	MEAN_E3 6.3368	ARIMA_E3 6.4020	ARIMA 7.5626	MEAN 7.5659
Merck & Co, Inc. (MRK)	HOLT 6.4003	NAÏVE 6.4005	ESM 6.4908	HOLT3_E3 6.5394	NAIVE3_E3 6.5396	ESM3_E3 6.5675	ARIMA 6.7005	ESM_E3 6.7535	ARIMA_E3 6.9892	M15_E3 7.3633	MEAN_E3 7.9130	MEAN 9.3193
Eli Lilly and Company (LLY)	NAÏVE 6.2333	HOLT 6.2335	NAIVE3_E3 6.2961	HOLT3_E3 6.2963	ESM3_E3 6.4081	ESM 6.4221	ESM_E3 6.4622	MEAN_E3 6.5699	M15_E3 6.7442	MEAN 7.3197	ARIMA_E3 7.3296	ARIMA 8.6850
Novo Nordisk A/S (NVO)	ESM3_E3 6.5868	ESM_E3 6.6045	ESM 6.6636	NAIVE3_E3 6.7534	HOLT3_E3 6.7753	NAÏVE 6.7830	HOLT 6.8134	M15_E3 6.9198	MEAN_E3 8.1182	ARIMA_E3 8.7535	MEAN 10.1549	ARIMA 14.7281
Bristol-Myers Squibb Company (BMY)	NAIVE3_E3 6.8909	HOLT3_E3 6.8913	NAÏVE 6.9362	HOLT 6.9368	M15_E3 6.9999	ESM_E3 7.1137	ARIMA_E3 7.1898	ESM3_E3 7.2501	ESM 7.4328	MEAN_E3 7.5152	ARIMA 8.0529	MEAN 9.2196
GlaxoSmithKline plc (GSK)	HOLT3_E3 5.1134	NAIVE3_E3 5.1139	HOLT 5.1157	NAÏVE 5.1182	ARIMA 5.1994	ARIMA_E3 5.2232	ESM_E3 5.2865	M15_E3 5.3507	ESM3_E3 5.3624	MEAN_E3 5.5002	ESM 5.5089	MEAN 6.4856
BIOMEDICAL AND GENETICS STOCKS												
Amgen Inc. (AMGN)	HOLT 6.5226	NAÏVE 6.5282	HOLT3_E3 6.5571	NAIVE3_E3 6.5599	ESM_E3 6.8460	M15_E3 6.9828	ESM3_E3 7.1111	ARIMA_E3 7.1492	MEAN_E3 7.2949	ESM 7.5602	ARIMA 8.5789	MEAN 8.7022
Celgene Corporation (CELG)	HOLT3_E3 11.0667	NAIVE3_E3 11.0910	HOLT 11.1161	NAÏVE 11.1425	ESM_E3 11.3208	M15_E3 11.4322	ESM3_E3 11.6289	ESM 12.1320	ARIMA_E3 12.8417	MEAN_E3 13.0177	MEAN 16.6021	ARIMA 18.5271
Bio-Techne Corporation (TECH)	NAIVE3_E3 7.7362	NAÏVE 7.8399	M15_E3 7.9063	HOLT3_E3 7.9150	HOLT 8.0702	ESM_E3 8.2479	MEAN_E3 8.4715	ESM3_E3 8.8033	ARIMA_E3 9.1192	ESM 9.5918	MEAN 10.3245	ARIMA 11.8157
Biogen Inc. (BIIB)	HOLT 10.3891	NAÏVE 10.4168	HOLT3_E3 10.5251	NAIVE3_E3 10.5478	ESM3_E3 10.9594	ESM_E3 10.9616	ARIMA_E3 11.0531	ESM 11.1447	M15_E3 11.4780	MEAN_E3 11.8664	ARIMA 13.7750	MEAN 14.0290
Gilead Sciences, Inc. (GILD)	NAÏVE 8.4141	HOLT 8.4175	NAIVE3_E3 8.4775	HOLT3_E3 8.4810	ESM3_E3 8.5138	ESM 8.5570	ESM_E3 8.5786	M15_E3 9.0514	MEAN_E3 9.5496	ARIMA_E3 9.8083	MEAN 11.3241	ARIMA 15.3191
Vertex Pharmaceuticals Inc. (VRTX)	NAÏVE 14.2582	HOLT 14.3077	ARIMA 14.3713	NAIVE3_E3 14.3811	HOLT3_E3 14.4192	ARIMA_E3 14.6824	ESM3_E3 14.8907	ESM_E3 14.9250	ESM 15.0751	M15_E3 15.4108	MEAN_E3 15.7330	MEAN 19.4844
Regeneron Pharmaceuticals, Inc. (REGN)	HOLT 14.3914	ESM 14.4472	NAÏVE 14.4862	HOLT3_E3 14.6615	ESM3_E3 14.7506	NAIVE3_E3 14.7914	ESM_E3 15.2515	MEAN_E3 16.5749	M15_E3 16.7482	ARIMA_E3 17.0375	MEAN 19.0453	ARIMA 22.3456

APPENDIX B: E3 Forecast Study Results

SECURITY	MAPE #1	MAPE #2	MAPE #3	MAPE #4	MAPE #5	MAPE #6	MAPE #7	MAPE #8	MAPE #9	MAPE #10	MAPE #11	MAPE #12
MEDICAL PRODUCTS MANUFACTURING												
Abbott Laboratories (ABT)	ESM 5.3968	ESM3_E3 5.4077	HOLT 5.4902	NAÏVE 5.4937	ESM_E3 5.5014	HOLT3_E3 5.5245	NAIVE3_E3 5.5256	M15_E3 5.8733	ARIMA_E3 5.8756	ARIMA 6.0480	MEAN_E3 6.1958	MEAN 7.4354
Stryker Corporation (SYK)	ESM3_E3 6.3708	ESM_E3 6.3741	HOLT 6.5232	NAÏVE 6.5331	HOLT3_E3 6.5485	NAIVE3_E3 6.5554	ESM 6.5564	M15_E3 6.8512	MEAN_E3 7.4929	ARIMA_E3 7.9290	MEAN 8.8462	ARIMA 13.2360
Bio-Rad Laboratories, Inc. (BIO)	ESM3_E3 8.0362	ESM 8.0530	ESM_E3 8.0939	NAIVE3_E3 8.1993	HOLT3_E3 8.1998	NAÏVE 8.2048	HOLT 8.2089	M15_E3 8.4897	ARIMA_E3 8.7298	MEAN_E3 8.8788	MEAN 10.4944	ARIMA 10.8066
Hill-Rom Holdings, Inc. (HRC)	ESM 7.8473	ESM3_E3 7.8864	NAÏVE 7.8911	HOLT 7.8921	ARIMA 7.9036	HOLT3_E3 7.9357	NAIVE3_E3 7.9363	ESM_E3 8.0006	ARIMA_E3 8.1228	M15_E3 8.3964	MEAN_E3 8.9578	MEAN 10.8332
Medtronic plc (MDT)	HOLT 6.0722	NAÏVE 6.0723	HOLT3_E3 6.1031	NAIVE3_E3 6.1032	ESM_E3 6.1234	ESM3_E3 6.1527	ESM 6.2940	M15_E3 6.4182	MEAN_E3 6.6108	ARIMA_E3 7.1180	MEAN 7.8621	ARIMA 9.2953
Baxter International Inc. (BAX)	ESM 6.2498	ESM3_E3 6.2562	NAIVE3_E3 6.3074	HOLT3_E3 6.3081	ESM_E3 6.3095	NAÏVE 6.3121	HOLT 6.3127	M15_E3 6.5191	ARIMA_E3 7.0631	MEAN_E3 7.1603	ARIMA 8.3735	MEAN 8.8698
Boston Scientific Corporation (BSX)	NAÏVE 9.3019	ESM 9.3026	HOLT 9.3028	ARIMA 9.3236	ESM3_E3 9.3599	NAIVE3_E3 9.3701	HOLT3_E3 9.3706	ESM_E3 9.4907	ARIMA_E3 9.5967	M15_E3 9.9206	MEAN_E3 10.7996	MEAN 13.3555
MEDICAL INSTRUMENTS MANUFACTURING												
Thermo Fisher Scientific Inc. (TMO)	ESM 7.4017	ESM3_E3 7.4584	ESM_E3 7.5939	NAÏVE 7.6494	HOLT 7.6594	NAIVE3_E3 7.7037	HOLT3_E3 7.7088	M15_E3 8.0649	ARIMA_E3 8.1024	MEAN_E3 8.3182	ARIMA 9.1425	MEAN 9.9160
Varian Medical Systems, Inc. (VAR)	NAÏVE 8.0680	NAIVE3_E3 8.0845	HOLT3_E3 8.1358	HOLT 8.1403	ESM3_E3 8.1660	ESM_E3 8.1781	ESM 8.2002	M15_E3 8.3594	MEAN_E3 8.8051	ARIMA_E3 9.1492	MEAN 10.6906	ARIMA 12.3175
ABIOMED, Inc. (ABMD)	NAIVE3_E3 17.0258	HOLT3_E3 17.0412	ESM_E3 17.0695	NAÏVE 17.1345	ESM3_E3 17.1424	HOLT 17.1567	M15_E3 17.2890	ESM 17.3386	ARIMA_E3 17.7803	MEAN_E3 18.0025	ARIMA 21.3786	MEAN 21.9347
Teleflex Inforporated (TFX)	ESM3_E3 7.0247	ESM_E3 7.0361	ESM 7.0899	NAIVE3_E3 7.1024	HOLT3_E3 7.1302	NAÏVE 7.1694	M15_E3 7.2229	HOLT 7.2375	ARIMA_E3 7.3082	MEAN_E3 7.8165	ARIMA 8.3606	MEAN 9.7989
Hologic, Inc. (HOLX)	NAÏVE 12.3865	NAIVE3_E3 12.4216	HOLT 12.4473	HOLT3_E3 12.4654	ESM3_E3 12.6452	ESM_E3 12.6558	ESM 12.7158	M15_E3 12.8171	MEAN_E3 13.0589	ARIMA_E3 13.7054	MEAN 14.8097	ARIMA 16.5273
IDEXX Laboratories (IDXX)	ESM 8.5667	ESM3_E3 8.5675	ESM_E3 8.7218	HOLT3_E3 9.0502	NAIVE3_E3 9.0509	HOLT 9.0677	NAÏVE 9.0692	M15_E3 9.3240	ARIMA_E3 9.4514	MEAN_E3 9.7031	ARIMA 10.7554	MEAN 11.0593
STERIS plc (STE)	ARIMA 6.8506	NAÏVE 6.8512	HOLT 6.8539	ESM3_E3 6.9355	ESM 6.9464	NAIVE3_E3 6.9492	HOLT3_E3 6.9523	ESM_E3 7.0440	ARIMA_E3 7.2248	M15_E3 7.6308	MEAN_E3 7.6550	MEAN 8.7946

Table B-38: Medical Sector E3 RMSE Accuracy Results

SECURITY	RMSE #1	RMSE #2	RMSE #3	RMSE #4	RMSE #5	RMSE #6	RMSE #7	RMSE #8	RMSE #9	RMSE #10	RMSE #11	RMSE #12
LARGE CAP PHARMACEUTICAL STOCKS												
Johnson & Johnson (JNJ)	ESM3_E3	ESM_E3	ESM	HOLT3_E3	NAIVE3_E3	HOLT	NAÏVE	M15_E3	MEAN_E3	MEAN	ARIMA_E3	ARIMA
	0.0622	0.0623	0.0625	0.0630	0.0630	0.0633	0.0633	0.0642	0.0676	0.0806	0.0925	0.1966
Pfizer Inc. (PFE)	NAÏVE	HOLT	NAIVE3_E3	HOLT3_E3	ESM_E3	ESM3_E3	ESM	M15_E3	MEAN_E3	MEAN	ARIMA_E3	ARIMA
	0.0324	0.0324	0.0326	0.0326	0.0332	0.0333	0.0340	0.0344	0.0345	0.0396	0.0432	0.0804
Merck & Co, Inc. (MRK)	HOLT	NAÏVE	HOLT3_E3	NAIVE3_E3	ESM3_E3	ESM_E3	ESM	M15_E3	ARIMA_E3	MEAN_E3	ARIMA	MEAN
	0.0665	0.0665	0.0671	0.0671	0.0688	0.0692	0.0695	0.0730	0.0733	0.0739	0.0796	0.0836
Eli Lilly and Company (LLY)	NAIVE3_E3	HOLT3_E3	NAÏVE	HOLT	MEAN_E3	M15_E3	ESM_E3	ESM3_E3	ESM	MEAN	ARIMA_E3	ARIMA
	0.0838	0.0838	0.0841	0.0841	0.0855	0.0863	0.0863	0.0879	0.0903	0.0958	0.1505	0.3222
Novo Nordisk A/S (NVO)	ESM	ESM3_E3	ESM_E3	NAÏVE	HOLT	NAIVE3_E3	HOLT3_E3	M15_E3	MEAN_E3	MEAN	ARIMA_E3	ARIMA
	0.0326	0.0326	0.0330	0.0331	0.0331	0.0333	0.0333	0.0347	0.0406	0.0502	0.0514	0.1096
Bristol-Myers Squibb Company (BMY)	M15_E3	NAIVE3_E3	HOLT3_E3	NAÏVE	HOLT	ESM_E3	MEAN_E3	ESM3_E3	ESM	ARIMA_E3	MEAN	ARIMA
	0.0604	0.0611	0.0611	0.0618	0.0619	0.0630	0.0641	0.0650	0.0673	0.0722	0.0764	0.1124
GlaxoSmithKline plc (GSK)	HOLT3_E3	NAIVE3_E3	HOLT	NAÏVE	ARIMA_E3	M15_E3	ARIMA	ESM_E3	ESM3_E3	MEAN_E3	ESM	MEAN
	0.0418	0.0418	0.0420	0.0421	0.0425	0.0430	0.0432	0.0435	0.0450	0.0450	0.0471	0.0520
BIOMEDICAL AND GENETICS STOCKS												
Amgen Inc. (AMGN)	NAIVE3_E3	HOLT3_E3	NAÏVE	HOLT	ESM_E3	MEAN_E3	ESM3_E3	M15_E3	ARIMA_E3	ESM	MEAN	ARIMA
	0.1209	0.1209	0.1209	0.1209	0.1212	0.1223	0.1237	0.1241	0.1269	0.1285	0.1366	0.1677
Celgene Corporation (CELG)	M15_E3	ESM_E3	HOLT3_E3	NAIVE3_E3	MEAN_E3	HOLT	NAÏVE	ESM3_E3	ESM	MEAN	ARIMA_E3	ARIMA
	0.1008	0.1017	0.1019	0.1020	0.1023	0.1030	0.1031	0.1036	0.1065	0.1151	0.1439	0.2742
Bio-Techne Corporation (TECH)	MEAN_E3	M15_E3	NAIVE3_E3	NAÏVE	HOLT3_E3	ESM_E3	HOLT	MEAN	ESM3_E3	ESM	ARIMA_E3	ARIMA
	0.1126	0.1157	0.1191	0.1210	0.1226	0.1255	0.1259	0.1277	0.1337	0.1437	0.1632	0.2693
Biogen Inc. (BIIB)	HOLT	NAÏVE	HOLT3_E3	NAIVE3_E3	M15_E3	ESM_E3	MEAN_E3	ESM3_E3	ESM	ARIMA_E3	MEAN	ARIMA
	0.3067	0.3068	0.3102	0.3102	0.3270	0.3296	0.3301	0.3376	0.3498	0.3528	0.3733	0.6870
Gilead Sciences, Inc. (GILD)	ESM3_E3	ESM	ESM_E3	NAÏVE	HOLT	NAIVE3_E3	HOLT3_E3	M15_E3	MEAN_E3	MEAN	ARIMA_E3	ARIMA
	0.0664	0.0666	0.0668	0.0672	0.0672	0.0673	0.0673	0.0693	0.0712	0.0837	0.0842	0.1810
Vertex Pharmaceuticals Inc. (VRTX)	NAÏVE	NAIVE3_E3	HOLT	HOLT3_E3	ARIMA_E3	ARIMA	ESM_E3	M15_E3	ESM3_E3	MEAN_E3	ESM	MEAN
	0.1483	0.1488	0.1492	0.1495	0.1501	0.1520	0.1557	0.1569	0.1598	0.1659	0.1666	0.1972
Regeneron Pharmaceuticals, Inc. (REGN)	ESM_E3	ESM3_E3	NAIVE3_E3	HOLT3_E3	ESM	NAÏVE	HOLT	M15_E3	MEAN_E3	MEAN	ARIMA_E3	ARIMA
	0.4264	0.4265	0.4282	0.4287	0.4292	0.4314	0.4321	0.4339	0.4381	0.5017	0.6128	1.3729

APPENDIX B: E3 Forecast Study Results

SECURITY	RMSE #1	RMSE #2	RMSE #3	RMSE #4	RMSE #5	RMSE #6	RMSE #7	RMSE #8	RMSE #9	RMSE #10	RMSE #11	RMSE #12
MEDICAL PRODUCTS MANUFACTURING												
Abbott Laboratories (ABT)	ESM_E3 0.0314	ESM3_E3 0.0314	HOLT 0.0317	NAIVE3_E3 0.0317	HOLT3_E3 0.0317	NAÏVE 0.0317	ESM 0.0319	M15_E3 0.0328	MEAN_E3 0.0338	ARIMA_E3 0.0394	MEAN 0.0402	ARIMA 0.0660
Stryker Corporation (SYK)	ESM3_E3 6.3708	ESM_E3 6.3741	HOLT 6.5232	NAÏVE 6.5331	HOLT3_E3 6.5485	NAIVE3_E3 6.5554	ESM 6.5564	M15_E3 6.8512	MEAN_E3 7.4929	ARIMA_E3 7.9290	MEAN 8.8462	ARIMA 13.2360
Bio-Rad Laboratories, Inc. (BIO)	ESM_E3 0.1610	ESM3_E3 0.1626	M15_E3 0.1638	HOLT3_E3 0.1651	NAIVE3_E3 0.1652	ESM 0.1661	HOLT 0.1665	NAÏVE 0.1666	MEAN_E3 0.1739	MEAN 0.1982	ARIMA_E3 0.2285	ARIMA 0.3830
Hill-Rom Holdings, Inc. (HRC)	ARIMA 0.0548	ESM 0.0549	ESM3_E3 0.0550	NAÏVE 0.0550	HOLT 0.0550	NAIVE3_E3 0.0551	HOLT3_E3 0.0551	ESM_E3 0.0554	ARIMA_E3 0.0558	M15_E3 0.0573	MEAN_E3 0.0584	MEAN 0.0673
Medtronic plc (MDT)	HOLT 0.0652	NAÏVE 0.0652	HOLT3_E3 0.0653	NAIVE3_E3 0.0653	MEAN_E3 0.0660	ESM_E3 0.0661	ESM3_E3 0.0670	M15_E3 0.0674	ESM 0.0689	MEAN 0.0751	ARIMA_E3 0.0823	ARIMA 0.1510
Baxter International Inc. (BAX)	ESM 0.0401	ESM3_E3 0.0401	HOLT 0.0403	NAÏVE 0.0403	NAIVE3_E3 0.0404	HOLT3_E3 0.0404	ESM_E3 0.0404	M15_E3 0.0415	MEAN_E3 0.0434	ARIMA_E3 0.0460	MEAN 0.0503	ARIMA 0.0690
Boston Scientific Corporation (BSX)	ESM3_E3 0.0302	ESM 0.0302	NAIVE3_E3 0.0304	HOLT3_E3 0.0304	ESM_E3 0.0304	NAÏVE 0.0304	HOLT 0.0304	ARIMA 0.0306	ARIMA_E3 0.0307	M15_E3 0.0313	MEAN_E3 0.0341	MEAN 0.0412
MEDICAL INSTRUMENTS MANUFACTURING												
Thermo Fisher Scientific Inc. (TMO)	ESM 0.1003	ESM3_E3 0.1004	ESM_E3 0.1015	NAÏVE 0.1037	HOLT 0.1037	NAIVE3_E3 0.1039	HOLT3_E3 0.1039	M15_E3 0.1064	MEAN_E3 0.1079	ARIMA_E3 0.1156	MEAN 0.1205	ARIMA 0.1703
Varian Medical Systems, Inc. (VAR)	NAÏVE 0.0715	HOLT 0.0715	ESM3_E3 0.0716	ESM 0.0717	NAIVE3_E3 0.0718	HOLT3_E3 0.0719	ESM_E3 0.0721	M15_E3 0.0746	MEAN_E3 0.0766	ARIMA_E3 0.0771	MEAN 0.0885	ARIMA 0.0968
ABIOMED, Inc. (ABMD)	M15_E3 0.2712	ESM_E3 0.2771	HOLT3_E3 0.2806	ESM3_E3 0.2811	NAIVE3_E3 0.2812	HOLT 0.2854	ESM 0.2857	NAÏVE 0.2857	MEAN_E3 0.2908	MEAN 0.3377	ARIMA_E3 0.5006	ARIMA 1.0287
Teleflex Inforporated (TFX)	ESM 0.1123	ESM3_E3 0.1131	ESM_E3 0.1145	HOLT 0.1150	HOLT3_E3 0.1152	NAÏVE 0.1158	NAIVE3_E3 0.1160	M15_E3 0.1192	ARIMA_E3 0.1325	MEAN_E3 0.1334	MEAN 0.1628	ARIMA 0.2535
Hologic, Inc. (HOLX)	NAÏVE 0.0353	HOLT 0.0355	NAIVE3_E3 0.0355	HOLT3_E3 0.0356	ESM_E3 0.0369	ESM3_E3 0.0371	M15_E3 0.0372	ESM 0.0376	MEAN_E3 0.0387	MEAN 0.0452	ARIMA_E3 0.0485	ARIMA 0.0926
IDEXX Laboratories (IDXX)	ESM 0.1199	ESM3_E3 0.1203	ESM_E3 0.1219	M15_E3 0.1283	NAIVE3_E3 0.1286	HOLT3_E3 0.1286	NAÏVE 0.1291	HOLT 0.1291	MEAN_E3 0.1370	MEAN 0.1539	ARIMA_E3 0.1763	ARIMA 0.2875
STERIS plc (STE)	ESM 0.0488	ESM3_E3 0.0494	ESM_E3 0.0507	NAÏVE 0.0517	HOLT 0.0518	ARIMA 0.0520	NAIVE3_E3 0.0522	HOLT3_E3 0.0523	ARIMA_E3 0.0533	M15_E3 0.0554	MEAN_E3 0.0561	MEAN 0.0636

B.18. Oil and Energy Sector

Table B-39: Oil and Energy Sector E3 MAPE Accuracy Results

SECURITY	MAPE #1	MAPE #2	MAPE #3	MAPE #4	MAPE #5	MAPE #6	MAPE #7	MAPE #8	MAPE #9	MAPE #10	MAPE #11	MAPE #12
OIL & GAS												
Exxon Mobile (XOM)	ESM_E3 5.1439	MEAN_E3 5.1453	HOLT3_E3 5.1521	NAIVE3_E3 5.1600	HOLT 5.1820	ESM3_E3 5.1864	NAÏVE 5.1978	M15_E3 5.2761	ESM 5.2993	ARIMA_E3 5.4263	MEAN 5.8862	ARIMA 6.3991
Chevron Corporation (CVX)	HOLT 5.7559	NAÏVE 5.7608	HOLT3_E3 5.7780	NAIVE3_E3 5.7804	ESM3_E3 5.8674	ESM 5.8903	ESM_E3 5.8965	ARIMA_E3 5.9435	ARIMA 5.9545	M15_E3 6.0748	MEAN_E3 6.2171	MEAN 7.1342
BP p.l.c. (BP)	ARIMA 6.7754	NAIVE3_E3 6.8904	HOLT3_E3 6.8915	NAÏVE 6.8959	HOLT 6.8968	ESM3_E3 6.9264	ARIMA_E3 6.9342	ESM 6.9518	ESM_E3 6.9529	MEAN_E3 7.0689	M15_E3 7.1356	MEAN 8.4503
Royal Dutch Shell plc (RDS-B)	HOLT3_E3 6.1773	HOLT 6.1793	NAIVE3_E3 6.1796	NAÏVE 6.1832	ARIMA 6.2592	ARIMA_E3 6.2758	ESM_E3 6.3621	M15_E3 6.3773	ESM3_E3 6.4204	MEAN_E3 6.4851	ESM 6.5337	MEAN 7.6336
Sasol Limited (SSL)	ESM_E3 8.5081	ESM3_E3 8.5207	HOLT3_E3 8.5584	NAIVE3_E3 8.5705	ESM 8.5914	HOLT 8.6319	M15_E3 8.6425	NAÏVE 8.6487	MEAN_E3 9.4893	ARIMA_E3 9.6871	MEAN 11.4806	ARIMA 11.9193
TOTAL S.A. (TOT)	ARIMA 6.1846	ARIMA_E3 6.2827	HOLT3_E3 6.3002	NAIVE3_E3 6.3005	HOLT 6.3251	NAÏVE 6.3261	ESM_E3 6.3909	ESM3_E3 6.4205	M15_E3 6.4875	ESM 6.5003	MEAN_E3 6.6233	MEAN 7.6109
YPF Sociedad Anonima (YPF)	NAIVE3_E3 11.0657	ESM3_E3 11.0798	HOLT3_E3 11.0936	NAÏVE 11.0983	ESM 11.1224	ESM_E3 11.1245	ARIMA 11.1450	HOLT 11.1504	ARIMA_E3 11.1891	M15_E3 11.4800	MEAN_E3 13.1636	MEAN 17.4219
OIL & GAS EXPLORATION												
Apache Corporation (APA)	HOLT3_E3 9.2601	NAIVE3_E3 9.2681	HOLT 9.2787	NAÏVE 9.2911	ESM3_E3 9.3012	ESM_E3 9.3230	ESM 9.3466	M15_E3 9.5391	ARIMA_E3 10.0038	MEAN_E3 10.2178	ARIMA 12.4211	MEAN 12.5393
Noble Energy, Inc. (NBL)	HOLT 8.7319	ESM 8.7320	NAÏVE 8.7656	ESM3_E3 8.8095	HOLT3_E3 8.8156	NAIVE3_E3 8.8409	ESM_E3 8.9470	M15_E3 9.3829	MEAN_E3 10.3504	ARIMA_E3 10.3908	MEAN 12.5354	ARIMA 13.3009
Murphy Oil Corporation (MUR)	NAIVE3_E3 9.0142	HOLT3_E3 9.0211	NAÏVE 9.0386	ESM3_E3 9.0419	HOLT 9.0487	ESM_E3 9.0579	ESM 9.0755	M15_E3 9.2011	MEAN_E3 9.7412	ARIMA_E3 10.5387	MEAN 11.9599	ARIMA 14.1332
Devon Energy Corporation (DVN)	ESM3_E3 9.8436	HOLT3_E3 9.8571	NAIVE3_E3 9.8648	ESM_E3 9.8653	ESM 9.8857	HOLT 9.8923	NAÏVE 9.8989	M15_E3 10.1103	ARIMA_E3 10.2546	MEAN_E3 11.1156	ARIMA 11.2469	MEAN 14.0379
Anadarko Petrolium Corp. (APC)	ESM 9.3708	ESM3_E3 9.4267	HOLT 9.4627	NAÏVE 9.4664	HOLT3_E3 9.4996	NAIVE3_E3 9.5023	ESM_E3 9.5401	ARIMA 9.6114	ARIMA_E3 9.7081	M15_E3 9.9065	MEAN_E3 11.1204	MEAN 14.0837
EOG Resources, Inc. (EOG)	ESM 8.9947	ESM3_E3 9.0675	HOLT 9.1625	NAÏVE 9.1797	ESM_E3 9.1943	HOLT3_E3 9.1992	NAIVE3_E3 9.2117	M15_E3 9.5828	ARIMA_E3 10.1309	MEAN_E3 10.4222	ARIMA 12.2930	MEAN 12.8960
Cabot Oil & Gas Corporation (COG)	ESM 9.7076	HOLT 9.7362	ESM3_E3 9.7394	HOLT3_E3 9.7781	NAÏVE 9.7960	NAIVE3_E3 9.8304	ESM_E3 9.8574	M15_E3 10.3491	ARIMA_E3 11.1615	MEAN_E3 11.4286	MEAN 14.1170	ARIMA 14.6148

APPENDIX B: E3 Forecast Study Results

SECURITY	MAPE #1	MAPE #2	MAPE #3	MAPE #4	MAPE #5	MAPE #6	MAPE #7	MAPE #8	MAPE #9	MAPE #10	MAPE #11	MAPE #12
OIL & GAS DRILLING												
Helmerich & Payne, Inc. (HP)	ARIMA	ESM	NAÏVE	ESM3_E3	NAIVE3_E3	HOLT	HOLT3_E3	ESM_E3	ARIMA_E3	M15_E3	MEAN_E3	MEAN
	11.1277	11.1445	11.1480	11.1874	11.1896	11.2513	11.2695	11.2960	11.3213	11.6996	13.2252	16.9647
Ensco plc (ESV)	ESM3_E3	HOLT3_E3	NAIVE3_E3	HOLT	ESM	NAÏVE	ESM_E3	ARIMA_E3	ARIMA	M15_E3	MEAN_E3	MEAN
	12.3764	12.3776	12.3901	12.3991	12.3991	12.4158	12.4317	12.6043	12.6636	12.7487	14.0038	17.3330
Nabors Industries Ltd. (NBR)	HOLT	NAÏVE	HOLT3_E3	NAIVE3_E3	ESM	ESM3_E3	ESM_E3	M15_E3	ARIMA_E3	ARIMA	MEAN_E3	MEAN
	13.4641	13.5651	13.6175	13.7010	13.7697	13.8297	13.9997	14.5879	14.8094	16.3919	17.1219	21.8887
Key Energy Services, Inc. (KEG)	ESM	ESM3_E3	HOLT	ARIMA	NAÏVE	ESM_E3	NAIVE3_E3	HOLT3_E3	ARIMA_E3	M15_E3	MEAN_E3	MEAN
	23.3437	23.7531	24.1263	24.1264	24.1264	24.2255	24.3189	24.3194	24.7570	25.3752	30.8240	40.1484
Rowan Companies, plc (RDC)	ESM_E3	ESM3_E3	HOLT3_E3	NAIVE3_E3	ARIMA_E3	ARIMA	ESM	HOLT	NAÏVE	M15_E3	MEAN_E3	MEAN
	12.5207	12.5210	12.5221	12.5311	12.5462	12.5752	12.5810	12.5842	12.5963	12.6792	13.2903	15.8393
Noble Corporation plc (NE)	ESM3_E3	HOLT3_E3	HOLT	NAIVE3_E3	ESM	NAÏVE	ESM_E3	M15_E3	ARIMA_E3	MEAN_E3	ARIMA	MEAN
	12.7509	12.7517	12.7567	12.7717	12.7721	12.7850	12.8141	13.1588	13.4547	13.8670	14.2772	16.7283
Transocean Ltd. (RIG)	ARIMA	HOLT	ESM	NAÏVE	HOLT3_E3	NAIVE3_E3	ESM3_E3	ESM_E3	ARIMA_E3	M15_E3	MEAN_E3	MEAN
	12.9505	12.9669	12.9698	12.9698	12.9911	12.9930	12.9930	13.1015	13.1986	13.5610	14.9641	19.1383
PRODUCTION & PIPELINE												
The Williams Companies, Inc. (WMB)	HOLT	NAÏVE	ESM	HOLT3_E3	NAIVE3_E3	ESM3_E3	ESM_E3	ARIMA	ARIMA_E3	M15_E3	MEAN_E3	MEAN
	13.2952	13.2976	13.3283	13.3434	13.3450	13.3482	13.4685	13.4733	13.6379	13.9393	18.9942	26.0318
TransCanada Corporation (TRP)	ESM_E3	NAIVE3_E3	HOLT3_E3	ESM3_E3	ARIMA_E3	NAÏVE	HOLT	ESM	ARIMA	M15_E3	MEAN_E3	MEAN
	5.2844	5.2874	5.2903	5.2922	5.3003	5.3254	5.3291	5.3319	5.3479	5.3858	6.2303	7.9742
Enbridge Inc. (ENB)	NAÏVE	ESM3_E3	NAIVE3_E3	ESM_E3	HOLT3_E3	HOLT	ESM	M15_E3	ARIMA_E3	MEAN_E3	MEAN	ARIMA
	5.2819	5.2825	5.2855	5.2946	5.3305	5.3428	5.3586	5.5520	5.9159	6.2594	7.7321	11.2827
OIL FIELD MACHINERY & EQUIPMENT												
Weatherford International plc (WFT)	HOLT	NAÏVE	ESM	ESM3_E3	HOLT3_E3	NAIVE3_E3	ESM_E3	M15_E3	ARIMA_E3	MEAN_E3	ARIMA	MEAN
	17.8523	17.8585	17.8587	17.9097	17.9249	17.9300	18.0350	18.4636	18.6653	19.9495	20.2118	24.0406
McDermott International, Inc. (MDR)	HOLT	NAÏVE	HOLT3_E3	NAIVE3_E3	ESM	ESM3_E3	ARIMA_E3	ESM_E3	ARIMA	M15_E3	MEAN_E3	MEAN
	18.3717	18.5631	18.5757	18.7726	18.8939	19.0033	19.1462	19.2097	19.2147	19.8389	23.5419	30.3820
Matrix Service Company (MTRX)	NAÏVE	HOLT	ESM	NAIVE3_E3	HOLT3_E3	ESM3_E3	ESM_E3	M15_E3	ARIMA_E3	ARIMA	MEAN_E3	MEAN
	15.6604	15.6867	15.7456	15.8130	15.8353	15.8821	16.1170	16.7985	16.9779	17.6675	18.5712	22.6869
ION Geophysical Corporation (IO)	NAIVE3_E3	NAÏVE	ARIMA	HOLT3_E3	HOLT	ARIMA_E3	M15_E3	ESM_E3	ESM3_E3	ESM	MEAN_E3	MEAN
	21.7439	21.7621	21.7676	21.7906	21.8264	21.9337	22.3877	23.2428	23.9165	24.7555	27.5577	35.3918
Superior Energy Services, Inc. (SPN)	NAIVE3_E3	HOLT3_E3	NAÏVE	HOLT	ESM3_E3	ESM	ESM_E3	M15_E3	ARIMA_E3	MEAN_E3	ARIMA	MEAN
	13.3904	13.4068	13.4128	13.4355	13.4791	13.5373	13.5523	14.0112	14.3118	15.3034	15.5016	18.9740

Table B-40: Oil and Energy Sector E3 RMSE Accuracy Results

SECURITY	RMSE #1	RMSE #2	RMSE #3	RMSE #4	RMSE #5	RMSE #6	RMSE #7	RMSE #8	RMSE #9	RMSE #10	RMSE #11	RMSE #12
OIL & GAS												
Exxon Mobile (XOM)	HOLT3_E3 0.0669	NAIVE3_E3 0.0670	HOLT 0.0671	MEAN_E3 0.0673	NAÏVE 0.0673	ESM_E3 0.0674	ESM3_E3 0.0682	M15_E3 0.0683	ESM 0.0699	ARIMA_E3 0.0725	MEAN 0.0760	ARIMA 0.0951
Chevron Corporation (CVX)	HOLT 0.0898	NAÏVE 0.0899	HOLT3_E3 0.0906	NAIVE3_E3 0.0906	ESM 0.0913	ESM3_E3 0.0915	ESM_E3 0.0922	MEAN_E3 0.0941	ARIMA_E3 0.0943	M15_E3 0.0951	ARIMA 0.0978	MEAN 0.1059
BP p.l.c. (BP)	ARIMA 0.0649	ARIMA_E3 0.0654	NAIVE3_E3 0.0659	HOLT3_E3 0.0659	MEAN_E3 0.0661	ESM_E3 0.0662	NAÏVE 0.0662	HOLT 0.0662	ESM3_E3 0.0665	M15_E3 0.0667	ESM 0.0671	MEAN 0.0763
Royal Dutch Shell plc (RDS-B)	HOLT3_E3 0.0722	NAIVE3_E3 0.0722	HOLT 0.0723	NAÏVE 0.0724	ARIMA_E3 0.0732	ARIMA 0.0736	ESM_E3 0.0740	M15_E3 0.0741	ESM3_E3 0.0750	MEAN_E3 0.0760	ESM 0.0765	MEAN 0.0877
Sasol Limited (SSL)	ESM_E3 0.0594	M15_E3 0.0595	HOLT3_E3 0.0596	ESM3_E3 0.0598	NAIVE3_E3 0.0598	HOLT 0.0602	NAÏVE 0.0605	ESM 0.0605	MEAN_E3 0.0673	ARIMA_E3 0.0745	MEAN 0.0807	ARIMA 0.1178
TOTAL S.A. (TOT)	ARIMA 0.0657	ARIMA_E3 0.0668	HOLT3_E3 0.0674	NAIVE3_E3 0.0674	HOLT 0.0677	NAÏVE 0.0678	M15_E3 0.0684	ESM_E3 0.0687	ESM3_E3 0.0696	MEAN_E3 0.0706	ESM 0.0709	MEAN 0.0803
YPF Sociedad Anonima (YPF)	NAIVE3_E3 0.0655	NAÏVE 0.0655	ESM3_E3 0.0655	HOLT3_E3 0.0656	ESM 0.0657	ARIMA 0.0657	ESM_E3 0.0657	HOLT 0.0657	ARIMA_E3 0.0659	M15_E3 0.0669	MEAN_E3 0.0707	MEAN 0.0857
OIL & GAS EXPLORATION												
Apache Corporation (APA)	HOLT 0.1304	HOLT3_E3 0.1307	NAÏVE 0.1309	NAIVE3_E3 0.1311	ESM3_E3 0.1315	ESM 0.1316	ESM_E3 0.1320	M15_E3 0.1348	MEAN_E3 0.1446	ARIMA_E3 0.1478	MEAN 0.1687	ARIMA 0.2420
Noble Energy, Inc. (NBL)	HOLT 0.0618	NAÏVE 0.0619	HOLT3_E3 0.0622	ESM 0.0622	NAIVE3_E3 0.0623	ESM3_E3 0.0624	ESM_E3 0.0629	M15_E3 0.0645	MEAN_E3 0.0676	MEAN 0.0782	ARIMA_E3 0.0881	ARIMA 0.1598
Murphy Oil Corporation (MUR)	NAIVE3_E3 0.0742	HOLT3_E3 0.0743	ESM_E3 0.0745	NAÏVE 0.0746	ESM3_E3 0.0748	HOLT 0.0748	M15_E3 0.0752	ESM 0.0753	MEAN_E3 0.0822	MEAN 0.0981	ARIMA_E3 0.1109	ARIMA 0.2153
Devon Energy Corporation (DVN)	HOLT 0.1036	NAÏVE 0.1038	HOLT3_E3 0.1039	NAIVE3_E3 0.1040	ESM 0.1042	ESM3_E3 0.1043	ESM_E3 0.1048	M15_E3 0.1072	MEAN_E3 0.1171	ARIMA_E3 0.1185	MEAN 0.1390	ARIMA 0.1601
Anadarko Petrolium Corp. (APC)	ESM 0.1026	ESM3_E3 0.1031	ESM_E3 0.1042	HOLT 0.1044	NAÏVE 0.1044	HOLT3_E3 0.1045	NAIVE3_E3 0.1045	ARIMA 0.1053	ARIMA_E3 0.1060	M15_E3 0.1079	MEAN_E3 0.1174	MEAN 0.1420
EOG Resources, Inc. (EOG)	ESM3_E3 0.0898	ESM 0.0899	ESM_E3 0.0902	HOLT3_E3 0.0911	NAIVE3_E3 0.0911	HOLT 0.0915	NAÏVE 0.0916	M15_E3 0.0925	ARIMA_E3 0.0986	MEAN_E3 0.0991	MEAN 0.1174	ARIMA 0.1313
Cabot Oil & Gas Corporation (COG)	ESM 0.0277	ESM3_E3 0.0279	HOLT 0.0281	NAÏVE 0.0282	HOLT3_E3 0.0284	ESM_E3 0.0284	NAIVE3_E3 0.0285	M15_E3 0.0300	ARIMA_E3 0.0329	MEAN_E3 0.0330	MEAN 0.0398	ARIMA 0.0502

APPENDIX B: E3 Forecast Study Results

SECURITY	RMSE #1	RMSE #2	RMSE #3	RMSE #4	RMSE #5	RMSE #6	RMSE #7	RMSE #8	RMSE #9	RMSE #10	RMSE #11	RMSE #12
OIL & GAS DRILLING												
Helmerich & Payne, Inc. (HP)	ESM3_E3 0.1014	ESM 0.1014	NAIVE3_E3 0.1014	NAÏVE 0.1014	ESM_E3 0.1018	ARIMA_E3 0.1026	HOLT3_E3 0.1027	HOLT 0.1032	M15_E3 0.1037	ARIMA 0.1117	MEAN_E3 0.1159	MEAN 0.1427
Ensco plc (ESV)	ESM 0.0789	HOLT 0.0792	NAÏVE 0.0793	ESM3_E3 0.0795	HOLT3_E3 0.0798	NAIVE3_E3 0.0798	ESM_E3 0.0805	M15_E3 0.0835	ARIMA_E3 0.0847	MEAN_E3 0.0898	ARIMA 0.0902	MEAN 0.1067
Nabors Industries Ltd. (NBR)	HOLT 0.0522	NAÏVE 0.0522	HOLT3_E3 0.0526	NAIVE3_E3 0.0527	ESM 0.0527	ESM3_E3 0.0530	ESM_E3 0.0535	M15_E3 0.0553	ARIMA_E3 0.0575	MEAN_E3 0.0610	MEAN 0.0748	ARIMA 0.0794
Key Energy Services, Inc. (KEG)	HOLT 5.5356	ARIMA 5.5367	NAÏVE 5.5367	ESM 5.5808	HOLT3_E3 5.5872	NAIVE3_E3 5.5879	ESM3_E3 5.6292	ESM_E3 5.6919	ARIMA_E3 5.7018	M15_E3 5.8580	MEAN_E3 6.1014	MEAN 7.2289
Rowan Companies, plc (RDC)	ESM3_E3 0.0554	HOLT3_E3 0.0555	NAIVE3_E3 0.0555	ESM_E3 0.0555	ARIMA 0.0556	ESM 0.0556	HOLT 0.0557	ARIMA_E3 0.0557	NAÏVE 0.0558	M15_E3 0.0566	MEAN_E3 0.0606	MEAN 0.0719
Noble Corporation plc (NE)	HOLT 0.0577	NAÏVE 0.0578	ESM 0.0580	HOLT3_E3 0.0581	NAIVE3_E3 0.0582	ESM3_E3 0.0583	ESM_E3 0.0588	M15_E3 0.0605	MEAN_E3 0.0621	ARIMA_E3 0.0661	MEAN 0.0703	ARIMA 0.0818
Transocean Ltd. (RIG)	HOLT 0.1457	ESM 0.1457	NAÏVE 0.1457	HOLT3_E3 0.1475	ESM3_E3 0.1476	NAIVE3_E3 0.1476	ARIMA 0.1480	ESM_E3 0.1501	ARIMA_E3 0.1526	M15_E3 0.1571	MEAN_E3 0.1736	MEAN 0.2100
PRODUCTION & PIPELINE												
The Williams Companies, Inc. (WMB)	HOLT3_E3 0.0513	NAIVE3_E3 0.0513	HOLT 0.0514	NAÏVE 0.0514	ESM3_E3 0.0515	ESM_E3 0.0516	ESM 0.0516	ARIMA_E3 0.0520	ARIMA 0.0522	M15_E3 0.0529	MEAN_E3 0.0600	MEAN 0.0773
TransCanada Corporation (TRP)	NAÏVE 0.0333	HOLT 0.0333	ESM 0.0334	NAIVE3_E3 0.0335	HOLT3_E3 0.0335	ARIMA 0.0335	ESM3_E3 0.0336	ESM_E3 0.0339	ARIMA_E3 0.0342	M15_E3 0.0351	MEAN_E3 0.0382	MEAN 0.0457
Enbridge Inc. (ENB)	NAÏVE 0.0260	HOLT 0.0261	NAIVE3_E3 0.0263	HOLT3_E3 0.0263	ESM_E3 0.0274	ESM3_E3 0.0277	M15_E3 0.0282	ESM 0.0284	MEAN_E3 0.0313	ARIMA_E3 0.0325	MEAN 0.0382	ARIMA 0.0747
OIL FIELD MACHINERY & EQUIPMENT												
Weatherford International plc (WFT)	HOLT 0.0462	NAÏVE 0.0462	HOLT3_E3 0.0465	NAIVE3_E3 0.0465	ESM3_E3 0.0472	ESM 0.0472	ESM_E3 0.0473	M15_E3 0.0483	MEAN_E3 0.0536	MEAN 0.0643	ARIMA_E3 0.0650	ARIMA 0.1156
McDermott International, Inc. (MDR)	ARIMA_E3 0.1001	HOLT 0.1029	NAÏVE 0.1043	HOLT3_E3 0.1043	NAIVE3_E3 0.1057	ESM 0.1062	ESM3_E3 0.1071	ESM_E3 0.1084	M15_E3 0.1118	MEAN_E3 0.1246	ARIMA 0.1268	MEAN 0.1501
Matrix Service Company (MTRX)	NAÏVE 0.0423	HOLT 0.0423	NAIVE3_E3 0.0424	HOLT3_E3 0.0424	ESM 0.0424	ESM3_E3 0.0425	ESM_E3 0.0428	M15_E3 0.0435	MEAN_E3 0.0470	ARIMA_E3 0.0470	MEAN 0.0552	ARIMA 0.0623
ION Geophysical Corporation (IO)	ARIMA 0.3652	NAÏVE 0.3655	HOLT 0.3663	NAIVE3_E3 0.3680	HOLT3_E3 0.3686	ARIMA_E3 0.3741	ESM3_E3 0.3747	ESM 0.3748	ESM_E3 0.3760	M15_E3 0.3829	MEAN_E3 0.4010	MEAN 0.4645
Superior Energy Services, Inc. (SPN)	NAÏVE 0.0572	ESM 0.0572	HOLT 0.0572	ESM3_E3 0.0576	NAIVE3_E3 0.0576	HOLT3_E3 0.0577	ESM_E3 0.0584	M15_E3 0.0607	MEAN_E3 0.0668	ARIMA_E3 0.0691	MEAN 0.0795	ARIMA 0.1134

B.19. Retail Sector

Table B-41: Retail Sector E3 MAPE Accuracy Results

SECURITY	MAPE #1	MAPE #2	MAPE #3	MAPE #4	MAPE #5	MAPE #6	MAPE #7	MAPE #8	MAPE #9	MAPE #10	MAPE #11	MAPE #12
FOOD & RESTAURANT												
McDonalds Corporation (MCD)	NAIVE3_E3 5.6350	HOLT3_E3 5.6354	NAÏVE 5.6420	HOLT 5.6425	ESM_E3 5.6516	ESM3_E3 5.6553	ESM 5.7277	M15_E3 5.8083	ARIMA_E3 6.3071	MEAN_E3 6.5244	MEAN 8.0219	ARIMA 8.5068
The Wendy's Company (WEN)	ESM 8.0171	ESM3_E3 8.0185	ARIMA 8.0656	NAÏVE 8.0656	HOLT 8.0669	NAIVE3_E3 8.0699	HOLT3_E3 8.0714	ESM_E3 8.1132	ARIMA_E3 8.2235	M15_E3 8.5313	MEAN_E3 10.2124	MEAN 13.2405
Cracker Barrel Old Country Store, Inc. (CBRL)	NAÏVE 8.5552	HOLT 8.5587	NAIVE3_E3 8.6532	HOLT3_E3 8.6560	ESM3_E3 8.7838	ESM_E3 8.8215	ESM 8.8712	ARIMA_E3 9.0706	M15_E3 9.2349	ARIMA 9.5184	MEAN_E3 9.9073	MEAN 12.2157
Jack in the Box Inc. (JACK)	ESM 7.6010	NAÏVE 7.6282	HOLT 7.6359	ESM3_E3 7.6475	NAIVE3_E3 7.6704	HOLT3_E3 7.6746	ESM_E3 7.7378	M15_E3 8.0333	ARIMA_E3 8.2231	MEAN_E3 9.2491	ARIMA 10.2792	MEAN 11.9245
Dine Brands Global, Inc. (DIN)	NAÏVE 11.0896	HOLT 11.1088	ESM3_E3 11.1196	NAIVE3_E3 11.1218	ESM 11.1288	HOLT3_E3 11.1363	ARIMA 11.1775	ESM_E3 11.2325	ARIMA_E3 11.3893	M15_E3 11.7470	MEAN_E3 13.3945	MEAN 16.9697
Starbucks Corporatino (SBUX)	ESM_E3 8.5601	ESM3_E3 8.6433	HOLT3_E3 8.6778	NAIVE3_E3 8.6791	HOLT 8.7547	NAÏVE 8.7556	M15_E3 8.8526	ESM 8.8917	ARIMA_E3 9.5720	MEAN_E3 9.7654	MEAN 12.1104	ARIMA 12.1675
The Cheescake Factory Inc. (CAKE)	NAÏVE 8.6367	NAIVE3_E3 8.6387	HOLT 8.6389	HOLT3_E3 8.6404	ESM3_E3 8.6943	ESM 8.7145	ESM_E3 8.7764	M15_E3 9.1948	ARIMA_E3 9.7098	MEAN_E3 9.9763	ARIMA 11.1805	MEAN 12.2231
APPAREL & SHOES												
Foot Locker, Inc. (FL)	ESM3_E3 10.1091	ESM 10.1212	NAIVE3_E3 10.1309	HOLT3_E3 10.1336	NAÏVE 10.1507	HOLT 10.1548	ESM_E3 10.1621	ARIMA 10.1669	ARIMA_E3 10.2329	M15_E3 10.4617	MEAN_E3 11.6503	MEAN 14.4031
Nordstrom, Inc. (JWN)	HOLT 10.3997	HOLT3_E3 10.4127	NAÏVE 10.4150	NAIVE3_E3 10.4229	ESM 10.4250	ESM3_E3 10.4290	ARIMA 10.5122	ESM_E3 10.5184	ARIMA_E3 10.5949	M15_E3 10.9412	MEAN_E3 12.2002	MEAN 14.8163
The Gap, Inc. (GPS)	NAÏVE 10.3026	HOLT 10.3029	NAIVE3_E3 10.3128	HOLT3_E3 10.3130	M15_E3 10.5906	ESM_E3 10.6602	ARIMA_E3 10.8738	ESM3_E3 10.8786	MEAN_E3 11.1046	ESM 11.1809	ARIMA 11.6711	MEAN 13.6382
Genesco Inc. (GCO)	ESM3_E3 12.0677	ESM 12.0798	NAÏVE 12.0802	HOLT 12.0815	NAIVE3_E3 12.0832	HOLT3_E3 12.0840	ESM_E3 12.1436	M15_E3 12.4768	ARIMA_E3 12.5008	ARIMA 12.8084	MEAN_E3 14.0087	MEAN 17.5594
L Brands, Inc. (LB)	NAÏVE 11.4003	HOLT 11.4035	NAIVE3_E3 11.4172	HOLT3_E3 11.4195	ESM_E3 11.5092	ESM3_E3 11.5203	ARIMA_E3 11.5766	ESM 11.5882	M15_E3 11.6556	ARIMA 12.0564	MEAN_E3 12.4980	MEAN 14.8944
Ascena Retail Group, Inc. (ASNA)	NAÏVE 12.1061	NAIVE3_E3 12.1267	HOLT 12.1549	HOLT3_E3 12.1640	ESM_E3 12.4334	ARIMA_E3 12.5041	ESM3_E3 12.5342	M15_E3 12.6229	ESM 12.7576	ARIMA 13.2720	MEAN_E3 14.0571	MEAN 17.2427
The Cato Corporation (CATO)	ESM3_E3 9.2652	ESM_E3 9.2690	HOLT3_E3 9.3060	NAIVE3_E3 9.3089	ARIMA_E3 9.3672	ARIMA 9.3700	ESM 9.3742	HOLT 9.3871	NAÏVE 9.3943	M15_E3 9.5843	MEAN_E3 9.9613	MEAN 11.6398

SECURITY	MAPE #1	MAPE #2	MAPE #3	MAPE #4	MAPE #5	MAPE #6	MAPE #7	MAPE #8	MAPE #9	MAPE #10	MAPE #11	MAPE #12
DISCOUNT & VARIETY												
Walmart Inc. (WMT)	HOLT3_E3 5.1493	NAIVE3_E3 5.1516	M15_E3 5.1850	HOLT 5.2027	NAÏVE 5.2061	ESM_E3 5.2111	ESM3_E3 5.3031	MEAN_E3 5.3588	ESM 5.4550	ARIMA_E3 5.7409	MEAN 6.2141	ARIMA 6.9363
Big Lots, Inc. (BIG)	HOLT 10.7854	NAÏVE 10.8011	HOLT3_E3 10.8157	ARIMA 10.8254	ESM 10.8298	NAIVE3_E3 10.8307	ESM3_E3 10.8357	ESM_E3 10.9298	ARIMA_E3 11.0208	M15_E3 11.3508	MEAN_E3 12.6701	MEAN 16.0077
Ross Stores, Inc. (ROST)	ESM_E3 8.5374	M15_E3 8.5821	HOLT3_E3 8.6086	NAIVE3_E3 8.6269	ESM3_E3 8.6402	HOLT 8.7067	NAÏVE 8.7300	ESM 8.8198	ARIMA_E3 9.5066	MEAN_E3 9.7327	MEAN 11.8985	ARIMA 14.2050
Costco Wholesale Corporation (COST)	ESM 6.0774	ESM3_E3 6.0816	HOLT3_E3 6.1274	NAIVE3_E3 6.1276	HOLT 6.1292	NAÏVE 6.1296	ESM_E3 6.1323	M15_E3 6.3685	ARIMA_E3 6.4926	MEAN_E3 6.8315	ARIMA 7.2504	MEAN 8.2985
The TJX Companies, Inc. (TJX)	NAÏVE 6.5732	ESM3_E3 6.5892	ESM 6.5942	NAIVE3_E3 6.5950	HOLT 6.6318	HOLT3_E3 6.6386	ESM_E3 6.6475	M15_E3 6.9156	ARIMA_E3 7.3992	MEAN_E3 7.7428	MEAN 9.4236	ARIMA 9.5903
Fred's Inc. (FRED)	HOLT 12.5920	HOLT3_E3 12.6054	ESM 12.6066	NAÏVE 12.6148	ESM3_E3 12.6214	NAIVE3_E3 12.6223	ESM_E3 12.6887	M15_E3 12.9921	ARIMA_E3 13.4381	MEAN_E3 14.7311	ARIMA 15.1309	MEAN 18.6206
Kohl's Corporation (KSS)	NAÏVE 7.9716	HOLT 7.9722	NAIVE3_E3 8.0145	HOLT3_E3 8.0149	ESM_E3 8.2857	ESM3_E3 8.3702	ARIMA_E3 8.4576	M15_E3 8.4840	ESM 8.5643	ARIMA 9.2061	MEAN_E3 9.2369	MEAN 11.1445

Table B-42: Retail Sector E3 RMSE Accuracy Results

SECURITY	RMSE #1	RMSE #2	RMSE #3	RMSE #4	RMSE #5	RMSE #6	RMSE #7	RMSE #8	RMSE #9	RMSE #10	RMSE #11	RMSE #12
FOOD & RESTAURANT												
McDonalds Corporation (MCD)	ESM3_E3 0.0722	ESM 0.0725	ESM_E3 0.0727	NAÏVE 0.0727	HOLT 0.0727	NAIVE3_E3 0.0729	HOLT3_E3 0.0730	M15_E3 0.0753	MEAN_E3 0.0813	ARIMA_E3 0.0916	MEAN 0.0959	ARIMA 0.1754
The Wendy's Company (WEN)	ARIMA_E3 0.0249	ESM_E3 0.0249	HOLT3_E3 0.0249	NAIVE3_E3 0.0249	ESM3_E3 0.0250	HOLT 0.0251	ARIMA 0.0251	NAÏVE 0.0251	M15_E3 0.0251	ESM 0.0252	MEAN_E3 0.0285	MEAN 0.0385
Cracker Barrel Old Country Store, Inc. (CBRL)	NAÏVE 0.1097	HOLT 0.1098	MEAN_E3 0.1099	NAIVE3_E3 0.1099	HOLT3_E3 0.1100	ESM_E3 0.1122	M15_E3 0.1136	ESM3_E3 0.1149	ESM 0.1198	ARIMA_E3 0.1211	MEAN 0.1297	ARIMA 0.1668
Jack in the Box Inc. (JACK)	ESM 0.0600	NAÏVE 0.0601	ESM3_E3 0.0606	HOLT 0.0606	NAIVE3_E3 0.0607	HOLT3_E3 0.0610	ESM_E3 0.0614	M15_E3 0.0640	MEAN_E3 0.0712	ARIMA_E3 0.0799	MEAN 0.0878	ARIMA 0.1737
Dine Brands Global, Inc. (DIN)	NAÏVE 0.0932	ESM 0.0933	HOLT 0.0933	ARIMA 0.0937	ESM3_E3 0.0937	NAIVE3_E3 0.0937	HOLT3_E3 0.0938	ESM_E3 0.0948	ARIMA_E3 0.0958	M15_E3 0.0991	MEAN_E3 0.1064	MEAN 0.1286
Starbucks Corporatino (SBUX)	ESM_E3 0.0297	ESM3_E3 0.0307	NAIVE3_E3 0.0309	HOLT3_E3 0.0309	M15_E3 0.0312	NAÏVE 0.0312	HOLT 0.0312	ESM 0.0327	MEAN_E3 0.0354	ARIMA_E3 0.0425	MEAN 0.0432	ARIMA 0.0871
The Cheesecake Factory Inc. (CAKE)	NAIVE3_E3 0.0460	HOLT3_E3 0.0460	ESM3_E3 0.0461	NAÏVE 0.0461	HOLT 0.0461	ESM_E3 0.0461	ESM 0.0464	M15_E3 0.0474	MEAN_E3 0.0507	ARIMA_E3 0.0530	MEAN 0.0614	ARIMA 0.0834

SECURITY	RMSE #1	RMSE #2	RMSE #3	RMSE #4	RMSE #5	RMSE #6	RMSE #7	RMSE #8	RMSE #9	RMSE #10	RMSE #11	RMSE #12
APPAREL & SHOES												
Foot Locker, Inc. (FL)	ESM 0.0537	ESM3_E3 0.0540	HOLT 0.0541	NAÏVE 0.0541	HOLT3_E3 0.0543	NAIVE3_E3 0.0543	ARIMA 0.0545	ESM_E3 0.0546	ARIMA_E3 0.0553	M15_E3 0.0564	MEAN_E3 0.0629	MEAN 0.0796
Nordstrom, Inc. (JWN)	HOLT 0.0674	NAÏVE 0.0674	ESM 0.0678	HOLT3_E3 0.0681	NAIVE3_E3 0.0681	ESM3_E3 0.0683	ARIMA 0.0690	ESM_E3 0.0692	ARIMA_E3 0.0697	M15_E3 0.0721	MEAN_E3 0.0751	MEAN 0.0869
The Gap, Inc. (GPS)	NAIVE3_E3 0.0504	HOLT3_E3 0.0504	NAÏVE 0.0505	HOLT 0.0505	M15_E3 0.0514	ESM_E3 0.0520	MEAN_E3 0.0521	ESM3_E3 0.0532	ESM 0.0548	ARIMA_E3 0.0619	MEAN 0.0625	ARIMA 0.0897
Genesco Inc. (GCO)	NAÏVE 0.0784	HOLT 0.0784	NAIVE3_E3 0.0786	HOLT3_E3 0.0786	ESM 0.0788	ESM3_E3 0.0788	ESM_E3 0.0792	M15_E3 0.0813	ARIMA_E3 0.0817	MEAN_E3 0.0860	ARIMA 0.0861	MEAN 0.1039
L Brands, Inc. (LB)	NAÏVE 0.0742	HOLT 0.0742	NAIVE3_E3 0.0746	HOLT3_E3 0.0746	M15_E3 0.0769	ESM_E3 0.0774	ARIMA_E3 0.0782	MEAN_E3 0.0783	ESM3_E3 0.0786	ESM 0.0805	MEAN 0.0906	ARIMA 0.0948
Ascena Retail Group, Inc. (ASNA)	NAÏVE 0.0204	NAIVE3_E3 0.0205	HOLT3_E3 0.0206	HOLT 0.0207	ESM_E3 0.0210	M15_E3 0.0210	ESM3_E3 0.0213	ARIMA_E3 0.0215	ESM 0.0218	MEAN_E3 0.0221	MEAN 0.0263	ARIMA 0.0263
The Cato Corporation (CATO)	HOLT3_E3 0.0352	NAIVE3_E3 0.0352	ESM3_E3 0.0353	HOLT 0.0353	NAÏVE 0.0353	ESM 0.0354	ESM_E3 0.0354	ARIMA 0.0354	ARIMA_E3 0.0356	M15_E3 0.0364	MEAN_E3 0.0380	MEAN 0.0453
DISCOUNT & VARIETY												
Walmart Inc. (WMT)	HOLT3_E3 0.0624	NAIVE3_E3 0.0624	M15_E3 0.0624	HOLT 0.0629	NAÏVE 0.0630	ESM_E3 0.0640	MEAN_E3 0.0641	ESM3_E3 0.0657	ESM 0.0680	MEAN 0.0757	ARIMA_E3 0.0784	ARIMA 0.1285
Big Lots, Inc. (BIG)	HOLT 0.0547	NAÏVE 0.0548	ARIMA 0.0549	ESM 0.0549	HOLT3_E3 0.0551	NAIVE3_E3 0.0552	ESM3_E3 0.0552	ESM_E3 0.0559	ARIMA_E3 0.0564	M15_E3 0.0580	MEAN_E3 0.0615	MEAN 0.0754
Ross Stores, Inc. (ROST)	ESM_E3 0.0400	ESM3_E3 0.0402	HOLT 0.0404	NAÏVE 0.0406	HOLT3_E3 0.0406	NAIVE3_E3 0.0407	ESM 0.0412	M15_E3 0.0421	MEAN_E3 0.0461	ARIMA_E3 0.0549	MEAN 0.0571	ARIMA 0.1185
Costco Wholesale Corporation (COST)	ESM3_E3 0.0926	ESM 0.0928	ESM_E3 0.0933	HOLT 0.0952	NAÏVE 0.0952	HOLT3_E3 0.0952	NAIVE3_E3 0.0952	M15_E3 0.0974	MEAN_E3 0.1058	ARIMA_E3 0.1118	MEAN 0.1273	ARIMA 0.1886
The TJX Companies, Inc. (TJX)	ESM_E3 0.0222	ESM3_E3 0.0224	NAIVE3_E3 0.0224	M15_E3 0.0225	NAÏVE 0.0226	HOLT3_E3 0.0227	ESM 0.0228	HOLT 0.0230	MEAN_E3 0.0245	MEAN 0.0299	ARIMA_E3 0.0323	ARIMA 0.0607
Fred's Inc. (FRED)	HOLT 0.0330	NAÏVE 0.0330	HOLT3_E3 0.0331	NAIVE3_E3 0.0331	ESM3_E3 0.0332	ESM 0.0333	ESM_E3 0.0333	M15_E3 0.0339	MEAN_E3 0.0373	ARIMA_E3 0.0393	MEAN 0.0456	ARIMA 0.0630
Kohl's Corporation (KSS)	NAÏVE 0.0815	HOLT 0.0815	NAIVE3_E3 0.0820	HOLT3_E3 0.0820	ESM_E3 0.0843	ESM3_E3 0.0852	M15_E3 0.0860	ARIMA_E3 0.0868	ESM 0.0871	MEAN_E3 0.0914	MEAN 0.1089	ARIMA 0.1128

B.20. Transportation Sector

Table B-43: Transportation Sector E3 MAPE Accuracy Results

SECURITY	MAPE #1	MAPE #2	MAPE #3	MAPE #4	MAPE #5	MAPE #6	MAPE #7	MAPE #8	MAPE #9	MAPE #10	MAPE #11	MAPE #12
AIRLINE STOCKS												
Southwest Airlines Co. (LUV)	ESM_E3 9.5228	ESM3_E3 9.5263	HOLT3_E3 9.5486	NAIVE3_E3 9.5534	HOLT 9.5832	ESM 9.5850	NAÏVE 9.5939	M15_E3 9.6897	ARIMA_E3 9.7222	MEAN_E3 10.5250	ARIMA 10.7034	MEAN 12.7673
Alaska Air Group, Inc. (ALK)	NAIVE3_E3 10.7405	HOLT3_E3 10.7447	NAÏVE 10.7770	HOLT 10.7844	ESM_E3 10.8011	ESM3_E3 10.8315	M15_E3 10.9125	ESM 10.9217	ARIMA_E3 11.2831	MEAN_E3 11.4226	ARIMA 13.0936	MEAN 14.1162
PHI, Inc. (PHII)	NAIVE3_E3 11.1954	NAÏVE 11.2373	HOLT3_E3 11.2413	HOLT 11.2435	ESM 11.2435	ESM3_E3 11.2499	ARIMA 11.2550	ESM_E3 11.3300	ARIMA_E3 11.4262	M15_E3 11.7330	MEAN_E3 12.6466	MEAN 15.2428
Skywest, Inc. (SKYW)	NAIVE3_E3 12.5531	HOLT3_E3 12.5650	NAÏVE 12.5777	HOLT 12.5924	ESM3_E3 12.7030	ESM_E3 12.7220	ARIMA 12.7242	ESM 12.7926	ARIMA_E3 12.7944	M15_E3 13.1298	MEAN_E3 13.5003	MEAN 16.3400
Bristow Group Inc. (BRS)	HOLT3_E3 13.5212	NAIVE3_E3 13.5213	ARIMA 13.5830	ESM3_E3 13.5965	ARIMA_E3 13.5966	ESM_E3 13.6044	HOLT 13.6140	NAÏVE 13.6143	ESM 13.7028	M15_E3 13.9009	MEAN_E3 15.0544	MEAN 18.8122
Hawaiian Holdings, Inc. (HA)	NAÏVE 15.8612	HOLT 15.8996	NAIVE3_E3 15.9018	HOLT3_E3 16.0097	ESM_E3 17.0517	M15_E3 17.3193	ESM3_E3 17.3454	ARIMA_E3 17.5167	ESM 17.9162	MEAN_E3 18.0720	ARIMA 19.5870	MEAN 22.0332
RAIL STOCKS												
Union Pacific Corporation (UNP)	ESM3_E3 6.6172	ESM_E3 6.6329	ESM 6.6418	HOLT3_E3 6.7043	NAIVE3_E3 6.7071	HOLT 6.7539	NAÏVE 6.7582	M15_E3 6.7755	ARIMA_E3 7.3967	MEAN_E3 7.7511	MEAN 9.4632	ARIMA 10.9407
CSX Coporation (CSX)	ESM3_E3 8.4490	ESM_E3 8.4709	ESM 8.4728	HOLT3_E3 8.4947	NAIVE3_E3 8.4998	HOLT 8.5170	NAÏVE 8.5242	M15_E3 8.6458	ARIMA_E3 8.7419	ARIMA 9.2999	MEAN_E3 9.9255	MEAN 12.3156
Kansas City Southern (KSU)	M15_E3 34.3471	NAIVE3_E3 34.3809	HOLT3_E3 34.3829	NAÏVE 34.6321	HOLT 34.6385	ESM_E3 34.9372	ESM3_E3 35.5256	ESM 36.3483	MEAN_E3 38.4240	MEAN 44.2145	ARIMA_E3 45.2224	ARIMA 64.7951
Norfolk Southern Corporatino (NSC)	ESM3_E3 8.6233	ESM 8.6412	ESM_E3 8.6582	ARIMA_E3 8.6856	HOLT3_E3 8.7092	NAIVE3_E3 8.7130	HOLT 8.7396	NAÏVE 8.7459	ARIMA 8.7537	M15_E3 8.8899	MEAN_E3 9.3883	MEAN 11.1151
Canadian Pacific Railway Limited (CP)	HOLT 7.9442	NAÏVE 7.9461	HOLT3_E3 7.9564	NAIVE3_E3 7.9582	ESM3_E3 8.0018	ESM_E3 8.0171	ESM 8.0541	M15_E3 8.2200	MEAN_E3 9.1791	ARIMA_E3 9.2191	MEAN 11.1816	ARIMA 12.6472

SECURITY	MAPE #1	MAPE #2	MAPE #3	MAPE #4	MAPE #5	MAPE #6	MAPE #7	MAPE #8	MAPE #9	MAPE #10	MAPE #11	MAPE #12
TRUCK TRANSPORTATION STOCKS												
J.. B. Hunt Transport Services, Inc. (JBHT)	ESM	ESM3_E3	NAÏVE	HOLT	NAIVE3_E3	ESM_E3	HOLT3_E3	M15_E3	MEAN_E3	ARIMA_E3	MEAN	ARIMA
	8.3931	8.4181	8.4746	8.4819	8.4913	8.4965	8.4971	8.8081	9.2339	10.0598	11.2105	14.8835
Werner Enterrises, Inc. (WERN)	MEAN_E3	ARIMA_E3	ESM_E3	ARIMA	HOLT3_E3	ESM3_E3	NAIVE3_E3	HOLT	M15_E3	ESM	NAÏVE	MEAN
	7.4335	7.4807	7.5131	7.5448	7.5533	7.5535	7.5645	7.6629	7.6643	7.6736	7.6812	8.5275
Heartland Express, Inc. (HTLD)	ARIMA_E3	ARIMA	ESM_E3	HOLT3_E3	ESM3_E3	NAIVE3_E3	M15_E3	HOLT	NAÏVE	ESM	MEAN_E3	MEAN
	7.0398	7.0709	7.0781	7.1175	7.1335	7.1455	7.1638	7.2143	7.2593	7.2595	7.3304	8.1874
Marten Transport, Ltd. (MRTN)	ESM_E3	ESM3_E3	NAIVE3_E3	HOLT3_E3	ESM	NAÏVE	HOLT	M15_E3	MEAN_E3	ARIMA_E3	MEAN	ARIMA
	8.6217	8.6224	8.6482	8.6679	8.7045	8.7539	8.7729	8.8628	9.3552	10.1061	11.0542	13.7303
Old Dominion Freight Line, Inc. (ODFL)	ESM	ESM3_E3	ESM_E3	HOLT	NAÏVE	NAIVE3_E3	HOLT3_E3	M15_E3	ARIMA_E3	MEAN_E3	MEAN	ARIMA
	8.6165	8.6875	8.8196	8.8459	8.8522	8.8530	8.8617	9.2410	9.8150	10.1793	12.3136	15.6395
Landstar Systems, Inc. (LSTR)	NAIVE3_E3	ESM3_E3	NAÏVE	ESM_E3	HOLT3_E3	ESM	HOLT	M15_E3	ARIMA_E3	MEAN_E3	MEAN	ARIMA
	7.0310	7.0335	7.0471	7.0535	7.0652	7.0889	7.0918	7.3361	7.6843	8.0074	9.7340	10.2701
Forward Air Corporation (FWRD)	ESM_E3	ESM3_E3	NAIVE3_E3	ARIMA_E3	HOLT3_E3	ESM	M15_E3	NAÏVE	HOLT	ARIMA	MEAN_E3	MEAN
	8.4409	8.4904	8.5646	8.6247	8.6337	8.6394	8.6776	8.6980	8.7923	9.0052	9.1564	11.0633

Table B-44: Transportation Sector E3 RMSE Accuracy Results

SECURITY	RMSE #1	RMSE #2	RMSE #3	RMSE #4	RMSE #5	RMSE #6	RMSE #7	RMSE #8	RMSE #9	RMSE #10	RMSE #11	RMSE #12
AIRLINE STOCKS												
Southwest Airlines Co. (LUV)	HOLT	NAÏVE	HOLT3_E3	NAIVE3_E3	MEAN_E3	M15_E3	ESM_E3	ARIMA_E3	ESM3_E3	ESM	MEAN	ARIMA
	0.0466	0.0467	0.0467	0.0467	0.0474	0.0475	0.0478	0.0481	0.0486	0.0498	0.0539	0.0593
Alaska Air Group, Inc. (ALK)	NAIVE3_E3	NAÏVE	HOLT3_E3	HOLT	ESM_E3	ESM3_E3	M15_E3	ESM	MEAN_E3	ARIMA_E3	MEAN	ARIMA
	0.0641	0.0641	0.0641	0.0642	0.0647	0.0650	0.0652	0.0658	0.0669	0.0772	0.0772	0.1263
PHI, Inc. (PHII)	NAIVE3_E3	NAÏVE	ARIMA	HOLT	ESM	HOLT3_E3	ESM3_E3	ESM_E3	ARIMA_E3	M15_E3	MEAN_E3	MEAN
	0.0521	0.0521	0.0522	0.0522	0.0522	0.0522	0.0523	0.0526	0.0530	0.0542	0.0555	0.0637
Skywest, Inc. (SKYW)	NAÏVE	NAIVE3_E3	HOLT	HOLT3_E3	ESM3_E3	ESM_E3	ARIMA	ESM	ARIMA_E3	M15_E3	MEAN_E3	MEAN
	0.0483	0.0483	0.0484	0.0484	0.0488	0.0489	0.0490	0.0492	0.0492	0.0502	0.0510	0.0600
Bristow Group Inc. (BRS)	HOLT3_E3	NAIVE3_E3	ARIMA	HOLT	NAÏVE	ESM3_E3	ESM_E3	ARIMA_E3	ESM	M15_E3	MEAN_E3	MEAN
	0.0700	0.0700	0.0700	0.0702	0.0702	0.0704	0.0705	0.0705	0.0708	0.0720	0.0772	0.0934
Hawaiian Holdings, Inc. (HA)	MEAN_E3	NAÏVE	HOLT	NAIVE3_E3	HOLT3_E3	ESM3_E3	ESM	ESM_E3	M15_E3	MEAN	ARIMA_E3	ARIMA
	0.0408	0.0420	0.0421	0.0421	0.0422	0.0422	0.0422	0.0425	0.0437	0.0453	0.0638	0.1322

APPENDIX B: E3 Forecast Study Results

SECURITY	RMSE #1	RMSE #2	RMSE #3	RMSE #4	RMSE #5	RMSE #6	RMSE #7	RMSE #8	RMSE #9	RMSE #10	RMSE #11	RMSE #12
RAIL STOCKS												
Union Pacific Corporation (UNP)	ESM 0.0641	ESM3_E3 0.0648	NAÏVE 0.0655	HOLT 0.0655	ESM_E3 0.0659	NAIVE3_E3 0.0659	HOLT3_E3 0.0659	M15_E3 0.0690	MEAN_E3 0.0782	ARIMA_E3 0.0808	MEAN 0.0975	ARIMA 0.1755
CSX Coporation (CSX)	HOLT3_E3 0.0385	NAIVE3_E3 0.0385	HOLT 0.0386	NAÏVE 0.0386	M15_E3 0.0388	ESM_E3 0.0389	ESM3_E3 0.0393	ESM 0.0399	ARIMA_E3 0.0433	MEAN_E3 0.0436	MEAN 0.0514	ARIMA 0.0614
Kansas City Southern (KSU)	MEAN_E3 0.2738	M15_E3 0.2796	HOLT3_E3 0.2825	NAIVE3_E3 0.2827	HOLT 0.2853	NAÏVE 0.2854	ESM_E3 0.2884	MEAN 0.2886	ESM3_E3 0.2952	ESM 0.3034	ARIMA_E3 0.4109	ARIMA 0.7372
Norfolk Southern Corporatino (NSC)	ARIMA 0.0946	ARIMA_E3 0.0933	ESM 0.0973	ESM_E3 0.0972	ESM3_E3 0.0969	HOLT 0.0965	HOLT3_E3 0.0968	M15_E3 0.0998	MEAN 0.1222	MEAN_E3 0.1049	NAÏVE 0.0965	NAIVE3_E3 0.0968
Canadian Pacific Railway Limited (CP)	HOLT 0.1215	NAÏVE 0.1216	ESM_E3 0.1220	HOLT3_E3 0.1221	ESM3_E3 0.1221	NAIVE3_E3 0.1222	ESM 0.1239	M15_E3 0.1268	MEAN_E3 0.1394	ARIMA_E3 0.1563	MEAN 0.1667	ARIMA 0.2802
TRUCK TRANSPORTATION STOCKS												
J.. B. Hunt Transport Services, Inc. (JBHT)	ESM 0.0602	ESM3_E3 0.0609	HOLT 0.0609	NAÏVE 0.0610	HOLT3_E3 0.0616	NAIVE3_E3 0.0616	ESM_E3 0.0621	M15_E3 0.0658	MEAN_E3 0.0693	MEAN 0.0823	ARIMA_E3 0.0903	ARIMA 0.2139
Werner Enterrises, Inc. (WERN)	ESM3_E3 0.0281	HOLT3_E3 0.0281	NAIVE3_E3 0.0281	ESM_E3 0.0281	HOLT 0.0282	ESM 0.0282	NAÏVE 0.0282	ARIMA 0.0283	ARIMA_E3 0.0283	MEAN_E3 0.0284	M15_E3 0.0290	MEAN 0.0327
Heartland Express, Inc. (HTLD)	ARIMA_E3 0.0190	M15_E3 0.0193	ESM_E3 0.0193	ARIMA 0.0193	HOLT3_E3 0.0194	NAIVE3_E3 0.0195	ESM3_E3 0.0197	HOLT 0.0197	NAÏVE 0.0198	MEAN_E3 0.0201	ESM 0.0203	MEAN 0.0234
Marten Transport, Ltd. (MRTN)	ESM_E3 0.0135	ESM3_E3 0.0135	HOLT3_E3 0.0136	NAIVE3_E3 0.0137	M15_E3 0.0137	ESM 0.0137	HOLT 0.0138	NAÏVE 0.0139	MEAN_E3 0.0150	ARIMA_E3 0.0165	MEAN 0.0180	ARIMA 0.0283
Old Dominion Freight Line, Inc. (ODFL)	ESM 0.0749	ESM3_E3 0.0752	ESM_E3 0.0760	HOLT 0.0762	NAÏVE 0.0763	HOLT3_E3 0.0763	NAIVE3_E3 0.0764	M15_E3 0.0785	MEAN_E3 0.0848	MEAN 0.0993	ARIMA_E3 0.1007	ARIMA 0.2029
Landstar Systems, Inc. (LSTR)	HOLT 0.0636	HOLT3_E3 0.0636	NAÏVE 0.0637	NAIVE3_E3 0.0637	ESM_E3 0.0643	ESM3_E3 0.0644	ESM 0.0651	M15_E3 0.0657	MEAN_E3 0.0695	ARIMA_E3 0.0708	MEAN 0.0815	ARIMA 0.1095
Forward Air Corporation (FWRD)	ESM_E3 0.0486	ARIMA_E3 0.0489	ESM3_E3 0.0491	NAIVE3_E3 0.0492	HOLT3_E3 0.0493	MEAN_E3 0.0494	M15_E3 0.0495	ESM 0.0500	NAÏVE 0.0500	HOLT 0.0502	ARIMA 0.0506	MEAN 0.0576

B.21. Utilities Sector

Table B-45: Utilities Sector E3 MAPE Accuracy Results

SECURITY	MAPE #1	MAPE #2	MAPE #3	MAPE #4	MAPE #5	MAPE #6	MAPE #7	MAPE #8	MAPE #9	MAPE #10	MAPE #11	MAPE #12
ELECTRICAL POWER DISTRIBUTION												
American Electric Power Co., Inc. (AEP)	HOLT3_E3 5.7561	ESM3_E3 5.7570	NAIVE3_E3 5.7575	HOLT 5.7588	ESM 5.7593	NAÏVE 5.7602	ARIMA 5.7654	ESM_E3 5.7988	ARIMA_E3 5.8466	M15_E3 6.0034	MEAN_E3 6.4720	MEAN 7.8660
Exelon Corporation (EXC)	ESM 5.7783	ESM3_E3 5.8099	NAÏVE 5.8506	HOLT 5.8723	NAIVE3_E3 5.8726	HOLT3_E3 5.8823	ESM_E3 5.8909	M15_E3 6.1699	ARIMA_E3 6.3657	MEAN_E3 6.9812	ARIMA 7.5969	MEAN 8.6097
NextEra Energy, Inc. (NEE)	ESM3_E3 4.8474	ESM 4.8489	ESM_E3 4.8930	NAIVE3_E3 4.9305	HOLT3_E3 4.9329	NAÏVE 4.9383	HOLT 4.9411	M15_E3 5.1142	MEAN_E3 5.7035	ARIMA_E3 5.7377	MEAN 6.9459	ARIMA 7.6994
Duke Energy Corporation (DUK)	HOLT 5.5112	NAÏVE 5.5121	HOLT3_E3 5.5316	NAIVE3_E3 5.5342	ARIMA 5.5383	ESM 5.5606	ESM3_E3 5.5613	ESM_E3 5.6131	ARIMA_E3 5.6473	M15_E3 5.8262	MEAN_E3 6.3470	MEAN 7.8802
Dominion Energy, Inc. (D)	ARIMA_E3 4.6836	ARIMA 4.6958	ESM_E3 4.7034	ESM3_E3 4.7255	HOLT3_E3 4.7296	NAIVE3_E3 4.7299	HOLT 4.7765	NAÏVE 4.7772	ESM 4.7909	M15_E3 4.7934	MEAN_E3 5.0196	MEAN 6.2570
Public Service Enterprise Group Inc. (PEG)	HOLT3_E3 5.6686	NAIVE3_E3 5.6734	ESM3_E3 5.6737	ESM_E3 5.6759	ARIMA_E3 5.6918	ARIMA 5.7033	HOLT 5.7062	ESM 5.7132	NAÏVE 5.7134	M15_E3 5.7977	MEAN_E3 6.3345	MEAN 7.8012
The Southern Comnpany (SO)	ARIMA 4.0939	HOLT3_E3 4.1031	NAIVE3_E3 4.1051	HOLT 4.1082	NAÏVE 4.1146	ARIMA_E3 4.1360	ESM_E3 4.1571	ESM3_E3 4.1745	ESM 4.2367	M15_E3 4.2469	MEAN_E3 4.3097	MEAN 5.1516
NATURAL GAS DISTRIBUTION												
National Fuel Gas Company (NFG)	HOLT3_E3 6.5857	NAIVE3_E3 6.5864	HOLT 6.5973	NAÏVE 6.5994	ESM3_E3 6.6087	ESM_E3 6.6187	ESM 6.6571	M15_E3 6.7984	ARIMA_E3 7.0023	MEAN_E3 7.4645	ARIMA 7.5542	MEAN 9.2197
ONEOK, Inc. (OKE)	HOLT 8.0865	NAÏVE 8.0871	ESM 8.0895	HOLT3_E3 8.1526	NAIVE3_E3 8.1530	ESM3_E3 8.1564	ESM_E3 8.2632	M15_E3 8.5714	MEAN_E3 9.8271	ARIMA_E3 10.2	MEAN 12.4	ARIMA 15.3
UGI Corporation (UGI)	ESM3_E3 5.1173	ESM 5.1338	NAÏVE 5.1378	NAIVE3_E3 5.1432	HOLT 5.1481	HOLT3_E3 5.1559	ESM_E3 5.1579	M15_E3 5.3864	MEAN_E3 5.9019	ARIMA_E3 6.5555	MEAN 7.1179	ARIMA 10.8308
MDU Resources Group, Inc. (MDU)	NAÏVE 6.5384	NAIVE3_E3 6.5463	HOLT 6.5552	HOLT3_E3 6.5583	ESM_E3 6.6240	ESM3_E3 6.6305	ESM 6.7113	M15_E3 6.8218	ARIMA_E3 7.5140	MEAN_E3 8	ARIMA 9	MEAN 10
New Jersey Resources Co. (NJR)	ESM_E3 4.3191	ESM3_E3 4.3343	NAIVE3_E3 4.3919	HOLT3_E3 4.3983	ESM 4.4302	NAÏVE 4.4533	HOLT 4.4565	M15_E3 4.5436	ARIMA_E3 4.5724	MEAN_E3 4.6761	ARIMA 5.0958	MEAN 5.4587
Atmos Energy Corporation (ATO)	ESM3_E3 4.8798	ESM_E3 4.8877	HOLT3_E3 4.9174	NAIVE3_E3 4.9247	ESM 4.9281	HOLT 4.9552	NAÏVE 4.9673	M15_E3 5.0351	MEAN_E3 5.3828	ARIMA_E3 5.5952	MEAN 6.3561	ARIMA 7.1297
Southwest Gas Holdings, Inc. (SWX)	ESM_E3 5.3857	HOLT3_E3 5.3865	ESM3_E3 5.3873	NAIVE3_E3 5.3976	HOLT 5.4370	ESM 5.4383	NAÏVE 5.4545	M15_E3 5.5042	MEAN_E3 5.8472	ARIMA_E3 5.8819	ARIMA 6.9412	MEAN 7.2077

SECURITY	MAPE #1	MAPE #2	MAPE #3	MAPE #4	MAPE #5	MAPE #6	MAPE #7	MAPE #8	MAPE #9	MAPE #10	MAPE #11	MAPE #12
WATER SUPPLY												
SJW Group (SJW)	NAIVE3_E3	HOLT3_E3	MEAN_E3	NAÏVE	HOLT	ESM_E3	ESM3_E3	M15_E3	ESM	ARIMA_E3	MEAN	ARIMA
	6.3444	6.3574	6.3837	6.4048	6.4270	6.4419	6.5288	6.5526	6.7250	7.2284	7.3532	8.8771
Aqua America, Inc. (WTR)	ESM_E3	HOLT3_E3	NAIVE3_E3	ESM3_E3	HOLT	NAÏVE	ESM	M15_E3	ARIMA_E3	MEAN_E3	ARIMA	MEAN
	5.4069	5.4102	5.4123	5.4315	5.4869	5.4917	5.5172	5.5203	5.8846	6.1351	6.7899	7.5481
American States Water Company (AWR)	ESM3_E3	ESM	ESM_E3	HOLT3_E3	NAIVE3_E3	HOLT	NAÏVE	MEAN_E3	M15_E3	ARIMA_E3	MEAN	ARIMA
	5.5601	5.5804	5.5958	5.6043	5.6055	5.6082	5.6157	5.6287	5.8598	5.9766	6.4189	6.7138
California Water Service Group (CWT)	ARIMA_E3	MEAN_E3	ARIMA	ESM_E3	HOLT3_E3	ESM3_E3	NAIVE3_E3	M15_E3	HOLT	ESM	NAÏVE	MEAN
	5.1769	5.1773	5.1869	5.2535	5.2852	5.2964	5.3159	5.3452	5.3748	5.4161	5.4260	6.0661
Middlesex Water Company (MSEX)	MEAN_E3	M15_E3	HOLT3_E3	ESM_E3	HOLT	ESM3_E3	NAIVE3_E3	ESM	NAÏVE	ARIMA_E3	MEAN	ARIMA
	4.9100	5.0371	5.0373	5.0432	5.1264	5.1292	5.1535	5.2735	5.3148	5.4469	5.6445	6.3689
Connecticut Water Service, Inc. (CTWS)	ARIMA_E3	ARIMA	ESM_E3	MEAN_E3	HOLT3_E3	M15_E3	ESM3_E3	NAIVE3_E3	HOLT	ESM	NAÏVE	MEAN
	4.8758	4.9005	4.9377	4.9545	4.9805	4.9807	5.0082	5.0569	5.0681	5.1425	5.2045	5.8428

Table B-46: Utilities Sector E3 RMSE Accuracy Results

SECURITY	RMSE #1	RMSE #2	RMSE #3	RMSE #4	RMSE #5	RMSE #6	RMSE #7	RMSE #8	RMSE #9	RMSE #10	RMSE #11	RMSE #12
ELECTRICAL POWER DISTRIBUTION												
American Electric Power Co., Inc. (AEP)	HOLT3_E3	ESM3_E3	NAIVE3_E3	ESM	HOLT	NAÏVE	ESM_E3	ARIMA	ARIMA_E3	M15_E3	MEAN_E3	MEAN
	0.0451	0.0451	0.0451	0.0453	0.0453	0.0453	0.0453	0.0454	0.0456	0.0465	0.0505	0.0601
Exelon Corporation (EXC)	ESM	ESM3_E3	NAÏVE	HOLT	HOLT3_E3	NAIVE3_E3	ESM_E3	M15_E3	MEAN_E3	ARIMA_E3	MEAN	ARIMA
	0.0516	0.0517	0.0520	0.0520	0.0520	0.0520	0.0520	0.0537	0.0615	0.0639	0.0750	0.1067
NextEra Energy, Inc. (NEE)	ESM	ESM3_E3	ESM_E3	NAIVE3_E3	HOLT3_E3	NAÏVE	HOLT	M15_E3	MEAN_E3	MEAN	ARIMA_E3	ARIMA
	0.0586	0.0589	0.0597	0.0612	0.0612	0.0615	0.0615	0.0626	0.0703	0.0857	0.0998	0.2165
Duke Energy Corporation (DUK)	HOLT3_E3	NAIVE3_E3	HOLT	NAÏVE	ARIMA	ESM3_E3	ESM_E3	ARIMA_E3	ESM	M15_E3	MEAN_E3	MEAN
	0.0608	0.0609	0.0609	0.0610	0.0611	0.0613	0.0614	0.0615	0.0616	0.0628	0.0692	0.0840
Dominion Energy, Inc. (D)	ARIMA_E3	ESM_E3	ARIMA	NAIVE3_E3	HOLT3_E3	ESM3_E3	HOLT	NAÏVE	M15_E3	ESM	MEAN_E3	MEAN
	0.0408	0.0409	0.0410	0.0411	0.0411	0.0411	0.0415	0.0415	0.0416	0.0417	0.0430	0.0514
Public Service Enterprise Group Inc. (PEG)	ESM_E3	ARIMA_E3	HOLT3_E3	ESM3_E3	NAIVE3_E3	HOLT	ESM	NAÏVE	ARIMA	M15_E3	MEAN_E3	MEAN
	0.0339	0.0339	0.0339	0.0339	0.0339	0.0342	0.0343	0.0343	0.0343	0.0344	0.0370	0.0457
The Southern Comnpany (SO)	HOLT3_E3	NAIVE3_E3	HOLT	ARIMA	NAÏVE	ARIMA_E3	ESM_E3	ESM3_E3	M15_E3	ESM	MEAN_E3	MEAN
	0.0268	0.0268	0.0268	0.0268	0.0269	0.0269	0.0272	0.0275	0.0276	0.0280	0.0282	0.0332

SECURITY	RMSE #1	RMSE #2	RMSE #3	RMSE #4	RMSE #5	RMSE #6	RMSE #7	RMSE #8	RMSE #9	RMSE #10	RMSE #11	RMSE #12
NATURAL GAS DISTRIBUTION												
National Fuel Gas Company (NFG)	HOLT3_E3 0.0611	NAIVE3_E3 0.0611	HOLT 0.0613	NAÏVE 0.0613	ESM_E3 0.0619	M15_E3 0.0620	ESM3_E3 0.0625	ESM 0.0635	MEAN_E3 0.0680	ARIMA_E3 0.0692	MEAN 0.0818	ARIMA 0.0947
ONEOK, Inc. (OKE)	HOLT 0.0498	NAÏVE 0.0499	ESM 0.0499	HOLT3_E3 0.0502	NAIVE3_E3 0.0502	ESM3_E3 0.0503	ESM_E3 0.0508	M15_E3 0.0524	MEAN_E3 0.0591	ARIMA_E3 0.0632	MEAN 0.0722	ARIMA 0.1313
UGI Corporation (UGI)	ESM3_E3 0.0216	ESM 0.0216	ESM_E3 0.0218	NAÏVE 0.0219	HOLT 0.0219	NAIVE3_E3 0.0220	HOLT3_E3 0.0220	M15_E3 0.0232	MEAN_E3 0.0256	MEAN 0.0312	ARIMA_E3 0.0342	ARIMA 0.0846
MDU Resources Group, Inc. (MDU)	M15_E3 0.0294	ESM_E3 0.0300	HOLT3_E3 0.0300	NAIVE3_E3 0.0300	HOLT 0.0304	NAÏVE 0.0305	ESM3_E3 0.0306	ESM 0.0315	MEAN_E3 0.0332	ARIMA_E3 0.0392	MEAN 0.0403	ARIMA 0.0721
New Jersey Resources Co. (NJR)	ESM_E3 0.0190	ESM3_E3 0.0192	HOLT3_E3 0.0193	NAIVE3_E3 0.0193	HOLT 0.0195	NAÏVE 0.0196	ESM 0.0196	M15_E3 0.0198	MEAN_E3 0.0213	MEAN 0.0260	ARIMA_E3 0.0264	ARIMA 0.0486
Atmos Energy Corporation (ATO)	ESM3_E3 0.0344	ESM_E3 0.0345	ESM 0.0347	HOLT3_E3 0.0351	NAIVE3_E3 0.0352	HOLT 0.0354	NAÏVE 0.0355	M15_E3 0.0359	MEAN_E3 0.0400	MEAN 0.0483	ARIMA_E3 0.0611	ARIMA 0.1373
Southwest Gas Holdings, Inc. (SWX)	ESM_E3 0.0450	M15_E3 0.0452	ESM3_E3 0.0453	HOLT3_E3 0.0453	NAIVE3_E3 0.0453	ESM 0.0457	HOLT 0.0457	NAÏVE 0.0458	MEAN_E3 0.0467	MEAN 0.0556	ARIMA_E3 0.0753	ARIMA 0.1614
WATER SUPPLY												
SJW Group (SJW)	MEAN_E3 0.0435	ESM_E3 0.0445	M15_E3 0.0446	HOLT3_E3 0.0446	NAIVE3_E3 0.0447	ESM3_E3 0.0450	HOLT 0.0451	NAÏVE 0.0452	ESM 0.0459	MEAN 0.0488	ARIMA_E3 0.0537	ARIMA 0.0847
Aqua America, Inc. (WTR)	M15_E3 0.0208	ESM_E3 0.0208	HOLT3_E3 0.0209	NAIVE3_E3 0.0210	ESM3_E3 0.0210	HOLT 0.0212	NAÏVE 0.0212	ESM 0.0213	MEAN_E3 0.0217	MEAN 0.0253	ARIMA_E3 0.0257	ARIMA 0.0393
American States Water Company (AWR)	ESM3_E3 0.0254	ESM_E3 0.0256	ESM 0.0256	HOLT3_E3 0.0266	NAIVE3_E3 0.0266	HOLT 0.0268	NAÏVE 0.0269	M15_E3 0.0270	MEAN_E3 0.0276	MEAN 0.0320	ARIMA_E3 0.0346	ARIMA 0.0676
California Water Service Group (CWT)	MEAN_E3 0.0233	ARIMA_E3 0.0233	ARIMA 0.0236	M15_E3 0.0236	ESM_E3 0.0238	HOLT3_E3 0.0239	NAIVE3_E3 0.0240	ESM3_E3 0.0242	HOLT 0.0243	NAÏVE 0.0245	ESM 0.0248	MEAN 0.0268
Middlesex Water Company (MSEX)	M15_E3 0.0276	MEAN_E3 0.0279	ESM_E3 0.0282	HOLT3_E3 0.0285	ESM3_E3 0.0287	NAIVE3_E3 0.0289	HOLT 0.0290	ESM 0.0295	NAÏVE 0.0297	MEAN 0.0321	ARIMA_E3 0.0419	ARIMA 0.0795
Connecticut Water Service, Inc. (CTWS)	ARIMA_E3 0.0304	ESM_E3 0.0306	M15_E3 0.0307	ARIMA 0.0308	ESM3_E3 0.0310	HOLT3_E3 0.0313	ESM 0.0317	NAIVE3_E3 0.0318	HOLT 0.0318	NAÏVE 0.0326	MEAN_E3 0.0328	MEAN 0.0395

B.22. Currency Exchange Rates

Table B-47: Currency Exchange Rates E3 MAPE Accuracy Results

SECURITY	MAPE #1	MAPE #2	MAPE #3	MAPE #4	MAPE #5	MAPE #6	MAPE #7	MAPE #8	MAPE #9	MAPE #10	MAPE #11	MAPE #12
CURRENCY EXCHANGE RATES												
Australian Dollar/US Dollar	HOLT 3.3233	ESM 3.3243	NAÏVE 3.3247	HOLT3_E3 3.3429	NAIVE3_E3 3.3434	ESM3_E3 3.3436	ESM_E3 3.3868	ARIMA_E3 3.5290	M15_E3 3.5368	ARIMA 3.7739	MEAN_E3 3.8676	MEAN 4.7465
Euro/US Dollar	ARIMA 2.5453	NAIVE3_E3 2.5726	ESM3_E3 2.5731	NAÏVE 2.5773	ESM 2.5777	HOLT3_E3 2.5905	ESM_E3 2.5924	ARIMA_E3 2.5935	HOLT 2.6068	M15_E3 2.6959	MEAN_E3 3.1160	MEAN 4.0742
Euro/Australian Dollar	NAÏVE 2.6272	ESM 2.6284	HOLT 2.6300	ARIMA 2.6312	NAIVE3_E3 2.6382	ESM3_E3 2.6391	HOLT3_E3 2.6408	ESM_E3 2.6731	ARIMA_E3 2.6993	M15_E3 2.8023	MEAN_E3 2.8321	MEAN 3.4187
British Pound/US Dollar	HOLT 2.2764	ARIMA 2.2783	NAÏVE 2.2789	ESM 2.2789	HOLT3_E3 2.3005	NAIVE3_E3 2.3043	ESM3_E3 2.3043	ESM_E3 2.3479	ARIMA_E3 2.3745	M15_E3 2.4773	MEAN_E3 2.8491	MEAN 3.5951
Euro/British Pound	NAÏVE 1.7864	ESM 1.7890	ARIMA 1.7952	NAIVE3_E3 1.7976	ESM3_E3 1.8012	HOLT 1.8199	HOLT3_E3 1.8234	ESM_E3 1.8375	ARIMA_E3 1.8664	M15_E3 1.9728	MEAN_E3 2.0599	MEAN 2.6722
US Dollar/ Canadian Dollar	NAÏVE 2.3485	ESM 2.3496	HOLT 2.3497	NAIVE3_E3 2.3567	HOLT3_E3 2.3577	ESM3_E3 2.3578	ESM_E3 2.3838	M15_E3 2.4837	ARIMA_E3 2.4950	MEAN_E3 2.6529	ARIMA 2.8207	MEAN 3.1668
US Dollar/Swiss Franc	ARIMA 2.5625	NAÏVE 2.5629	ESM 2.5631	HOLT 2.5632	NAIVE3_E3 2.5735	ESM3_E3 2.5738	HOLT3_E3 2.5738	ESM_E3 2.6071	ARIMA_E3 2.6334	M15_E3 2.7452	MEAN_E3 2.9440	MEAN 3.7531
US Dollar Currency Index	NAÏVE 2.1518	ESM 2.1528	HOLT 2.1555	NAIVE3_E3 2.1676	ESM3_E3 2.1686	HOLT3_E3 2.1704	ESM_E3 2.2024	M15_E3 2.3169	MEAN_E3 2.6480	ARIMA_E3 2.7950	MEAN 3.3887	ARIMA 3.8393
US Dollar/ Japanese Yen	ESM3_E3 2.6316	NAIVE3_E3 2.6325	HOLT3_E3 2.6364	ESM_E3 2.6390	ESM 2.6465	ARIMA 2.6481	NAÏVE 2.6481	ARIMA_E3 2.6510	HOLT 2.6539	M15_E3 2.7281	MEAN_E3 2.9530	MEAN 3.7211
British Pound/ Japanese Yen	NAÏVE 3.0400	NAIVE3_E3 3.0473	ESM 3.0506	HOLT3_E3 3.0508	ESM3_E3 3.0530	HOLT 3.0535	ESM_E3 3.0854	M15_E3 3.2280	MEAN_E3 3.5530	MEAN 4.5803	ARIMA_E3 5.1323	ARIMA 11.3804

Table B-48: Currency Exchange Rates E3 RMSE Accuracy Results

SECURITY	RMSE #1	RMSE #2	RMSE #3	RMSE #4	RMSE #5	RMSE #6	RMSE #7	RMSE #8	RMSE #9	RMSE #10	RMSE #11	RMSE #12
CURRENCY EXCHANGE RATES												
Australian Dollar/US Dollar	ESM 0.0005	HOLT 0.0005	NAÏVE 0.0005	NAIVE3_E3 0.0005	ESM3_E3 0.0005	HOLT3_E3 0.0005	ESM_E3 0.0005	M15_E3 0.0005	ARIMA_E3 0.0006	MEAN_E3 0.0006	MEAN 0.0008	ARIMA 0.0008
Euro/US Dollar	ARIMA 0.0006	NAÏVE 0.0006	ESM 0.0006	NAIVE3_E3 0.0006	ESM3_E3 0.0006	HOLT 0.0006	HOLT3_E3 0.0006	ARIMA_E3 0.0006	ESM_E3 0.0006	M15_E3 0.0007	MEAN_E3 0.0007	MEAN 0.0009
Euro/Australian Dollar	NAÏVE 0.0008	HOLT 0.0008	ESM 0.0008	ARIMA 0.0008	NAIVE3_E3 0.0008	ESM3_E3 0.0008	HOLT3_E3 0.0008	ESM_E3 0.0009	ARIMA_E3 0.0009	M15_E3 0.0009	MEAN_E3 0.0009	MEAN 0.0011
British Pound/US Dollar	HOLT 0.0008	ARIMA 0.0008	NAÏVE 0.0008	ESM 0.0008	HOLT3_E3 0.0008	NAIVE3_E3 0.0008	ESM3_E3 0.0008	ESM_E3 0.0008	ARIMA_E3 0.0008	M15_E3 0.0008	MEAN_E3 0.0009	MEAN 0.0011
Euro/British Pound	NAÏVE 0.0003	ESM 0.0003	NAIVE3_E3 0.0003	ARIMA 0.0003	ESM3_E3 0.0003	ESM_E3 0.0003	HOLT3_E3 0.0003	ARIMA_E3 0.0003	HOLT 0.0003	M15_E3 0.0003	MEAN_E3 0.0003	MEAN 0.0004
US Dollar/ Canadian Dollar	NAÏVE 0.0006	ESM 0.0006	HOLT 0.0006	NAIVE3_E3 0.0006	ESM3_E3 0.0006	HOLT3_E3 0.0006	ESM_E3 0.0006	ARIMA_E3 0.0006	M15_E3 0.0006	MEAN_E3 0.0007	ARIMA 0.0007	MEAN 0.0008
US Dollar/Swiss Franc	ARIMA 0.0006	NAÏVE 0.0006	ESM 0.0006	HOLT 0.0006	NAIVE3_E3 0.0006	HOLT3_E3 0.0006	ESM3_E3 0.0006	ESM_E3 0.0006	ARIMA_E3 0.0006	M15_E3 0.0006	MEAN_E3 0.0007	MEAN 0.0008
US Dollar Currency Index	NAÏVE 0.0387	ESM 0.0387	HOLT 0.0387	NAIVE3_E3 0.0390	ESM3_E3 0.0390	HOLT3_E3 0.0390	ESM_E3 0.0395	M15_E3 0.0411	MEAN_E3 0.0458	MEAN 0.0560	ARIMA_E3 0.0621	ARIMA 0.1283
US Dollar/ Japanese Yen	ESM 0.0553	ARIMA 0.0553	NAÏVE 0.0553	HOLT 0.0554	ESM3_E3 0.0554	NAIVE3_E3 0.0554	HOLT3_E3 0.0554	ESM_E3 0.0558	ARIMA_E3 0.0562	M15_E3 0.0577	MEAN_E3 0.0617	MEAN 0.0757
British Pound/ Japanese Yen	HOLT 0.1121	NAÏVE 0.1123	ESM 0.1125	HOLT3_E3 0.1132	NAIVE3_E3 0.1135	ESM3_E3 0.1136	ESM_E3 0.1153	M15_E3 0.1202	MEAN_E3 0.1329	MEAN 0.1625	ARIMA_E3 0.1689	ARIMA 0.4105

APPENDIX C: M3 Forecast Study Results

This appendix contains the full accuracy results for both **M3 MAPE** and **M3 RMSE** for the financial forecasts, organized by the sector. Each table includes the results of all 12 forecast models, ranked in descending order of accuracy. The summary results by sector and category are presented first. Significant results (60% or greater) are highlighted.

Source files, raw data, R scripts and Excel files used to generate the results of this study are available by request at **TheScienceofAstrology.com**.

C.1. Summary Results of M3 Forecast Study

Table C-1: M3 MAPE Accuracy Ranked by Sector

SECTOR	TOTAL	MAPE #1	% MAPE #1
STOCK MARKET INDEXES	10	10	100%
AEROSPACE	12	11	91.67%
INDUSTRIAL PRODUCTS	27	23	85.19%
TRANSPORTATION	18	15	83.33%
FINANCE (INSURANCE)	11	9	81.82%
FINANCE (BANKS)	28	21	75.00%
RETAIL/WHOLESALE	21	15	71.43%
AUTO/TIRES/TRUCKS	12	8	66.67%
BUSINESS SERVICES	24	15	62.50%
FINANCE (REIT/REAL ESTATE)	21	13	61.90%
UTILITIES	20	12	60.00%
COMPUTER & TECHNOLOGY	43	25	58.14%
CONSUMER DISCRETIONARY	22	12	54.55%
CONSUMER STAPLES	22	12	54.55%
RATES & BONDS	10	5	50.00%
CONSTRUCTION	14	7	50.00%
MEDICAL SECTOR	28	14	50.00%
BASIC MATERIALS	27	11	40.74%
COMMODITIES	21	8	38.10%
OIL & ENERGY	29	11	37.93%
CURRENCY EXCHANGE RATES	10	2	20.00%
TOTAL STOCKS	430	259	60.23%

SECTOR	TOTAL	MAPE #1/#2	% MAPE #1/#2
STOCK MARKET INDEXES	10	10	100%
AEROSPACE	12	12	100%
FINANCE (INSURANCE)	11	11	100%
TRANSPORTATION	18	18	100%
INDUSTRIAL PRODUCTS	27	24	88.89%
FINANCE (BANKS)	28	24	85.71%
FINANCE (REIT/REAL ESTATE)	21	18	85.71%
RETAIL/WHOLESALE	21	18	85.71%
AUTO/TIRES/TRUCKS	12	10	83.33%
RATES & BONDS	10	8	80.00%
UTILITIES	20	16	80.00%
COMPUTER & TECHNOLOGY	43	33	76.74%
BUSINESS SERVICES	24	16	66.67%
OIL & ENERGY	29	19	65.52%
MEDICAL SECTOR	28	18	64.29%
CONSUMER DISCRETIONARY	22	14	63.64%
CONSUMER STAPLES	22	14	63.64%
COMMODITIES	21	13	61.90%
CONSTRUCTION	14	8	57.14%
BASIC MATERIALS	27	14	51.85%
CURRENCY EXCHANGE RATES	10	3	30.00%
TOTAL STOCKS	430	321	74.65%

Table C-2: M3 RMSE Accuracy Ranked by Sector

SECTOR	TOTAL	RMSE #1	% RMSE #1	SECTOR	TOTAL	RMSE #1/#2	% RMSE #1/#2
FINANCE (INSURANCE)	11	10	90.91%	STOCK MARKET INDEXES	10	10	100%
RATES & BONDS	10	8	80.00%	TRANSPORTATION	18	18	100%
UTILITIES	20	15	75.00%	FINANCE (INSURANCE)	11	10	90.91%
TRANSPORTATION	18	13	72.22%	RATES & BONDS	10	9	90.00%
FINANCE (REIT/REAL ESTATE)	21	15	71.43%	UTILITIES	20	18	90.00%
RETAIL/WHOLESALE	21	15	71.43%	MEDICAL SECTOR	28	24	85.71%
BUSINESS SERVICES	24	17	70.83%	BUSINESS SERVICES	24	20	83.33%
AEROSPACE	12	8	66.67%	AEROSPACE	12	9	75.00%
MEDICAL SECTOR	28	18	64.29%	INDUSTRIAL PRODUCTS	27	20	74.07%
CONSUMER DISCRETIONARY	22	14	63.64%	CONSUMER DISCRETIONARY	22	16	72.73%
STOCK MARKET INDEXES	10	6	60.00%	FINANCE (REIT/REAL ESTATE)	21	15	71.43%
INDUSTRIAL PRODUCTS	27	16	59.26%	AUTO/TIRES/TRUCKS	12	8	66.67%
FINANCE (BANKS)	28	16	57.14%	BASIC MATERIALS	27	18	66.67%
BASIC MATERIALS	27	15	55.56%	RETAIL/WHOLESALE	21	14	66.67%
CONSUMER STAPLES	22	12	54.55%	COMPUTER & TECHNOLOGY	43	28	65.12%
AUTO/TIRES/TRUCKS	12	6	50.00%	FINANCE (BANKS)	28	18	64.29%
COMPUTER & TECHNOLOGY	43	19	44.19%	CONSUMER STAPLES	22	14	63.64%
COMMODITIES	21	8	38.10%	COMMODITIES	21	12	57.14%
OIL & ENERGY	29	8	27.59%	CONSTRUCTION	14	7	50.00%
CONSTRUCTION	14	3	21.43%	OIL & ENERGY	29	13	44.83%
CURRENCY EXCHANGE RATES	10	0	0.00%	CURRENCY EXCHANGE RATES	10	0	0.00%
TOTAL STOCKS	430	242	56.28%	TOTAL STOCKS	430	301	70.00%

Table C-3: M3 MAPE Accuracy Ranked by Category

CATEGORY	#	MAPE #1	% MAPE #1	CATEGORY	#	MAPE #1	% MAPE #1
Stock Market Indexes	10	10	100%	Semiconductor Stocks	7	4	57.14%
Energy (Commodity)	3	3	100%	Building & Construction	7	4	57.14%
Meat (Commodity)	2	2	100%	Real Estate Operations Stocks	7	4	57.14%
Aerospace Equipment	7	7	100%	Large Cap Pharmaceutical	7	4	57.14%
Automotive Stocks	5	5	100%	Medical Instruments Manufacturing	7	4	57.14%
Financial Transaction Services	4	4	100%	Food & Restaurant	7	4	57.14%
Schools	3	3	100%	Rates & Bonds	10	5	50.00%
Bank Stocks	7	7	100%	Communication Components	4	2	50.00%
General Industrial Machinery	7	7	100%	OEM Manufacturer Stocks	7	3	42.86%
Electrical Machinery	7	7	100%	Computer Hardware	7	3	42.86%
Protection — Safety Equipment	7	7	100%	Residential & Commercial	7	3	42.86%
Steel Producers	7	6	85.71%	Furniture Manufacturing	7	3	42.86%
Food Item Stocks	7	6	85.71%	Investment Management Stocks	7	3	42.86%
Midwest Bank Stocks	7	6	85.71%	REIT—Retail Equity Trusts Stocks	7	3	42.86%
Insurance Stocks	7	6	85.71%	Biomedical and Genetics	7	3	42.86%
REIT—Other Equity Trust Stocks	7	6	85.71%	Medical Products Manufacturing	7	3	42.86%
Apparel & Shoes	7	6	85.71%	Oil & Gas Stocks	7	3	42.86%
Truck Transportation	7	6	85.71%	Oil & Gas Drilling	7	3	42.86%
Airlines	6	5	83.33%	Oil Field Machinery & Equipment	5	2	40.00%
Water Supply	6	5	83.33%	Pollution Control Equipment	6	2	33.33%
Aerospace General	5	4	80.00%	Oil & Gas Production & Pipeline	3	1	33.33%
Leisure & Recreation	5	4	80.00%	Textile	7	2	28.57%
Alcoholic Beverages	5	4	80.00%	Oil & Gas U.S. Exploration	7	2	28.57%
Rail	5	4	80.00%	Electric Power Distribution	7	2	28.57%
Staffing Services	4	3	75.00%	Outsourcing Services	4	1	25.00%
Insurance Brokers Stocks	4	3	75.00%	Scientific Instruments	4	1	25.00%
Computer Software	7	5	71.43%	Metals (Commodity)	5	1	20.00%
Wireless Providers	7	5	71.43%	Softs (Commodity)	5	1	20.00%
Electrical Products	7	5	71.43%	Soft Drinks	5	1	20.00%
Investment Bank Stocks	7	5	71.43%	Tobacco Products	5	1	20.00%
Discount & Variety	7	5	71.43%	Currency Exchange Rates	10	2	20.00%
Natural Gas Distribution	7	5	71.43%	Grains (Commodity)	6	1	16.67%
Advertising and Marketing	5	3	60.00%	Gold Mining Stocks	7	1	14.29%
Chemicals	7	4	57.14%	Mining	6	0	0.00%
Business Services	7	4	57.14%				

Table C-4: #1/#2 M3 MAPE Accuracy Ranked by Category

CATEGORY	#	MAPE #1/#2	% MAPE #1/#2	CATEGORY	#	MAPE #1/#2	% MAPE #1/#2
Stock Market Indexes	10	10	100%	Communication Components	4	3	75.00%
Energy (Commodity)	3	3	100%	OEM Manufacturer Stocks	7	5	71.43%
Meat (Commodity)	2	2	100%	Chemicals	7	5	71.43%
Aerospace General	5	5	100%	Computer Hardware	7	5	71.43%
Aerospace Equipment	7	7	100%	Semiconductor Stocks	7	5	71.43%
Automotive Stocks	5	5	100%	Electrical Products	7	5	71.43%
Financial Transaction Services	4	4	100%	Building & Construction	7	5	71.43%
Computer Software	7	7	100%	Investment Bank Stocks	7	5	71.43%
Schools	3	3	100%	Investment Management Stocks	7	5	71.43%
Bank Stocks	7	7	100%	Real Estate Operations Stocks	7	5	71.43%
Midwest Bank Stocks	7	7	100%	Medical Products Manufacturing	7	5	71.43%
Insurance Stocks	7	7	100%	Oil & Gas U.S. Exploration	7	5	71.43%
Insurance Brokers Stocks	4	4	100%	Oil & Gas Drilling	7	5	71.43%
REIT—Other Equity Trust Stocks	7	7	100%	Discount & Variety	7	5	71.43%
General Industrial Machinery	7	7	100%	Oil & Gas Production & Pipeline	3	2	66.67%
Electrical Machinery	7	7	100%	Oil Field Machinery & Equipment	5	3	60.00%
Protection — Safety Equipment	7	7	100%	Business Services	7	4	57.14%
Apparel & Shoes	7	7	100%	Furniture Manufacturing	7	4	57.14%
Airlines	6	6	100%	Large Cap Pharmaceutical	7	4	57.14%
Rail	5	5	100%	Oil & Gas Stocks	7	4	57.14%
Truck Transportation	7	7	100%	Electric Power Distribution	7	4	57.14%
Water Supply	6	6	100%	Scientific Instruments	4	2	50.00%
Steel Producers	7	6	85.71%	Pollution Control Equipment	6	3	50.00%
Wireless Providers	7	6	85.71%	Residential & Commercial	7	3	42.86%
Food Item Stocks	7	6	85.71%	Textile	7	3	42.86%
REIT—Retail Equity Trusts Stocks	7	6	85.71%	Biomedical and Genetics	7	3	42.86%
Medical Instruments Manufacturing	7	6	85.71%	Metals (Commodity)	5	2	40.00%
Food & Restaurant	7	6	85.71%	Soft Drinks	5	2	40.00%
Natural Gas Distribution	7	6	85.71%	Tobacco Products	5	2	40.00%
Softs (Commodity)	5	4	80.00%	Grains (Commodity)	6	2	33.33%
Rates & Bonds	10	8	80.00%	Currency Exchange Rates	10	3	30.00%
Advertising and Marketing	5	4	80.00%	Gold Mining Stocks	7	2	28.57%
Leisure & Recreation	5	4	80.00%	Outsourcing Services	4	1	25.00%
Alcoholic Beverages	5	4	80.00%	Mining	6	1	16.67%
Staffing Services	4	3	75.00%				

Table C-5: #1 M3 RMSE Accuracy Ranked by Category

CATEGORY	#	RMSE #1	% RMSE #1	CATEGORY	#	RMSE #1	% RMSE #1
Meat (Commodity)	2	2	100%	Textile	7	4	57.14%
Insurance Brokers Stocks	4	4	100%	Furniture Manufacturing	7	4	57.14%
Medical Products Manufacturing	7	7	100%	Investment Bank Stocks	7	4	57.14%
Steel Producers	7	6	85.71%	REIT—Retail Equity Trusts Stocks	7	4	57.14%
Bank Stocks	7	6	85.71%	Electrical Machinery	7	4	57.14%
Midwest Bank Stocks	7	6	85.71%	Protection — Safety Equipment	7	4	57.14%
Insurance Stocks	7	6	85.71%	Large Cap Pharmaceutical	7	4	57.14%
REIT—Other Equity Trust Stocks	7	6	85.71%	Biomedical and Genetics	7	4	57.14%
Apparel & Shoes	7	6	85.71%	Oil & Gas Stocks	7	4	57.14%
Natural Gas Distribution	7	6	85.71%	Discount & Variety	7	4	57.14%
Airlines	6	5	83.33%	Truck Transportation	7	4	57.14%
Water Supply	6	5	83.33%	Electric Power Distribution	7	4	57.14%
Rates & Bonds	10	8	80.00%	Mining	6	3	50.00%
Aerospace General	5	4	80.00%	Staffing Services	4	2	50.00%
Advertising and Marketing	5	4	80.00%	Pollution Control Equipment	6	3	50.00%
Leisure & Recreation	5	4	80.00%	OEM Manufacturer Stocks	7	3	42.86%
Alcoholic Beverages	5	4	80.00%	Medical Instruments Manufacturing	7	3	42.86%
Rail	5	4	80.00%	Grains (Commodity)	6	2	33.33%
Financial Transaction Services	4	3	75.00%	Gold Mining Stocks	7	2	28.57%
Outsourcing Services	4	3	75.00%	Computer Hardware	7	2	28.57%
Business Services	7	5	71.43%	Semiconductor Stocks	7	2	28.57%
Electrical Products	7	5	71.43%	Building & Construction	7	2	28.57%
Food Item Stocks	7	5	71.43%	Oil & Gas U.S. Exploration	7	2	28.57%
Real Estate Operations Stocks	7	5	71.43%	Communication Components	4	1	25.00%
General Industrial Machinery	7	5	71.43%	Scientific Instruments	4	1	25.00%
Food & Restaurant	7	5	71.43%	Metals (Commodity)	5	1	20.00%
Schools	3	2	66.67%	Oil Field Machinery & Equipment	5	1	20.00%
Stock Market Indexes	10	6	60.00%	Residential & Commercial	7	1	14.29%
Softs (Commodity)	5	3	60.00%	Oil & Gas Drilling	7	1	14.29%
Automotive Stocks	5	3	60.00%	Energy (Commodity)	3	0	0.00%
Soft Drinks	5	3	60.00%	Tobacco Products	5	0	0.00%
Aerospace Equipment	7	4	57.14%	Investment Management Stocks	7	0	0.00%
Chemicals	7	4	57.14%	Oil & Gas Production & Pipeline	3	0	0.00%
Computer Software	7	4	57.14%	Currency Exchange Rates	10	0	0.00%
Wireless Providers	7	4	57.14%				

Table C-6: #1/#2 M3 RMSE Accuracy Ranked by Category

CATEGORY	#	RMSE #1/#2	% RMSE #1/#2	CATEGORY	#	RMSE #1/#2	% RMSE #1/#2
Stock Market Indexes	10	10	100%	Scientific Instruments	4	3	75.00%
Meat (Commodity)	2	2	100%	Aerospace Equipment	7	5	71.43%
Business Services	7	7	100%	Building & Construction	7	5	71.43%
Insurance Brokers Stocks	4	4	100%	Food Item Stocks	7	5	71.43%
Medical Products Manufacturing	7	7	100%	Real Estate Operations Stocks	7	5	71.43%
Airlines	6	6	100%	Protection — Safety Equipment	7	5	71.43%
Rail	5	5	100%	Large Cap Pharmaceutical	7	5	71.43%
Truck Transportation	7	7	100%	Oil & Gas U.S. Exploration	7	5	71.43%
Water Supply	6	6	100%	Food & Restaurant	7	5	71.43%
Rates & Bonds	10	9	90.00%	Discount & Variety	7	5	71.43%
Chemicals	7	6	85.71%	Schools	3	2	66.67%
Steel Producers	7	6	85.71%	Soft Drinks	5	3	60.00%
Computer Software	7	6	85.71%	OEM Manufacturer Stocks	7	4	57.14%
Electrical Products	7	6	85.71%	Wireless Providers	7	4	57.14%
Textile	7	6	85.71%	Semiconductor Stocks	7	4	57.14%
Bank Stocks	7	6	85.71%	Furniture Manufacturing	7	4	57.14%
Midwest Bank Stocks	7	6	85.71%	Investment Bank Stocks	7	4	57.14%
Insurance Stocks	7	6	85.71%	REIT—Retail Equity Trusts Stocks	7	4	57.14%
REIT—Other Equity Trust Stocks	7	6	85.71%	Apparel & Shoes	7	4	57.14%
General Industrial Machinery	7	6	85.71%	Grains (Commodity)	6	3	50.00%
Electrical Machinery	7	6	85.71%	Mining	6	3	50.00%
Biomedical and Genetics	7	6	85.71%	Communication Components	4	2	50.00%
Medical Instruments Manufacturing	7	6	85.71%	Pollution Control Equipment	6	3	50.00%
Oil & Gas Stocks	7	6	85.71%	Gold Mining Stocks	7	3	42.86%
Electric Power Distribution	7	6	85.71%	Computer Hardware	7	3	42.86%
Natural Gas Distribution	7	6	85.71%	Metals (Commodity)	5	2	40.00%
Softs (Commodity)	5	4	80.00%	Tobacco Products	5	2	40.00%
Aerospace General	5	4	80.00%	Energy (Commodity)	3	1	33.33%
Automotive Stocks	5	4	80.00%	Residential & Commercial	7	2	28.57%
Advertising and Marketing	5	4	80.00%	Investment Management Stocks	7	2	28.57%
Leisure & Recreation	5	4	80.00%	Oil Field Machinery & Equipment	5	1	20.00%
Alcoholic Beverages	5	4	80.00%	Oil & Gas Drilling	7	1	14.29%
Financial Transaction Services	4	3	75.00%	Oil & Gas Production & Pipeline	3	0	0.00%
Staffing Services	4	3	75.00%	Currency Exchange Rates	10	0	0.00%
Outsourcing Services	4	3	75.00%				

C.2. Stock Market Indexes

Table C-7: M3 MAPE Accuracy Results: Stock Market Indexes

SECURITY	MAPE #1	MAPE #2	MAPE #3	MAPE #4	MAPE #5	MAPE #6	MAPE #7	MAPE #8	MAPE #9	MAPE #10	MAPE #11	MAPE #12
STOCK MARKET INDEXES												
S&P 500 (^GSPC)	ESM_M3 4.3851	ESM3_M3 4.3996	ESM 4.4770	HOLT3_M3 4.5127	NAIVM3_M3 4.5164	ARIMA_M3 4.5274	M15_M3 4.5297	HOLT 4.5473	NAÏVE 4.5530	ARIMA 4.8700	MEAN_M3 4.8803	MEAN 6.0296
NASDAQ Composite (^IXIC)	ESM3_M3 6.4528	ESM_M3 6.4812	ESM 6.4818	ARIMA_M3 6.5120	HOLT3_M3 6.6540	NAIVM3_M3 6.6766	M15_M3 6.6894	HOLT 6.7195	NAÏVE 6.7380	MEAN_M3 7.2977	ARIMA 7.7890	MEAN 8.9988
Dow Jones Industrial (^DJI)	ESM_M3 4.2538	M15_M3 4.2739	HOLT3_M3 4.2775	NAIVM3_M3 4.2790	ESM3_M3 4.2799	ARIMA_M3 4.2942	HOLT 4.3194	NAÏVE 4.3220	ESM 4.3345	ARIMA 4.4405	MEAN_M3 4.4991	MEAN 5.4157
Dow Jones Transportation (^DJT)	ARIMA_M3 6.0854	ESM_M3 6.0941	ESM3_M3 6.1609	M15_M3 6.1630	ARIMA 6.1699	NAIVM3_M3 6.1771	HOLT3_M3 6.1831	NAÏVE 6.2505	HOLT 6.2646	ESM 6.3006	MEAN_M3 6.5217	MEAN 7.8809
Dow Jones Utility (^DJU)	ESM3_M3 4.4886	ESM_M3 4.5039	ESM 4.5102	HOLT3_M3 4.5117	NAIVM3_M3 4.5128	HOLT 4.5396	NAÏVE 4.5409	ARIMA_M3 4.5473	ARIMA 4.5711	M15_M3 4.6200	MEAN_M3 5.1758	MEAN 6.5040
Dow Jones Composite (^DJA)	ESM_M3 4.0383	ESM3_M3 4.0728	ARIMA_M3 4.1556	M15_M3 4.1567	HOLT3_M3 4.1692	NAIVM3_M3 4.1692	ESM 4.1708	HOLT 4.2189	NAÏVE 4.2199	ARIMA 4.3157	MEAN_M3 4.5397	MEAN 5.6715
NYSE Composite (^NYA)	ESM_M3 4.3751	ESM3_M3 4.4058	NAIVM3_M3 4.4780	HOLT3_M3 4.4788	NAÏVE 4.5042	ESM 4.5047	HOLT 4.5048	M15_M3 4.5314	ARIMA_M3 4.7565	MEAN_M3 4.9134	ARIMA 5.3086	MEAN 6.0735
Russell 2000 (^RUT)	ARIMA_M3 6.0351	ESM_M3 6.0468	HOLT3_M3 6.0489	NAIVE3_M3 6.0560	ESM3_M3 6.0690	ARIMA 6.0857	HOLT 6.0938	NAÏVE 6.1000	M15_M3 6.1196	ESM 6.1293	MEAN_M3 6.5075	MEAN 7.8769
Nikkei 225 (^N225)	HOLT3_M3 6.0098	HOLT 6.0260	NAIVE3_M3 6.0292	NAÏVE 6.0480	ARIMA 6.0504	ESM3_M3 6.0544	ESM_M3 6.0723	ESM 6.0814	ARIMA_M3 6.0874	M15_M3 6.2219	MEAN_M3 6.9770	MEAN 8.6752
CBOE Volatility Index (^VIX)	MEAN_M3 18.0726	ARIMA 18.1236	ARIMA_M3 18.3800	NAIVE3_M3 18.4704	HOLT3_M3 18.4856	NAÏVE 18.6753	HOLT 18.6762	M15_M3 18.9192	MEAN 19.7358	ESM_M3 19.8360	ESM3_M3 20.8108	ESM 22.0285

Table C-8: M3 RMSE Accuracy Results: Stock Market Indexes

SECURITY	RMSE #1	RMSE #2	RMSE #3	RMSE #4	RMSE #5	RMSE #6	RMSE #7	RMSE #8	RMSE #9	RMSE #10	RMSE #11	RMSE #12
STOCK MARKET INDEXES												
S&P 500 (^GSPC)	ESM_M3	HOLT3_M3	NAIVM3_M3	HOLT	NAÏVE	ESM3_M3	ARIMA_M3	M15_M3	ESM	MEAN_M3	ARIMA	MEAN
	1.1841	1.1850	1.1860	1.1883	1.1897	1.1960	1.1965	1.2051	1.2225	1.2815	1.3422	1.5163
NASDAQ Composite (^IXIC)	ESM	ESM3_M3	HOLT	ESM_M3	HOLT3_M3	NAÏVE	NAIVM3_M3	M15_M3	ARIMA_M3	MEAN_M3	MEAN	ARIMA
	3.8871	3.9020	3.9442	3.9503	3.9605	3.9792	3.9998	4.1420	4.1628	4.3868	5.1930	7.1975
Dow Jones Industrial (^DJI)	ESM_M3	ESM3_M3	HOLT3_M3	NAIVM3_M3	HOLT	M15_M3	NAÏVE	ESM	ARIMA_M3	MEAN_M3	ARIMA	MEAN
	10.3173	10.3527	10.3584	10.3615	10.4205	10.4226	10.4258	10.4464	10.4596	10.9409	10.9845	12.9940
Dow Jones Transportation (^DJT)	ARIMA	ARIMA_M3	ESM_M3	NAIVM3_M3	HOLT3_M3	ESM3_M3	M15_M3	NAÏVE	HOLT	ESM	MEAN_M3	MEAN
	5.4146	5.5111	5.6072	5.6600	5.6639	5.6644	5.6912	5.6996	5.7058	5.7859	6.0809	7.1776
Dow Jones Utility (^DJU)	ESM3_M3	ESM_M3	HOLT3_M3	NAIVM3_M3	ARIMA_M3	ESM	HOLT	NAÏVE	ARIMA	M15_M3	MEAN_M3	MEAN
	0.3606	0.3607	0.3611	0.3612	0.3623	0.3628	0.3632	0.3633	0.3645	0.3676	0.4100	0.5025
Dow Jones Composite (^DJA)	ESM_M3	ESM3_M3	M15_M3	NAIVM3_M3	HOLT3_M3	ESM	NAÏVE	HOLT	ARIMA_M3	MEAN_M3	ARIMA	MEAN
	3.7629	3.7933	3.8443	3.8710	3.8742	3.9020	3.9066	3.9097	3.9110	3.9698	4.1159	4.5358
NYSE Composite (^NYA)	HOLT	HOLT3_M3	NAÏVE	NAIVM3_M3	ESM_M3	ESM3_M3	M15_M3	ESM	ARIMA_M3	MEAN_M3	ARIMA	MEAN
	6.7368	6.7395	6.7406	6.7428	6.8286	6.9098	6.9299	7.0760	7.2256	7.5776	8.7339	8.9923
Russell 2000 (^RUT)	HOLT3_M3	HOLT	ESM3_M3	NAÏVE	NAIVE3_M3	ARIMA	ESM	ESM_M3	ARIMA_M3	M15_M3	MEAN_M3	MEAN
	0.8654	0.8656	0.8659	0.8662	0.8662	0.8663	0.8670	0.8699	0.8741	0.8926	0.9430	1.0872
Nikkei 225 (^N225)	NAIVE3_M3	HOLT3_M3	ARIMA_M3	NAÏVE	HOLT	ESM_M3	ARIMA	ESM3_M3	ESM	M15_M3	MEAN_M3	MEAN
	15.8875	15.8886	15.9315	15.9467	15.9596	15.9615	15.9673	15.9924	16.0908	16.1033	17.3444	20.8469
CBOE Volatility Index (^VIX)	ARIMA	ARIMA_M3	HOLT3_M3	NAIVE3_M3	HOLT	ESM_M3	NAÏVE	M15_M3	ESM3_M3	MEAN_M3	ESM	MEAN
	0.0858	0.0859	0.0861	0.0865	0.0868	0.0869	0.0873	0.0875	0.0881	0.0893	0.0902	0.1013

C.3. Commodities

Table C-9: M3 MAPE Accuracy Results: Commodities

SECURITY	MAPE #1	MAPE #2	MAPE #3	MAPE #4	MAPE #5	MAPE #6	MAPE #7	MAPE #8	MAPE #9	MAPE #10	MAPE #11	MAPE #12
METALS												
Gold Futures	ESM_M3	ESM3_M3	NAÏVE	HOLT	NAIVE3_M3	HOLT3_M3	ESM	M15_M3	MEAN_M3	ARIMA_M3	MEAN	ARIMA
	4.3195	4.3412	4.3573	4.3601	4.3617	4.3646	4.4388	4.5169	4.6988	4.7590	5.5928	5.8827
Platinum Futures	HOLT	NAÏVE	HOLT3_M3	ESM	ESM3_M3	NAIVE3_M3	ESM_M3	M15_M3	ARIMA_M3	ARIMA	MEAN_M3	MEAN
	5.8462	5.9000	5.9012	5.9062	5.9663	5.9699	6.0745	6.3807	6.3821	6.6274	7.6090	9.5002
Silver Futures	ESM	ESM3_M3	NAÏVE	HOLT	NAIVE3_M3	HOLT3_M3	ESM_M3	ARIMA	ARIMA_M3	M15_M3	MEAN_M3	MEAN
	8.7322	8.7473	8.8066	8.8295	8.8510	8.8696	8.8816	8.9386	9.1131	9.4105	9.7764	11.5908
Copper Futures	NAÏVE	ESM	NAIVE3_M3	ESM3_M3	ESM_M3	HOLT3_M3	HOLT	ARIMA_M3	M15_M3	ARIMA	MEAN_M3	MEAN
	7.5696	7.5717	7.5987	7.6009	7.6712	7.7069	7.7242	7.9053	7.9360	8.2911	8.7691	10.9439
Palladium Futures	ESM	HOLT	NAÏVE	HOLT3_M3	ESM3_M3	NAIVE3_M3	ARIMA	ESM_M3	ARIMA_M3	M15_M3	MEAN_M3	MEAN
	8.3682	8.4074	8.4262	8.4652	8.4868	8.5745	8.7083	8.7180	9.0112	9.4393	11.3923	14.5746

APPENDIX C: M3 Forecast Study Results

SECURITY	MAPE #1	MAPE #2	MAPE #3	MAPE #4	MAPE #5	MAPE #6	MAPE #7	MAPE #8	MAPE #9	MAPE #10	MAPE #11	MAPE #12
SOFTS												
Coffee Futures	ESM3_M3 7.2435	NAIVE3_M3 7.2436	HOLT3_M3 7.2436	ARIMA 7.2553	ESM 7.2640	HOLT 7.2641	NAÏVE 7.2642	ESM_M3 7.2864	ARIMA_M3 7.3277	M15_M3 7.5330	MEAN_M3 8.5628	MEAN 10.9282
Lumber Futures	HOLT 8.7711	HOLT3_M3 8.7742	ESM 8.7825	NAÏVE 8.7825	ESM3_M3 8.7898	NAIVE3_M3 8.7898	ARIMA 8.7972	ESM_M3 8.8578	ARIMA_M3 8.9181	M15_M3 9.1446	MEAN_M3 10.2392	MEAN 13.0247
Cocoa Futures	ARIMA 7.3340	ESM3_M3 7.3636	NAIVE3_M3 7.3644	HOLT3_M3 7.3646	ESM 7.4019	NAÏVE 7.4031	HOLT 7.4035	ESM_M3 7.4079	ARIMA_M3 7.4391	M15_M3 7.6886	MEAN_M3 8.1051	MEAN 9.5613
Sugar Futures	NAÏVE 8.9223	NAIVE3_M3 8.9394	ESM 8.9409	ESM3_M3 8.9438	HOLT3_M3 8.9837	ESM_M3 9.1003	HOLT 9.1022	M15_M3 9.7781	MEAN_M3 10.9843	ARIMA_M3 12.0290	MEAN 13.9239	ARIMA 16.6941
Cotton Futures	HOLT 7.0340	ARIMA 7.0695	NAÏVE 7.0972	ESM 7.0998	HOLT3_M3 7.1120	ESM3_M3 7.1678	NAIVE3_M3 7.2029	ESM_M3 7.3355	ARIMA_M3 7.4775	M15_M3 7.9290	MEAN_M3 9.4661	MEAN 11.9867
GRAINS												
Corn Futures	ARIMA 7.0312	HOLT 7.2033	NAÏVE 7.2816	ESM 7.2816	HOLT3_M3 7.4074	NAIVE3_M3 7.4892	ESM3_M3 7.4892	ESM_M3 7.7743	ARIMA_M3 7.8358	M15_M3 8.5198	MEAN_M3 9.4687	MEAN 11.2777
Oats Futures	ESM 8.2940	ESM3_M3 8.3168	ARIMA 8.3176	NAÏVE 8.3284	HOLT3_M3 8.3537	NAIVE3_M3 8.3602	HOLT 8.3845	ESM_M3 8.4157	ARIMA_M3 8.5052	M15_M3 8.8461	MEAN_M3 9.6319	MEAN 11.6271
Soybean Futures	ARIMA 6.9217	HOLT 6.9496	NAÏVE 6.9664	ESM 6.9694	HOLT3_M3 7.0056	ESM3_M3 7.0165	NAIVE3_M3 7.0447	ESM_M3 7.1530	ARIMA_M3 7.2638	M15_M3 7.6857	MEAN_M3 8.4906	MEAN 10.3551
Soybean Oil Futures	ESM 6.3800	NAÏVE 6.3800	HOLT 6.3887	HOLT3_M3 6.4013	NAIVE3_M3 6.4062	ESM3_M3 6.4062	ESM_M3 6.4959	ARIMA_M3 6.7909	M15_M3 6.8431	ARIMA 6.9886	MEAN_M3 7.8879	MEAN 9.5350
Soybean Meal Futures	HOLT3_M3 8.2056	ESM3_M3 8.2065	NAIVE3_M3 8.2065	ESM 8.2371	NAÏVE 8.2371	HOLT 8.2431	ESM_M3 8.2489	ARIMA 8.2493	ARIMA_M3 8.3003	M15_M3 8.5400	MEAN_M3 9.3711	MEAN 11.5485
Wheat Futures	NAÏVE 7.5016	ESM 7.5026	HOLT 7.5107	ESM3_M3 7.5605	NAIVE3_M3 7.5608	HOLT3_M3 7.5663	ESM_M3 7.7052	ARIMA_M3 8.0308	ARIMA 8.1350	M15_M3 8.1996	MEAN_M3 8.4172	MEAN 9.7449
ENERGY												
Brent Oil Futures	NAIVE3_M3 9.3362	NAÏVE 9.3512	ESM3_M3 9.3777	ESM 9.4067	ESM_M3 9.4099	HOLT3_M3 9.5029	HOLT 9.5927	M15_M3 9.6227	ARIMA_M3 9.7247	ARIMA 10.1397	MEAN_M3 10.9150	MEAN 13.8578
Crude Oil WTI Futures	NAIVE3_M3 9.7444	HOLT3_M3 9.7488	ESM3_M3 9.7574	NAÏVE 9.7596	ESM 9.7759	HOLT 9.7943	ESM_M3 9.7999	M15_M3 10.0453	ARIMA_M3 10.3017	ARIMA 11.1985	MEAN_M3 11.3050	MEAN 14.2917
Heating Oil Futures	HOLT3_M3 8.4506	NAIVE3_M3 8.4542	ESM3_M3 8.4548	HOLT 8.4761	NAÏVE 8.4836	ESM 8.4838	ESM_M3 8.4877	ARIMA_M3 8.6751	M15_M3 8.7359	ARIMA 8.8972	MEAN_M3 10.2035	MEAN 13.1985
MEATS												
Cattle Futures	HOLT3_M3 4.3142	NAIVE3_M3 4.3165	ESM3_M3 4.3167	ESM_M3 4.3351	NAÏVE 4.3446	ESM 4.3485	ARIMA 4.3515	HOLT 4.3521	ARIMA_M3 4.3637	M15_M3 4.5291	MEAN_M3 5.1197	MEAN 6.4061
Hog Futures	NAIVE3_M3 10.5212	HOLT3_M3 10.5322	ESM3_M3 10.5352	NAÏVE 10.5444	ESM_M3 10.5475	ARIMA 10.5584	HOLT 10.5622	ESM 10.5639	ARIMA_M3 10.5695	M15_M3 10.6968	MEAN_M3 11.1466	MEAN 13.1042

Table C-10: M3 RMSE Accuracy Results: Commodities

SECURITY	RMSE #1	RMSE #2	RMSE #3	RMSE #4	RMSE #5	RMSE #6	RMSE #7	RMSE #8	RMSE #9	RMSE #10	RMSE #11	RMSE #12
METALS												
Gold Futures	ESM_M3 0.9424	NAIVE3_M3 0.9452	HOLT3_M3 0.9460	NAÏVE 0.9505	HOLT 0.9517	M15_M3 0.9523	ESM3_M3 0.9683	MEAN_M3 0.9856	ESM 1.0131	ARIMA_M3 1.1406	MEAN 1.1414	ARIMA 2.0937
Platinum Futures	HOLT 1.6349	HOLT3_M3 1.6379	NAÏVE 1.6411	NAIVE3_M3 1.6466	ESM_M3 1.6807	ESM3_M3 1.6819	ESM 1.6912	M15_M3 1.7027	ARIMA_M3 1.8780	MEAN_M3 1.9813	MEAN 2.4434	ARIMA 2.5574
Silver Futures	NAÏVE 0.0359	ARIMA 0.0360	HOLT 0.0360	NAIVE3_M3 0.0360	ESM3_M3 0.0361	HOLT3_M3 0.0361	ESM 0.0362	ESM_M3 0.0363	ARIMA_M3 0.0366	M15_M3 0.0374	MEAN_M3 0.0374	MEAN 0.0457
Copper Futures	NAÏVE 0.0040	ESM 0.0040	NAIVE3_M3 0.0041	ESM3_M3 0.0041	ESM_M3 0.0041	HOLT3_M3 0.0041	HOLT 0.0041	M15_M3 0.0042	ARIMA_M3 0.0043	MEAN_M3 0.0048	ARIMA 0.0053	MEAN 0.0057
Palladium Futures	ESM 0.8666	HOLT 0.8673	NAÏVE 0.8685	HOLT3_M3 0.8694	ESM3_M3 0.8701	NAIVE3_M3 0.8739	ESM_M3 0.8807	ARIMA_M3 0.9208	M15_M3 0.9219	ARIMA 0.9826	MEAN_M3 1.0937	MEAN 1.3746
SOFTS												
Coffee Futures	HOLT3_M3 0.1861	ESM3_M3 0.1861	NAIVE3_M3 0.1861	ESM_M3 0.1864	ARIMA 0.1868	HOLT 0.1869	ESM 0.1869	NAÏVE 0.1870	ARIMA_M3 0.1871	M15_M3 0.1907	MEAN_M3 0.2263	MEAN 0.2891
Lumber Futures	HOLT3_M3 0.5178	HOLT 0.5182	ESM 0.5182	NAÏVE 0.5182	ESM3_M3 0.5182	NAIVE3_M3 0.5182	ARIMA 0.5190	ESM_M3 0.5204	ARIMA_M3 0.5226	M15_M3 0.5314	MEAN_M3 0.5790	MEAN 0.7097
Cocoa Futures	ARIMA 3.1025	ESM3_M3 3.1180	NAIVE3_M3 3.1182	HOLT3_M3 3.1183	ESM 3.1214	NAÏVE 3.1217	HOLT 3.1218	ESM_M3 3.1378	ARIMA_M3 3.1506	M15_M3 3.2444	MEAN_M3 3.3706	MEAN 3.9238
Sugar Futures	NAIVE3_M3 0.0003	ESM3_M3 0.0003	ESM_M3 0.0003	HOLT3_M3 0.0003	NAÏVE 0.0003	ESM 0.0003	HOLT 0.0003	M15_M3 0.0003	MEAN_M3 0.0003	MEAN 0.0004	ARIMA_M3 0.0008	ARIMA 0.0018
Cotton Futures	HOLT 0.0014	ESM 0.0014	ARIMA 0.0014	HOLT3_M3 0.0014	NAÏVE 0.0014	ESM3_M3 0.0014	NAIVE3_M3 0.0014	ESM_M3 0.0015	ARIMA_M3 0.0015	M15_M3 0.0016	MEAN_M3 0.0018	MEAN 0.0022
GRAINS												
Corn Futures	ARIMA 0.6390	HOLT 0.6789	NAÏVE 0.6835	ESM 0.6835	HOLT3_M3 0.6907	NAIVE3_M3 0.6959	ESM3_M3 0.6959	ARIMA_M3 0.7027	ESM_M3 0.7142	M15_M3 0.7666	MEAN_M3 0.8332	MEAN 0.9765
Oats Futures	ESM 0.0042	ESM3_M3 0.0042	NAÏVE 0.0042	ARIMA 0.0042	HOLT3_M3 0.0042	NAIVE3_M3 0.0042	HOLT 0.0043	ESM_M3 0.0043	ARIMA_M3 0.0043	M15_M3 0.0045	MEAN_M3 0.0049	MEAN 0.0058
Soybean Futures	ARIMA 0.0141	NAÏVE 0.0142	HOLT 0.0142	ESM 0.0142	HOLT3_M3 0.0143	ESM3_M3 0.0143	NAIVE3_M3 0.0143	ESM_M3 0.0145	ARIMA_M3 0.0146	M15_M3 0.0153	MEAN_M3 0.0167	MEAN 0.0199
Soybean Oil Futures	ESM3_M3 0.0505	NAIVE3_M3 0.0505	ESM 0.0506	NAÏVE 0.0506	HOLT3_M3 0.0507	ESM_M3 0.0508	HOLT 0.0509	M15_M3 0.0523	ARIMA_M3 0.0599	MEAN_M3 0.0605	MEAN 0.0729	ARIMA 0.0808
Soybean Meal Futures	ARIMA_M3 0.5145	ESM_M3 0.5146	ESM3_M3 0.5168	NAIVE3_M3 0.5168	HOLT3_M3 0.5172	M15_M3 0.5188	ESM 0.5219	NAÏVE 0.5219	ARIMA 0.5222	HOLT 0.5228	MEAN_M3 0.5392	MEAN 0.6415
Wheat Futures	ESM 0.0090	NAÏVE 0.0090	HOLT 0.0090	ESM3_M3 0.0090	NAIVE3_M3 0.0090	HOLT3_M3 0.0090	ESM_M3 0.0092	M15_M3 0.0097	ARIMA_M3 0.0100	MEAN_M3 0.0103	MEAN 0.0121	ARIMA 0.0126

SECURITY	RMSE #1	RMSE #2	RMSE #3	RMSE #4	RMSE #5	RMSE #6	RMSE #7	RMSE #8	RMSE #9	RMSE #10	RMSE #11	RMSE #12
ENERGY												
Brent Oil Futures	NAÏVE 0.1367	NAIVE3_M3 0.1373	HOLT3_M3 0.1387	HOLT 0.1390	ESM3_M3 0.1396	ESM_M3 0.1398	ESM 0.1399	M15_M3 0.1417	MEAN_M3 0.1547	ARIMA_M3 0.1562	MEAN 0.1858	ARIMA 0.1935
Crude Oil WTI Futures	NAÏVE 0.1361	ESM 0.1365	HOLT 0.1371	NAIVE3_M3 0.1372	ESM3_M3 0.1374	HOLT3_M3 0.1377	ESM_M3 0.1389	M15_M3 0.1432	MEAN_M3 0.1559	ARIMA_M3 0.1583	MEAN 0.1861	ARIMA 0.2028
Heating Oil Futures	HOLT 0.0034	ESM 0.0034	NAÏVE 0.0034	HOLT3_M3 0.0034	ESM3_M3 0.0034	NAIVE3_M3 0.0034	ESM_M3 0.0035	M15_M3 0.0036	ARIMA_M3 0.0038	MEAN_M3 0.0040	ARIMA 0.0046	MEAN 0.0049
MEATS												
Cattle Futures	ESM_M3 0.0825	NAIVE3_M3 0.0825	ESM3_M3 0.0825	HOLT3_M3 0.0825	ARIMA_M3 0.0829	NAÏVE 0.0834	ESM 0.0835	ARIMA 0.0836	HOLT 0.0836	M15_M3 0.0852	MEAN_M3 0.0953	MEAN 0.1182
Hog Futures	M15_M3 0.1424	ARIMA_M3 0.1425	MEAN_M3 0.1425	ESM_M3 0.1430	NAIVE3_M3 0.1434	HOLT3_M3 0.1437	ESM3_M3 0.1439	NAÏVE 0.1445	ARIMA 0.1447	HOLT 0.1450	ESM 0.1452	MEAN 0.1643

C.4. Rates & Bonds

Table C-11: Rates & Bonds M3 MAPE Accuracy Results

SECURITY	MAPE #1	MAPE #2	MAPE #3	MAPE #4	MAPE #5	MAPE #6	MAPE #7	MAPE #8	MAPE #9	MAPE #10	MAPE #11	MAPE #12
RATES & BONDS												
U.S. 30-Year Bond	NAIVE3_M3 5.8095	ESM3_M3 5.8102	HOLT3_M3 5.8109	NAÏVE 5.8191	ESM 5.8198	HOLT 5.8209	ESM_M3 5.8271	ARIMA_M3 5.8603	ARIMA 5.8662	M15_M3 5.9454	MEAN_M3 6.1190	MEAN 7.3915
U.S. 10-Year Bond	NAIVE3_M3 8.0782	NAÏVE 8.0904	ARIMA 8.0981	HOLT3_M3 8.1296	ARIMA_M3 8.1299	ESM3_M3 8.1371	ESM_M3 8.1377	HOLT 8.1670	ESM 8.1742	M15_M3 8.2694	MEAN_M3 8.4487	MEAN 10.3773
U.S. 5-Year Bond	ESM_M3 11.6339	HOLT3_M3 11.6358	ESM3_M3 11.6376	NAIVE3_M3 11.6472	HOLT 11.7008	ESM 11.7028	NAÏVE 11.7185	ARIMA_M3 11.7322	M15_M3 11.7754	ARIMA 12.1120	MEAN_M3 12.3432	MEAN 15.1648
U.S. 2-Year Bond	ARIMA_M3 14.7178	M15_M3 14.8327	NAIVE3_M3 14.8787	ESM_M3 14.8885	ESM3_M3 15.1010	NAÏVE 15.1278	HOLT3_M3 15.1596	ESM 15.4402	HOLT 15.5404	ARIMA 15.8344	MEAN_M3 16.2619	MEAN 20.7118
U.S. 30-Year T-Bond Futures	ARIMA 2.8115	ESM 2.8134	NAÏVE 2.8134	HOLT 2.8184	NAIVE3_M3 2.8197	ESM3_M3 2.8197	HOLT3_M3 2.8234	ESM_M3 2.8456	ARIMA_M3 2.8651	M15_M3 2.9574	MEAN_M3 2.9660	MEAN 3.6398
U.S. 10-Year T-Bond Futures	ESM 1.6661	NAÏVE 1.6661	ESM3_M3 1.6673	NAIVE3_M3 1.6673	HOLT3_M3 1.6697	HOLT 1.6699	ARIMA 1.6765	ESM_M3 1.6815	ARIMA_M3 1.6963	M15_M3 1.7460	MEAN_M3 1.7828	MEAN 2.1844
Federal Funds Rate	M15_M3 18.5630	ARIMA_M3 18.7890	ARIMA 20.5169	ESM_M3 21.3828	HOLT3_M3 22.7166	ESM3_M3 23.9558	HOLT 24.9032	MEAN_M3 25.4175	NAIVE3_M3 26.0578	ESM 26.9142	NAÏVE 29.6013	MEAN 34.8371
1 Year LIBOR Rate	ESM 8.0831	ESM3_M3 8.1563	HOLT 8.1746	HOLT3_M3 8.2466	ESM_M3 8.3196	NAÏVE 8.3597	NAIVE3_M3 8.4245	ARIMA_M3 8.6094	ARIMA 8.6312	M15_M3 8.8330	MEAN_M3 11.5495	MEAN 15.5246
6 Month LIBOR Rate	HOLT 8.6441	ESM3_M3 8.7120	ESM 8.7244	HOLT3_M3 8.7470	ESM_M3 8.8860	ARIMA 8.9739	NAÏVE 9.1381	ARIMA_M3 9.1947	NAIVE3_M3 9.2137	M15_M3 9.6601	MEAN_M3 13.7282	MEAN 18.9925
3 Month LIBOR Rate	ESM 9.6205	ESM3_M3 9.7116	ESM_M3 9.8800	NAÏVE 9.9104	NAIVE3_M3 9.9944	HOLT 10.0407	HOLT3_M3 10.0740	ARIMA 10.1436	ARIMA_M3 10.2009	M15_M3 10.4617	MEAN_M3 15.4351	MEAN 21.4382

Table C-12: Rates & Bonds M3 RMSE Accuracy Results

SECURITY	RMSE #1	RMSE #2	RMSE #3	RMSE #4	RMSE #5	RMSE #6	RMSE #7	RMSE #8	RMSE #9	RMSE #10	RMSE #11	RMSE #12
RATES & BONDS												
U.S. 30-Year Bond	MEAN_M3	NAIVE3_M3	ESM3_M3	HOLT3_M3	NAÏVE	ESM	HOLT	ESM_M3	ARIMA_M3	ARIMA	M15_M3	MEAN
	0.0046	0.0046	0.0046	0.0046	0.0046	0.0046	0.0046	0.0046	0.0047	0.0047	0.0047	0.0053
U.S. 10-Year Bond	NAIVE3_M3	NAÏVE	ARIMA	HOLT3_M3	ESM3_M3	ESM_M3	ARIMA_M3	HOLT	ESM	MEAN_M3	M15_M3	MEAN
	0.0051	0.0051	0.0051	0.0051	0.0051	0.0051	0.0051	0.0051	0.0051	0.0051	0.0052	0.0060
U.S. 5-Year Bond	HOLT3_M3	ESM3_M3	HOLT	NAIVE3_M3	ESM	NAÏVE	ESM_M3	ARIMA_M3	ARIMA	M15_M3	MEAN_M3	MEAN
	0.0054	0.0054	0.0054	0.0054	0.0054	0.0054	0.0054	0.0055	0.0055	0.0055	0.0057	0.0067
U.S. 2-Year Bond	ESM	HOLT	ESM3_M3	HOLT3_M3	NAÏVE	NAIVE3_M3	ESM_M3	ARIMA	ARIMA_M3	M15_M3	MEAN_M3	MEAN
	0.0048	0.0048	0.0048	0.0048	0.0049	0.0049	0.0049	0.0049	0.0050	0.0051	0.0057	0.0070
U.S. 30-Year T-Bond Futures	NAIVE3_M3	ESM3_M3	ARIMA	NAÏVE	ESM	HOLT3_M3	HOLT	ESM_M3	ARIMA_M3	MEAN_M3	M15_M3	MEAN
	0.0726	0.0726	0.0727	0.0727	0.0727	0.0727	0.0728	0.0728	0.0731	0.0734	0.0743	0.0892
U.S. 10-Year T-Bond Futures	ESM3_M3	NAIVE3_M3	HOLT3_M3	ESM	NAÏVE	ESM_M3	HOLT	ARIMA	ARIMA_M3	M15_M3	MEAN_M3	MEAN
	0.0392	0.0392	0.0393	0.0393	0.0393	0.0393	0.0394	0.0395	0.0395	0.0402	0.0409	0.0501
Federal Funds Rate	ARIMA_M3	ARIMA	M15_M3	ESM_M3	ESM3_M3	HOLT3_M3	ESM	HOLT	NAIVE3_M3	NAÏVE	MEAN_M3	MEAN
	0.0044	0.0044	0.0045	0.0045	0.0046	0.0047	0.0048	0.0049	0.0050	0.0055	0.0059	0.0077
1 Year LIBOR Rate	ESM	ESM3_M3	ESM_M3	NAÏVE	HOLT3_M3	HOLT	ARIMA	NAIVE3_M3	ARIMA_M3	M15_M3	MEAN_M3	MEAN
	0.0047	0.0047	0.0048	0.0048	0.0048	0.0048	0.0048	0.0048	0.0048	0.0049	0.0058	0.0075
6 Month LIBOR Rate	ESM_M3	HOLT3_M3	NAIVE3_M3	NAÏVE	ESM3_M3	M15_M3	HOLT	ARIMA_M3	ESM	ARIMA	MEAN_M3	MEAN
	0.0048	0.0048	0.0048	0.0048	0.0048	0.0049	0.0049	0.0049	0.0050	0.0051	0.0060	0.0077
3 Month LIBOR Rate	M15_M3	ESM_M3	ESM3_M3	NAIVE3_M3	NAÏVE	ESM	ARIMA_M3	HOLT3_M3	HOLT	ARIMA	MEAN_M3	MEAN
	0.0050	0.0050	0.0050	0.0050	0.0051	0.0051	0.0051	0.0053	0.0054	0.0056	0.0061	0.0080

C.5. Aerospace Sector

Table C-13: Aerospace Sector M3 MAPE Accuracy Results

SECURITY	MAPE #1	MAPE #2	MAPE #3	MAPE #4	MAPE #5	MAPE #6	MAPE #7	MAPE #8	MAPE #9	MAPE #10	MAPE #11	MAPE #12
AEROSPACE AND DEFENSE GENERAL STOCKS												
The Boeing Company (BA)	ESM3_M3 7.5804	ESM 7.6103	ESM_M3 7.6327	NAIVE3_M3 7.7231	HOLT3_M3 7.7347	NAÏVE 7.7613	HOLT 7.7765	ARIMA_M3 7.8664	M15_M3 7.9362	ARIMA 8.1593	MEAN_M3 9.3223	MEAN 11.8609
Lockheed Martin Corporation (LMT)	ESM 6.4349	ESM3_M3 6.4421	ESM_M3 6.5382	NAÏVE 6.5948	HOLT 6.6052	NAIVE3_M3 6.6156	HOLT3_M3 6.6247	M15_M3 6.9245	MEAN_M3 7.5531	ARIMA_M3 7.7152	MEAN 8.9634	ARIMA 10.1463
General Dynamics Corporation (GD)	ESM3_M3 7.2971	ESM_M3 7.3218	ESM 7.3361	NAIVE3_M3 7.3933	HOLT3_M3 7.3999	NAÏVE 7.4198	HOLT 7.4363	M15_M3 7.5121	MEAN_M3 8.0298	ARIMA_M3 8.4437	MEAN 9.5658	ARIMA 10.7653
Textron Inc. (TXT)	HOLT3_M3 12.1336	HOLT 12.1364	NAIVE3_M3 12.2424	NAÏVE 12.2533	ARIMA 12.2562	ARIMA_M3 12.3315	ESM3_M3 12.3359	ESM_M3 12.3386	ESM 12.4038	M15_M3 12.5323	MEAN_M3 14.0720	MEAN 17.2741
Northrop Grumman Corporation (NOC)	ESM3_M3 6.4201	HOLT3_M3 6.4211	HOLT 6.4275	NAIVE3_M3 6.4281	NAÏVE 6.4307	ESM 6.4309	ESM_M3 6.4541	M15_M3 6.6611	ARIMA_M3 7.2080	MEAN_M3 7.4721	ARIMA 8.8857	MEAN 9.2746
AEROSPACE AND DEFENSE EQUIPMENT STOCKS												
Esterline Technologies Corporation (ESL)	ESM_M3 8.7395	ESM3_M3 8.7680	HOLT3_M3 8.7786	NAIVE3_M3 8.7825	ARIMA_M3 8.8393	HOLT 8.8864	NAÏVE 8.8909	ESM 8.9000	M15_M3 8.9850	ARIMA 9.1009	MEAN_M3 10.2349	MEAN 13.5833
HEICO Corporation (HEI)	NAIVE3_M3 8.4748	ESM3_M3 8.4935	ESM_M3 8.4949	HOLT3_M3 8.5234	NAÏVE 8.5578	ESM 8.6466	HOLT 8.7373	M15_M3 8.8674	ARIMA_M3 9.5168	MEAN_M3 9.8110	MEAN 12.6701	ARIMA 12.8070
Hexcel Corporation (HXL)	HOLT3_M3 11.4461	HOLT 11.5224	NAIVE3_M3 11.5650	ARIMA_M3 11.5900	ESM_M3 11.6599	NAÏVE 11.6697	ARIMA 11.7171	M15_M3 11.7656	ESM3_M3 11.7947	ESM 12.0478	MEAN_M3 15.2677	MEAN 20.0097
Curtiss-Wright Corporation (CW)	NAIVE3_M3 7.8398	NAÏVE 7.8402	ESM3_M3 7.8618	HOLT3_M3 7.8680	ESM_M3 7.8941	HOLT 7.8993	ESM 7.9070	ARIMA_M3 8.1740	M15_M3 8.2119	MEAN_M3 8.6995	ARIMA 9.9104	MEAN 10.3230
Moog Inc. (MOG-A)	ESM_M3 8.7973	NAIVE3_M3 8.8414	ESM3_M3 8.8923	HOLT3_M3 8.9065	M15_M3 8.9156	NAÏVE 8.9584	MEAN_M3 9.0483	HOLT 9.0529	ESM 9.0713	ARIMA_M3 9.6797	MEAN 10.6506	ARIMA 12.5834
Raytheon Company (RTN)	ESM3_M3 6.4505	NAIVE3_M3 6.4559	HOLT3_M3 6.4701	ESM 6.4737	NAÏVE 6.4809	ESM_M3 6.4874	HOLT 6.5180	M15_M3 6.7052	ARIMA_M3 6.7497	ARIMA 7.1079	MEAN_M3 7.6869	MEAN 9.8807
Aerojet Rocketdyne Holdings, Inc. (AJRD)	NAIVE3_M3 12.0342	NAÏVE 12.0784	ESM3_M3 12.0893	HOLT3_M3 12.0917	ARIMA 12.0986	ESM_M3 12.1292	ESM 12.1494	ARIMA_M3 12.1580	HOLT 12.1931	M15_M3 12.4480	MEAN_M3 13.6320	MEAN 16.9416

Table C-14: Aerospace Sector M3 RMSE Accuracy Results

SECURITY	RMSE #1	RMSE #2	RMSE #3	RMSE #4	RMSE #5	RMSE #6	RMSE #7	RMSE #8	RMSE #9	RMSE #10	RMSE #11	RMSE #12
AEROSPACE AND DEFENSE GENERAL STOCKS												
The Boeing Company (BA)	ESM_M3 0.1635	ESM3_M3 0.1641	M15_M3 0.1654	ESM 0.1656	NAIVE3_M3 0.1675	NAÏVE 0.1690	HOLT3_M3 0.1691	HOLT 0.1712	ARIMA_M3 0.1876	MEAN_M3 0.1952	MEAN 0.2414	ARIMA 0.2474
Lockheed Martin Corporation (LMT)	MEAN_M3 0.1591	M15_M3 0.1591	NAIVE3_M3 0.1604	HOLT3_M3 0.1606	ESM_M3 0.1610	NAÏVE 0.1618	HOLT 0.1621	ESM3_M3 0.1645	ESM 0.1695	MEAN 0.1768	ARIMA_M3 0.2404	ARIMA 0.4585
General Dynamics Corporation (GD)	ESM3_M3 0.1123	ESM 0.1125	ESM_M3 0.1126	HOLT3_M3 0.1135	NAIVE3_M3 0.1135	HOLT 0.1138	NAÏVE 0.1139	M15_M3 0.1146	MEAN_M3 0.1208	ARIMA_M3 0.1380	MEAN 0.1400	ARIMA 0.2191
Textron Inc. (TXT)	HOLT 0.0694	NAÏVE 0.0694	ARIMA 0.0696	HOLT3_M3 0.0699	NAIVE3_M3 0.0700	ARIMA_M3 0.0715	ESM3_M3 0.0724	ESM_M3 0.0724	ESM 0.0728	M15_M3 0.0735	MEAN_M3 0.0786	MEAN 0.0927
Northrop Grumman Corporation (NOC)	M15_M3 0.1433	HOLT3_M3 0.1436	ESM_M3 0.1441	HOLT 0.1445	NAIVE3_M3 0.1448	ESM3_M3 0.1452	NAÏVE 0.1461	ESM 0.1467	MEAN_M3 0.1595	MEAN 0.1884	ARIMA_M3 0.2180	ARIMA 0.4143
AEROSPACE AND DEFENSE EQUIPMENT STOCKS												
Esterline Technologies Corporation (ESL)	HOLT 0.0980	NAÏVE 0.0980	HOLT3_M3 0.0981	NAIVE3_M3 0.0982	ESM3_M3 0.0992	ESM_M3 0.0992	ESM 0.0998	ARIMA_M3 0.1007	M15_M3 0.1011	MEAN_M3 0.1112	ARIMA 0.1112	MEAN 0.1356
HEICO Corporation (HEI)	ESM3_M3 0.0354	ESM 0.0354	ESM_M3 0.0357	M15_M3 0.0371	NAIVE3_M3 0.0372	HOLT3_M3 0.0372	NAÏVE 0.0374	HOLT 0.0375	MEAN_M3 0.0388	MEAN 0.0476	ARIMA_M3 0.0519	ARIMA 0.0952
Hexcel Corporation (HXL)	ESM 0.0376	ESM3_M3 0.0376	HOLT 0.0377	HOLT3_M3 0.0378	NAÏVE 0.0378	ARIMA 0.0379	NAIVE3_M3 0.0379	ESM_M3 0.0380	ARIMA_M3 0.0386	M15_M3 0.0399	MEAN_M3 0.0447	MEAN 0.0542
Curtiss-Wright Corporation (CW)	ESM 0.0808	NAÏVE 0.0810	HOLT 0.0811	ESM3_M3 0.0814	NAIVE3_M3 0.0816	HOLT3_M3 0.0816	ESM_M3 0.0823	M15_M3 0.0851	ARIMA_M3 0.0857	MEAN_M3 0.0864	MEAN 0.0977	ARIMA 0.1168
Moog Inc. (MOG-A)	NAIVE3_M3 0.0636	HOLT3_M3 0.0637	NAÏVE 0.0637	HOLT 0.0638	ESM3_M3 0.0640	ESM_M3 0.0641	ESM 0.0644	MEAN_M3 0.0648	M15_M3 0.0656	ARIMA_M3 0.0692	MEAN 0.0736	ARIMA 0.0992
Raytheon Company (RTN)	ESM_M3 0.0961	M15_M3 0.0964	ESM3_M3 0.0964	NAIVE3_M3 0.0966	HOLT3_M3 0.0966	ESM 0.0970	NAÏVE 0.0972	HOLT 0.0973	MEAN_M3 0.1031	ARIMA_M3 0.1168	MEAN 0.1219	ARIMA 0.1672
Aerojet Rocketdyne Holdings, Inc. (AJRD)	NAIVE3_M3 0.0271	NAÏVE 0.0272	ESM3_M3 0.0272	ARIMA 0.0272	ESM_M3 0.0272	ARIMA_M3 0.0272	ESM 0.0273	HOLT3_M3 0.0273	HOLT 0.0275	M15_M3 0.0276	MEAN_M3 0.0291	MEAN 0.0355

C.6. Auto/Tires/Trucks Sector

Table C-15: Auto/Tires/Trucks Sector M3 MAPE Accuracy Results

SECURITY	MAPE #1	MAPE #2	MAPE #3	MAPE #4	MAPE #5	MAPE #6	MAPE #7	MAPE #8	MAPE #9	MAPE #10	MAPE #11	MAPE #12
AUTOMOTIVE STOCKS												
Ford Motor Company (F)	HOLT3_M3	HOLT	NAIVE3_M3	ARIMA	NAÏVE	ARIMA_M3	ESM_M3	ESM3_M3	ESM	M15_M3	MEAN_M3	MEAN
	11.5713	11.6043	11.6250	11.6398	11.6728	11.6802	11.6883	11.6995	11.7898	11.9076	13.1794	15.9354
Toyota Motor Corporation (TM)	NAIVE3_M3	HOLT3_M3	NAÏVE	HOLT	ESM_M3	M15_M3	ESM3_M3	ESM	ARIMA_M3	MEAN_M3	ARIMA	MEAN
	6.1220	6.1230	6.1579	6.1586	6.3591	6.3611	6.4898	6.7189	6.8989	7.1487	8.1450	9.4176
Honda Motor Co., Ltd (HMC)	ESM3_M3	ESM	HOLT	HOLT3_M3	NAÏVE	NAIVE3_M3	ESM_M3	ARIMA	ARIMA_M3	M15_M3	MEAN_M3	MEAN
	6.7234	6.7336	6.7581	6.7590	6.7607	6.7607	6.7850	6.9322	6.9486	7.0990	7.3479	8.8451
PACCAR Inc. (PCAR)	HOLT3_M3	NAIVE3_M3	ESM_M3	HOLT	M15_M3	NAÏVE	ESM3_M3	ESM	MEAN_M3	ARIMA_M3	MEAN	ARIMA
	8.4030	8.4220	8.4582	8.4788	8.4903	8.5059	8.5359	8.6653	8.7962	8.8272	10.4019	11.6289
Harley-Davidson, Inc. (HOG)	ARIMA_M3	HOLT3_M3	NAIVE3_M3	HOLT	NAÏVE	M15_M3	ESM_M3	ESM3_M3	MEAN_M3	ESM	ARIMA	MEAN
	10.0009	10.0992	10.0997	10.1562	10.1570	10.2167	10.4127	10.6669	11.0165	11.0321	11.2199	13.1090
AUTO & TRUCK ORIGINAL EQUIPMENT MANUFACTURERS STOCKS												
Gentex Corporation (GNTX)	ARIMA	ARIMA_M3	HOLT	NAÏVE	HOLT3_M3	NAIVE3_M3	ESM_M3	M15_M3	ESM3_M3	ESM	MEAN_M3	MEAN
	9.6025	9.8181	9.8632	9.8638	9.8729	9.8733	10.0962	10.2136	10.2567	10.5666	10.8361	12.8868
Tenneco Inc. (TEN)	HOLT3_M3	ESM3_M3	NAIVE3_M3	HOLT	ESM	NAÏVE	ESM_M3	ARIMA	ARIMA_M3	M15_M3	MEAN_M3	MEAN
	17.5515	17.5518	17.5518	17.5557	17.5558	17.5559	17.6799	17.6941	17.9069	18.4081	23.2195	30.9881
Modine Manufacturing Company (MOD)	ARIMA	HOLT	HOLT3_M3	NAÏVE	NAIVE3_M3	ESM3_M3	ARIMA_M3	ESM	ESM_M3	M15_M3	MEAN_M3	MEAN
	15.9458	15.9564	16.0024	16.0173	16.0431	16.2029	16.2200	16.2456	16.2567	16.6018	19.4118	24.1208
Magna International, Inc. (MGA)	HOLT	ESM	NAÏVE	ARIMA	ESM3_M3	HOLT3_M3	NAIVE3_M3	ESM_M3	ARIMA_M3	M15_M3	MEAN_M3	MEAN
	8.7945	8.8026	8.8031	8.8058	8.8498	8.8641	8.8797	8.9626	9.0732	9.3622	10.6190	13.1190
LCI Industries (LC II)	NAIVE3_M3	NAÏVE	ESM3_M3	ESM	HOLT	HOLT3_M3	ESM_M3	M15_M3	MEAN_M3	ARIMA_M3	MEAN	ARIMA
	10.3230	10.3253	10.3312	10.3353	10.3416	10.3448	10.4125	10.7544	12.4622	12.4679	15.3174	17.7164
Oshkosh Corporation (OSK)	NAÏVE	NAIVE3_M3	HOLT3_M3	HOLT	ESM3_M3	ESM_M3	ESM	M15_M3	ARIMA_M3	MEAN_M3	ARIMA	MEAN
	12.0411	12.0769	12.1210	12.1222	12.1941	12.2209	12.2902	12.6184	13.5355	15.0220	15.8696	19.1566
Wabash National Corporation (WNC)	ESM3_M3	NAIVE3_M3	HOLT3_M3	ESM	NAÏVE	ARIMA	ESM_M3	HOLT	ARIMA_M3	M15_M3	MEAN_M3	MEAN
	17.5268	17.5452	17.5575	17.5654	17.5922	17.5932	17.6131	17.6163	17.6999	18.0206	20.2501	25.1145

Table C-16: Auto/Tires/Trucks Sector M3 RMSE Accuracy Results

SECURITY	RMSE #1	RMSE #2	RMSE #3	RMSE #4	RMSE #5	RMSE #6	RMSE #7	RMSE #8	RMSE #9	RMSE #10	RMSE #11	RMSE #12
AUTOMOTIVE STOCKS												
Ford Motor Company (F)	HOLT 0.0250	HOLT3_M3 0.0250	NAIVE3_M3 0.0251	NAÏVE 0.0251	ARIMA 0.0251	ESM3_M3 0.0253	ESM_M3 0.0253	ARIMA_M3 0.0253	ESM 0.0254	M15_M3 0.0259	MEAN_M3 0.0266	MEAN 0.0317
Toyota Motor Corporation (TM)	HOLT3_M3 0.1052	NAIVE3_M3 0.1052	HOLT 0.1058	NAÏVE 0.1058	ESM_M3 0.1069	M15_M3 0.1081	ESM3_M3 0.1085	ESM 0.1113	MEAN_M3 0.1204	ARIMA_M3 0.1349	MEAN 0.1508	ARIMA 0.2361
Honda Motor Co., Ltd (HMC)	HOLT3_M3 0.0382	NAIVE3_M3 0.0382	HOLT 0.0382	NAÏVE 0.0382	ESM3_M3 0.0383	ESM_M3 0.0384	ESM 0.0385	ARIMA_M3 0.0389	ARIMA 0.0391	M15_M3 0.0393	MEAN_M3 0.0404	MEAN 0.0469
PACCAR Inc. (PCAR)	HOLT3_M3 0.0625	HOLT 0.0626	NAIVE3_M3 0.0626	NAÏVE 0.0627	ESM_M3 0.0637	ESM3_M3 0.0639	M15_M3 0.0644	ESM 0.0646	MEAN_M3 0.0657	ARIMA_M3 0.0658	MEAN 0.0753	ARIMA 0.0954
Harley-Davidson, Inc. (HOG)	HOLT 0.0752	NAÏVE 0.0752	HOLT3_M3 0.0755	NAIVE3_M3 0.0755	ARIMA_M3 0.0759	ESM_M3 0.0786	M15_M3 0.0787	ESM3_M3 0.0808	MEAN_M3 0.0809	ESM 0.0844	ARIMA 0.0926	MEAN 0.0927
AUTO & TRUCK ORIGINAL EQUIPMENT MANUFACTURERS STOCKS												
Gentex Corporation (GNTX)	ARIMA_M3 0.0183	ARIMA 0.0183	NAIVE3_M3 0.0185	HOLT3_M3 0.0185	NAÏVE 0.0186	HOLT 0.0186	ESM_M3 0.0188	M15_M3 0.0188	ESM3_M3 0.0193	MEAN_M3 0.0201	ESM 0.0203	MEAN 0.0239
Tenneco Inc. (TEN)	ESM3_M3 0.0599	NAIVE3_M3 0.0599	HOLT3_M3 0.0599	ESM 0.0602	NAÏVE 0.0602	HOLT 0.0602	ESM_M3 0.0602	ARIMA 0.0603	ARIMA_M3 0.0607	M15_M3 0.0626	MEAN_M3 0.0673	MEAN 0.0821
Modine Manufacturing Company (MOD)	HOLT 0.0383	ARIMA 0.0383	NAÏVE 0.0383	HOLT3_M3 0.0384	NAIVE3_M3 0.0385	ESM3_M3 0.0390	ESM 0.0391	ARIMA_M3 0.0391	ESM_M3 0.0392	M15_M3 0.0403	MEAN_M3 0.0427	MEAN 0.0506
Magna International, Inc. (MGA)	ARIMA 0.0402	NAÏVE 0.0402	ESM 0.0404	HOLT 0.0404	ESM3_M3 0.0406	NAIVE3_M3 0.0406	HOLT3_M3 0.0407	ESM_M3 0.0412	ARIMA_M3 0.0417	M15_M3 0.0435	MEAN_M3 0.0480	MEAN 0.0582
LCI Industries (LC II)	NAÏVE 0.0705	HOLT 0.0706	NAIVE3_M3 0.0713	HOLT3_M3 0.0714	ESM3_M3 0.0739	ESM_M3 0.0741	ESM 0.0743	M15_M3 0.0758	MEAN_M3 0.0884	ARIMA_M3 0.0976	MEAN 0.1099	ARIMA 0.2100
Oshkosh Corporation (OSK)	NAÏVE 0.0726	NAIVE3_M3 0.0733	HOLT 0.0736	HOLT3_M3 0.0738	ESM_M3 0.0751	ESM3_M3 0.0751	ESM 0.0762	M15_M3 0.0781	MEAN_M3 0.0881	ARIMA_M3 0.0984	MEAN 0.1079	ARIMA 0.1655
Wabash National Corporation (WNC)	NAIVE3_M3 0.0328	ARIMA 0.0328	ESM3_M3 0.0328	NAÏVE 0.0328	HOLT3_M3 0.0328	ESM 0.0329	HOLT 0.0329	ESM_M3 0.0330	ARIMA_M3 0.0332	M15_M3 0.0341	MEAN_M3 0.0358	MEAN 0.0423

C.7. Basic Materials Sector

Table C-17: Basic Materials Sector M3 MAPE Accuracy Results

SECURITY	MAPE #1	MAPE #2	MAPE #3	MAPE #4	MAPE #5	MAPE #6	MAPE #7	MAPE #8	MAPE #9	MAPE #10	MAPE #11	MAPE #12
CHEMICALS												
Valhi, Inc. (VHI)	HOLT	NAÏVE	ESM	HOLT3_M3	NAIVE3_M3	ESM3_M3	ESM_M3	ARIMA_M3	M15_M3	ARIMA	MEAN_M3	MEAN
	15.1211	15.1815	15.2130	15.2456	15.3079	15.3323	15.6053	15.6636	16.4490	17.4846	19.2955	24.8044
Air Products and Chemicals, Inc. (APD)	ARIMA	ARIMA_M3	ESM_M3	NAIVE3_M3	HOLT3_M3	ESM3_M3	NAÏVE	HOLT	M15_M3	ESM	MEAN_M3	MEAN
	6.1771	6.2195	6.2450	6.2487	6.2494	6.2793	6.2982	6.2983	6.3535	6.3766	6.6734	8.2110
FMC Corporation (FMC)	NAIVE3_M3	NAÏVE	HOLT3_M3	HOLT	ESM_M3	ARIMA_M3	ESM3_M3	ESM	ARIMA	M15_M3	MEAN_M3	MEAN
	8.8460	8.8509	8.8744	8.9070	8.9886	8.9943	9.0143	9.1136	9.1564	9.1886	10.6187	13.5957
Olin Corporation (OLN)	NAIVE3_M3	HOLT3_M3	ARIMA	NAÏVE	ESM3_M3	HOLT	ESM_M3	ARIMA_M3	ESM	M15_M3	MEAN_M3	MEAN
	8.8526	8.8651	8.8878	8.8919	8.9071	8.9115	8.9321	8.9506	8.9704	9.2311	9.5192	11.8832
Cabot Corporation (CBT)	ARIMA_M3	ESM_M3	NAIVE3_M3	M15_M3	ESM3_M3	HOLT3_M3	NAÏVE	ARIMA	ESM	HOLT	MEAN_M3	MEAN
	11.0532	11.0675	11.1062	11.1104	11.1269	11.1455	11.2124	11.2239	11.2405	11.2681	11.8593	14.2242
PPG Industries, Inc. (PPG)	ESM_M3	ESM3_M3	M15_M3	HOLT3_M3	NAIVE3_M3	ESM	ARIMA_M3	HOLT	NAÏVE	MEAN_M3	ARIMA	MEAN
	6.9107	6.9362	6.9938	7.0013	7.0058	7.0085	7.0207	7.0814	7.0878	7.4267	8.1473	8.9133
Methanex Corporation (MEOH)	ESM	NAÏVE	ESM3_M3	NAIVE3_M3	HOLT	HOLT3_M3	ESM_M3	ARIMA_M3	M15_M3	ARIMA	MEAN_M3	MEAN
	11.5683	11.5973	11.6246	11.6438	11.6885	11.7011	11.7393	11.7723	12.2357	13.0546	13.5224	16.8716
MINING STOCKS												
McEwen Mining, Inc. (MUX)	HOLT	NAÏVE	HOLT3_M3	NAIVE3_M3	M15_M3	ARIMA_M3	ESM_M3	ARIMA	MEAN_M3	ESM3_M3	ESM	MEAN
	19.6592	19.6630	19.6900	19.6968	20.7476	20.9932	22.3179	22.4446	23.8421	23.9920	25.9059	29.7466
BHP Group (BHP)	NAÏVE	HOLT	NAIVE3_M3	HOLT3_M3	ESM3_M3	ESM_M3	ESM	M15_M3	ARIMA_M3	MEAN_M3	ARIMA	MEAN
	8.7807	8.7867	8.7916	8.7959	8.8868	8.9202	8.9304	9.1782	9.7000	10.2933	11.9643	12.8216
Cleveland-Cliffs, Inc. (CLF)	NAÏVE	NAIVE3_M3	HOLT	HOLT3_M3	ESM_M3	M15_M3	ESM3_M3	ESM	ARIMA_M3	MEAN_M3	ARIMA	MEAN
	17.8959	17.9780	18.1207	18.1411	18.5721	18.6529	18.6672	18.8852	19.2998	22.2160	22.3218	28.6194
Materion Corporation (MTRN)	ARIMA	NAÏVE	HOLT	HOLT3_M3	NAIVE3_M3	ESM	ESM3_M3	ESM_M3	ARIMA_M3	M15_M3	MEAN_M3	MEAN
	12.7652	12.7869	12.7887	12.8580	12.8657	12.9304	12.9586	13.0543	13.0829	13.4451	14.9484	18.1814
Rio Tinto plc (RIO)	ESM	HOLT	ESM3_M3	HOLT3_M3	NAÏVE	NAIVE3_M3	ESM_M3	M15_M3	ARIMA_M3	MEAN_M3	ARIMA	MEAN
	10.3186	10.3315	10.3319	10.3418	10.3479	10.3537	10.4078	10.7523	11.2939	12.2952	13.6841	15.5787
Taseko Mines Limited (TGB)	NAÏVE	HOLT	NAIVE3_M3	HOLT3_M3	ESM3_M3	ESM_M3	ESM	M15_M3	MEAN_M3	ARIMA_M3	MEAN	ARIMA
	20.1960	20.2222	20.3705	20.3924	21.0973	21.1451	21.1855	21.6360	25.8738	27.6634	33.0341	47.4164

SECURITY	MAPE #1	MAPE #2	MAPE #3	MAPE #4	MAPE #5	MAPE #6	MAPE #7	MAPE #8	MAPE #9	MAPE #10	MAPE #11	MAPE #12
GOLD MINER STOCKS												
U.S. Gold Corp (USAU)	NAÏVE 16.9962	NAIVE3_M3 17.1702	HOLT3_M3 17.2975	HOLT 17.2989	ARIMA_M3 18.6468	M15_M3 18.6973	ESM_M3 19.5846	ARIMA 20.1032	ESM3_M3 20.6483	ESM 21.8515	MEAN_M3 23.1739	MEAN 30.5156
Newmont Mining Corporation (NEM)	ESM 9.3630	NAÏVE 9.3630	HOLT 9.3717	ARIMA 9.3821	ESM3_M3 9.4111	NAIVE3_M3 9.4111	HOLT3_M3 9.4149	ESM_M3 9.5338	ARIMA_M3 9.6177	M15_M3 9.9250	MEAN_M3 10.5311	MEAN 12.7347
Agnico Eagle Mines Limited (AEM)	NAÏVE 11.7903	HOLT 11.8059	ESM 11.8088	NAIVE3_M3 11.8686	HOLT3_M3 11.8710	ESM3_M3 11.8825	ESM_M3 12.0282	M15_M3 12.5405	ARIMA_M3 12.8947	MEAN_M3 13.3082	ARIMA 14.0037	MEAN 15.5416
Gold Fields Limited (GFI)	HOLT 10.9949	NAÏVE 11.0046	HOLT3_M3 11.0649	NAIVE3_M3 11.0705	ARIMA 11.0902	ESM3_M3 11.1762	ESM 11.1873	ESM_M3 11.2549	ARIMA_M3 11.3515	M15_M3 11.6895	MEAN_M3 13.0181	MEAN 16.5002
Royal Gold, Inc. (RGLD)	NAIVE3_M3 11.6196	NAÏVE 11.6292	HOLT3_M3 11.6369	HOLT 11.6655	ESM3_M3 11.7499	ESM_M3 11.7520	ESM 11.8463	ARIMA_M3 12.0511	M15_M3 12.0615	MEAN_M3 12.0944	ARIMA 12.4504	MEAN 14.6906
Barrick Gold Corporation (GOLD)	NAÏVE 10.9457	HOLT 10.9521	NAIVE3_M3 11.0166	HOLT3_M3 11.0216	ARIMA 11.0636	ESM 11.1693	ESM3_M3 11.1927	ESM_M3 11.2681	ARIMA_M3 11.2808	M15_M3 11.5731	MEAN_M3 11.6111	MEAN 13.3653
Vista Gold Corp. (VGZ)	HOLT 18.6508	NAÏVE 18.6794	ESM 18.6964	HOLT3_M3 18.8342	NAIVE3_M3 18.8367	ESM3_M3 18.8425	ARIMA 18.9057	ESM_M3 19.1061	ARIMA_M3 19.4279	M15_M3 20.0171	MEAN_M3 21.9471	MEAN 26.4962
STEEL PRODUCERS												
Nucor Corporation (NUE)	ESM_M3 8.8160	HOLT3_M3 8.8297	NAIVE3_M3 8.8360	ESM3_M3 8.8667	HOLT 8.9049	NAÏVE 8.9163	M15_M3 8.9385	ESM 9.0010	ARIMA_M3 9.1550	MEAN_M3 9.7383	ARIMA 10.3274	MEAN 12.0849
Commercial Metals Company (CMC)	HOLT3_M3 10.7399	NAIVE3_M3 10.7449	HOLT 10.7569	NAÏVE 10.7619	ESM3_M3 10.8412	ESM_M3 10.8607	ESM 10.9056	M15_M3 11.1420	MEAN_M3 12.3282	ARIMA_M3 12.3997	MEAN 14.9830	ARIMA 15.5348
L. B. Foster Company (FSTR)	HOLT3_M3 11.5506	ESM3_M3 11.5760	ESM_M3 11.5773	NAIVE3_M3 11.6033	HOLT 11.6395	ESM 11.6616	NAÏVE 11.7247	M15_M3 11.8462	ARIMA_M3 12.3964	MEAN_M3 13.2689	ARIMA 15.6943	MEAN 16.6893
United States Steel Corporation (X)	ARIMA_M3 17.2214	HOLT3_M3 17.3249	HOLT 17.3596	ARIMA 17.3977	NAIVE3_M3 17.4536	ESM3_M3 17.4589	NAÏVE 17.4769	ESM 17.4871	ESM_M3 17.5099	M15_M3 17.8245	MEAN_M3 20.9409	MEAN 27.2162
Shiloh Industries Inc. (SHLO)	NAÏVE 16.9777	ARIMA 17.0064	HOLT 17.0068	NAIVE3_M3 17.1633	HOLT3_M3 17.2026	ESM 17.5326	ESM3_M3 17.5738	ARIMA_M3 17.7426	ESM_M3 17.7484	M15_M3 18.5065	MEAN_M3 21.7400	MEAN 27.4445
Schnitzer Steel Industries, Inc. (SCHN)	HOLT3_M3 13.6199	NAÏVE 13.6212	NAIVE3_M3 13.6532	HOLT 13.6643	ESM_M3 14.0834	ESM3_M3 14.1334	M15_M3 14.2195	ESM 14.2555	MEAN_M3 14.6168	ARIMA_M3 15.9581	MEAN 17.7771	ARIMA 19.3458
Olympic Steel, Inc. (ZEUS)	NAIVE3_M3 15.1062	HOLT3_M3 15.1791	NAÏVE 15.1801	ESM_M3 15.2575	ESM3_M3 15.2740	HOLT 15.2953	ESM 15.4122	ARIMA_M3 15.4545	M15_M3 15.6012	ARIMA 15.7650	MEAN_M3 17.9813	MEAN 23.4836

Table C-18: Basic Materials Sector M3 RMSE Accuracy Results

SECURITY	RMSE #1	RMSE #2	RMSE #3	RMSE #4	RMSE #5	RMSE #6	RMSE #7	RMSE #8	RMSE #9	RMSE #10	RMSE #11	RMSE #12
CHEMICALS												
Valhi, Inc. (VHI)	NAÏVE 0.0221	NAIVE3_M3 0.0221	HOLT3_M3 0.0221	HOLT 0.0221	ESM3_M3 0.0222	ESM 0.0222	ESM_M3 0.0223	M15_M3 0.0229	ARIMA_M3 0.0246	MEAN_M3 0.0253	MEAN 0.0309	ARIMA 0.0421
Air Products and Chemicals, Inc. (APD)	ARIMA 0.0849	ARIMA_M3 0.0857	HOLT3_M3 0.0858	NAIVE3_M3 0.0858	ESM_M3 0.0859	ESM3_M3 0.0862	HOLT 0.0863	NAÏVE 0.0863	ESM 0.0872	M15_M3 0.0876	MEAN_M3 0.0965	MEAN 0.1192
FMC Corporation (FMC)	NAÏVE 0.0500	HOLT 0.0502	NAIVE3_M3 0.0504	HOLT3_M3 0.0506	ESM 0.0506	ESM3_M3 0.0507	ESM_M3 0.0513	ARIMA_M3 0.0515	M15_M3 0.0541	ARIMA 0.0556	MEAN_M3 0.0593	MEAN 0.0725
Olin Corporation (OLN)	NAIVE3_M3 0.0352	ARIMA 0.0352	NAÏVE 0.0352	HOLT3_M3 0.0352	HOLT 0.0352	ESM3_M3 0.0353	ESM 0.0354	ESM_M3 0.0355	ARIMA_M3 0.0356	M15_M3 0.0366	MEAN_M3 0.0387	MEAN 0.0470
Cabot Corporation (CBT)	NAIVE3_M3 0.0693	ESM3_M3 0.0694	ESM_M3 0.0694	HOLT3_M3 0.0694	NAÏVE 0.0695	ARIMA_M3 0.0696	ARIMA 0.0696	ESM 0.0697	HOLT 0.0698	M15_M3 0.0705	MEAN_M3 0.0729	MEAN 0.0841
PPG Industries, Inc. (PPG)	ESM_M3 0.0667	ESM3_M3 0.0668	HOLT3_M3 0.0673	ESM 0.0673	NAIVE3_M3 0.0673	M15_M3 0.0675	HOLT 0.0679	NAÏVE 0.0679	MEAN_M3 0.0687	ARIMA_M3 0.0717	MEAN 0.0806	ARIMA 0.1250
Methanex Corporation (MEOH)	ARIMA_M3 0.0657	NAÏVE 0.0676	HOLT 0.0678	ESM 0.0678	NAIVE3_M3 0.0683	HOLT3_M3 0.0684	ESM3_M3 0.0684	ESM_M3 0.0693	M15_M3 0.0716	MEAN_M3 0.0729	ARIMA 0.0788	MEAN 0.0831
MINING STOCKS												
McEwen Mining, Inc. (MUX)	NAIVE3_M3 0.0112	HOLT3_M3 0.0112	ESM_M3 0.0112	ESM3_M3 0.0113	NAÏVE 0.0113	HOLT 0.0114	ESM 0.0115	M15_M3 0.0116	MEAN_M3 0.0125	ARIMA_M3 0.0154	MEAN 0.0154	ARIMA 0.0265
BHP Group (BHP)	NAÏVE 0.0846	HOLT 0.0847	HOLT3_M3 0.0848	NAIVE3_M3 0.0848	ESM_M3 0.0870	ESM3_M3 0.0874	ESM 0.0886	M15_M3 0.0887	MEAN_M3 0.0993	ARIMA_M3 0.0994	MEAN 0.1209	ARIMA 0.1597
Cleveland-Cliffs, Inc. (CLF)	NAIVE3_M3 0.1224	NAÏVE 0.1231	M15_M3 0.1244	HOLT3_M3 0.1309	ARIMA_M3 0.1314	ESM_M3 0.1342	HOLT 0.1353	MEAN_M3 0.1423	ESM3_M3 0.1426	ESM 0.1528	ARIMA 0.1605	MEAN 0.1766
Materion Corporation (MTRN)	ARIMA 0.0600	NAÏVE 0.0600	HOLT 0.0600	ESM 0.0600	ESM3_M3 0.0603	HOLT3_M3 0.0604	NAIVE3_M3 0.0604	ESM_M3 0.0610	ARIMA_M3 0.0614	M15_M3 0.0633	MEAN_M3 0.0696	MEAN 0.0839
Rio Tinto plc (RIO)	ESM 0.1136	HOLT 0.1139	ESM3_M3 0.1142	HOLT3_M3 0.1146	NAÏVE 0.1147	NAIVE3_M3 0.1151	ESM_M3 0.1154	M15_M3 0.1191	MEAN_M3 0.1365	ARIMA_M3 0.1591	MEAN 0.1659	ARIMA 0.2817
Taseko Mines Limited (TGB)	NAIVE3_M3 0.0076	HOLT3_M3 0.0076	HOLT 0.0076	NAÏVE 0.0076	ESM3_M3 0.0076	ESM_M3 0.0077	ESM 0.0077	M15_M3 0.0078	MEAN_M3 0.0090	MEAN 0.0110	ARIMA_M3 0.0166	ARIMA 0.0392

SECURITY	RMSE #1	RMSE #2	RMSE #3	RMSE #4	RMSE #5	RMSE #6	RMSE #7	RMSE #8	RMSE #9	RMSE #10	RMSE #11	RMSE #12
GOLD MINER STOCKS												
U.S. Gold Corp (USAU)	NAÏVE 1.4374	ESM 1.4470	NAIVE3_M3 1.4557	ESM3_M3 1.4636	ESM_M3 1.4979	HOLT3_M3 1.5990	M15_M3 1.6145	HOLT 1.7433	MEAN_M3 1.9452	ARIMA_M3 2.0434	MEAN 2.6727	ARIMA 4.4900
Newmont Mining Corporation (NEM)	NAÏVE 0.0608	ESM 0.0608	HOLT 0.0608	ARIMA 0.0609	NAIVE3_M3 0.0611	ESM3_M3 0.0611	HOLT3_M3 0.0611	ESM_M3 0.0617	ARIMA_M3 0.0622	M15_M3 0.0641	MEAN_M3 0.0693	MEAN 0.0832
Agnico Eagle Mines Limited (AEM)	HOLT 0.0859	NAÏVE 0.0862	ESM 0.0865	HOLT3_M3 0.0868	NAIVE3_M3 0.0871	ESM3_M3 0.0873	ESM_M3 0.0885	M15_M3 0.0922	MEAN_M3 0.0983	MEAN 0.1139	ARIMA_M3 0.1186	ARIMA 0.2029
Gold Fields Limited (GFI)	ESM3_M3 0.0174	HOLT3_M3 0.0174	NAIVE3_M3 0.0174	ESM_M3 0.0175	HOLT 0.0175	NAÏVE 0.0175	ESM 0.0175	ARIMA 0.0175	ARIMA_M3 0.0177	M15_M3 0.0181	MEAN_M3 0.0207	MEAN 0.0257
Royal Gold, Inc. (RGLD)	NAIVE3_M3 0.0937	NAÏVE 0.0937	HOLT3_M3 0.0938	HOLT 0.0940	ESM_M3 0.0947	ESM3_M3 0.0948	ESM 0.0955	M15_M3 0.0966	MEAN_M3 0.0999	ARIMA_M3 0.1095	MEAN 0.1181	ARIMA 0.1535
Barrick Gold Corporation (GOLD)	NAÏVE 0.0506	NAIVE3_M3 0.0506	HOLT3_M3 0.0507	HOLT 0.0507	ARIMA 0.0512	ARIMA_M3 0.0513	ESM_M3 0.0515	ESM3_M3 0.0515	ESM 0.0517	M15_M3 0.0521	MEAN_M3 0.0542	MEAN 0.0615
Vista Gold Corp. (VGZ)	NAÏVE 0.0110	HOLT 0.0110	NAIVE3_M3 0.0111	ESM 0.0111	HOLT3_M3 0.0111	ARIMA 0.0112	ESM3_M3 0.0112	ESM_M3 0.0113	ARIMA_M3 0.0115	M15_M3 0.0119	MEAN_M3 0.0127	MEAN 0.0146
STEEL PRODUCERS												
Nucor Corporation (NUE)	HOLT3_M3 0.0694	ESM_M3 0.0694	NAIVE3_M3 0.0695	ESM3_M3 0.0695	HOLT 0.0697	NAÏVE 0.0698	ESM 0.0700	M15_M3 0.0706	ARIMA_M3 0.0768	MEAN_M3 0.0771	MEAN 0.0930	ARIMA 0.1209
Commercial Metals Company (CMC)	NAIVE3_M3 0.0382	HOLT3_M3 0.0382	NAÏVE 0.0383	HOLT 0.0383	M15_M3 0.0390	ESM_M3 0.0393	ESM3_M3 0.0396	ESM 0.0402	MEAN_M3 0.0422	ARIMA_M3 0.0484	MEAN 0.0501	ARIMA 0.0890
L. B. Foster Company (FSTR)	HOLT3_M3 0.0586	NAIVE3_M3 0.0587	HOLT 0.0587	ESM3_M3 0.0588	ESM_M3 0.0589	NAÏVE 0.0589	ESM 0.0590	M15_M3 0.0597	MEAN_M3 0.0651	ARIMA_M3 0.0685	MEAN 0.0789	ARIMA 0.1205
United States Steel Corporation (X)	ARIMA_M3 0.1520	NAÏVE 0.1729	ESM 0.1729	HOLT 0.1748	NAIVE3_M3 0.1753	ESM3_M3 0.1753	HOLT3_M3 0.1762	ESM_M3 0.1781	M15_M3 0.1846	ARIMA 0.1851	MEAN_M3 0.2087	MEAN 0.2543
Shiloh Industries Inc. (SHLO)	NAÏVE 0.0287	HOLT 0.0287	ARIMA 0.0287	NAIVE3_M3 0.0287	HOLT3_M3 0.0288	ESM3_M3 0.0292	ARIMA_M3 0.0293	ESM_M3 0.0293	ESM 0.0293	M15_M3 0.0301	MEAN_M3 0.0316	MEAN 0.0366
Schnitzer Steel Industries, Inc. (SCHN)	M15_M3 0.1184	NAIVE3_M3 0.1191	NAÏVE 0.1202	HOLT3_M3 0.1203	ESM_M3 0.1206	MEAN_M3 0.1217	HOLT 0.1224	ESM3_M3 0.1230	ESM 0.1261	MEAN 0.1437	ARIMA_M3 0.1688	ARIMA 0.2800
Olympic Steel, Inc. (ZEUS)	M15_M3 0.0748	NAIVE3_M3 0.0751	ESM_M3 0.0751	ESM3_M3 0.0758	NAÏVE 0.0758	ESM 0.0767	HOLT3_M3 0.0792	MEAN_M3 0.0803	HOLT 0.0816	ARIMA_M3 0.0837	MEAN 0.0985	ARIMA 0.1033

C.8. Business Services Sector

Table C-19: Business Services M3 MAPE Accuracy Results

SECURITY	MAPE #1	MAPE #2	MAPE #3	MAPE #4	MAPE #5	MAPE #6	MAPE #7	MAPE #8	MAPE #9	MAPE #10	MAPE #11	MAPE #12
FINANCIAL TRANSACTIONS												
Equifax Inc. (EFX)	HOLT3_M3 7.7623	HOLT 7.7677	NAIVE3_M3 7.7691	ESM_M3 7.7710	ESM3_M3 7.7763	NAÏVE 7.7776	ESM 7.8679	ARIMA_M3 7.9357	M15_M3 7.9510	MEAN_M3 8.3251	ARIMA 8.7821	MEAN 9.9536
Diebold Nixdorf, Incorporated (DBD)	ARIMA_M3 10.0503	ESM_M3 10.0820	M15_M3 10.1032	HOLT3_M3 10.1198	NAIVE3_M3 10.1271	ESM3_M3 10.1511	ARIMA 10.2170	HOLT 10.2354	NAÏVE 10.2456	ESM 10.2784	MEAN_M3 11.7670	MEAN 14.6414
Total System Services, Inc (TSS)	ESM3_M3 8.2389	HOLT3_M3 8.2622	HOLT 8.2624	ESM 8.2675	ESM_M3 8.2895	NAÏVE 8.2933	NAIVE3_M3 8.2986	ARIMA_M3 8.5194	ARIMA 8.6410	M15_M3 8.6633	MEAN_M3 9.9056	MEAN 12.0264
Fiserv, Inc. (FISV)	ESM_M3 6.5713	ESM3_M3 6.6757	M15_M3 6.7549	HOLT3_M3 6.7560	NAIVE3_M3 6.7819	HOLT 6.8419	NAÏVE 6.8831	ESM 6.8960	MEAN_M3 7.0209	ARIMA_M3 7.9152	MEAN 8.4563	ARIMA 10.4831
STAFFING												
Kelly Services, Inc. (KELYA)	ARIMA_M3 9.1652	M15_M3 9.1860	ESM_M3 9.2559	HOLT3_M3 9.2636	NAIVE3_M3 9.3112	ESM3_M3 9.3833	ARIMA 9.4039	HOLT 9.4118	NAÏVE 9.4818	ESM 9.5977	MEAN_M3 9.8183	MEAN 12.0018
ManpowerGroup Inc. (MAN)	ESM_M3 10.1212	ARIMA_M3 10.1344	NAIVE3_M3 10.1373	ESM3_M3 10.1377	HOLT3_M3 10.1430	NAÏVE 10.2024	ESM 10.2028	ARIMA 10.2099	HOLT 10.2113	M15_M3 10.2368	MEAN_M3 10.9833	MEAN 13.4925
GEE Group., Inc. (JOB)	HOLT3_M3 18.0139	NAIVE3_M3 18.0569	ARIMA_M3 18.0757	ARIMA 18.1214	HOLT 18.1756	NAÏVE 18.2294	M15_M3 18.4074	ESM_M3 20.7927	MEAN_M3 22.6880	ESM3_M3 22.7376	ESM 24.9935	MEAN 29.9891
Robert Half International, Inc. (RHI)	HOLT 9.9912	NAÏVE 9.9917	HOLT3_M3 9.9972	NAIVE3_M3 9.9973	ARIMA 10.0746	ARIMA_M3 10.1384	M15_M3 10.3201	ESM_M3 10.3226	ESM3_M3 10.4643	ESM 10.7081	MEAN_M3 11.0638	MEAN 12.8875
OUTSOURCING												
Automatic Data Processing, Inc. (ADP)	NAIVE3_M3 5.6858	HOLT3_M3 5.6888	M15_M3 5.6929	NAÏVE 5.7519	HOLT 5.7534	ESM_M3 5.7591	ESM3_M3 5.8785	MEAN_M3 5.9631	ESM 6.0563	ARIMA_M3 6.4509	MEAN 7.0391	ARIMA 8.2441
R.R. Donnelley & Sons Company (RRD)	HOLT 10.3291	ARIMA 10.3300	NAÏVE 10.3319	HOLT3_M3 10.3483	NAIVE3_M3 10.3512	ESM 10.3559	ESM3_M3 10.3602	ESM_M3 10.4314	ARIMA_M3 10.4881	M15_M3 10.7485	MEAN_M3 12.2375	MEAN 14.8914
Paychex, Inc. (PAYX)	NAÏVE 6.7251	HOLT 6.7291	NAIVE3_M3 6.7611	HOLT3_M3 6.7688	ESM_M3 7.0267	ESM3_M3 7.1326	M15_M3 7.1755	ESM 7.3982	ARIMA_M3 7.4638	MEAN_M3 7.7777	ARIMA 8.6891	MEAN 9.1645
Barrett Business Services, Inc. (BBSI)	NAÏVE 11.5892	HOLT 11.5928	NAIVE3_M3 11.6111	HOLT3_M3 11.6188	ESM3_M3 11.6538	ESM 11.6679	ESM_M3 11.7802	M15_M3 12.3399	ARIMA_M3 12.6286	MEAN_M3 13.9028	ARIMA 14.7379	MEAN 17.1869

SECURITY	MAPE #1	MAPE #2	MAPE #3	MAPE #4	MAPE #5	MAPE #6	MAPE #7	MAPE #8	MAPE #9	MAPE #10	MAPE #11	MAPE #12
BUSINESS SERVICES												
Xerox Corporation (XRX)	ESM_M3 11.2385	ARIMA_M3 11.2457	ESM3_M3 11.3334	NAIVE3_M3 11.3376	M15_M3 11.4340	HOLT3_M3 11.4342	ARIMA 11.5362	ESM 11.5776	NAÏVE 11.5861	HOLT 11.7364	MEAN_M3 12.4048	MEAN 15.4212
Volt Information Services, Inc. (VISI)	ARIMA 12.7116	NAÏVE 12.7175	NAIVE3_M3 12.7731	HOLT 12.7924	HOLT3_M3 12.7965	ESM3_M3 13.0811	ESM 13.1114	ARIMA_M3 13.1600	ESM_M3 13.1961	M15_M3 13.7747	MEAN_M3 15.6254	MEAN 19.9618
Healthcare Services Group, Inc. (HCSG)	HOLT3_M3 7.6860	NAIVE3_M3 7.6952	HOLT 7.7064	NAÏVE 7.7444	ESM3_M3 7.7470	ESM_M3 7.7548	ESM 7.8664	M15_M3 8.1282	MEAN_M3 8.6131	ARIMA_M3 10.3084	MEAN 10.3358	ARIMA 18.6814
Avis Budget Group, Inc. (CAR)	HOLT 20.3083	NAÏVE 20.3086	HOLT3_M3 20.3684	NAIVE3_M3 20.3687	ARIMA 20.3831	ARIMA_M3 20.6818	ESM3_M3 20.7083	ESM_M3 20.7321	ESM 20.8142	M15_M3 21.1972	MEAN_M3 27.4381	MEAN 36.1545
Spherix Incorporated (SPEX)	NAIVE3_M3 25.3848	HOLT 25.3934	NAÏVE 25.3944	HOLT3_M3 25.4262	ARIMA 25.8764	ESM_M3 26.1533	ESM3_M3 26.2267	ARIMA_M3 26.2975	ESM 26.5787	M15_M3 26.9996	MEAN_M3 32.8035	MEAN 42.2901
Crawford & Company (CRD-A)	ESM 10.1372	NAÏVE 10.1374	ARIMA 10.1447	HOLT 10.1449	HOLT3_M3 10.1663	ESM3_M3 10.1686	NAIVE3_M3 10.1687	ESM_M3 10.3082	ARIMA_M3 10.4191	M15_M3 10.8869	MEAN_M3 12.5120	MEAN 15.8346
CorVel Corporation (CRVL	NAIVE3_M3 9.0995	NAÏVE 9.1390	HOLT3_M3 9.1952	ESM_M3 9.2103	ESM3_M3 9.2228	HOLT 9.2585	ESM 9.3196	ARIMA_M3 9.4309	M15_M3 9.4417	ARIMA 10.1512	MEAN_M3 10.6856	MEAN 13.2917
ADVERTISING & MARKETING												
Omnicom Group, Inc (OMC)	HOLT3_M3 7.0509	NAIVE3_M3 7.0566	ARIMA_M3 7.0763	M15_M3 7.1098	HOLT 7.1209	NAÏVE 7.1284	ESM_M3 7.1966	MEAN_M3 7.2451	ESM3_M3 7.3947	ESM 7.6732	ARIMA 8.2965	MEAN 8.7662
The Interpublic Group of Co., Inc. (IPG)	NAIVE3_M3 10.4541	M15_M3 10.4669	HOLT3_M3 10.4730	ESM_M3 10.5235	NAÏVE 10.5601	HOLT 10.5891	ESM3_M3 10.6651	ESM 10.8657	ARIMA_M3 11.2570	MEAN_M3 11.6120	ARIMA 12.8952	MEAN 14.6869
WPP plc (WPP)	NAIVE3_M3 8.3295	HOLT3_M3 8.3451	ESM_M3 8.3585	NAÏVE 8.3742	ESM3_M3 8.3786	HOLT 8.4005	ESM 8.4741	M15_M3 8.5269	MEAN_M3 9.4713	ARIMA_M3 9.7443	MEAN 11.6831	ARIMA 12.4003
Insignia Systems, Inc. (ISIG)	ARIMA 18.2666	HOLT3_M3 18.3116	HOLT 18.3349	NAIVE3_M3 18.3477	NAÏVE 18.3861	ARIMA_M3 18.5473	ESM_M3 18.5872	ESM3_M3 18.7008	ESM 19.1015	M15_M3 19.1165	MEAN_M3 21.5277	MEAN 27.6159
Harte Hanks, Inc. (HHS)	HOLT 11.1890	NAÏVE 11.1904	NAIVE3_M3 11.2354	HOLT3_M3 11.2395	ESM3_M3 11.4841	ESM_M3 11.5053	ESM 11.5889	ARIMA_M3 11.7478	M15_M3 11.9079	ARIMA 13.7468	MEAN_M3 14.2692	MEAN 17.8338

APPENDIX C: M3 Forecast Study Results

Table C-20: Business Services M3 RMSE Accuracy Results

SECURITY	RMSE #1	RMSE #2	RMSE #3	RMSE #4	RMSE #5	RMSE #6	RMSE #7	RMSE #8	RMSE #9	RMSE #10	RMSE #11	RMSE #12
FINANCIAL TRANSACTIONS												
Equifax Inc. (EFX)	HOLT	NAÏVE	HOLT3_M3	NAIVE3_M3	M15_M3	MEAN_M3	ESM_M3	ESM3_M3	ESM	ARIMA_M3	MEAN	ARIMA
	0.0860	0.0861	0.0864	0.0865	0.0892	0.0896	0.0903	0.0917	0.0936	0.0970	0.1031	0.1638
Diebold Nixdorf, Incorporated (DBD)	ARIMA_M3	ESM_M3	M15_M3	HOLT3_M3	NAIVE3_M3	ESM3_M3	ARIMA	HOLT	NAÏVE	ESM	MEAN_M3	MEAN
	0.0517	0.0519	0.0520	0.0522	0.0522	0.0523	0.0528	0.0529	0.0529	0.0530	0.0563	0.0690
Total System Services, Inc (TSS)	ESM_M3	ESM3_M3	HOLT3_M3	NAIVE3_M3	ESM	NAÏVE	HOLT	M15_M3	MEAN_M3	ARIMA_M3	MEAN	ARIMA
	0.0518	0.0520	0.0526	0.0527	0.0527	0.0528	0.0528	0.0532	0.0570	0.0579	0.0660	0.0696
Fiserv, Inc. (FISV)	ESM_M3	ESM3_M3	M15_M3	HOLT3_M3	NAIVE3_M3	ESM	HOLT	NAÏVE	MEAN_M3	MEAN	ARIMA_M3	ARIMA
	0.0236	0.0245	0.0251	0.0260	0.0261	0.0263	0.0266	0.0267	0.0282	0.0345	0.0480	0.1017
STAFFING												
Kelly Services, Inc. (KELYA)	ARIMA_M3	M15_M3	HOLT3_M3	ESM_M3	NAIVE3_M3	HOLT	ARIMA	ESM3_M3	NAÏVE	ESM	MEAN_M3	MEAN
	0.0337	0.0338	0.0340	0.0340	0.0342	0.0344	0.0344	0.0345	0.0347	0.0352	0.0353	0.0418
ManpowerGroup Inc. (MAN)	ESM3_M3	NAIVE3_M3	HOLT3_M3	ESM_M3	ESM	NAÏVE	HOLT	ARIMA_M3	ARIMA	M15_M3	MEAN_M3	MEAN
	0.1063	0.1063	0.1064	0.1065	0.1066	0.1066	0.1066	0.1070	0.1070	0.1081	0.1182	0.1430
GEE Group., Inc. (JOB)	ARIMA	HOLT3_M3	NAIVE3_M3	HOLT	NAÏVE	ARIMA_M3	ESM_M3	M15_M3	ESM3_M3	ESM	MEAN_M3	MEAN
	0.0359	0.0359	0.0360	0.0360	0.0360	0.0364	0.0374	0.0375	0.0383	0.0398	0.0413	0.0524
Robert Half International, Inc. (RHI)	NAÏVE	HOLT	NAIVE3_M3	HOLT3_M3	ARIMA	ARIMA_M3	ESM_M3	M15_M3	ESM3_M3	ESM	MEAN_M3	MEAN
	0.0581	0.0581	0.0581	0.0581	0.0584	0.0588	0.0595	0.0598	0.0606	0.0624	0.0637	0.0736
OUTSOURCING												
Automatic Data Processing, Inc. (ADP)	M15_M3	NAIVE3_M3	HOLT3_M3	ESM_M3	NAÏVE	HOLT	ESM3_M3	ESM	MEAN_M3	MEAN	ARIMA_M3	ARIMA
	0.0607	0.0610	0.0611	0.0611	0.0616	0.0617	0.0620	0.0633	0.0635	0.0755	0.0755	0.1237
R.R. Donnelley & Sons Company (RRD)	HOLT3_M3	NAIVE3_M3	HOLT	NAÏVE	ARIMA	ARIMA_M3	ESM_M3	ESM3_M3	ESM	M15_M3	MEAN_M3	MEAN
	0.0388	0.0388	0.0389	0.0389	0.0389	0.0389	0.0390	0.0390	0.0393	0.0394	0.0428	0.0504
Paychex, Inc. (PAYX)	NAÏVE	HOLT	NAIVE3_M3	HOLT3_M3	M15_M3	ESM_M3	ESM3_M3	MEAN_M3	ESM	ARIMA_M3	MEAN	ARIMA
	0.0469	0.0470	0.0472	0.0473	0.0503	0.0504	0.0524	0.0543	0.0556	0.0583	0.0649	0.0971
Barrett Business Services, Inc. (BBSI)	ESM_M3	ESM3_M3	NAIVE3_M3	M15_M3	ESM	NAÏVE	HOLT3_M3	HOLT	MEAN_M3	ARIMA_M3	MEAN	ARIMA
	0.0858	0.0861	0.0862	0.0863	0.0867	0.0869	0.0869	0.0879	0.0943	0.1058	0.1168	0.1740

SECURITY	RMSE #1	RMSE #2	RMSE #3	RMSE #4	RMSE #5	RMSE #6	RMSE #7	RMSE #8	RMSE #9	RMSE #10	RMSE #11	RMSE #12
BUSINESS SERVICES												
Xerox Corporation (XRX)	ESM3_M3 0.0633	NAIVE3_M3 0.0633	HOLT3_M3 0.0633	ESM_M3 0.0633	ARIMA 0.0634	ARIMA_M3 0.0635	ESM 0.0636	NAÏVE 0.0636	HOLT 0.0637	M15_M3 0.0646	MEAN_M3 0.0749	MEAN 0.0976
Volt Information Services, Inc. (VISI)	ARIMA 0.0299	NAIVE3_M3 0.0299	NAÏVE 0.0300	ESM3_M3 0.0301	ARIMA_M3 0.0302	ESM_M3 0.0302	HOLT3_M3 0.0303	ESM 0.0303	HOLT 0.0306	M15_M3 0.0310	MEAN_M3 0.0347	MEAN 0.0461
Healthcare Services Group, Inc. (HCSG)	NAIVE3_M3 0.0246	HOLT3_M3 0.0246	ESM3_M3 0.0247	ESM_M3 0.0247	NAÏVE 0.0247	HOLT 0.0247	ESM 0.0248	M15_M3 0.0252	MEAN_M3 0.0278	MEAN 0.0333	ARIMA_M3 0.0376	ARIMA 0.0870
Avis Budget Group, Inc. (CAR)	NAIVE3_M3 0.0607	HOLT3_M3 0.0607	NAÏVE 0.0610	HOLT 0.0610	ARIMA 0.0610	ARIMA_M3 0.0614	ESM_M3 0.0618	ESM3_M3 0.0621	ESM 0.0632	M15_M3 0.0634	MEAN_M3 0.0712	MEAN 0.0894
Spherix Incorporated (SPEX)	NAÏVE 51.4268	NAIVE3_M3 51.6055	HOLT 51.6388	HOLT3_M3 51.8095	ARIMA 51.8677	MEAN_M3 52.4161	ARIMA_M3 52.7859	M15_M3 53.8868	ESM_M3 55.8480	MEAN 57.8041	ESM3_M3 57.8820	ESM 60.5321
Crawford & Company (CRD-A)	ESM3_M3 0.0143	NAIVE3_M3 0.0143	ARIMA 0.0143	ESM 0.0143	NAÏVE 0.0143	HOLT3_M3 0.0143	HOLT 0.0144	ESM_M3 0.0144	ARIMA_M3 0.0144	M15_M3 0.0148	MEAN_M3 0.0153	MEAN 0.0183
CorVel Corporation (CRVL	ESM3_M3 0.0388	ESM 0.0389	ESM_M3 0.0390	NAÏVE 0.0391	NAIVE3_M3 0.0392	HOLT3_M3 0.0399	HOLT 0.0401	M15_M3 0.0407	ARIMA_M3 0.0417	MEAN_M3 0.0468	ARIMA 0.0518	MEAN 0.0584
ADVERTISING & MARKETING												
Omnicom Group, Inc (OMC)	M15_M3 0.0612	HOLT3_M3 0.0613	NAIVE3_M3 0.0614	MEAN_M3 0.0616	ARIMA_M3 0.0619	ESM_M3 0.0620	HOLT 0.0621	NAÏVE 0.0621	ESM3_M3 0.0634	ESM 0.0656	MEAN 0.0740	ARIMA 0.0844
The Interpublic Group of Co.s, Inc. (IPG)	MEAN_M3 0.0362	M15_M3 0.0364	NAIVE3_M3 0.0380	HOLT3_M3 0.0384	ESM_M3 0.0384	NAÏVE 0.0389	HOLT 0.0394	ESM3_M3 0.0399	ESM 0.0417	MEAN 0.0454	ARIMA_M3 0.0684	ARIMA 0.1385
WPP plc (WPP)	NAIVE3_M3 0.0957	HOLT3_M3 0.0959	ESM_M3 0.0960	NAÏVE 0.0961	ESM3_M3 0.0962	HOLT 0.0964	ESM 0.0971	M15_M3 0.0977	MEAN_M3 0.1063	MEAN 0.1293	ARIMA_M3 0.1668	ARIMA 0.3569
Insignia Systems, Inc. (ISIG)	MEAN_M3 0.0141	ARIMA 0.0145	HOLT 0.0145	NAÏVE 0.0145	ESM3_M3 0.0145	ESM 0.0145	NAIVE3_M3 0.0145	HOLT3_M3 0.0145	ESM_M3 0.0146	ARIMA_M3 0.0147	M15_M3 0.0151	MEAN 0.0163
Harte Hanks, Inc. (HHS)	HOLT 0.1993	NAÏVE 0.1996	HOLT3_M3 0.2011	NAIVE3_M3 0.2012	ARIMA_M3 0.2031	ESM3_M3 0.2054	ESM_M3 0.2062	ESM 0.2064	M15_M3 0.2132	MEAN_M3 0.2414	ARIMA 0.2657	MEAN 0.2936

C.9. Computer & Technology Sector

Table C-21: Computer & Technology Sector M3 MAPE Accuracy Results

SECURITY	MAPE #1	MAPE #2	MAPE #3	MAPE #4	MAPE #5	MAPE #6	MAPE #7	MAPE #8	MAPE #9	MAPE #10	MAPE #11	MAPE #12
COMPUTER SOFTWARE												
Autodesk, Inc. (ADSK)	ESM_M3 12.2645	ESM3_M3 12.3426	NAIVE3_M3 12.3472	HOLT3_M3 12.3531	M15_M3 12.4076	NAÏVE 12.4471	HOLT 12.4573	ESM 12.5145	MEAN_M3 13.0853	ARIMA_M3 13.1120	ARIMA 15.2136	MEAN 15.5680
Cadence Design Systems, Inc. (CDNS)	ARIMA 10.3507	NAIVE3_M3 10.3572	HOLT3_M3 10.3574	NAÏVE 10.3752	HOLT 10.3754	ARIMA_M3 10.4363	ESM_M3 10.4727	ESM3_M3 10.4902	ESM 10.5687	M15_M3 10.6429	MEAN_M3 11.7405	MEAN 14.5574
Microsoft (MSFT)	ARIMA 8.0481	MEAN_M3 8.1138	ARIMA_M3 8.2902	NAIVE3_M3 8.4527	HOLT3_M3 8.4639	NAÏVE 8.4780	HOLT 8.4927	M15_M3 8.6094	ESM_M3 8.9568	MEAN 9.1435	ESM3_M3 9.1970	ESM 9.4858
Oracle Corporation (ORCL)	HOLT3_M3 7.6106	NAIVE3_M3 7.6701	HOLT 7.6820	NAÏVE 7.7307	ARIMA_M3 7.8218	M15_M3 7.9389	ESM_M3 8.0027	ESM3_M3 8.4014	ARIMA 8.5315	MEAN_M3 8.5578	ESM 9.0369	MEAN 11.2493
Adobe, Inc. (ADBE)	NAIVE3_M3 10.5261	HOLT3_M3 10.5282	ESM_M3 10.5298	ESM3_M3 10.5340	HOLT 10.5998	NAÏVE 10.6040	ESM 10.6314	M15_M3 10.7379	ARIMA_M3 11.1083	MEAN_M3 11.9356	ARIMA 12.7309	MEAN 15.1365
Symantec (SYMC)	HOLT3_M3 10.2469	NAIVE3_M3 10.2650	HOLT 10.2761	NAÏVE 10.3026	M15_M3 10.4552	ARIMA_M3 10.5368	ESM_M3 10.5746	ESM3_M3 10.8016	ESM 11.1301	MEAN_M3 11.2130	ARIMA 11.7582	MEAN 13.7008
PTC, Inc. (PTC)	ESM3_M3 14.9004	HOLT3_M3 14.9280	NAIVE3_M3 14.9280	ARIMA 14.9423	ESM_M3 14.9533	HOLT 14.9643	NAÏVE 14.9650	ESM 14.9689	ARIMA_M3 15.0513	M15_M3 15.3722	MEAN_M3 15.5243	MEAN 18.0233
COMPUTERS												
International Business Machines (IBM)	HOLT 7.2533	HOLT3_M3 7.2543	NAIVE3_M3 7.2548	NAÏVE 7.2548	ARIMA_M3 7.3311	ESM_M3 7.3421	ESM3_M3 7.3570	ESM 7.4140	M15_M3 7.4548	MEAN_M3 7.4955	ARIMA 7.5073	MEAN 8.3999
HP Inc. (HPQ)	HOLT3_M3 9.2711	NAIVE3_M3 9.2786	HOLT 9.3163	NAÏVE 9.3280	ESM_M3 9.4404	M15_M3 9.4954	ESM3_M3 9.5264	ESM 9.6954	ARIMA_M3 9.7530	MEAN_M3 10.6701	ARIMA 11.2630	MEAN 14.0009
Agilsys, Inc. (AGYS)	ARIMA 12.6197	HOLT 12.6271	NAÏVE 12.6366	NAIVE3_M3 12.6720	HOLT3_M3 12.6783	ESM 12.7136	ESM3_M3 12.7409	ESM_M3 12.8906	ARIMA_M3 12.9897	M15_M3 13.5265	MEAN_M3 14.4233	MEAN 16.9250
CSP Inc. (CSPI)	ARIMA 10.8455	NAIVE3_M3 10.8473	HOLT3_M3 10.8581	HOLT 10.8735	NAÏVE 10.8776	ESM3_M3 10.8977	ESM 10.9187	ESM_M3 10.9645	ARIMA_M3 11.0109	M15_M3 11.3402	MEAN_M3 11.9776	MEAN 14.5891
Apple Inc. (AAPL)	HOLT 11.2023	NAÏVE 11.3290	ESM 11.5225	HOLT3_M3 11.6484	NAIVE3_M3 11.7596	ESM3_M3 11.8731	ESM_M3 12.2829	M15_M3 13.3445	ARIMA_M3 13.8553	MEAN_M3 14.8596	MEAN 18.5417	ARIMA 19.8208
3D Systems Corporation (DDD)	NAIVE3_M3 18.4691	HOLT3_M3 18.5537	ESM_M3 18.5621	M15_M3 18.6403	NAÏVE 18.6628	HOLT 18.7773	ESM3_M3 18.8032	ESM 19.1868	ARIMA_M3 19.3260	MEAN_M3 20.2630	ARIMA 21.3402	MEAN 24.6898
PAR Technology Corporation (PAR)	NAIVE3_M3 12.9886	NAÏVE 13.0044	HOLT3_M3 13.0089	HOLT 13.0554	ESM3_M3 13.0700	ESM 13.1144	ESM_M3 13.1522	ARIMA 13.2653	ARIMA_M3 13.3033	M15_M3 13.6002	MEAN_M3 14.9870	MEAN 18.3670

SECURITY	MAPE #1	MAPE #2	MAPE #3	MAPE #4	MAPE #5	MAPE #6	MAPE #7	MAPE #8	MAPE #9	MAPE #10	MAPE #11	MAPE #12
WIRELESS PROVIDERS												
CenturyLink, Inc (CTL)	NAIVE3_M3 7.8184	HOLT3_M3 7.8202	NAÏVE 7.8257	HOLT 7.8303	M15_M3 8.0586	ESM_M3 8.0779	ARIMA_M3 8.1203	ESM3_M3 8.2055	ESM 8.4023	ARIMA 8.4198	MEAN_M3 8.4249	MEAN 9.9794
Verizon Communications Inc. (VZ)	HOLT 5.9636	NAÏVE 5.9691	HOLT3_M3 5.9723	NAIVE3_M3 5.9748	ARIMA 6.0227	ESM3_M3 6.0363	ESM 6.0556	ESM_M3 6.0672	ARIMA_M3 6.1004	MEAN_M3 6.2511	M15_M3 6.2548	MEAN 7.0969
AT&T Inc. (T)	HOLT3_M3 5.9648	NAIVE3_M3 5.9669	HOLT 5.9811	NAÏVE 5.9849	ARIMA 5.9976	ESM3_M3 6.0423	ESM_M3 6.0505	ARIMA_M3 6.0595	ESM 6.0875	M15_M3 6.2013	MEAN_M3 6.2041	MEAN 7.3076
Sprint Corporation (S)	HOLT 14.7902	HOLT3_M3 14.8116	NAÏVE 14.8362	NAIVE3_M3 14.8612	ARIMA_M3 14.9428	ARIMA 15.0612	ESM_M3 15.2436	M15_M3 15.3102	ESM3_M3 15.4183	ESM 15.7333	MEAN_M3 17.4171	MEAN 22.3875
Vodafone Group PLC (VOD)	ARIMA_M3 7.1579	HOLT3_M3 7.3193	NAIVE3_M3 7.3251	HOLT 7.3776	NAÏVE 7.3869	M15_M3 7.4379	ESM_M3 7.4652	ESM3_M3 7.5659	ESM 7.7197	ARIMA 7.7625	MEAN_M3 8.6668	MEAN 11.1029
United States Cellular Corp. (USM)	ESM_M3 8.1941	HOLT3_M3 8.2047	NAIVE3_M3 8.2065	ESM3_M3 8.2330	ARIMA_M3 8.2673	NAÏVE 8.2914	HOLT 8.3076	M15_M3 8.3354	ESM 8.3538	ARIMA 8.4476	MEAN_M3 8.9025	MEAN 10.7071
ATN International, Inc. (ATNI)	ESM3_M3 9.4530	ESM 9.4536	HOLT3_M3 9.4678	HOLT 9.4718	NAIVE3_M3 9.4841	NAÏVE 9.4923	ESM_M3 9.5181	MEAN_M3 9.8076	M15_M3 9.9767	ARIMA_M3 10.9961	MEAN 11.3348	ARIMA 14.6015
SEMICONDUCTORS												
Texas Instruments Inc. (TXN)	NAIVE3_M3 9.0805	HOLT3_M3 9.0807	NAÏVE 9.0909	HOLT 9.0912	ESM_M3 9.2243	ESM3_M3 9.2432	ESM 9.3438	M15_M3 9.4330	ARIMA_M3 10.0416	MEAN_M3 10.5954	ARIMA 12.8284	MEAN 13.1634
Intel Corporation (INTC)	NAÏVE 8.5304	NAIVE3_M3 8.5326	HOLT3_M3 8.5381	HOLT 8.5528	ESM_M3 8.6027	ESM3_M3 8.6107	ESM 8.7036	ARIMA_M3 8.8171	M15_M3 8.8907	ARIMA 9.2934	MEAN_M3 10.2587	MEAN 13.2636
Analog Devices, Inc. (ADI)	HOLT3_M3 9.2588	HOLT 9.2812	NAIVE3_M3 9.3112	NAÏVE 9.3388	ESM_M3 9.5818	M15_M3 9.6019	ESM3_M3 9.7790	ARIMA_M3 9.9070	ESM 10.1187	MEAN_M3 10.1886	MEAN 12.1147	ARIMA 12.4356
Semtech Corporation (SMTC)	HOLT 11.9062	NAÏVE 11.9065	HOLT3_M3 12.0377	NAIVE3_M3 12.0380	ARIMA_M3 12.1467	ESM 12.1620	ESM3_M3 12.1861	ESM_M3 12.3119	M15_M3 12.8275	MEAN_M3 13.3219	MEAN 16.1003	ARIMA 16.6362
Amtech Systems, Inc. (ASYS)	NAIVE3_M3 18.1981	ESM3_M3 18.2089	HOLT3_M3 18.2089	ESM 18.2627	HOLT 18.2636	NAÏVE 18.2648	ARIMA 18.3206	ESM_M3 18.3532	ARIMA_M3 18.6050	M15_M3 19.1574	MEAN_M3 21.7233	MEAN 27.7400
Maxim Integrated Products, Inc. (MXIM)	HOLT3_M3 9.8144	NAIVE3_M3 9.8466	M15_M3 9.8877	HOLT 9.9112	NAÏVE 9.9598	ARIMA_M3 10.0770	ESM_M3 10.0940	MEAN_M3 10.1025	ESM3_M3 10.3474	ESM 10.6964	ARIMA 10.6988	MEAN 11.7529
Microchip Technology Inc. (MCHP)	HOLT 8.6922	NAÏVE 8.7122	HOLT3_M3 8.7661	ARIMA 8.7712	NAIVE3_M3 8.7744	ESM3_M3 8.7871	ESM 8.7929	ESM_M3 8.8825	ARIMA_M3 9.0760	M15_M3 9.4231	MEAN_M3 9.6431	MEAN 11.2663

APPENDIX C: M3 Forecast Study Results

SECURITY	MAPE #1	MAPE #2	MAPE #3	MAPE #4	MAPE #5	MAPE #6	MAPE #7	MAPE #8	MAPE #9	MAPE #10	MAPE #11	MAPE #12
ELECTRICAL PRODUCTS												
Koninkijke Philips N.V. (PHG)	M15_M3 9.3276	ESM_M3 9.3469	HOLT3_M3 9.3739	NAIVE3_M3 9.3761	ESM3_M3 9.4466	HOLT 9.4993	NAÏVE 9.5032	ARIMA_M3 9.5111	ESM 9.6001	MEAN_M3 9.9712	ARIMA 10.5737	MEAN 12.4082
Cubic Corporation (CUB)	NAIVE3_M3 9.5512	HOLT3_M3 9.5595	NAÏVE 9.5932	HOLT 9.6059	ARIMA_M3 9.6568	ESM_M3 9.7011	ARIMA 9.7847	M15_M3 9.7847	ESM3_M3 9.7941	ESM 9.9540	MEAN_M3 10.4452	MEAN 12.3719
Bell Fuse Inc. (BELFA)	NAIVE3_M3 11.1919	ESM_M3 11.2237	HOLT3_M3 11.2258	ESM3_M3 11.2425	NAÏVE 11.2838	HOLT 11.3307	ESM 11.3461	M15_M3 11.4658	MEAN_M3 11.8400	ARIMA_M3 12.3691	ARIMA 14.3479	MEAN 14.4172
Bonso Electronics International, Inc. (BNSO)	ARIMA 14.8268	HOLT 14.8475	NAÏVE 14.9080	HOLT3_M3 14.9484	NAIVE3_M3 14.9853	ARIMA_M3 15.3419	ESM3_M3 15.5140	ESM_M3 15.5338	ESM 15.6357	M15_M3 16.0253	MEAN_M3 17.5410	MEAN 21.8215
Trimble Inc. (TRMB)	ESM_M3 11.2182	ESM3_M3 11.2218	ESM 11.2999	NAIVE3_M3 11.3171	HOLT3_M3 11.3285	NAÏVE 11.3865	HOLT 11.4012	ARIMA_M3 11.4064	M15_M3 11.4668	ARIMA 12.0019	MEAN_M3 13.0211	MEAN 16.5463
Kopin Corporation (KOPN)	HOLT 14.7506	NAÏVE 14.8216	HOLT3_M3 14.9920	NAIVE3_M3 15.0490	ARIMA 15.3765	ESM3_M3 15.5992	ESM 15.6004	ESM_M3 15.7538	ARIMA_M3 15.7905	M15_M3 16.5880	MEAN_M3 18.3103	MEAN 22.6991
Flex Ltd. (FLEX)	ARIMA_M3 12.7648	NAIVE3_M3 13.0303	HOLT3_M3 13.0303	NAÏVE 13.0778	HOLT 13.0779	ESM_M3 13.3260	ESM3_M3 13.3856	M15_M3 13.5871	ESM 13.5936	ARIMA 13.7295	MEAN_M3 15.1276	MEAN 18.8526
COMMUNICATIONS COMPONENT												
Corning Incorporated (GLW)	ARIMA_M3 12.6670	HOLT 12.7791	HOLT3_M3 12.8369	NAÏVE 12.9103	NAIVE3_M3 12.9362	ARIMA 13.2602	ESM_M3 13.4579	M15_M3 13.4835	ESM3_M3 13.7281	ESM 14.1851	MEAN_M3 17.4592	MEAN 23.3375
Communications Systems, Inc. (JCS)	HOLT 8.2018	NAÏVE 8.2208	HOLT3_M3 8.2506	NAIVE3_M3 8.2582	ARIMA 8.2842	ARIMA_M3 8.5591	ESM_M3 8.6301	ESM3_M3 8.6551	ESM 8.7893	M15_M3 8.8794	MEAN_M3 9.7775	MEAN 12.1522
Plantronics (PLT)	NAIVE3_M3 12.0911	NAÏVE 12.1236	HOLT3_M3 12.1257	ARIMA 12.1899	ARIMA_M3 12.1988	HOLT 12.1999	ESM_M3 12.3967	M15_M3 12.4703	ESM3_M3 12.5067	ESM 12.7053	MEAN_M3 14.1626	MEAN 17.3219
Technical Communications Corp. (TCCO)	ARIMA 17.3917	NAIVE3_M3 17.7315	NAÏVE 17.7342	HOLT3_M3 17.8034	HOLT 17.8148	ARIMA_M3 17.8440	M15_M3 18.4287	MEAN_M3 20.4624	ESM_M3 22.7460	MEAN 25.3780	ESM3_M3 26.5078	ESM 30.7622
SCIENTIFIC INSTRUMENTS												
MTS Systems Corporation (MTSC)	ESM3_M3 9.5497	ESM_M3 9.5505	HOLT3_M3 9.5866	NAIVE3_M3 9.6190	ESM 9.6447	HOLT 9.6742	NAÏVE 9.7487	ARIMA_M3 9.7787	M15_M3 9.8138	MEAN_M3 9.9738	ARIMA 10.2474	MEAN 11.9715
Kewaunee Scientific Corp. (KEQU)	NAÏVE 6.6090	HOLT 6.6147	ARIMA 6.6395	ESM 6.6630	NAIVE3_M3 6.6814	HOLT3_M3 6.6978	ESM3_M3 6.7135	ESM_M3 6.8837	ARIMA_M3 7.0413	M15_M3 7.5428	MEAN_M3 8.6556	MEAN 11.4645
PerkinElmer, Inc. (PKI)	NAÏVE 11.0089	NAIVE3_M3 11.0152	HOLT3_M3 11.0479	HOLT 11.0582	ESM3_M3 11.0791	ESM 11.1013	ESM_M3 11.1088	M15_M3 11.3286	ARIMA_M3 11.6455	ARIMA 12.4711	MEAN_M3 12.9090	MEAN 16.4499
Mixonix, Inc. (MSON)	ARIMA 15.0436	HOLT 15.0684	NAÏVE 15.0942	NAIVE3_M3 15.2493	HOLT3_M3 15.2502	ARIMA_M3 15.7474	ESM_M3 16.0466	ESM3_M3 16.1265	ESM 16.4063	M15_M3 16.4684	MEAN_M3 17.5287	MEAN 21.6729

Table C-22: Computer & Technology Sector M3 RMSE Accuracy Results

SECURITY	RMSE #1	RMSE #2	RMSE #3	RMSE #4	RMSE #5	RMSE #6	RMSE #7	RMSE #8	RMSE #9	RMSE #10	RMSE #11	RMSE #12
COMPUTER SOFTWARE												
Autodesk, Inc. (ADSK)	MEAN_M3 0.0913	ESM_M3 0.0931	ESM3_M3 0.0934	M15_M3 0.0941	ESM 0.0942	NAIVE3_M3 0.0948	HOLT3_M3 0.0948	NAÏVE 0.0957	HOLT 0.0958	MEAN 0.1018	ARIMA_M3 0.1217	ARIMA 0.2066
Cadence Design Systems, Inc. (CDNS)	ARIMA_M3 0.0320	NAIVE3_M3 0.0320	HOLT3_M3 0.0320	ARIMA 0.0321	NAÏVE 0.0322	HOLT 0.0322	ESM_M3 0.0323	M15_M3 0.0324	ESM3_M3 0.0326	ESM 0.0332	MEAN_M3 0.0342	MEAN 0.0415
Microsoft (MSFT)	ARIMA 0.0586	ARIMA_M3 0.0613	MEAN_M3 0.0618	NAIVE3_M3 0.0642	HOLT3_M3 0.0643	NAÏVE 0.0643	HOLT 0.0644	M15_M3 0.0649	MEAN 0.0674	ESM_M3 0.0678	ESM3_M3 0.0703	ESM 0.0733
Oracle Corporation (ORCL)	HOLT 0.0377	HOLT3_M3 0.0378	NAÏVE 0.0391	NAIVE3_M3 0.0394	ARIMA_M3 0.0405	M15_M3 0.0419	ESM_M3 0.0439	MEAN_M3 0.0455	ESM3_M3 0.0463	ESM 0.0495	ARIMA 0.0558	MEAN 0.0588
Adobe, Inc. (ADBE)	NAIVE3_M3 0.1185	HOLT3_M3 0.1187	M15_M3 0.1189	NAÏVE 0.1190	HOLT 0.1193	ESM_M3 0.1226	ESM3_M3 0.1259	ESM 0.1302	MEAN_M3 0.1331	ARIMA_M3 0.1478	MEAN 0.1577	ARIMA 0.2370
Symantec (SYMC)	HOLT3_M3 0.0317	NAIVE3_M3 0.0318	HOLT 0.0319	NAÏVE 0.0319	M15_M3 0.0322	ESM_M3 0.0331	ARIMA_M3 0.0335	MEAN_M3 0.0339	ESM3_M3 0.0341	ESM 0.0356	MEAN 0.0406	ARIMA 0.0490
PTC, Inc. (PTC)	ARIMA 0.0849	NAÏVE 0.0850	HOLT 0.0851	NAIVE3_M3 0.0866	HOLT3_M3 0.0866	ESM 0.0869	ESM3_M3 0.0876	ESM_M3 0.0889	ARIMA_M3 0.0895	MEAN_M3 0.0930	M15_M3 0.0934	MEAN 0.1018
COMPUTERS												
International Business Machines (IBM)	HOLT 0.1645	NAÏVE 0.1645	HOLT3_M3 0.1649	NAIVE3_M3 0.1649	ARIMA_M3 0.1673	MEAN_M3 0.1682	ESM_M3 0.1683	ESM3_M3 0.1691	M15_M3 0.1698	ESM 0.1710	ARIMA 0.1732	MEAN 0.1851
HP Inc. (HPQ)	HOLT 0.0262	NAÏVE 0.0262	HOLT3_M3 0.0263	NAIVE3_M3 0.0263	ESM_M3 0.0274	M15_M3 0.0276	ESM3_M3 0.0276	ESM 0.0281	MEAN_M3 0.0307	ARIMA_M3 0.0327	MEAN 0.0380	ARIMA 0.0638
Agilsys, Inc. (AGYS)	ARIMA 0.0260	HOLT 0.0260	NAÏVE 0.0260	NAIVE3_M3 0.0261	HOLT3_M3 0.0261	ESM3_M3 0.0265	ESM 0.0265	ARIMA_M3 0.0266	ESM_M3 0.0266	M15_M3 0.0274	MEAN_M3 0.0282	MEAN 0.0323
CSP Inc. (CSPI)	NAÏVE 0.0166	ARIMA 0.0166	NAIVE3_M3 0.0166	HOLT 0.0167	HOLT3_M3 0.0167	ESM 0.0168	ESM3_M3 0.0168	ESM_M3 0.0169	ARIMA_M3 0.0170	M15_M3 0.0175	MEAN_M3 0.0177	MEAN 0.0201
Apple Inc. (AAPL)	M15_M3 0.1235	NAIVE3_M3 0.1236	HOLT3_M3 0.1236	NAÏVE 0.1243	HOLT 0.1244	MEAN_M3 0.1251	ESM_M3 0.1282	ESM3_M3 0.1324	ESM 0.1378	MEAN 0.1399	ARIMA_M3 0.1697	ARIMA 0.3118
3D Systems Corporation (DDD)	NAIVE3_M3 0.0696	NAÏVE 0.0698	ESM_M3 0.0699	ESM3_M3 0.0700	ESM 0.0703	M15_M3 0.0706	MEAN_M3 0.0738	HOLT3_M3 0.0750	HOLT 0.0773	MEAN 0.0903	ARIMA_M3 0.0994	ARIMA 0.1704
PAR Technology Corporation (PAR)	NAÏVE 0.0227	NAIVE3_M3 0.0228	ESM 0.0228	HOLT 0.0228	ESM3_M3 0.0229	HOLT3_M3 0.0229	ESM_M3 0.0230	M15_M3 0.0237	ARIMA_M3 0.0239	ARIMA 0.0255	MEAN_M3 0.0262	MEAN 0.0319

APPENDIX C: M3 Forecast Study Results

SECURITY	RMSE #1	RMSE #2	RMSE #3	RMSE #4	RMSE #5	RMSE #6	RMSE #7	RMSE #8	RMSE #9	RMSE #10	RMSE #11	RMSE #12
WIRELESS PROVIDERS												
CenturyLink, Inc (CTL)	NAIVE3_M3 0.0486	HOLT3_M3 0.0487	NAÏVE 0.0488	HOLT 0.0488	M15_M3 0.0493	ESM_M3 0.0503	MEAN_M3 0.0503	ESM3_M3 0.0514	ESM 0.0529	ARIMA_M3 0.0532	MEAN 0.0570	ARIMA 0.0643
Verizon Communications Inc. (VZ)	HOLT 0.0448	NAÏVE 0.0448	NAIVE3_M3 0.0451	HOLT3_M3 0.0451	ARIMA 0.0455	ESM 0.0456	ESM3_M3 0.0456	ESM_M3 0.0460	ARIMA_M3 0.0465	MEAN_M3 0.0467	M15_M3 0.0478	MEAN 0.0517
AT&T Inc. (T)	HOLT3_M3 0.0387	NAIVE3_M3 0.0388	HOLT 0.0390	NAÏVE 0.0390	ARIMA_M3 0.0392	ESM_M3 0.0393	M15_M3 0.0395	ESM3_M3 0.0397	ARIMA 0.0397	ESM 0.0403	MEAN_M3 0.0403	MEAN 0.0476
Sprint Corporation (S)	NAÏVE 0.0405	HOLT 0.0406	NAIVE3_M3 0.0421	HOLT3_M3 0.0422	ARIMA_M3 0.0444	ESM 0.0453	ESM3_M3 0.0456	ESM_M3 0.0462	M15_M3 0.0483	ARIMA 0.0497	MEAN_M3 0.0526	MEAN 0.0630
Vodafone Group PLC (VOD)	ARIMA_M3 0.0408	HOLT 0.0446	NAÏVE 0.0447	HOLT3_M3 0.0450	NAIVE3_M3 0.0450	M15_M3 0.0478	ESM_M3 0.0487	ESM3_M3 0.0497	ESM 0.0511	MEAN_M3 0.0515	MEAN 0.0619	ARIMA 0.0680
United States Cellular Corp. (USM)	HOLT3_M3 0.0884	NAIVE3_M3 0.0888	ESM_M3 0.0888	HOLT 0.0889	ESM3_M3 0.0890	NAÏVE 0.0892	M15_M3 0.0897	ESM 0.0897	ARIMA_M3 0.0919	MEAN_M3 0.0920	ARIMA 0.0976	MEAN 0.1075
ATN International, Inc. (ATNI)	ESM 0.0721	HOLT 0.0724	ESM3_M3 0.0724	NAÏVE 0.0724	HOLT3_M3 0.0726	NAIVE3_M3 0.0726	ESM_M3 0.0730	MEAN_M3 0.0734	M15_M3 0.0750	ARIMA_M3 0.0844	MEAN 0.0844	ARIMA 0.1415
SEMICONDUCTORS												
Texas Instruments Incorporated (TXN)	HOLT 0.0751	NAÏVE 0.0751	HOLT3_M3 0.0753	NAIVE3_M3 0.0753	ESM_M3 0.0770	ESM3_M3 0.0770	ESM 0.0778	M15_M3 0.0791	MEAN_M3 0.0879	MEAN 0.1065	ARIMA_M3 0.1247	ARIMA 0.2795
Intel Corporation (INTC)	NAÏVE 0.0519	ESM3_M3 0.0520	ESM 0.0522	NAIVE3_M3 0.0522	HOLT 0.0523	HOLT3_M3 0.0524	ESM_M3 0.0524	M15_M3 0.0547	ARIMA_M3 0.0591	MEAN_M3 0.0635	MEAN 0.0836	ARIMA 0.0855
Analog Devices, Inc. (ADI)	HOLT 0.0875	HOLT3_M3 0.0890	NAÏVE 0.0905	NAIVE3_M3 0.0914	M15_M3 0.0981	ESM_M3 0.0983	ESM3_M3 0.1002	ESM 0.1031	MEAN_M3 0.1033	MEAN 0.1225	ARIMA_M3 0.1417	ARIMA 0.3283
Semtech Corporation (SMTC)	NAÏVE 0.0607	HOLT 0.0607	NAIVE3_M3 0.0623	HOLT3_M3 0.0624	ARIMA_M3 0.0660	ESM 0.0662	ESM3_M3 0.0663	ESM_M3 0.0669	MEAN_M3 0.0681	M15_M3 0.0696	MEAN 0.0765	ARIMA 0.1765
Amtech Systems, Inc. (ASYS)	ARIMA_M3 0.0346	ESM_M3 0.0346	NAIVE3_M3 0.0347	ESM3_M3 0.0347	HOLT3_M3 0.0347	ARIMA 0.0347	M15_M3 0.0349	NAÏVE 0.0349	ESM 0.0349	HOLT 0.0349	MEAN_M3 0.0363	MEAN 0.0435
Maxim Integrated Products, Inc. (MXIM)	MEAN_M3 0.0772	M15_M3 0.0790	NAIVE3_M3 0.0795	HOLT3_M3 0.0797	NAÏVE 0.0804	HOLT 0.0805	ESM_M3 0.0823	ESM3_M3 0.0859	MEAN 0.0878	ESM 0.0905	ARIMA_M3 0.0921	ARIMA 0.1241
Microchip Technology Inc. (MCHP)	ARIMA 0.0572	ESM 0.0574	HOLT 0.0575	ESM3_M3 0.0576	NAÏVE 0.0576	HOLT3_M3 0.0580	NAIVE3_M3 0.0580	ESM_M3 0.0584	ARIMA_M3 0.0594	M15_M3 0.0615	MEAN_M3 0.0663	MEAN 0.0768

SECURITY	RMSE #1	RMSE #2	RMSE #3	RMSE #4	RMSE #5	RMSE #6	RMSE #7	RMSE #8	RMSE #9	RMSE #10	RMSE #11	RMSE #12
ELECTRICAL PRODUCTS												
Koninkijke Philips N.V. (PHG)	HOLT3_M3 0.0503	NAIVE3_M3 0.0503	HOLT 0.0505	NAÏVE 0.0505	ESM_M3 0.0506	ESM3_M3 0.0507	M15_M3 0.0512	ESM 0.0512	ARIMA_M3 0.0555	MEAN_M3 0.0557	MEAN 0.0674	ARIMA 0.0877
Cubic Corporation (CUB)	HOLT3_M3 0.0610	NAIVE3_M3 0.0610	ARIMA_M3 0.0613	HOLT 0.0614	NAÏVE 0.0614	M15_M3 0.0614	ESM_M3 0.0619	ARIMA 0.0625	ESM3_M3 0.0629	ESM 0.0643	MEAN_M3 0.0652	MEAN 0.0762
Bell Fuse Inc. (BELFA)	ESM_M3 0.0464	ESM3_M3 0.0467	NAIVE3_M3 0.0469	M15_M3 0.0469	HOLT3_M3 0.0473	ESM 0.0474	NAÏVE 0.0477	HOLT 0.0484	MEAN_M3 0.0487	MEAN 0.0594	ARIMA_M3 0.0827	ARIMA 0.1818
Bonso Electronics International, Inc. (BNSO)	NAÏVE 0.0159	HOLT 0.0159	ARIMA 0.0160	NAIVE3_M3 0.0162	HOLT3_M3 0.0163	ESM 0.0165	ESM3_M3 0.0167	ESM_M3 0.0171	ARIMA_M3 0.0171	M15_M3 0.0183	MEAN_M3 0.0195	MEAN 0.0222
Trimble Inc. (TRMB)	ESM_M3 0.0345	NAIVE3_M3 0.0345	HOLT3_M3 0.0345	ESM3_M3 0.0346	NAÏVE 0.0347	HOLT 0.0347	ARIMA_M3 0.0348	M15_M3 0.0348	ESM 0.0350	MEAN_M3 0.0368	ARIMA 0.0390	MEAN 0.0446
Kopin Corporation (KOPN)	HOLT 0.0362	HOLT3_M3 0.0377	NAÏVE 0.0388	NAIVE3_M3 0.0397	ARIMA_M3 0.0427	M15_M3 0.0432	ESM_M3 0.0450	ESM3_M3 0.0468	ESM 0.0491	MEAN_M3 0.0501	ARIMA 0.0586	MEAN 0.0604
Flex Ltd. (FLEX)	ARIMA_M3 0.0375	MEAN_M3 0.0411	NAÏVE 0.0425	HOLT 0.0425	NAIVE3_M3 0.0425	HOLT3_M3 0.0425	ESM_M3 0.0438	ESM3_M3 0.0441	M15_M3 0.0445	ESM 0.0448	MEAN 0.0468	ARIMA 0.0532
COMMUNICATIONS COMPONENT												
Corning Incorporated (GLW)	ARIMA_M3 0.0617	HOLT 0.0729	HOLT3_M3 0.0760	NAÏVE 0.0771	NAIVE3_M3 0.0794	M15_M3 0.0877	ESM_M3 0.0968	MEAN_M3 0.1012	ESM3_M3 0.1032	ARIMA 0.1057	ESM 0.1105	MEAN 0.1300
Communications Systems, Inc. (JCS)	HOLT 0.0184	NAÏVE 0.0185	HOLT3_M3 0.0185	NAIVE3_M3 0.0186	ARIMA 0.0187	ESM3_M3 0.0188	ESM 0.0188	ESM_M3 0.0189	ARIMA_M3 0.0191	M15_M3 0.0195	MEAN_M3 0.0202	MEAN 0.0235
Plantronics (PLT)	NAÏVE 0.0748	HOLT 0.0752	ARIMA 0.0753	NAIVE3_M3 0.0753	HOLT3_M3 0.0754	ARIMA_M3 0.0768	ESM_M3 0.0779	ESM3_M3 0.0783	M15_M3 0.0789	ESM 0.0793	MEAN_M3 0.0853	MEAN 0.1011
Technical Communications Corp. (TCCO)	ARIMA 0.0173	ARIMA_M3 0.0177	NAIVE3_M3 0.0182	HOLT3_M3 0.0183	NAÏVE 0.0184	M15_M3 0.0184	HOLT 0.0184	MEAN_M3 0.0185	ESM_M3 0.0191	ESM3_M3 0.0202	ESM 0.0217	MEAN 0.0218
SCIENTIFIC INSTRUMENTS												
MTS Systems Corporation (MTSC)	MEAN_M3 0.0676	HOLT3_M3 0.0700	NAIVE3_M3 0.0701	ESM_M3 0.0701	ESM3_M3 0.0703	HOLT 0.0703	NAÏVE 0.0704	M15_M3 0.0708	ESM 0.0708	ARIMA_M3 0.0718	MEAN 0.0744	ARIMA 0.0814
Kewaunee Scientific Corp. (KEQU)	HOLT 0.0200	NAÏVE 0.0200	ARIMA 0.0200	ESM 0.0201	HOLT3_M3 0.0201	NAIVE3_M3 0.0201	ESM3_M3 0.0201	ESM_M3 0.0204	ARIMA_M3 0.0207	M15_M3 0.0217	MEAN_M3 0.0246	MEAN 0.0312
PerkinElmer, Inc. (PKI)	NAÏVE 0.0673	NAIVE3_M3 0.0673	ESM_M3 0.0676	ESM3_M3 0.0678	HOLT3_M3 0.0679	MEAN_M3 0.0679	M15_M3 0.0681	ESM 0.0682	HOLT 0.0682	ARIMA_M3 0.0781	MEAN 0.0807	ARIMA 0.1035
Mixonix, Inc. (MSON)	ARIMA 0.0186	HOLT3_M3 0.0187	NAIVE3_M3 0.0187	HOLT 0.0187	NAÏVE 0.0187	ESM3_M3 0.0188	ESM_M3 0.0188	ARIMA_M3 0.0189	ESM 0.0189	M15_M3 0.0195	MEAN_M3 0.0203	MEAN 0.0242

C.10. Construction Sector

Table C-23: Construction Sector M3 MAPE Accuracy Results

SECURITY	MAPE #1	MAPE #2	MAPE #3	MAPE #4	MAPE #5	MAPE #6	MAPE #7	MAPE #8	MAPE #9	MAPE #10	MAPE #11	MAPE #12
BUILDING AND CONSTRUCTION PRODUCTS												
Masco Corporation (MAS)	HOLT3_M3	NAIVE3_M3	ARIMA	HOLT	NAÏVE	ESM3_M3	ESM_M3	ARIMA_M3	ESM	M15_M3	MEAN_M3	MEAN
	9.5045	9.5048	9.5372	9.5612	9.5613	9.5827	9.6056	9.6176	9.6653	9.8765	10.7161	13.3594
USG Corporation (USG)	ESM3_M3	NAIVE3_M3	HOLT3_M3	ESM	NAÏVE	ARIMA	HOLT	ESM_M3	ARIMA_M3	M15_M3	MEAN_M3	MEAN
	17.7690	17.7773	17.7857	17.8015	17.8151	17.8349	17.8427	17.8447	17.9315	18.2957	20.7395	26.9389
Patrick Industries, Inc. (PATK)	NAÏVE	HOLT	NAIVE3_M3	HOLT3_M3	ESM	ESM3_M3	ESM_M3	M15_M3	ARIMA_M3	MEAN_M3	ARIMA	MEAN
	17.8209	17.9053	17.9214	17.9973	18.0234	18.0890	18.3089	19.1372	20.5625	24.4357	25.3927	32.1665
Aegion Corporation (AEGN)	ARIMA	NAÏVE	HOLT	NAIVE3_M3	HOLT3_M3	ARIMA_M3	ESM3_M3	ESM_M3	ESM	M15_M3	MEAN_M3	MEAN
	11.2127	11.2349	11.2357	11.2829	11.2836	11.4848	11.5141	11.5339	11.5926	11.7930	11.8039	14.3282
CRH plc (CRH)	NAIVE3_M3	HOLT3_M3	NAÏVE	HOLT	ARIMA	ARIMA_M3	ESM_M3	ESM3_M3	ESM	M15_M3	MEAN_M3	MEAN
	7.1779	7.1890	7.1895	7.1944	7.1945	7.3252	7.3603	7.3695	7.4588	7.5745	8.1764	10.0378
NCI Building Systems (NCS)	ARIMA	HOLT3_M3	NAÏVE	HOLT	NAIVE3_M3	ARIMA_M3	ESM_M3	M15_M3	ESM3_M3	ESM	MEAN_M3	MEAN
	15.2338	15.3397	15.3435	15.3466	15.3504	15.5502	15.9761	16.0334	16.5012	17.2938	18.0274	23.5704
Gibraltar Industries (ROCK)	NAIVE3_M3	NAÏVE	HOLT3_M3	HOLT	ARIMA	ARIMA_M3	ESM_M3	ESM3_M3	ESM	M15_M3	MEAN_M3	MEAN
	13.4122	13.4336	13.4412	13.4579	13.5437	13.6103	13.6768	13.7179	13.8513	13.8682	14.2800	17.0860
RESIDENTIAL AND COMMERCIAL BUILDING												
Lennar Corporation (LEN)	NAIVE3_M3	HOLT3_M3	NAÏVE	HOLT	ESM_M3	ESM3_M3	ESM	ARIMA_M3	M15_M3	ARIMA	MEAN_M3	MEAN
	10.0006	10.0010	10.0325	10.0337	10.1678	10.1711	10.3508	10.5378	10.6246	11.1496	11.9813	15.1575
M.D.C. Holdings, Inc. (MDC)	NAÏVE	HOLT	NAIVE3_M3	HOLT3_M3	ESM_M3	ESM3_M3	ESM	M15_M3	ARIMA_M3	MEAN_M3	MEAN	ARIMA
	8.9443	8.9469	9.0218	9.0243	9.2981	9.3275	9.5244	9.6794	10.4264	10.5559	12.8455	13.6869
NVR, Inc. (NVR)	ESM_M3	ESM3_M3	HOLT3_M3	NAIVE3_M3	HOLT	ESM	NAÏVE	M15_M3	MEAN_M3	ARIMA_M3	MEAN	ARIMA
	8.2404	8.2599	8.2698	8.3460	8.3548	8.3804	8.4041	8.5726	10.2599	10.4705	13.0967	18.8004
PulteGroup, Inc. (PHM)	NAÏVE	HOLT	ESM	NAIVE3_M3	HOLT3_M3	ESM3_M3	ESM_M3	ARIMA_M3	M15_M3	MEAN_M3	ARIMA	MEAN
	11.3518	11.3590	11.3648	11.4580	11.4633	11.4672	11.6429	11.8781	12.1894	12.7356	13.1188	15.1063
Toll Brothers, Inc. (TOL)	HOLT3_M3	HOLT	NAIVE3_M3	NAÏVE	ESM_M3	ESM3_M3	ESM	ARIMA_M3	M15_M3	MEAN_M3	ARIMA	MEAN
	9.2240	9.2300	9.2515	9.2624	9.4786	9.4923	9.6018	9.6235	9.6609	10.5331	10.5737	13.0062
KB Home (KBH)	NAÏVE	HOLT	ESM	NAIVE3_M3	HOLT3_M3	ESM3_M3	ESM_M3	ARIMA_M3	ARIMA	M15_M3	MEAN_M3	MEAN
	11.5299	11.5391	11.5512	11.7188	11.7250	11.7399	11.9975	12.2641	12.4757	12.7399	14.0094	17.3160
D.R. Horton, Inc. (DHI)	NAÏVE	HOLT	HOLT3_M3	NAIVE3_M3	ESM3_M3	ESM	ESM_M3	M15_M3	ARIMA_M3	MEAN_M3	ARIMA	MEAN
	10.7533	10.7538	10.7748	10.7750	10.8641	10.8787	10.9515	11.3459	11.5707	12.3628	12.5433	15.0436

Table C-24: Construction Sector M3 RMSE Accuracy Results

SECURITY	RMSE #1	RMSE #2	RMSE #3	RMSE #4	RMSE #5	RMSE #6	RMSE #7	RMSE #8	RMSE #9	RMSE #10	RMSE #11	RMSE #12
BUILDING AND CONSTRUCTION PRODUCTS												
Masco Corporation (MAS)	HOLT3_M3 0.0318	NAIVE3_M3 0.0318	HOLT 0.0318	NAÏVE 0.0318	ARIMA 0.0319	ESM3_M3 0.0320	ESM 0.0320	ESM_M3 0.0321	ARIMA_M3 0.0322	M15_M3 0.0329	MEAN_M3 0.0353	MEAN 0.0420
USG Corporation (USG)	ESM3_M3 0.0906	NAIVE3_M3 0.0907	ESM_M3 0.0907	HOLT3_M3 0.0908	ESM 0.0909	NAÏVE 0.0909	ARIMA_M3 0.0909	ARIMA 0.0910	HOLT 0.0911	M15_M3 0.0919	MEAN_M3 0.1045	MEAN 0.1271
Patrick Industries, Inc. (PATK)	HOLT 0.0432	NAÏVE 0.0433	HOLT3_M3 0.0435	NAIVE3_M3 0.0436	ESM 0.0437	ESM3_M3 0.0439	ESM_M3 0.0444	M15_M3 0.0461	MEAN_M3 0.0488	MEAN 0.0575	ARIMA_M3 0.0613	ARIMA 0.1398
Aegion Corporation (AEGN)	ARIMA 0.0436	NAIVE3_M3 0.0438	HOLT3_M3 0.0438	NAÏVE 0.0440	HOLT 0.0440	ARIMA_M3 0.0440	ESM_M3 0.0444	ESM3_M3 0.0446	M15_M3 0.0449	MEAN_M3 0.0450	ESM 0.0452	MEAN 0.0549
CRH plc (CRH)	ARIMA 0.0337	NAIVE3_M3 0.0342	NAÏVE 0.0342	HOLT3_M3 0.0342	HOLT 0.0343	ARIMA_M3 0.0345	ESM_M3 0.0348	ESM3_M3 0.0349	ESM 0.0353	M15_M3 0.0355	MEAN_M3 0.0389	MEAN 0.0480
NCI Building Systems (NCS)	ARIMA 0.2530	HOLT 0.2542	NAÏVE 0.2547	HOLT3_M3 0.2561	NAIVE3_M3 0.2565	ARIMA_M3 0.2606	ESM_M3 0.2619	ESM3_M3 0.2623	ESM 0.2650	M15_M3 0.2678	MEAN_M3 0.2938	MEAN 0.3585
Gibraltar Industries (ROCK)	NAÏVE 0.0448	NAIVE3_M3 0.0448	HOLT 0.0448	HOLT3_M3 0.0448	ARIMA 0.0450	ARIMA_M3 0.0454	ESM_M3 0.0456	ESM3_M3 0.0457	ESM 0.0460	M15_M3 0.0461	MEAN_M3 0.0474	MEAN 0.0538
RESIDENTIAL AND COMMERCIAL BUILDING												
Lennar Corporation (LEN)	NAÏVE 0.0539	HOLT 0.0539	HOLT3_M3 0.0540	NAIVE3_M3 0.0540	ESM_M3 0.0553	ESM3_M3 0.0556	ESM 0.0567	M15_M3 0.0571	ARIMA_M3 0.0600	MEAN_M3 0.0639	ARIMA 0.0792	MEAN 0.0797
M.D.C. Holdings, Inc. (MDC)	NAÏVE 0.0550	HOLT 0.0551	NAIVE3_M3 0.0551	HOLT3_M3 0.0551	ESM_M3 0.0561	ESM3_M3 0.0564	M15_M3 0.0572	ESM 0.0574	MEAN_M3 0.0620	ARIMA_M3 0.0718	MEAN 0.0742	ARIMA 0.1311
NVR, Inc. (NVR)	M15_M3 1.6959	NAIVE3_M3 1.6970	HOLT3_M3 1.7163	NAÏVE 1.7185	ESM_M3 1.7268	HOLT 1.7500	ESM3_M3 1.7689	ESM 1.8272	MEAN_M3 1.9712	MEAN 2.4519	ARIMA_M3 3.1952	ARIMA 7.5166
PulteGroup, Inc. (PHM)	NAÏVE 0.0342	HOLT 0.0342	ESM 0.0343	NAIVE3_M3 0.0347	HOLT3_M3 0.0347	ESM3_M3 0.0348	ESM_M3 0.0354	ARIMA_M3 0.0365	M15_M3 0.0374	MEAN_M3 0.0400	MEAN 0.0483	ARIMA 0.0584
Toll Brothers, Inc. (TOL)	HOLT 0.0424	HOLT3_M3 0.0426	NAÏVE 0.0427	NAIVE3_M3 0.0429	ESM3_M3 0.0435	ESM_M3 0.0436	ESM 0.0438	M15_M3 0.0451	ARIMA_M3 0.0475	MEAN_M3 0.0512	MEAN 0.0631	ARIMA 0.0669
KB Home (KBH)	HOLT 0.0517	NAÏVE 0.0517	ESM 0.0519	HOLT3_M3 0.0523	NAIVE3_M3 0.0523	ESM3_M3 0.0525	ESM_M3 0.0534	M15_M3 0.0562	ARIMA_M3 0.0587	MEAN_M3 0.0651	MEAN 0.0823	ARIMA 0.0887
D.R. Horton, Inc. (DHI)	HOLT 0.0391	NAÏVE 0.0391	HOLT3_M3 0.0395	NAIVE3_M3 0.0395	ESM 0.0398	ESM3_M3 0.0398	ESM_M3 0.0402	M15_M3 0.0418	MEAN_M3 0.0443	ARIMA_M3 0.0446	MEAN 0.0527	ARIMA 0.0584

C.11. Consumer Discretionary Sector

Table C-25: Consumer Discretionary M3 MAPE Accuracy Results

SECURITY	MAPE #1	MAPE #2	MAPE #3	MAPE #4	MAPE #5	MAPE #6	MAPE #7	MAPE #8	MAPE #9	MAPE #10	MAPE #11	MAPE #12
LEISURE												
The Marcus Corporation (MCS)	ARIMA_M3 8.7658	ESM_M3 8.8055	HOLT3_M3 8.8610	ARIMA 8.8790	NAIVE3_M3 8.8896	M15_M3 8.9181	ESM3_M3 8.9261	HOLT 9.0077	NAÏVE 9.0577	ESM 9.1531	MEAN_M3 9.1543	MEAN 10.9104
Cedar Fair, L.P. (FUN)	HOLT3_M3 6.5663	HOLT 6.5667	NAIVE3_M3 6.5804	NAÏVE 6.5884	ESM3_M3 6.6099	ESM_M3 6.6330	ESM 6.6398	ARIMA 6.6736	ARIMA_M3 6.6864	M15_M3 6.8124	MEAN_M3 7.6380	MEAN 9.4845
Carnival Corporation (CCL)	ESM_M3 7.6520	ARIMA_M3 7.6695	NAIVE3_M3 7.7003	HOLT3_M3 7.7005	ESM3_M3 7.7135	M15_M3 7.7687	NAÏVE 7.8178	HOLT 7.8210	ESM 7.8503	ARIMA 7.8586	MEAN_M3 8.9089	MEAN 11.6124
Reading International, Inc. (RDI)	ESM 7.6651	HOLT 7.6659	NAÏVE 7.6660	HOLT3_M3 7.7502	NAIVE3_M3 7.7506	ESM3_M3 7.7605	ESM_M3 7.9218	M15_M3 8.4166	MEAN_M3 9.2314	ARIMA_M3 10.7506	MEAN 11.5887	ARIMA 19.6823
Royal Caribbean Cruises, Ltd. (RCL)	ESM_M3 13.1901	ARIMA_M3 13.2121	NAIVE3_M3 13.2714	ESM3_M3 13.2818	HOLT3_M3 13.2879	M15_M3 13.3443	NAÏVE 13.4333	ARIMA 13.4550	HOLT 13.4631	ESM 13.4952	MEAN_M3 14.9684	MEAN 19.1067
TEXTILES												
V.F. Corporation (VFC)	ESM3_M3 6.3644	ESM_M3 6.3655	ESM 6.4203	HOLT3_M3 6.4662	NAIVE3_M3 6.4687	M15_M3 6.5202	HOLT 6.5320	NAÏVE 6.5360	ARIMA_M3 6.8938	MEAN_M3 7.3090	ARIMA 8.4428	MEAN 9.1405
Oxford Industries, Inc. (OXM)	HOLT 13.7225	HOLT3_M3 13.7360	NAÏVE 13.7404	ESM3_M3 13.7866	NAIVE3_M3 13.7930	ESM 13.8588	ESM_M3 13.9347	ARIMA_M3 14.3656	M15_M3 14.6483	ARIMA 14.6712	MEAN_M3 15.7935	MEAN 18.2599
Interface, Inc. (TILE)	HOLT 13.3819	ARIMA 13.3833	NAÏVE 13.3848	ESM 13.3932	ESM3_M3 13.4082	NAIVE3_M3 13.4120	HOLT3_M3 13.4133	ESM_M3 13.5495	ARIMA_M3 13.6697	M15_M3 14.1168	MEAN_M3 16.4996	MEAN 20.5301
Culp, Inc. (CULP)	ESM 13.1837	ARIMA 13.2488	ESM3_M3 13.2644	NAÏVE 13.2918	NAIVE3_M3 13.3360	HOLT 13.3978	HOLT3_M3 13.4262	ESM_M3 13.4418	ARIMA_M3 13.5914	M15_M3 14.0592	MEAN_M3 16.1678	MEAN 20.6370
PVH Corp (PVH)	ESM3_M3 9.4422	ESM_M3 9.4443	NAIVE3_M3 9.4960	HOLT3_M3 9.5077	ESM 9.5356	NAÏVE 9.5922	HOLT 9.6133	M15_M3 9.6879	ARIMA_M3 9.9120	ARIMA 10.9674	MEAN_M3 11.5127	MEAN 14.7761
G-III Apparel Group, Ltd. (GIII)	NAÏVE 12.9340	HOLT 12.9633	ESM 13.0078	NAIVE3_M3 13.0166	HOLT3_M3 13.0956	ESM3_M3 13.1440	ESM_M3 13.4503	ARIMA_M3 14.3103	M15_M3 14.3768	ARIMA 16.6041	MEAN_M3 16.7117	MEAN 21.5521
The Dixie Group (DXYN)	NAÏVE 15.4309	ARIMA 15.4410	HOLT 15.4411	NAIVE3_M3 15.6399	HOLT3_M3 15.6569	ARIMA_M3 16.2831	ESM3_M3 16.3158	ESM 16.3247	ESM_M3 16.4728	M15_M3 17.1320	MEAN_M3 20.9706	MEAN 26.9382
SCHOOLS												
Graham Holdings Company (GHC)	NAIVE3_M3 5.8074	HOLT3_M3 5.8176	NAÏVE 5.8296	HOLT 5.8446	ESM_M3 5.8789	ARIMA_M3 5.8890	ARIMA 5.9084	ESM3_M3 5.9272	ESM 6.0450	M15_M3 6.0696	MEAN_M3 6.8606	MEAN 8.5837
GP Strategies Corporation (GPX)	NAIVE3_M3 10.1382	HOLT3_M3 10.1383	NAÏVE 10.1732	HOLT 10.1736	ARIMA 10.1752	ARIMA_M3 10.2912	ESM_M3 10.4515	M15_M3 10.6730	ESM3_M3 10.7107	ESM 11.2499	MEAN_M3 11.7135	MEAN 14.5517
Adtalem Global Education (ATGE)	NAIVE3_M3 10.9682	HOLT3_M3 10.9852	NAÏVE 11.0202	HOLT 11.0320	ARIMA_M3 11.1812	ARIMA 11.1826	ESM_M3 11.4192	M15_M3 11.4240	ESM3_M3 11.6154	ESM 11.9069	MEAN_M3 12.8617	MEAN 15.4043

SECURITY	MAPE #1	MAPE #2	MAPE #3	MAPE #4	MAPE #5	MAPE #6	MAPE #7	MAPE #8	MAPE #9	MAPE #10	MAPE #11	MAPE #12
FURNITURE MANUFACTURERS												
Leggett & Platt, Incorporated (LEG)	NAIVE3_M3 7.1246	NAÏVE 7.1270	HOLT3_M3 7.1284	ARIMA 7.1358	HOLT 7.1358	ESM3_M3 7.1816	ESM_M3 7.2036	ARIMA_M3 7.2290	ESM 7.2365	M15_M3 7.4376	MEAN_M3 7.7889	MEAN 9.4163
La-Z-Boy Incorporated (LZB)	NAÏVE 13.2283	HOLT 13.2574	ARIMA 13.2837	NAIVE3_M3 13.3708	ESM 13.3833	HOLT3_M3 13.3978	ESM3_M3 13.4335	ESM_M3 13.6245	ARIMA_M3 13.8330	M15_M3 14.4237	MEAN_M3 16.7993	MEAN 21.0107
Kimball International, Inc. (KBAL)	ARIMA_M3 9.9903	ESM_M3 10.0668	M15_M3 10.0754	HOLT3_M3 10.0977	NAIVE3_M3 10.1667	ESM3_M3 10.1986	ARIMA 10.2465	HOLT 10.2684	NAÏVE 10.3797	ESM 10.4096	MEAN_M3 10.6927	MEAN 12.9975
Bassett Furniture Industries Inc. (BSET)	NAÏVE 13.1292	ESM 13.1546	HOLT 13.1627	ARIMA 13.1888	ESM3_M3 13.2032	NAIVE3_M3 13.2072	HOLT3_M3 13.2471	ESM_M3 13.4139	ARIMA_M3 13.6349	M15_M3 14.3248	MEAN_M3 15.7909	MEAN 19.0029
Flexsteel Industies, Inc. (FLXS)	HOLT 8.8737	ARIMA 8.8878	NAÏVE 8.9047	NAIVE3_M3 8.9216	HOLT3_M3 8.9257	ESM 8.9466	ESM3_M3 8.9733	ESM_M3 9.0739	ARIMA_M3 9.1370	M15_M3 9.4415	MEAN_M3 9.6559	MEAN 11.2930
Virco Mfg. Corporation (VIRC)	NAIVE3_M3 9.1739	HOLT3_M3 9.2050	NAÏVE 9.2358	ESM3_M3 9.2612	ESM_M3 9.2828	HOLT 9.3061	ESM 9.3528	ARIMA_M3 9.3798	ARIMA 9.4241	M15_M3 9.6705	MEAN_M3 11.4331	MEAN 15.0312
American Woodmark Corp. (AMWD)	NAÏVE 11.4845	NAIVE3_M3 11.5012	HOLT 11.5248	ESM 11.5305	HOLT3_M3 11.5357	ESM3_M3 11.5499	ESM_M3 11.6435	M15_M3 12.0466	ARIMA_M3 12.3657	MEAN_M3 13.1912	ARIMA 13.4789	MEAN 15.9270

Table C-26: Consumer Discretionary M3 RMSE Accuracy Results

SECURITY	RMSE #1	RMSE #2	RMSE #3	RMSE #4	RMSE #5	RMSE #6	RMSE #7	RMSE #8	RMSE #9	RMSE #10	RMSE #11	RMSE #12
LEISURE												
The Marcus Corporation (MCS)	ESM_M3 0.0294	ARIMA_M3 0.0294	HOLT3_M3 0.0296	NAIVE3_M3 0.0297	ARIMA 0.0297	ESM3_M3 0.0297	M15_M3 0.0298	HOLT 0.0300	NAÏVE 0.0301	ESM 0.0304	MEAN_M3 0.0315	MEAN 0.0377
Cedar Fair, L.P. (FUN)	ESM3_M3 0.0373	ESM 0.0374	HOLT3_M3 0.0374	ESM_M3 0.0374	NAIVE3_M3 0.0375	HOLT 0.0375	NAÏVE 0.0376	ARIMA 0.0377	ARIMA_M3 0.0378	M15_M3 0.0387	MEAN_M3 0.0423	MEAN 0.0509
Carnival Corporation (CCL)	NAIVE3_M3 0.0576	HOLT3_M3 0.0576	ESM_M3 0.0577	ARIMA_M3 0.0579	ESM3_M3 0.0579	NAÏVE 0.0581	HOLT 0.0581	M15_M3 0.0584	ESM 0.0586	ARIMA 0.0588	MEAN_M3 0.0627	MEAN 0.0772
Reading International, Inc. (RDI)	ESM 0.0113	HOLT 0.0113	NAÏVE 0.0114	HOLT3_M3 0.0114	NAIVE3_M3 0.0114	ESM3_M3 0.0114	ESM_M3 0.0115	M15_M3 0.0119	MEAN_M3 0.0130	MEAN 0.0155	ARIMA_M3 0.0198	ARIMA 0.0482
Royal Caribbean Cruises, Ltd. (RCL)	M15_M3 0.0981	ARIMA_M3 0.0985	NAIVE3_M3 0.0992	HOLT3_M3 0.0993	ESM_M3 0.0996	NAÏVE 0.1007	HOLT 0.1008	ESM3_M3 0.1019	ARIMA 0.1020	MEAN_M3 0.1022	ESM 0.1052	MEAN 0.1204
TEXTILES												
V.F. Corporation (VFC)	ESM3_M3 0.0458	ESM_M3 0.0458	ESM 0.0460	NAIVE3_M3 0.0463	HOLT3_M3 0.0463	M15_M3 0.0465	NAÏVE 0.0465	HOLT 0.0465	MEAN_M3 0.0500	ARIMA_M3 0.0583	MEAN 0.0590	ARIMA 0.0942
Oxford Industries, Inc. (OXM)	NAÏVE 0.0808	ESM3_M3 0.0809	HOLT 0.0809	HOLT3_M3 0.0809	NAIVE3_M3 0.0811	ESM 0.0813	ESM_M3 0.0818	ARIMA_M3 0.0861	M15_M3 0.0873	MEAN_M3 0.0924	ARIMA 0.0953	MEAN 0.1085
Interface, Inc. (TILE)	HOLT 0.0271	HOLT3_M3 0.0271	NAÏVE 0.0271	NAIVE3_M3 0.0272	ARIMA 0.0272	ESM3_M3 0.0272	ESM 0.0272	ESM_M3 0.0274	ARIMA_M3 0.0276	M15_M3 0.0284	MEAN_M3 0.0305	MEAN 0.0362
Culp, Inc. (CULP)	NAIVE3_M3 0.0261	ESM3_M3 0.0262	ARIMA 0.0262	NAÏVE 0.0263	ESM_M3 0.0263	ARIMA_M3 0.0263	ESM 0.0264	HOLT3_M3 0.0266	HOLT 0.0270	M15_M3 0.0270	MEAN_M3 0.0297	MEAN 0.0373
PVH Corp (PVH)	NAIVE3_M3 0.1140	HOLT3_M3 0.1141	ESM3_M3 0.1142	ESM_M3 0.1144	NAÏVE 0.1145	HOLT 0.1146	ESM 0.1149	M15_M3 0.1175	ARIMA_M3 0.1234	MEAN_M3 0.1287	MEAN 0.1544	ARIMA 0.1577

APPENDIX C: M3 Forecast Study Results

SECURITY	RMSE #1	RMSE #2	RMSE #3	RMSE #4	RMSE #5	RMSE #6	RMSE #7	RMSE #8	RMSE #9	RMSE #10	RMSE #11	RMSE #12
G-III Apparel Group, Ltd. (GIII)	ARIMA_M3 0.0464	NAÏVE 0.0477	HOLT 0.0477	NAIVE3_M3 0.0482	HOLT3_M3 0.0483	ESM 0.0483	ESM3_M3 0.0487	ESM_M3 0.0493	M15_M3 0.0517	MEAN_M3 0.0559	MEAN 0.0691	ARIMA 0.0784
The Dixie Group (DXYN)	ARIMA 0.0190	NAÏVE 0.0190	HOLT 0.0191	ESM 0.0191	NAIVE3_M3 0.0192	HOLT3_M3 0.0192	ESM3_M3 0.0193	ESM_M3 0.0196	ARIMA_M3 0.0198	M15_M3 0.0207	MEAN_M3 0.0225	MEAN 0.0272
SCHOOLS												
Graham Holdings Company (GHC)	NAIVE3_M3 0.4597	HOLT3_M3 0.4601	NAÏVE 0.4612	HOLT 0.4618	ARIMA_M3 0.4620	ESM_M3 0.4623	ESM3_M3 0.4642	ARIMA 0.4645	M15_M3 0.4681	ESM 0.4693	MEAN_M3 0.5329	MEAN 0.6505
GP Strategies Corporation (GPX)	NAÏVE 0.0271	HOLT 0.0271	ARIMA 0.0272	NAIVE3_M3 0.0272	HOLT3_M3 0.0272	ARIMA_M3 0.0277	ESM_M3 0.0279	ESM3_M3 0.0281	M15_M3 0.0286	ESM 0.0286	MEAN_M3 0.0289	MEAN 0.0339
Adtalem Global Education (ATGE)	NAIVE3_M3 0.0723	HOLT3_M3 0.0724	ARIMA_M3 0.0728	NAÏVE 0.0729	HOLT 0.0730	M15_M3 0.0736	ARIMA 0.0736	ESM_M3 0.0750	ESM3_M3 0.0768	MEAN_M3 0.0782	ESM 0.0793	MEAN 0.0914
FURNITURE MANUFACTURERS												
Leggett & Platt, Incorporated (LEG)	NAIVE3_M3 0.0344	HOLT3_M3 0.0344	NAÏVE 0.0345	HOLT 0.0345	ARIMA 0.0346	ESM_M3 0.0348	ARIMA_M3 0.0348	ESM3_M3 0.0348	ESM 0.0352	M15_M3 0.0357	MEAN_M3 0.0369	MEAN 0.0433
La-Z-Boy Incorporated (LZB)	ARIMA 0.0333	HOLT 0.0333	NAÏVE 0.0333	HOLT3_M3 0.0335	NAIVE3_M3 0.0335	ESM3_M3 0.0340	ARIMA_M3 0.0341	ESM 0.0341	ESM_M3 0.0341	MEAN_M3 0.0350	M15_M3 0.0351	MEAN 0.0413
Kimball International, Inc. (KBAL)	ARIMA_M3 0.0197	M15_M3 0.0198	ESM_M3 0.0198	HOLT3_M3 0.0199	NAIVE3_M3 0.0200	ESM3_M3 0.0200	ARIMA 0.0202	HOLT 0.0202	NAÏVE 0.0203	ESM 0.0204	MEAN_M3 0.0211	MEAN 0.0254
Bassett Furniture Industries Inc. (BSET)	NAÏVE 0.0333	ESM 0.0333	HOLT 0.0334	NAIVE3_M3 0.0334	ESM3_M3 0.0334	ARIMA 0.0334	HOLT3_M3 0.0335	ESM_M3 0.0340	ARIMA_M3 0.0345	M15_M3 0.0365	MEAN_M3 0.0377	MEAN 0.0434
Flexsteel Industies, Inc. (FLXS)	MEAN_M3 0.0458	NAÏVE 0.0479	NAIVE3_M3 0.0482	HOLT 0.0483	ARIMA 0.0485	HOLT3_M3 0.0485	ESM 0.0488	ESM3_M3 0.0489	ESM_M3 0.0492	ARIMA_M3 0.0493	M15_M3 0.0504	MEAN 0.0507
Virco Mfg. Corporation (VIRC)	NAIVE3_M3 0.0104	HOLT3_M3 0.0105	NAÏVE 0.0105	ESM3_M3 0.0105	ESM_M3 0.0105	ARIMA_M3 0.0106	HOLT 0.0106	ARIMA 0.0107	ESM 0.0107	M15_M3 0.0109	MEAN_M3 0.0120	MEAN 0.0150
American Woodmark Corp. (AMWD)	NAÏVE 0.0825	HOLT 0.0826	ESM 0.0831	NAIVE3_M3 0.0835	HOLT3_M3 0.0836	ESM3_M3 0.0840	ESM_M3 0.0852	M15_M3 0.0886	ARIMA_M3 0.0948	MEAN_M3 0.0973	MEAN 0.1218	ARIMA 0.1272

C.12. Consumer Staples Sector

Table C-27: Consumer Staples Sector M3 MAPE Accuracy Results

SECURITY	MAPE #1	MAPE #2	MAPE #3	MAPE #4	MAPE #5	MAPE #6	MAPE #7	MAPE #8	MAPE #9	MAPE #10	MAPE #11	MAPE #12
SOFT DRINKS												
The Coca-Cola Company (KO)	HOLT 4.4990	NAÏVE 4.5004	HOLT3_M3 4.5053	NAIVE3_M3 4.5065	ARIMA 4.5230	ARIMA_M3 4.5734	ESM_M3 4.5983	M15_M3 4.6850	ESM3_M3 4.6914	ESM 4.8797	MEAN_M3 4.9980	MEAN 6.2218
PepsiCo, Inc. (PEP)	ARIMA 4.4485	HOLT 4.5005	NAÏVE 4.5089	HOLT3_M3 4.5303	NAIVE3_M3 4.5335	ESM_M3 4.5370	ESM3_M3 4.5606	ARIMA_M3 4.6051	ESM 4.6773	M15_M3 4.7657	MEAN_M3 4.9010	MEAN 5.5439
Coca-Cola European Partners (CCEP)	HOLT 7.9007	ESM 7.9092	NAÏVE 7.9103	ARIMA 7.9170	HOLT3_M3 7.9498	ESM3_M3 7.9660	NAIVE3_M3 7.9664	ESM_M3 8.0835	ARIMA_M3 8.1612	M15_M3 8.4376	MEAN_M3 8.7664	MEAN 10.5138
National Beverage Corp. (FIZZ)	ESM3_M3 8.3161	ESM_M3 8.3495	NAIVE3_M3 8.3600	ESM 8.3687	NAÏVE 8.4196	HOLT3_M3 8.4380	HOLT 8.5748	M15_M3 8.6478	MEAN_M3 9.2391	ARIMA_M3 9.6513	MEAN 11.4322	ARIMA 12.8036
Cott Corporation (COT)	HOLT 12.2732	HOLT3_M3 12.2808	ARIMA 12.3270	NAIVE3_M3 12.3595	NAÏVE 12.3686	ARIMA_M3 12.5251	ESM_M3 12.6539	ESM3_M3 12.9083	M15_M3 12.9146	ESM 13.3788	MEAN_M3 16.0635	MEAN 20.6629
ALCOHOLIC BEVERAGES												
Brown-Forman Corporation (BF-A)	ESM3_M3 5.3695	ESM_M3 5.3713	HOLT3_M3 5.3739	NAIVE3_M3 5.3933	HOLT 5.4079	ESM 5.4125	NAÏVE 5.4271	M15_M3 5.5068	MEAN_M3 5.9895	ARIMA_M3 7.0652	MEAN 7.3936	ARIMA 11.8273
Molson Coors Brewing Company (TAP)	HOLT3_M3 6.7476	NAIVE3_M3 6.7520	ESM_M3 6.7530	ESM3_M3 6.7846	HOLT 6.8055	M15_M3 6.8094	NAÏVE 6.8127	ESM 6.8634	ARIMA_M3 7.1001	MEAN_M3 7.2850	ARIMA 7.7671	MEAN 8.8696
Constellation Brans, Inc. (STZ)	ESM 7.8615	NAÏVE 7.8679	HOLT 7.8783	ESM3_M3 7.8880	NAIVE3_M3 7.9147	HOLT3_M3 7.9245	ESM_M3 7.9845	M15_M3 8.4466	MEAN_M3 9.2646	ARIMA_M3 9.5878	MEAN 11.0387	ARIMA 13.1527
Compañia Cervecerias Unidas S.A. (CCU)	NAIVE3_M3 7.0096	ESM_M3 7.0401	ARIMA_M3 7.0492	HOLT3_M3 7.0549	NAÏVE 7.0762	ESM3_M3 7.0947	ARIMA 7.1120	HOLT 7.1758	M15_M3 7.1958	ESM 7.2437	MEAN_M3 7.6288	MEAN 9.3280
Willamette Valley Vineyards, Inc. (WVVI)	HOLT3_M3 7.6871	HOLT 7.7168	ESM3_M3 7.8063	NAIVE3_M3 7.8237	ESM_M3 7.8386	ESM 7.8954	NAÏVE 7.9455	ARIMA_M3 8.1258	M15_M3 8.2238	ARIMA 8.5869	MEAN_M3 8.6511	MEAN 10.6927
TOBACCO PRODUCTS												
Altria Group, Inc. (MO)	ESM 6.1163	NAÏVE 6.2874	HOLT 6.4090	ARIMA 6.6549	ESM3_M3 6.7648	NAIVE3_M3 6.9339	HOLT3_M3 7.0298	ESM_M3 7.5084	ARIMA_M3 8.1566	M15_M3 9.2052	MEAN_M3 10.8292	MEAN 13.5035
British American Tobacco p.l.c. (BTI)	ESM 5.7321	NAÏVE 5.7653	HOLT 5.7660	ESM3_M3 5.8060	NAIVE3_M3 5.8298	HOLT3_M3 5.8326	ESM_M3 5.9288	M15_M3 6.3004	MEAN_M3 7.0247	ARIMA_M3 7.1069	MEAN 8.7161	ARIMA 11.0103
Vector Group Ltd. (VGR)	NAÏVE 7.5813	NAIVE3_M3 7.5993	HOLT3_M3 7.6169	HOLT 7.6200	ESM3_M3 7.6295	ESM 7.6524	ESM_M3 7.6874	M15_M3 7.9762	ARIMA_M3 8.1648	MEAN_M3 8.8192	ARIMA 8.8841	MEAN 10.8477
Universal Corporation (UVV)	NAIVE3_M3 7.2787	HOLT3_M3 7.2884	ESM3_M3 7.3005	NAÏVE 7.3117	ARIMA 7.3198	HOLT 7.3240	ESM 7.3392	ESM_M3 7.3396	ARIMA_M3 7.3857	M15_M3 7.6461	MEAN_M3 8.6329	MEAN 11.0832
Schweitzer-Mauduit Intl., Inc. (SWM)	ARIMA 9.7176	NAÏVE 9.7388	HOLT 9.7599	ESM 9.7704	NAIVE3_M3 9.8685	HOLT3_M3 9.8733	ESM3_M3 9.8924	ARIMA_M3 9.9950	ESM_M3 10.0985	MEAN_M3 10.5588	M15_M3 10.7012	MEAN 12.1291

SECURITY	MAPE #1	MAPE #2	MAPE #3	MAPE #4	MAPE #5	MAPE #6	MAPE #7	MAPE #8	MAPE #9	MAPE #10	MAPE #11	MAPE #12
FOOD ITEMS												
Sysco Corporation (SYY)	ESM_M3 5.5978	HOLT3_M3 5.6010	NAIVE3_M3 5.6058	ESM3_M3 5.6239	HOLT 5.6639	NAÏVE 5.6742	M15_M3 5.6928	ESM 5.7004	MEAN_M3 5.7608	MEAN 6.9035	ARIMA_M3 7.5695	ARIMA 13.4759
General Mills, Inc. (GIS)	NAIVE3_M3 3.6956	HOLT3_M3 3.6959	HOLT 3.7031	NAÏVE 3.7046	ESM3_M3 3.7206	ESM_M3 3.7616	ESM 3.7760	M15_M3 4.0121	ARIMA_M3 4.1966	MEAN_M3 4.6310	ARIMA 4.9050	MEAN 5.8997
Kellogg Company (K)	ESM3_M3 4.5671	ESM 4.5905	ESM_M3 4.6053	HOLT 4.6078	HOLT3_M3 4.6084	NAIVE3_M3 4.6085	NAÏVE 4.6145	ARIMA 4.6187	ARIMA_M3 4.6919	M15_M3 4.8320	MEAN_M3 5.1818	MEAN 6.1757
McCormick & Company, Inc. (MKC)	ESM_M3 4.4085	ESM3_M3 4.4286	NAIVE3_M3 4.4400	HOLT3_M3 4.4407	HOLT 4.4838	NAÏVE 4.4857	ESM 4.5443	M15_M3 4.6092	MEAN_M3 5.3264	ARIMA_M3 6.5512	MEAN 6.7833	ARIMA 11.6031
Conagra Brands, Inc. (CAG)	HOLT 5.6467	NAÏVE 5.6468	ESM3_M3 5.6473	NAIVE3_M3 5.6497	ARIMA 5.6499	HOLT3_M3 5.6501	ESM 5.6638	ESM_M3 5.7006	ARIMA_M3 5.7595	M15_M3 5.9814	MEAN_M3 6.1385	MEAN 7.6344
Campbell Soup Company (CPG)	HOLT3_M3 5.0503	NAIVE3_M3 5.0561	ARIMA_M3 5.0582	HOLT 5.1079	NAÏVE 5.1202	ARIMA 5.1229	ESM_M3 5.1369	M15_M3 5.1411	ESM3_M3 5.2150	ESM 5.3430	MEAN_M3 5.6191	MEAN 7.1908
Seaboard Corporation (SEB)	NAIVE3_M3 8.5404	HOLT3_M3 8.5561	NAÏVE 8.5863	HOLT 8.6764	M15_M3 8.9330	ESM_M3 8.9465	ESM3_M3 9.1558	ESM 9.4858	ARIMA_M3 9.7997	MEAN_M3 10.1549	MEAN 12.8411	ARIMA 13.4104

Table C-28: Consumer Staples Sector M3 RMSE Accuracy Results

SECURITY	RMSE #1	RMSE #2	RMSE #3	RMSE #4	RMSE #5	RMSE #6	RMSE #7	RMSE #8	RMSE #9	RMSE #10	RMSE #11	RMSE #12
SOFT DRINKS												
The Coca-Cola Company (KO)	HOLT3_M3 0.0251	NAIVE3_M3 0.0251	HOLT 0.0252	NAÏVE 0.0252	ARIMA_M3 0.0253	ARIMA 0.0254	M15_M3 0.0257	ESM_M3 0.0258	ESM3_M3 0.0266	MEAN_M3 0.0270	ESM 0.0280	MEAN 0.0330
PepsiCo, Inc. (PEP)	ARIMA 0.0576	HOLT 0.0577	HOLT3_M3 0.0578	NAÏVE 0.0578	NAIVE3_M3 0.0579	ARIMA_M3 0.0584	ESM_M3 0.0591	M15_M3 0.0597	ESM3_M3 0.0606	MEAN_M3 0.0607	ESM 0.0632	MEAN 0.0684
Coca-Cola European Partners (CCEP)	HOLT3_M3 0.0404	ESM3_M3 0.0404	NAIVE3_M3 0.0404	ARIMA 0.0404	ESM 0.0404	HOLT 0.0405	NAÏVE 0.0405	ESM_M3 0.0405	ARIMA_M3 0.0406	M15_M3 0.0414	MEAN_M3 0.0415	MEAN 0.0489
National Beverage Corp. (FIZZ)	M15_M3 0.0785	ESM_M3 0.0798	ESM3_M3 0.0810	NAIVE3_M3 0.0817	ESM 0.0825	NAÏVE 0.0833	HOLT3_M3 0.0837	MEAN_M3 0.0841	HOLT 0.0870	MEAN 0.1029	ARIMA_M3 0.1379	ARIMA 0.2791
Cott Corporation (COT)	HOLT 0.0238	ARIMA 0.0239	HOLT3_M3 0.0239	NAÏVE 0.0239	NAIVE3_M3 0.0240	ESM_M3 0.0242	ESM3_M3 0.0242	ARIMA_M3 0.0244	ESM 0.0245	M15_M3 0.0252	MEAN_M3 0.0285	MEAN 0.0353
ALCOHOLIC BEVERAGES												
Brown-Forman Corporation (BF-A)	HOLT3_M3 0.0251	HOLT 0.0251	ESM3_M3 0.0251	NAIVE3_M3 0.0251	ESM_M3 0.0251	NAÏVE 0.0252	ESM 0.0252	M15_M3 0.0254	MEAN_M3 0.0292	ARIMA_M3 0.0348	MEAN 0.0362	ARIMA 0.0716
Molson Coors Brewing Company (TAP)	M15_M3 0.0594	ESM_M3 0.0594	HOLT3_M3 0.0595	NAIVE3_M3 0.0596	ESM3_M3 0.0599	HOLT 0.0601	NAÏVE 0.0601	ESM 0.0607	MEAN_M3 0.0647	ARIMA_M3 0.0708	MEAN 0.0796	ARIMA 0.1052
Constellation Brans, Inc. (STZ)	ESM 0.0869	HOLT 0.0871	NAÏVE 0.0872	ESM3_M3 0.0875	NAIVE3_M3 0.0878	HOLT3_M3 0.0878	ESM_M3 0.0889	M15_M3 0.0939	MEAN_M3 0.1056	MEAN 0.1245	ARIMA_M3 0.1930	ARIMA 0.5012
Compañia Cervecerias Unidas S.A. (CCU)	M15_M3 0.0220	ARIMA_M3 0.0220	ESM_M3 0.0222	NAIVE3_M3 0.0223	HOLT3_M3 0.0225	ESM3_M3 0.0227	NAÏVE 0.0227	ARIMA 0.0229	HOLT 0.0230	MEAN_M3 0.0231	ESM 0.0233	MEAN 0.0282
Willamette Valley Vineyards, Inc. (WVVI)	HOLT3_M3 0.0077	MEAN_M3 0.0077	M15_M3 0.0077	HOLT 0.0077	NAIVE3_M3 0.0078	ESM_M3 0.0078	NAÏVE 0.0079	ESM3_M3 0.0079	ESM 0.0081	ARIMA_M3 0.0087	MEAN 0.0090	ARIMA 0.0114

SECURITY	RMSE #1	RMSE #2	RMSE #3	RMSE #4	RMSE #5	RMSE #6	RMSE #7	RMSE #8	RMSE #9	RMSE #10	RMSE #11	RMSE #12
TOBACCO PRODUCTS												
Altria Group, Inc. (MO)	ESM 0.0625	NAÏVE 0.0632	HOLT 0.0633	ARIMA 0.0642	ESM3_M3 0.0643	NAIVE3_M3 0.0654	HOLT3_M3 0.0661	ESM_M3 0.0712	ARIMA_M3 0.0792	M15_M3 0.0951	MEAN_M3 0.1050	MEAN 0.1218
British American Tobacco p.l.c. (BTI)	HOLT 0.0396	NAÏVE 0.0397	ESM 0.0397	HOLT3_M3 0.0399	ESM3_M3 0.0399	NAIVE3_M3 0.0399	ESM_M3 0.0404	M15_M3 0.0421	MEAN_M3 0.0457	MEAN 0.0539	ARIMA_M3 0.0556	ARIMA 0.1222
Vector Group Ltd. (VGR)	NAÏVE 0.0180	HOLT 0.0180	ESM 0.0180	ESM3_M3 0.0181	NAIVE3_M3 0.0181	HOLT3_M3 0.0181	ESM_M3 0.0183	M15_M3 0.0191	MEAN_M3 0.0212	ARIMA_M3 0.0229	MEAN 0.0257	ARIMA 0.0382
Universal Corporation (UVV)	NAÏVE 0.0672	NAIVE3_M3 0.0673	HOLT 0.0673	HOLT3_M3 0.0674	ARIMA 0.0675	ESM3_M3 0.0677	ESM 0.0678	ESM_M3 0.0680	ARIMA_M3 0.0682	M15_M3 0.0695	MEAN_M3 0.0737	MEAN 0.0904
Schweitzer-Mauduit Intl., Inc. (SWM)	ARIMA 0.0490	ARIMA_M3 0.0491	NAÏVE 0.0492	HOLT 0.0495	ESM 0.0496	NAIVE3_M3 0.0498	HOLT3_M3 0.0500	ESM3_M3 0.0501	ESM_M3 0.0508	M15_M3 0.0529	MEAN_M3 0.0546	MEAN 0.0639
FOOD ITEMS												
Sysco Corporation (SYY)	HOLT3_M3 0.0369	NAIVE3_M3 0.0370	HOLT 0.0372	ESM_M3 0.0372	NAÏVE 0.0372	ESM3_M3 0.0374	M15_M3 0.0375	MEAN_M3 0.0378	ESM 0.0379	MEAN 0.0442	ARIMA_M3 0.0505	ARIMA 0.1028
General Mills, Inc. (GIS)	HOLT3_M3 0.0265	NAIVE3_M3 0.0265	HOLT 0.0266	ESM3_M3 0.0266	NAÏVE 0.0266	ESM_M3 0.0267	ESM 0.0269	M15_M3 0.0281	MEAN_M3 0.0326	MEAN 0.0430	ARIMA_M3 0.0527	ARIMA 0.1259
Kellogg Company (K)	HOLT3_M3 0.0433	HOLT 0.0433	NAIVE3_M3 0.0434	NAÏVE 0.0434	ESM3_M3 0.0436	ESM_M3 0.0436	ARIMA 0.0437	ARIMA_M3 0.0440	ESM 0.0441	M15_M3 0.0450	MEAN_M3 0.0476	MEAN 0.0562
McCormick & Company, Inc. (MKC)	HOLT 0.0473	NAÏVE 0.0474	HOLT3_M3 0.0475	NAIVE3_M3 0.0475	ESM_M3 0.0484	ESM3_M3 0.0486	ESM 0.0498	M15_M3 0.0515	MEAN_M3 0.0610	MEAN 0.0756	ARIMA_M3 0.1189	ARIMA 0.3284
Conagra Brands, Inc. (CAG)	HOLT 0.0236	NAÏVE 0.0236	ARIMA 0.0236	HOLT3_M3 0.0237	NAIVE3_M3 0.0237	ESM3_M3 0.0239	ESM 0.0240	ESM_M3 0.0240	ARIMA_M3 0.0243	M15_M3 0.0253	MEAN_M3 0.0258	MEAN 0.0308
Campbell Soup Company (CPG)	M15_M3 0.0381	ARIMA_M3 0.0381	HOLT3_M3 0.0383	NAIVE3_M3 0.0384	ESM_M3 0.0386	HOLT 0.0388	NAÏVE 0.0389	ARIMA 0.0392	ESM3_M3 0.0394	ESM 0.0404	MEAN_M3 0.0420	MEAN 0.0521
Seaboard Corporation (SEB)	NAIVE3_M3 3.1412	NAÏVE 3.1605	HOLT3_M3 3.1609	HOLT 3.1901	M15_M3 3.2306	ESM_M3 3.2859	ESM3_M3 3.4292	MEAN_M3 3.4473	ESM 3.6392	ARIMA_M3 3.9706	MEAN 4.1647	ARIMA 7.3415

C.13. Financial Sector (Banks)

Table C-29: Financial Sector (Banks) M3 MAPE Accuracy Results

SECURITY	MAPE #1	MAPE #2	MAPE #3	MAPE #4	MAPE #5	MAPE #6	MAPE #7	MAPE #8	MAPE #9	MAPE #10	MAPE #11	MAPE #12
BANKS												
US Bancorp (USB)	ARIMA_M3 6.5149	M15_M3 6.5473	ESM_M3 6.6135	NAIVE3_M3 6.6378	HOLT3_M3 6.6537	ESM3_M3 6.7577	MEAN_M3 6.7639	ARIMA 6.7689	NAÏVE 6.7837	HOLT 6.8074	ESM 6.9591	MEAN 8.1143
Wells Fargo & Company (WFC)	HOLT3_M3 6.8541	NAIVE3_M3 6.8659	M15_M3 6.8953	ESM_M3 6.9114	ARIMA_M3 6.9519	HOLT 6.9650	NAÏVE 6.9833	MEAN_M3 7.0027	ESM3_M3 7.0728	ARIMA 7.3146	ESM 7.3636	MEAN 8.3135
The Bank of New York Mellon Corp. (BK)	HOLT3_M3 6.6803	NAIVE3_M3 6.6873	HOLT 6.6889	NAÏVE 6.7037	ESM_M3 6.7614	ESM3_M3 6.7731	ESM 6.8565	M15_M3 6.9524	ARIMA_M3 7.3362	MEAN_M3 7.6326	MEAN 9.5313	ARIMA 9.9136
Citigroup Inc. (C)	NAIVE3_M3 12.1734	HOLT3_M3 12.1736	M15_M3 12.2263	NAÏVE 12.2970	HOLT 12.2974	ESM_M3 12.3060	ESM3_M3 12.4942	ESM 12.7755	ARIMA_M3 12.7796	MEAN_M3 13.8922	ARIMA 14.2046	MEAN 17.6316
JP Morgan Chase & Co (JPM)	ESM_M3 8.2001	HOLT3_M3 8.2195	NAIVE3_M3 8.2258	ESM3_M3 8.2319	ARIMA_M3 8.2367	HOLT 8.3038	NAÏVE 8.3126	M15_M3 8.3193	ESM 8.3284	ARIMA 8.3801	MEAN_M3 8.4477	MEAN 10.1802
Bank of America Corporation (BAC)	ESM_M3 11.2466	ESM3_M3 11.2726	HOLT3_M3 11.2774	NAIVE3_M3 11.2877	ARIMA_M3 11.3064	ESM 11.3572	HOLT 11.3603	NAÏVE 11.3742	M15_M3 11.4071	ARIMA 11.4568	MEAN_M3 12.7868	MEAN 15.5536
The PNC Financial Services Group (PNC)	ESM_M3 7.0008	HOLT3_M3 7.0081	NAIVE3_M3 7.0119	ESM3_M3 7.0179	ARIMA_M3 7.0253	HOLT 7.0899	NAÏVE 7.0964	ESM 7.1031	ARIMA 7.1153	M15_M3 7.1447	MEAN_M3 7.8399	MEAN 9.5629
MIDWEST BANKS												
UMB Financial Corporation (UMBF)	HOLT3_M3 6.0361	ESM_M3 6.0469	M15_M3 6.0884	ESM3_M3 6.1039	HOLT 6.1139	NAIVE3_M3 6.1186	ESM 6.2212	NAÏVE 6.2469	ARIMA_M3 6.2926	MEAN_M3 6.3241	ARIMA 6.9048	MEAN 7.4640
Huntington Bancshares Inc. (HBAN)	ESM_M3 11.4744	HOLT3_M3 11.4791	ARIMA_M3 11.4908	M15_M3 11.5011	NAIVE3_M3 11.5174	ESM3_M3 11.5768	HOLT 11.6074	NAÏVE 11.6762	ARIMA 11.7654	ESM 11.7734	MEAN_M3 11.9680	MEAN 14.2711
Commerce Bancshares, Inc. (CBSH)	ARIMA_M3 4.9494	M15_M3 4.9939	HOLT3_M3 5.0130	ARIMA 5.0245	NAIVE3_M3 5.0299	ESM_M3 5.0714	HOLT 5.0790	NAÏVE 5.1038	MEAN_M3 5.1648	ESM3_M3 5.1930	ESM 5.3609	MEAN 6.0447
Associated Banc-Corp (ASB)	ARIMA 6.7592	ESM3_M3 6.8003	ESM 6.8115	NAIVE3_M3 6.8144	HOLT3_M3 6.8160	NAÏVE 6.8305	HOLT 6.8311	ESM_M3 6.8524	ARIMA_M3 6.8852	M15_M3 7.1165	MEAN_M3 7.3671	MEAN 8.8603
First Financial Bancorp (FFBC)	HOLT3_M3 7.2459	ARIMA 7.2499	MEAN_M3 7.2690	HOLT 7.2788	ARIMA_M3 7.3077	NAIVE3_M3 7.3127	ESM_M3 7.3398	ESM3_M3 7.3488	NAÏVE 7.4145	ESM 7.4175	M15_M3 7.4732	MEAN 8.2879
Old National Bancorp (ONB)	MEAN_M3 6.3869	ARIMA_M3 6.6151	M15_M3 6.6324	ESM_M3 6.6338	ESM3_M3 6.7122	HOLT3_M3 6.7224	NAIVE3_M3 6.7349	ESM 6.8304	ARIMA 6.8410	HOLT 6.8476	NAÏVE 6.8714	MEAN 7.2023
TCF Financial Corporation (TCF)	ARIMA_M3 7.7815	HOLT 7.8163	HOLT3_M3 7.8174	NAÏVE 7.8182	NAIVE3_M3 7.8208	ESM 7.8252	ESM3_M3 7.8260	ESM_M3 7.8782	ARIMA 8.1215	M15_M3 8.1563	MEAN_M3 8.5795	MEAN 10.4540

SECURITY	MAPE #1	MAPE #2	MAPE #3	MAPE #4	MAPE #5	MAPE #6	MAPE #7	MAPE #8	MAPE #9	MAPE #10	MAPE #11	MAPE #12
INVESTMENT BANKS												
Siebert Financial Corp. (SIEB)	NAÏVE 12.7973	HOLT 12.7975	NAIVE3_M3 12.8945	HOLT3_M3 12.8947	ARIMA 13.0250	ARIMA_M3 13.6523	M15_M3 14.2704	MEAN_M3 14.3215	ESM_M3 15.2000	ESM3_M3 16.3666	MEAN 17.3765	ESM 17.8691
Stifel Financial Corp. (SF)	NAIVE3_M3 8.8918	ESM_M3 8.8919	HOLT3_M3 8.8966	ESM3_M3 8.8970	NAÏVE 8.9544	ESM 8.9609	HOLT 8.9653	M15_M3 9.0403	MEAN_M3 9.7957	ARIMA_M3 10.9593	MEAN 12.1075	ARIMA 15.1268
Raymond James Financial, Inc. (RJF)	ARIMA_M3 9.0788	ESM_M3 9.0874	HOLT3_M3 9.1235	M15_M3 9.1606	NAIVE3_M3 9.2016	HOLT 9.2394	ESM3_M3 9.2563	ARIMA 9.2675	NAÏVE 9.3511	MEAN_M3 9.5353	ESM 9.5397	MEAN 11.0160
The Charles Schwab Corp. (SCHW)	HOLT3_M3 10.4709	NAIVE3_M3 10.4751	ARIMA 10.4831	ARIMA_M3 10.5019	HOLT 10.5306	NAÏVE 10.5398	M15_M3 10.6513	MEAN_M3 10.8836	ESM_M3 11.0696	ESM3_M3 11.5139	ESM 12.1091	MEAN 13.1911
WisdomTree Investments, Inc. (WETF)	NAÏVE 22.6224	HOLT 22.6478	ARIMA 22.6939	ESM 22.8337	NAIVE3_M3 23.3213	HOLT3_M3 23.3375	ESM3_M3 23.3995	ESM_M3 24.1798	ARIMA_M3 24.5998	M15_M3 26.1484	MEAN_M3 29.9931	MEAN 37.6617
Oppenheimer Holdings, Inc. (OPY)	NAIVE3_M3 10.6226	HOLT3_M3 10.6527	ARIMA_M3 10.6651	NAÏVE 10.7039	ESM_M3 10.7125	HOLT 10.7454	ESM3_M3 10.7622	ARIMA 10.7706	M15_M3 10.8528	ESM 10.8866	MEAN_M3 12.0793	MEAN 14.8277
National Holdings Corp. (NHLD)	ARIMA_M3 21.5273	ARIMA 21.6422	HOLT3_M3 22.1036	M15_M3 22.4233	NAIVE3_M3 22.4430	HOLT 23.0283	NAÏVE 23.5797	MEAN_M3 24.3556	MEAN 30.0433	ESM_M3 37.1946	ESM3_M3 48.3173	ESM 59.9233
INVESTMENT MANAGEMENT												
Eaton Vance Corp. (EV)	ESM_M3 8.3711	HOLT3_M3 8.4595	NAIVE3_M3 8.4933	HOLT 8.5222	ESM3_M3 8.5642	NAÏVE 8.5696	M15_M3 8.5927	ESM 8.9990	MEAN_M3 9.4267	ARIMA_M3 9.8999	MEAN 11.4185	ARIMA 15.1013
Barings Corporate Investors (MCI)	ESM 4.0111	ESM3_M3 4.0254	ESM_M3 4.1040	NAIVE3_M3 4.1111	NAÏVE 4.1127	HOLT3_M3 4.1132	HOLT 4.1145	ARIMA 4.1391	ARIMA_M3 4.2330	M15_M3 4.3796	MEAN_M3 4.6580	MEAN 5.7397
SEI Investments Co. (SEIC)	HOLT 10.3462	NAÏVE 10.3674	HOLT3_M3 10.3760	NAIVE3_M3 10.3910	M15_M3 10.6591	ESM_M3 10.6990	ESM3_M3 10.8180	MEAN_M3 10.8677	ESM 11.0066	ARIMA_M3 11.2192	MEAN 12.5418	ARIMA 13.5138
Franklin Resources, Inc. (BEN)	NAIVE3_M3 8.2181	HOLT3_M3 8.2232	ARIMA_M3 8.2441	NAÏVE 8.2508	HOLT 8.2552	ESM3_M3 8.2554	ESM_M3 8.2582	ARIMA 8.2825	ESM 8.3003	M15_M3 8.4287	MEAN_M3 9.2609	MEAN 11.3217
Legg Mason, Inc. (LM)	HOLT 10.1744	HOLT3_M3 10.2055	NAÏVE 10.2333	NAIVE3_M3 10.2481	ESM3_M3 10.4734	ESM_M3 10.4913	ESM 10.5563	M15_M3 10.7326	ARIMA_M3 11.1617	MEAN_M3 11.9614	ARIMA 13.2399	MEAN 14.5445
T. Rowe Price Group, Inc. (TROW)	HOLT3_M3 8.4118	NAIVE3_M3 8.4330	HOLT 8.4410	ARIMA_M3 8.4445	ESM_M3 8.4617	ESM3_M3 8.4634	NAÏVE 8.4713	ESM 8.5381	M15_M3 8.6532	ARIMA 8.6791	MEAN_M3 9.1089	MEAN 11.0461
Alliance- Bernstein Holding L.P. (AB)	HOLT 9.0092	NAÏVE 9.0097	HOLT3_M3 9.0644	NAIVE3_M3 9.0647	ESM_M3 9.3733	ESM3_M3 9.3740	ESM 9.4808	ARIMA_M3 9.5528	M15_M3 9.6585	ARIMA 10.0130	MEAN_M3 11.6451	MEAN 14.5893

Table C-30: Financial Sector (Banks) M3 RMSE Accuracy Results

SECURITY	RMSE #1	RMSE #2	RMSE #3	RMSE #4	RMSE #5	RMSE #6	RMSE #7	RMSE #8	RMSE #9	RMSE #10	RMSE #11	RMSE #12
BANKS												
US Bancorp (USB)	M15_M3 0.0364	ARIMA_M3 0.0366	ESM_M3 0.0371	NAIVE3_M3 0.0371	HOLT3_M3 0.0372	MEAN_M3 0.0378	NAÏVE 0.0379	ESM3_M3 0.0380	HOLT 0.0380	ARIMA 0.0383	ESM 0.0391	MEAN 0.0446
Wells Fargo & Company (WFC)	MEAN_M3 0.0445	M15_M3 0.0448	HOLT3_M3 0.0454	NAIVE3_M3 0.0455	ARIMA_M3 0.0455	ESM_M3 0.0459	HOLT 0.0462	NAÏVE 0.0464	ESM3_M3 0.0473	ESM 0.0494	ARIMA 0.0502	MEAN 0.0516
The Bank of New York Mellon Corp. (BK)	HOLT 0.0452	NAÏVE 0.0452	HOLT3_M3 0.0453	NAIVE3_M3 0.0454	ESM_M3 0.0469	ESM3_M3 0.0470	ESM 0.0477	M15_M3 0.0479	MEAN_M3 0.0506	MEAN 0.0614	ARIMA_M3 0.0626	ARIMA 0.1416
Citigroup Inc. (C)	NAIVE3_M3 0.4869	HOLT3_M3 0.4869	M15_M3 0.4894	NAÏVE 0.4894	HOLT 0.4894	MEAN_M3 0.5047	ESM_M3 0.5058	ESM3_M3 0.5230	ESM 0.5455	MEAN 0.5978	ARIMA_M3 0.6178	ARIMA 0.9745
JP Morgan Chase & Co (JPM)	ESM3_M3 0.0712	ESM_M3 0.0713	ESM 0.0715	HOLT3_M3 0.0718	NAIVE3_M3 0.0718	ARIMA_M3 0.0721	HOLT 0.0722	NAÏVE 0.0723	ARIMA 0.0727	M15_M3 0.0730	MEAN_M3 0.0731	MEAN 0.0851
Bank of America Corporation (BAC)	ARIMA_M3 0.0460	ESM_M3 0.0461	M15_M3 0.0461	HOLT3_M3 0.0463	ESM3_M3 0.0464	NAIVE3_M3 0.0464	HOLT 0.0468	ARIMA 0.0469	ESM 0.0469	NAÏVE 0.0470	MEAN_M3 0.0479	MEAN 0.0561
The PNC Financial Services Group (PNC)	ARIMA_M3 0.0917	ESM_M3 0.0917	HOLT3_M3 0.0919	NAIVE3_M3 0.0920	ESM3_M3 0.0921	ARIMA 0.0927	M15_M3 0.0927	HOLT 0.0928	NAÏVE 0.0929	ESM 0.0930	MEAN_M3 0.1011	MEAN 0.1230
MIDWEST BANKS												
UMB Financial Corporation (UMBF)	M15_M3 0.0542	HOLT3_M3 0.0543	ESM_M3 0.0543	NAIVE3_M3 0.0547	HOLT 0.0549	ESM3_M3 0.0549	NAÏVE 0.0555	ESM 0.0559	MEAN_M3 0.0571	ARIMA_M3 0.0600	MEAN 0.0666	ARIMA 0.0875
Huntington Bancshares Inc. (HBAN)	ARIMA_M3 0.0203	M15_M3 0.0204	HOLT3_M3 0.0204	NAIVE3_M3 0.0205	ESM_M3 0.0206	HOLT 0.0207	ARIMA 0.0208	NAÏVE 0.0208	ESM3_M3 0.0210	ESM 0.0215	MEAN_M3 0.0222	MEAN 0.0266
Commerce Bancshares, Inc. (CBSH)	ARIMA_M3 0.0308	ESM_M3 0.0310	ARIMA 0.0311	M15_M3 0.0311	HOLT3_M3 0.0311	NAIVE3_M3 0.0312	HOLT 0.0314	ESM3_M3 0.0316	NAÏVE 0.0316	MEAN_M3 0.0324	ESM 0.0325	MEAN 0.0370
Associated Banc-Corp (ASB)	ESM3_M3 0.0256	HOLT3_M3 0.0257	ESM 0.0257	ARIMA 0.0257	NAIVE3_M3 0.0257	ESM_M3 0.0257	HOLT 0.0258	NAÏVE 0.0258	ARIMA_M3 0.0258	M15_M3 0.0265	MEAN_M3 0.0284	MEAN 0.0337
First Financial Bancorp (FFBC)	HOLT3_M3 0.0231	ARIMA 0.0231	HOLT 0.0231	NAIVE3_M3 0.0232	ARIMA_M3 0.0233	ESM_M3 0.0233	ESM3_M3 0.0233	NAÏVE 0.0235	ESM 0.0235	M15_M3 0.0238	MEAN_M3 0.0240	MEAN 0.0277
Old National Bancorp (ONB)	MEAN_M3 0.0201	M15_M3 0.0209	ARIMA_M3 0.0213	ESM_M3 0.0214	ESM3_M3 0.0219	HOLT3_M3 0.0219	NAIVE3_M3 0.0220	MEAN 0.0222	ESM 0.0225	ARIMA 0.0225	HOLT 0.0225	NAÏVE 0.0227
TCF Financial Corporation (TCF)	NAÏVE 0.0274	HOLT 0.0274	ESM 0.0274	ARIMA_M3 0.0274	ESM3_M3 0.0274	HOLT3_M3 0.0274	NAIVE3_M3 0.0274	ESM_M3 0.0276	M15_M3 0.0285	MEAN_M3 0.0298	ARIMA 0.0313	MEAN 0.0354

SECURITY	RMSE #1	RMSE #2	RMSE #3	RMSE #4	RMSE #5	RMSE #6	RMSE #7	RMSE #8	RMSE #9	RMSE #10	RMSE #11	RMSE #12
INVESTMENT BANKS												
Siebert Financial Corp. (SIEB)	NAÏVE 0.0178	HOLT 0.0178	NAIVE3_M3 0.0185	HOLT3_M3 0.0186	ARIMA 0.0187	MEAN_M3 0.0195	ARIMA_M3 0.0203	M15_M3 0.0215	MEAN 0.0220	ESM_M3 0.0250	ESM3_M3 0.0282	ESM 0.0319
Stifel Financial Corp. (SF)	HOLT3_M3 0.0530	NAIVE3_M3 0.0530	ESM3_M3 0.0531	HOLT 0.0531	NAÏVE 0.0532	ESM_M3 0.0532	ESM 0.0533	M15_M3 0.0541	MEAN_M3 0.0585	ARIMA_M3 0.0660	MEAN 0.0711	ARIMA 0.1035
Raymond James Financial, Inc. (RJF)	ARIMA_M3 0.0621	HOLT3_M3 0.0622	ESM_M3 0.0623	M15_M3 0.0624	NAIVE3_M3 0.0627	HOLT 0.0628	ARIMA 0.0629	ESM3_M3 0.0632	NAÏVE 0.0634	MEAN_M3 0.0637	ESM 0.0648	MEAN 0.0711
The Charles Schwab Corp. (SCHW)	HOLT3_M3 0.0432	NAIVE3_M3 0.0432	HOLT 0.0433	NAÏVE 0.0433	ARIMA 0.0433	ARIMA_M3 0.0436	MEAN_M3 0.0442	M15_M3 0.0443	ESM_M3 0.0470	ESM3_M3 0.0496	MEAN 0.0522	ESM 0.0528
WisdomTree Investments, Inc. (WETF)	HOLT 0.0204	NAÏVE 0.0204	ESM 0.0204	ARIMA 0.0204	HOLT3_M3 0.0206	NAIVE3_M3 0.0206	ESM3_M3 0.0207	ESM_M3 0.0210	ARIMA_M3 0.0213	M15_M3 0.0221	MEAN_M3 0.0228	MEAN 0.0271
Oppenheimer Holdings, Inc. (OPY)	NAIVE3_M3 0.0529	ARIMA_M3 0.0529	HOLT3_M3 0.0531	NAÏVE 0.0532	ESM_M3 0.0532	M15_M3 0.0533	HOLT 0.0534	ESM3_M3 0.0535	ARIMA 0.0536	ESM 0.0540	MEAN_M3 0.0576	MEAN 0.0691
National Holdings Corp. (NHLD)	NAÏVE 0.0923	HOLT 0.0928	ARIMA 0.0948	NAIVE3_M3 0.0962	HOLT3_M3 0.0967	ESM 0.1006	MEAN_M3 0.1008	ESM3_M3 0.1010	ESM_M3 0.1050	ARIMA_M3 0.1084	MEAN 0.1087	M15_M3 0.1217
INVESTMENT MANAGEMENT												
Eaton Vance Corp. (EV)	HOLT 0.0541	HOLT3_M3 0.0541	NAIVE3_M3 0.0544	NAÏVE 0.0545	ARIMA_M3 0.0546	M15_M3 0.0553	ESM_M3 0.0566	MEAN_M3 0.0567	ESM3_M3 0.0583	ESM 0.0608	MEAN 0.0637	ARIMA 0.0777
Barings Corporate Investors (MCI)	ESM 0.0112	ESM3_M3 0.0113	ESM_M3 0.0114	NAIVE3_M3 0.0114	NAÏVE 0.0114	HOLT3_M3 0.0114	HOLT 0.0114	ARIMA 0.0115	ARIMA_M3 0.0116	M15_M3 0.0119	MEAN_M3 0.0126	MEAN 0.0152
SEI Investments Co. (SEIC)	HOLT 0.0522	NAÏVE 0.0523	HOLT3_M3 0.0524	NAIVE3_M3 0.0525	M15_M3 0.0538	ESM_M3 0.0550	MEAN_M3 0.0552	ESM3_M3 0.0565	ARIMA_M3 0.0583	ESM 0.0585	MEAN 0.0629	ARIMA 0.0742
Franklin Resources, Inc. (BEN)	HOLT 0.0455	NAÏVE 0.0456	HOLT3_M3 0.0456	NAIVE3_M3 0.0456	ESM3_M3 0.0461	ESM 0.0461	ESM_M3 0.0463	ARIMA_M3 0.0465	ARIMA 0.0466	M15_M3 0.0476	MEAN_M3 0.0516	MEAN 0.0622
Legg Mason, Inc. (LM)	HOLT 0.0864	NAÏVE 0.0868	HOLT3_M3 0.0873	NAIVE3_M3 0.0876	ESM 0.0884	ESM3_M3 0.0887	ESM_M3 0.0896	M15_M3 0.0930	MEAN_M3 0.1074	ARIMA_M3 0.1120	MEAN 0.1303	ARIMA 0.2335
T. Rowe Price Group, Inc. (TROW)	HOLT 0.0783	NAÏVE 0.0785	ESM 0.0786	HOLT3_M3 0.0788	NAIVE3_M3 0.0789	ESM3_M3 0.0789	ESM_M3 0.0797	ARIMA_M3 0.0802	M15_M3 0.0823	ARIMA 0.0827	MEAN_M3 0.0885	MEAN 0.1033
Alliance- Bernstein Holding L.P. (AB)	HOLT 0.0630	NAÏVE 0.0630	HOLT3_M3 0.0641	NAIVE3_M3 0.0641	ESM_M3 0.0667	ESM3_M3 0.0669	ESM 0.0679	M15_M3 0.0692	ARIMA_M3 0.0727	MEAN_M3 0.0818	ARIMA 0.0994	MEAN 0.1001

C.14. Financial Sector (Insurance)

Table C-31: Financial Sector (Insurance) M3 MAPE Accuracy Results

SECURITY	MAPE #1	MAPE #2	MAPE #3	MAPE #4	MAPE #5	MAPE #6	MAPE #7	MAPE #8	MAPE #9	MAPE #10	MAPE #11	MAPE #12
INSURANCE STOCKS												
W. R. Berkley Corporation (WRB)	HOLT3_M3 5.8436	NAIVE3_M3 5.8545	ESM3_M3 5.8772	ESM_M3 5.8796	HOLT 5.8820	NAÏVE 5.8943	ESM 5.9738	M15_M3 6.3766	MEAN_M3 6.8378	ARIMA_M3 7.1435	MEAN 8.2270	ARIMA 10.8211
Alleghany Corporation (Y)	HOLT3_M3 4.3993	NAIVE3_M3 4.4195	ARIMA_M3 4.4523	HOLT 4.4528	ARIMA 4.4651	ESM_M3 4.4824	NAÏVE 4.4832	M15_M3 4.6102	ESM3_M3 4.6147	ESM 4.8606	MEAN_M3 4.9419	MEAN 5.9173
The Progressive Corporation (PGR)	ESM_M3 6.6057	ESM3_M3 6.6090	HOLT3_M3 6.6157	NAIVE3_M3 6.6508	HOLT 6.6691	ESM 6.6808	NAÏVE 6.6978	ARIMA_M3 6.7400	M15_M3 6.7898	MEAN_M3 7.3480	ARIMA 7.4831	MEAN 9.0234
The Travelers Companies, Inc. (TRV)	ESM3_M3 6.7126	HOLT3_M3 6.7166	NAIVE3_M3 6.7386	HOLT 6.7420	ESM_M3 6.7470	ESM 6.7513	ARIMA 6.7538	NAÏVE 6.7633	ARIMA_M3 6.8042	MEAN_M3 6.8195	M15_M3 6.9742	MEAN 7.6028
Cincinnati Financial Corp. (CINF)	HOLT3_M3 5.5991	ESM_M3 5.6137	NAIVE3_M3 5.6161	HOLT 5.6340	ARIMA_M3 5.6369	ESM3_M3 5.6535	ARIMA 5.6614	NAÏVE 5.6644	M15_M3 5.7542	ESM 5.7869	MEAN_M3 6.0320	MEAN 7.0352
CNA Financial Corporation (CNA)	HOLT3_M3 8.6264	NAIVE3_M3 8.6300	NAÏVE 8.6338	HOLT 8.6391	ARIMA 8.6511	ESM3_M3 8.6649	ESM 8.6825	ESM_M3 8.6950	ARIMA_M3 8.7100	M15_M3 8.9055	MEAN_M3 9.5052	MEAN 11.4471
Markel Corporation (MKL)	HOLT 4.6442	HOLT3_M3 4.6489	NAÏVE 4.6600	NAIVE3_M3 4.6604	ESM_M3 4.6809	ESM3_M3 4.7131	ESM 4.8709	M15_M3 5.0453	ARIMA_M3 5.6827	MEAN_M3 5.7144	MEAN 7.0191	ARIMA 8.3080
INSURANCE BROKERS												
March & McLennan Co, Inc. (MMC)	NAIVE3_M3 6.3543	HOLT3_M3 6.3652	NAÏVE 6.4222	ESM_M3 6.4237	HOLT 6.4440	ESM3_M3 6.4562	ESM 6.5789	M15_M3 6.5962	MEAN_M3 6.7576	ARIMA_M3 7.7783	MEAN 7.9841	ARIMA 10.2731
Aon plc (AON)	ESM3_M3 7.4268	NAIVE3_M3 7.4436	HOLT3_M3 7.4462	ESM_M3 7.4510	ESM 7.4646	NAÏVE 7.4688	HOLT 7.4731	ARIMA_M3 7.5700	ARIMA 7.7007	M15_M3 7.7017	MEAN_M3 8.0299	MEAN 9.2922
Brown & Brown, Inc. (BRO)	HOLT 6.0230	HOLT3_M3 6.0451	NAÏVE 6.0761	NAIVE3_M3 6.0881	ESM_M3 6.2352	ESM3_M3 6.3374	M15_M3 6.4470	ESM 6.5559	MEAN_M3 6.9462	ARIMA_M3 7.1794	MEAN 8.4400	ARIMA 13.1718
Arthur J. Gallagher & Co (AJG)	HOLT3_M3 6.1716	NAIVE3_M3 6.1731	HOLT 6.2133	NAÏVE 6.2143	ESM_M3 6.2724	M15_M3 6.2996	ESM3_M3 6.3926	ARIMA_M3 6.4060	MEAN_M3 6.5390	ESM 6.5840	MEAN 7.7122	ARIMA 7.7993

Table C-32: Financial Sector (Insurance) M3 RMSE Accuracy Results

SECURITY	RMSE #1	RMSE #2	RMSE #3	RMSE #4	RMSE #5	RMSE #6	RMSE #7	RMSE #8	RMSE #9	RMSE #10	RMSE #11	RMSE #12
INSURANCE STOCKS												
W. R. Berkley Corporation (WRB)	ESM3_M3 0.0341	ESM_M3 0.0341	NAIVE3_M3 0.0341	HOLT3_M3 0.0342	HOLT 0.0343	NAÏVE 0.0343	ESM 0.0344	M15_M3 0.0352	MEAN_M3 0.0363	MEAN 0.0423	ARIMA_M3 0.0570	ARIMA 0.1284
Alleghany Corporation (Y)	ARIMA_M3 0.3171	HOLT3_M3 0.3180	ARIMA 0.3192	ESM_M3 0.3202	HOLT 0.3203	NAIVE3_M3 0.3205	M15_M3 0.3215	NAÏVE 0.3238	MEAN_M3 0.3240	ESM3_M3 0.3284	ESM 0.3419	MEAN 0.3724
The Progressive Corporation (PGR)	M15_M3 0.0294	ESM_M3 0.0294	NAIVE3_M3 0.0295	HOLT3_M3 0.0297	NAÏVE 0.0298	ESM3_M3 0.0298	HOLT 0.0301	ESM 0.0304	MEAN_M3 0.0315	ARIMA_M3 0.0332	MEAN 0.0387	ARIMA 0.0468
The Travelers Companies, Inc. (TRV)	MEAN_M3 0.0727	ESM3_M3 0.0734	HOLT3_M3 0.0734	NAIVE3_M3 0.0735	ESM_M3 0.0735	ARIMA 0.0736	ESM 0.0737	HOLT 0.0737	NAÏVE 0.0738	ARIMA_M3 0.0739	M15_M3 0.0754	MEAN 0.0809
Cincinnati Financial Co. (CINF)	HOLT3_M3 0.0432	NAIVE3_M3 0.0433	ESM_M3 0.0433	ARIMA_M3 0.0433	ARIMA 0.0434	HOLT 0.0435	NAÏVE 0.0437	ESM3_M3 0.0438	M15_M3 0.0443	ESM 0.0449	MEAN_M3 0.0454	MEAN 0.0522
CNA Financial Corporation (CNA)	HOLT 0.0477	NAÏVE 0.0478	ARIMA 0.0479	HOLT3_M3 0.0480	NAIVE3_M3 0.0481	ESM3_M3 0.0487	ESM 0.0487	ARIMA_M3 0.0489	ESM_M3 0.0489	M15_M3 0.0501	MEAN_M3 0.0527	MEAN 0.0615
Markel Corporation (MKL)	ESM_M3 0.4690	HOLT 0.4701	NAÏVE 0.4710	HOLT3_M3 0.4730	NAIVE3_M3 0.4735	ESM3_M3 0.4760	ESM 0.4974	M15_M3 0.4997	MEAN_M3 0.5112	MEAN 0.5819	ARIMA_M3 0.6584	ARIMA 1.3620
INSURANCE BROKERS												
March & McLennan Co., Inc. (MMC)	NAIVE3_M3 0.0528	HOLT3_M3 0.0529	NAÏVE 0.0531	HOLT 0.0533	ESM_M3 0.0535	ESM3_M3 0.0539	M15_M3 0.0543	MEAN_M3 0.0545	ESM 0.0549	MEAN 0.0627	ARIMA_M3 0.0912	ARIMA 0.2018
Aon plc (AON)	NAIVE3_M3 0.0666	HOLT3_M3 0.0666	NAÏVE 0.0666	HOLT 0.0667	ESM3_M3 0.0670	ESM_M3 0.0671	ESM 0.0673	M15_M3 0.0686	ARIMA_M3 0.0696	MEAN_M3 0.0715	ARIMA 0.0822	MEAN 0.0849
Brown & Brown, Inc. (BRO)	HOLT3_M3 0.0141	NAIVE3_M3 0.0141	HOLT 0.0141	NAÏVE 0.0142	ESM_M3 0.0142	M15_M3 0.0144	ESM3_M3 0.0145	ESM 0.0151	MEAN_M3 0.0153	ARIMA_M3 0.0170	MEAN 0.0183	ARIMA 0.0330
Arthur J. Gallagher & Co (AJG)	HOLT3_M3 0.0379	NAIVE3_M3 0.0380	HOLT 0.0381	NAÏVE 0.0381	ESM_M3 0.0384	M15_M3 0.0387	ESM3_M3 0.0389	ESM 0.0399	MEAN_M3 0.0408	ARIMA_M3 0.0416	MEAN 0.0484	ARIMA 0.0710

C.15. Financial Sector (REIT/Real Estate)

Table C-33: Financial Sector (REIT/Real Estate) M3 MAPE Accuracy Results

SECURITY	MAPE #1	MAPE #2	MAPE #3	MAPE #4	MAPE #5	MAPE #6	MAPE #7	MAPE #8	MAPE #9	MAPE #10	MAPE #11	MAPE #12
REIT—OTHER EQUITY TRUSTS												
Host Hotels & Resorts (HST)	ESM3_M3 9.1294	ESM_M3 9.1496	ESM 9.1728	HOLT3_M3 9.1932	NAIVE3_M3 9.1933	HOLT 9.2213	NAÏVE 9.2214	ARIMA 9.2300	ARIMA_M3 9.2450	M15_M3 9.3974	MEAN_M3 10.5951	MEAN 13.4545
Public Storage (PSA)	ESM_M3 6.3054	M15_M3 6.3106	HOLT3_M3 6.3206	NAIVE3_M3 6.3365	ESM3_M3 6.3544	HOLT 6.3907	NAÏVE 6.4132	ESM 6.4363	MEAN_M3 6.8641	ARIMA_M3 7.9371	MEAN 8.1151	ARIMA 13.2784
Welltower Inc. (WELL)	NAIVE3_M3 6.5944	HOLT3_M3 6.5965	ARIMA 6.6011	ESM_M3 6.6056	ARIMA_M3 6.6091	ESM3_M3 6.6153	HOLT 6.6204	NAÏVE 6.6219	ESM 6.6737	M15_M3 6.7038	MEAN_M3 7.1662	MEAN 8.4427
Vornado Realty Trust (VNO)	ESM3_M3 7.1036	ESM 7.1124	ESM_M3 7.1422	HOLT3_M3 7.2141	NAIVE3_M3 7.2278	HOLT 7.2296	NAÏVE 7.2506	M15_M3 7.3537	ARIMA_M3 7.5604	MEAN_M3 8.0182	ARIMA 8.6769	MEAN 9.5240
EastGroup Properties, Inc. (EGP)	ESM 5.5654	ESM3_M3 5.5689	HOLT 5.5810	HOLT3_M3 5.5818	NAIVE3_M3 5.5837	NAÏVE 5.5903	ARIMA 5.6152	ESM_M3 5.6233	ARIMA_M3 5.6831	M15_M3 5.8311	MEAN_M3 6.2817	MEAN 7.3692
HCP, Inc. (HCP)	ESM3_M3 7.3110	NAIVE3_M3 7.3277	HOLT3_M3 7.3321	ESM 7.3328	ESM_M3 7.3374	NAÏVE 7.3436	HOLT 7.3482	ARIMA 7.3654	ARIMA_M3 7.3889	M15_M3 7.5241	MEAN_M3 7.8024	MEAN 9.0297
Duke Realty Corporation (DRE)	ESM3_M3 8.6432	HOLT3_M3 8.6435	NAIVE3_M3 8.6512	ESM 8.6680	HOLT 8.6684	ARIMA 8.6735	NAÏVE 8.6830	ESM_M3 8.6984	ARIMA_M3 8.7630	M15_M3 9.0096	MEAN_M3 9.6507	MEAN 11.1878
REIT—RETAIL EQUITY TRUSTS												
Federal Realty Investment Trust (FRT)	ESM 6.1443	ESM3_M3 6.1587	ESM_M3 6.2261	HOLT3_M3 6.2818	NAIVE3_M3 6.2847	HOLT 6.3178	NAÏVE 6.3262	M15_M3 6.4674	MEAN_M3 6.5465	ARIMA_M3 7.0323	MEAN 7.4641	ARIMA 9.0607
Pennsylvania REIT (PEI)	ESM 12.1037	NAÏVE 12.1166	ARIMA 12.1186	HOLT 12.1188	ESM3_M3 12.1741	NAIVE3_M3 12.1848	HOLT3_M3 12.1865	ESM_M3 12.3339	ARIMA_M3 12.4524	M15_M3 12.8727	MEAN_M3 14.1597	MEAN 17.0285
Urstadt Biddle Properties, Inc. (UBP)	ESM_M3 4.4706	ARIMA_M3 4.4793	ESM3_M3 4.5211	HOLT3_M3 4.5215	NAIVE3_M3 4.5560	ARIMA 4.5906	M15_M3 4.6330	ESM 4.6393	HOLT 4.6401	NAÏVE 4.7017	MEAN_M3 4.8678	MEAN 5.7340
National Retail Properties, Inc. (NNN)	ESM 6.1552	NAIVE3_M3 6.1562	HOLT 6.1563	ESM3_M3 6.1563	HOLT3_M3 6.1567	NAÏVE 6.1750	ARIMA 6.1796	ESM_M3 6.1992	ARIMA_M3 6.2475	M15_M3 6.3950	MEAN_M3 6.4275	MEAN 7.3536
Weingarten Realty Investors (WRI)	ARIMA 7.8138	HOLT3_M3 7.8355	ESM3_M3 7.8377	NAIVE3_M3 7.8442	HOLT 7.8662	ESM_M3 7.8669	ESM 7.8691	ARIMA_M3 7.8761	NAÏVE 7.8788	M15_M3 8.0521	MEAN_M3 8.2154	MEAN 9.3366
Kimco Realty Corporation (KIM)	NAIVE3_M3 8.6345	HOLT3_M3 8.6381	NAÏVE 8.6531	HOLT 8.6558	ESM3_M3 8.6792	ESM_M3 8.6915	ESM 8.7330	M15_M3 8.8689	ARIMA_M3 8.8862	MEAN_M3 9.2136	ARIMA 9.6053	MEAN 11.3338
Taubman Centes, Inc. (TCO)	ESM3_M3 7.7725	NAIVE3_M3 7.7863	HOLT3_M3 7.7883	ESM_M3 7.7902	ESM 7.8001	NAÏVE 7.8269	HOLT 7.8275	M15_M3 7.9456	ARIMA_M3 8.5313	MEAN_M3 8.6816	MEAN 10.4381	ARIMA 10.6035

SECURITY	MAPE #1	MAPE #2	MAPE #3	MAPE #4	MAPE #5	MAPE #6	MAPE #7	MAPE #8	MAPE #9	MAPE #10	MAPE #11	MAPE #12
REAL ESTATE OPERATIONS STOCKS												
Texas Pacific Land Trust (TPL)	ESM3_M3 9.4976	ESM_M3 9.5256	ESM 9.5519	NAIVE3_M3 9.5665	NAÏVE 9.6007	HOLT3_M3 9.6100	HOLT 9.6953	M15_M3 9.8953	MEAN_M3 11.5646	ARIMA_M3 12.2672	MEAN 14.5043	ARIMA 21.0625
Tejon Ranch Co. (TRC)	NAÏVE 8.7923	NAIVE3_M3 8.7927	ARIMA 8.7995	HOLT3_M3 8.8054	HOLT 8.8112	ESM3_M3 8.8460	ESM 8.8637	ESM_M3 8.8713	ARIMA_M3 8.8821	M15_M3 9.0547	MEAN_M3 9.4890	MEAN 11.4011
American Realty Investors, Inc. (ARL)	ARIMA 12.7004	HOLT 12.7318	NAÏVE 12.7348	HOLT3_M3 12.8021	NAIVE3_M3 12.8089	ESM 12.8677	ESM3_M3 12.9039	ESM_M3 13.1143	ARIMA_M3 13.2248	M15_M3 13.9916	MEAN_M3 14.1640	MEAN 17.5852
Brookfield Asset Management, Inc. (BAM)	NAIVE3_M3 7.2758	HOLT3_M3 7.3266	NAÏVE 7.3299	ESM_M3 7.3354	ESM3_M3 7.3842	HOLT 7.4101	M15_M3 7.4187	ESM 7.4926	MEAN_M3 8.5267	ARIMA_M3 8.5480	MEAN 10.5997	ARIMA 13.1632
FRP Holdings, Inc. (FRPH)	ESM_M3 8.7133	ESM3_M3 8.7180	HOLT3_M3 8.7255	NAIVE3_M3 8.7648	ESM 8.7914	HOLT 8.8157	MEAN_M3 8.8811	NAÏVE 8.8914	M15_M3 8.9144	ARIMA_M3 9.0991	ARIMA 9.9085	MEAN 10.3390
Income Opportunity Realty Investors, Inc. (IOR)	NAIVE3_M3 11.1015	NAÏVE 11.1446	ESM3_M3 11.1712	HOLT3_M3 11.1740	ESM_M3 11.2279	ESM 11.2489	HOLT 11.2630	ARIMA_M3 11.4538	M15_M3 11.6283	MEAN_M3 11.7642	ARIMA 11.9588	MEAN 14.1795
J W Mays (MAYS)	NAÏVE 9.0293	HOLT 9.0469	ESM 9.0822	NAIVE3_M3 9.1018	HOLT3_M3 9.1215	ESM3_M3 9.1441	ESM_M3 9.3234	ARIMA_M3 9.4386	M15_M3 9.8869	MEAN_M3 10.3973	ARIMA 10.4396	MEAN 12.7140

Table C-34: Financial Sector (REIT/Real Estate) M3 RMSE Accuracy Results

SECURITY	RMSE #1	RMSE #2	RMSE #3	RMSE #4	RMSE #5	RMSE #6	RMSE #7	RMSE #8	RMSE #9	RMSE #10	RMSE #11	RMSE #12
REIT—OTHER EQUITY TRUSTS												
Host Hotels & Resorts (HST)	HOLT3_M3 0.0232	NAIVE3_M3 0.0232	HOLT 0.0233	NAÏVE 0.0233	ARIMA 0.0233	ESM_M3 0.0234	ESM3_M3 0.0234	ARIMA_M3 0.0234	ESM 0.0236	M15_M3 0.0238	MEAN_M3 0.0257	MEAN 0.0314
Public Storage (PSA)	M15_M3 0.1319	ESM_M3 0.1332	HOLT3_M3 0.1340	NAIVE3_M3 0.1345	ESM3_M3 0.1350	HOLT 0.1361	NAÏVE 0.1368	MEAN_M3 0.1373	ESM 0.1374	MEAN 0.1616	ARIMA_M3 0.2300	ARIMA 0.4924
Welltower Inc. (WELL)	HOLT3_M3 0.0582	NAIVE3_M3 0.0582	ARIMA_M3 0.0583	HOLT 0.0585	ESM_M3 0.0585	NAÏVE 0.0586	ARIMA 0.0588	M15_M3 0.0589	ESM3_M3 0.0589	ESM 0.0597	MEAN_M3 0.0626	MEAN 0.0726
Vornado Realty Trust (VNO)	ESM3_M3 0.0781	ESM_M3 0.0782	HOLT3_M3 0.0784	ESM 0.0785	HOLT 0.0787	NAIVE3_M3 0.0787	NAÏVE 0.0790	M15_M3 0.0794	ARIMA_M3 0.0842	MEAN_M3 0.0857	MEAN 0.0999	ARIMA 0.1089
EastGroup Properties, Inc. (EGP)	HOLT 0.0523	ESM 0.0523	HOLT3_M3 0.0523	ESM3_M3 0.0523	NAÏVE 0.0524	NAIVE3_M3 0.0524	ESM_M3 0.0526	ARIMA 0.0528	ARIMA_M3 0.0529	M15_M3 0.0539	MEAN_M3 0.0570	MEAN 0.0655
HCP, Inc. (HCP)	ARIMA_M3 0.0406	M15_M3 0.0408	ESM_M3 0.0408	HOLT3_M3 0.0409	NAIVE3_M3 0.0410	ESM3_M3 0.0411	ARIMA 0.0412	HOLT 0.0413	MEAN_M3 0.0413	NAÏVE 0.0414	ESM 0.0416	MEAN 0.0463
Duke Realty Corporation (DRE)	ESM3_M3 0.0358	HOLT3_M3 0.0358	ESM 0.0359	HOLT 0.0359	ARIMA 0.0359	NAIVE3_M3 0.0359	ESM_M3 0.0359	NAÏVE 0.0359	ARIMA_M3 0.0361	M15_M3 0.0365	MEAN_M3 0.0388	MEAN 0.0442

APPENDIX C: M3 Forecast Study Results

SECURITY	RMSE #1	RMSE #2	RMSE #3	RMSE #4	RMSE #5	RMSE #6	RMSE #7	RMSE #8	RMSE #9	RMSE #10	RMSE #11	RMSE #12
REIT—RETAIL EQUITY TRUSTS												
Federal Realty Investment Trust (FRT)	ESM_M3	ESM3_M3	HOLT3_M3	NAIVE3_M3	ESM	HOLT	MEAN_M3	NAÏVE	M15_M3	MEAN	ARIMA_M3	ARIMA
	0.0961	0.0962	0.0962	0.0963	0.0967	0.0968	0.0968	0.0969	0.0975	0.1073	0.1236	0.2330
Pennsylvania REIT (PEI)	NAÏVE	HOLT	ESM	ARIMA	NAIVE3_M3	HOLT3_M3	ESM3_M3	ESM_M3	ARIMA_M3	M15_M3	MEAN_M3	MEAN
	0.0395	0.0395	0.0395	0.0396	0.0397	0.0397	0.0397	0.0400	0.0403	0.0412	0.0447	0.0523
Urstadt Biddle Properties, Inc. (UBP)	ESM_M3	ARIMA	ESM3_M3	HOLT3_M3	ARIMA_M3	NAIVE3_M3	ESM	HOLT	NAÏVE	M15_M3	MEAN_M3	MEAN
	0.0138	0.0138	0.0138	0.0138	0.0138	0.0139	0.0140	0.0140	0.0142	0.0143	0.0143	0.0165
National Retail Properties, Inc. (NNN)	ESM	HOLT	ESM3_M3	HOLT3_M3	NAIVE3_M3	NAÏVE	ARIMA	ESM_M3	ARIMA_M3	M15_M3	MEAN_M3	MEAN
	0.0322	0.0322	0.0323	0.0323	0.0324	0.0324	0.0324	0.0325	0.0328	0.0334	0.0338	0.0381
Weingarten Realty Investors (WRI)	ARIMA_M3	ESM_M3	HOLT3_M3	NAIVE3_M3	ESM3_M3	ARIMA	M15_M3	HOLT	MEAN_M3	NAÏVE	ESM	MEAN
	0.0445	0.0447	0.0447	0.0448	0.0448	0.0449	0.0449	0.0450	0.0450	0.0451	0.0452	0.0499
Kimco Realty Corporation (KIM)	HOLT	NAÏVE	HOLT3_M3	NAIVE3_M3	ESM_M3	M15_M3	ESM3_M3	ESM	MEAN_M3	ARIMA_M3	MEAN	ARIMA
	0.0442	0.0442	0.0443	0.0443	0.0452	0.0453	0.0455	0.0458	0.0466	0.0475	0.0527	0.0635
Taubman Centes, Inc. (TCO)	ESM_M3	ESM3_M3	HOLT3_M3	NAIVE3_M3	ESM	HOLT	NAÏVE	M15_M3	MEAN_M3	ARIMA_M3	MEAN	ARIMA
	0.0665	0.0665	0.0667	0.0668	0.0669	0.0672	0.0673	0.0673	0.0719	0.0732	0.0845	0.1117
REAL ESTATE OPERATIONS STOCKS												
Texas Pacific Land Trust (TPL)	M15_M3	NAIVE3_M3	NAÏVE	HOLT3_M3	ESM_M3	HOLT	MEAN_M3	ESM3_M3	ESM	MEAN	ARIMA_M3	ARIMA
	0.5635	0.5650	0.5663	0.5780	0.5825	0.5845	0.5857	0.5980	0.6172	0.6477	0.6557	0.9608
Tejon Ranch Co. (TRC)	ARIMA_M3	NAIVE3_M3	ESM_M3	HOLT3_M3	M15_M3	ESM3_M3	NAÏVE	ARIMA	HOLT	ESM	MEAN_M3	MEAN
	0.0487	0.0488	0.0489	0.0489	0.0491	0.0491	0.0491	0.0492	0.0493	0.0495	0.0504	0.0591
American Realty Investors, Inc. (ARL)	HOLT3_M3	NAIVE3_M3	ESM3_M3	ESM_M3	ARIMA	HOLT	NAÏVE	ARIMA_M3	ESM	MEAN_M3	M15_M3	MEAN
	0.0212	0.0212	0.0213	0.0213	0.0214	0.0214	0.0214	0.0214	0.0214	0.0215	0.0221	0.0263
Brookfield Asset Management, Inc. (BAM)	NAIVE3_M3	NAÏVE	MEAN_M3	MEAN	M15_M3	HOLT3_M3	HOLT	ESM3_M3	ESM_M3	ESM	ARIMA_M3	ARIMA
	0.0245	0.0245	0.0267	0.0310	0.0252	0.0246	0.0246	0.0250	0.0249	0.0253	0.0293	0.0545
FRP Holdings, Inc. (FRPH)	NAÏVE	HOLT	NAIVE3_M3	ESM	HOLT3_M3	ESM3_M3	ESM_M3	MEAN_M3	M15_M3	ARIMA_M3	MEAN	ARIMA
	0.0456	0.0457	0.0458	0.0459	0.0459	0.0461	0.0465	0.0473	0.0479	0.0494	0.0521	0.0616
Income Opportunity Realty Investors, Inc. (IOR)	MEAN_M3	ESM3_M3	ESM_M3	NAIVE3_M3	ESM	NAÏVE	ARIMA_M3	M15_M3	HOLT3_M3	HOLT	ARIMA	MEAN
	0.0123	0.0128	0.0128	0.0129	0.0129	0.0130	0.0130	0.0131	0.0133	0.0135	0.0135	0.0138
J W Mays (MAYS)	NAÏVE	HOLT	NAIVE3_M3	ESM	HOLT3_M3	ESM3_M3	ESM_M3	ARIMA_M3	M15_M3	MEAN_M3	MEAN	ARIMA
	0.0436	0.0445	0.0446	0.0447	0.0452	0.0454	0.0464	0.0469	0.0493	0.0499	0.0576	0.0984

C.16. Industrial Products Sector

Table C-35: Industrial Products Sector M3 MAPE Accuracy Results

SECURITY	MAPE #1	MAPE #2	MAPE #3	MAPE #4	MAPE #5	MAPE #6	MAPE #7	MAPE #8	MAPE #9	MAPE #10	MAPE #11	MAPE #12
GENERAL INDUSTRIAL MACHINERY												
Ingersoll-Rand Plc (IR)	ARIMA_M3 9.2266	ESM_M3 9.2708	M15_M3 9.3025	HOLT3_M3 9.3329	NAIVE3_M3 9.3379	ESM3_M3 9.3412	HOLT 9.4563	ARIMA 9.4569	NAÏVE 9.4627	ESM 9.4718	MEAN_M3 10.2367	MEAN 12.5942
Dover Corporation (DOV)	ESM_M3 7.9938	HOLT3_M3 8.0378	NAIVE3_M3 8.0414	ESM3_M3 8.0465	M15_M3 8.0585	HOLT 8.0964	NAÏVE 8.1016	ARIMA_M3 8.1020	ESM 8.1587	MEAN_M3 8.3755	ARIMA 8.3839	MEAN 9.8603
Nordson Corporation (NDSN)	HOLT3_M3 8.7453	ESM_M3 8.7646	NAIVE3_M3 8.7673	HOLT 8.8010	ESM3_M3 8.8061	ARIMA_M3 8.8321	NAÏVE 8.8379	M15_M3 8.8544	ESM 8.9002	MEAN_M3 9.4637	ARIMA 9.6813	MEAN 11.7960
Graco Inc. (GGG)	ESM_M3 7.5889	ESM3_M3 7.6203	NAIVE3_M3 7.7343	HOLT3_M3 7.7358	NAÏVE 7.7869	HOLT 7.7962	ESM 7.8000	M15_M3 7.9381	MEAN_M3 8.5226	ARIMA_M3 9.3949	MEAN 10.0694	ARIMA 14.3069
Illinois Tools Works, Inc. (ITW)	ARIMA_M3 6.3534	ESM_M3 6.3949	ESM3_M3 6.4649	M15_M3 6.4819	HOLT3_M3 6.5067	NAIVE3_M3 6.5100	HOLT 6.5789	NAÏVE 6.5834	ESM 6.6015	ARIMA 6.7037	MEAN_M3 6.8319	MEAN 8.3587
Parker-Hannifn Corporation (PH)	M15_M3 8.7729	ARIMA_M3 8.7860	ESM_M3 8.8624	NAIVE3_M3 8.8665	HOLT3_M3 8.8825	NAÏVE 8.9811	ESM3_M3 8.9934	HOLT 9.0017	ARIMA 9.0279	MEAN_M3 9.0993	ESM 9.1693	MEAN 10.7132
Roper Technologies, Inc. (ROP)	HOLT3_M3 6.9397	NAIVE3_M3 6.9406	ESM_M3 6.9696	HOLT 7.0313	NAÏVE 7.0336	ESM3_M3 7.0602	M15_M3 7.1340	ESM 7.2596	ARIMA_M3 7.4131	MEAN_M3 7.6310	MEAN 9.0835	ARIMA 9.4383
ELECTRICAL MACHINERY												
Emerson Electric Co (EMR)	ESM_M3 6.8163	ARIMA_M3 6.8453	ESM3_M3 6.8462	HOLT3_M3 6.8496	NAIVE3_M3 6.8588	M15_M3 6.8751	ESM 6.9176	HOLT 6.9189	NAÏVE 6.9328	ARIMA 6.9583	MEAN_M3 7.0870	MEAN 8.4490
Eaton Corporation plc (ETN)	ESM_M3 8.2746	ESM3_M3 8.2839	ARIMA_M3 8.3296	ESM 8.3388	NAIVE3_M3 8.3599	M15_M3 8.3653	HOLT3_M3 8.3720	NAÏVE 8.4149	HOLT 8.4316	ARIMA 8.5398	MEAN_M3 8.8729	MEAN 10.5685
Franlin Electric Co. Inc. (FELE)	ESM_M3 7.2937	HOLT3_M3 7.3126	ESM3_M3 7.3398	NAIVE3_M3 7.3424	HOLT 7.3680	NAÏVE 7.4170	M15_M3 7.4451	ESM 7.4618	ARIMA_M3 7.8148	MEAN_M3 7.9154	ARIMA 9.0918	MEAN 9.3392
A. O. Smith Corporation (AOS)	ESM_M3 9.6026	ESM3_M3 9.6401	NAIVE3_M3 9.6850	M15_M3 9.7342	HOLT3_M3 9.7407	ESM 9.7432	NAÏVE 9.7755	HOLT 9.8628	ARIMA_M3 9.9854	MEAN_M3 10.1504	MEAN 11.7995	ARIMA 13.0693
Regal Beloit Corporation (RBC)	ARIMA_M3 8.3291	ESM_M3 8.3486	M15_M3 8.3748	HOLT3_M3 8.4123	NAIVE3_M3 8.4156	ESM3_M3 8.4430	HOLT 8.5469	NAÏVE 8.5514	ARIMA 8.5886	ESM 8.5986	MEAN_M3 8.8840	MEAN 10.6831
II-VI Incorporated (IIVI)	HOLT3_M3 11.6555	NAIVE3_M3 11.6831	ARIMA_M3 11.7233	ARIMA 11.7349	HOLT 11.7946	NAÏVE 11.7995	M15_M3 12.0077	ESM_M3 12.2322	MEAN_M3 12.4469	ESM3_M3 12.5547	ESM 12.9866	MEAN 14.7872
ESCO Technologies, Inc. (ESE)	ESM3_M3 7.9609	ESM_M3 7.9853	ESM 8.0023	NAIVE3_M3 8.0142	HOLT3_M3 8.0219	NAÏVE 8.0704	HOLT 8.0805	M15_M3 8.2269	ARIMA_M3 8.6244	MEAN_M3 8.7794	ARIMA 10.4028	MEAN 10.6552

APPENDIX C: M3 Forecast Study Results

SECURITY	MAPE #1	MAPE #2	MAPE #3	MAPE #4	MAPE #5	MAPE #6	MAPE #7	MAPE #8	MAPE #9	MAPE #10	MAPE #11	MAPE #12
PROTECTION—SAFETY EQUIPMENT & SERVICES												
MSA Safety Incorporated (MSA)	NAIVE3_M3 8.4089	HOLT3_M3 8.4160	HOLT 8.4161	NAÏVE 8.4374	ESM_M3 8.6880	M15_M3 8.7932	ARIMA_M3 8.8381	ESM3_M3 8.8628	ESM 9.1860	MEAN_M3 9.5318	ARIMA 10.8797	MEAN 11.5215
The Eastern Company (EML)	NAIVE3_M3 8.0875	HOLT3_M3 8.0896	ESM3_M3 8.1069	NAÏVE 8.1125	HOLT 8.1398	ESM 8.1436	ESM_M3 8.1828	ARIMA_M3 8.3603	ARIMA 8.4203	M15_M3 8.6068	MEAN_M3 9.0755	MEAN 10.7177
Napco Security Technologies, Inc. (NSSC)	ESM3_M3 10.4241	ESM 10.4491	NAIVE3_M3 10.4873	ESM_M3 10.4929	HOLT3_M3 10.4944	NAÏVE 10.5203	HOLT 10.5330	M15_M3 10.8732	ARIMA_M3 11.6072	MEAN_M3 13.3952	ARIMA 13.7007	MEAN 17.6654
Brady Corporation (BRC)	ESM_M3 7.7896	ARIMA_M3 7.8016	HOLT3_M3 7.8089	NAIVE3_M3 7.8270	M15_M3 7.8541	ESM3_M3 7.8776	HOLT 7.9483	NAÏVE 7.9826	ESM 8.0560	ARIMA 8.0788	MEAN_M3 8.3184	MEAN 9.8766
Johnson Controls Intl. plc (JCI)	NAIVE3_M3 11.5922	HOLT3_M3 11.5969	NAÏVE 11.6207	HOLT 11.6289	ARIMA_M3 11.6365	ARIMA 11.6515	ESM_M3 11.6585	ESM3_M3 11.6694	ESM 11.7650	M15_M3 11.8721	MEAN_M3 13.3479	MEAN 16.3377
Lakeland Industries, Inc. (LAKE)	NAIVE3_M3 10.8167	HOLT3_M3 10.8234	ESM_M3 10.8420	ESM3_M3 10.8984	NAÏVE 10.9092	HOLT 10.9199	ESM 11.0945	M15_M3 11.2103	MEAN_M3 12.1525	ARIMA_M3 13.2115	MEAN 15.0239	ARIMA 17.1903
Magal Security Systems Ltd. (MAGS)	NAIVE3_M3 12.0191	HOLT3_M3 12.0903	NAÏVE 12.1621	HOLT 12.2657	M15_M3 12.3625	ESM_M3 12.3712	ESM3_M3 12.6207	MEAN_M3 12.9798	ESM 13.0164	ARIMA_M3 15.0817	MEAN 15.4071	ARIMA 20.6161
POLLUTION CONTROL EQUIPMENT & SERVICES												
Donaldson Company, Inc. (DCI)	NAIVE3_M3 6.4646	HOLT3_M3 6.4702	NAÏVE 6.5017	HOLT 6.5077	ESM_M3 6.5317	ESM3_M3 6.5561	ESM 6.6417	M15_M3 6.6862	MEAN_M3 6.8604	ARIMA_M3 7.2927	MEAN 8.2179	ARIMA 9.0341
CECO Environmental Corp. (CECE)	ARIMA 13.7024	HOLT 13.7403	HOLT3_M3 13.7960	NAÏVE 13.8012	ESM 13.8107	NAIVE3_M3 13.8519	ESM3_M3 13.8609	ESM_M3 13.9978	ARIMA_M3 14.0634	M15_M3 14.5592	MEAN_M3 16.0781	MEAN 19.9006
Ecology & Environment, Inc. (EEI)	ESM3_M3 6.6060	HOLT3_M3 6.6134	NAIVE3_M3 6.6157	ESM 6.6201	HOLT 6.6308	NAÏVE 6.6346	ARIMA 6.6408	ESM_M3 6.6601	ARIMA_M3 6.7257	M15_M3 6.9492	MEAN_M3 7.5632	MEAN 9.6600
Tetra Tech, Inc. (TTEK)	ARIMA 10.3585	NAÏVE 10.3670	HOLT 10.3684	ESM 10.4110	NAIVE3_M3 10.4227	HOLT3_M3 10.4252	ESM3_M3 10.4560	ESM_M3 10.5856	ARIMA_M3 10.6792	M15_M3 11.0362	MEAN_M3 11.0868	MEAN 12.7417
Appliance Recycling Centers of America (ARCI)	NAÏVE 18.7880	HOLT 18.9017	NAIVE3_M3 19.0764	HOLT3_M3 19.1991	ARIMA 19.2445	ARIMA_M3 19.7694	M15_M3 20.8259	ESM_M3 21.0652	ESM3_M3 21.7402	ESM 22.7908	MEAN_M3 22.8728	MEAN 28.0183
Fuel Tech, Inc. (FTEK)	HOLT 13.6975	HOLT3_M3 13.7028	NAIVE3_M3 13.7107	NAÏVE 13.7109	ARIMA_M3 14.1639	M15_M3 14.4335	ESM_M3 14.7242	ARIMA 15.0240	ESM3_M3 15.1737	MEAN_M3 15.5774	ESM 15.8090	MEAN 19.7447

Table C-36: Industrial Products Sector M3 RMSE Accuracy Results

SECURITY	RMSE #1	RMSE #2	RMSE #3	RMSE #4	RMSE #5	RMSE #6	RMSE #7	RMSE #8	RMSE #9	RMSE #10	RMSE #11	RMSE #12
GENERAL INDUSTRIAL MACHINERY												
Ingersoll-Rand Plc (IR)	ARIMA_M3	ESM_M3	HOLT3_M3	NAIVE3_M3	ESM3_M3	M15_M3	ARIMA	HOLT	NAÏVE	ESM	MEAN_M3	MEAN
	0.0571	0.0578	0.0580	0.0580	0.0581	0.0583	0.0583	0.0585	0.0585	0.0588	0.0632	0.0758
Dover Corporation (DOV)	HOLT3_M3	NAIVE3_M3	HOLT	ESM_M3	NAÏVE	ESM3_M3	ARIMA_M3	ESM	M15_M3	ARIMA	MEAN_M3	MEAN
	0.0544	0.0545	0.0545	0.0546	0.0546	0.0547	0.0551	0.0552	0.0555	0.0564	0.0572	0.0670
Nordson Corporation (NDSN)	ESM	ESM3_M3	HOLT	NAÏVE	HOLT3_M3	NAIVE3_M3	ESM_M3	M15_M3	ARIMA_M3	MEAN_M3	MEAN	ARIMA
	0.0764	0.0765	0.0766	0.0766	0.0768	0.0768	0.0771	0.0795	0.0803	0.0816	0.0978	0.1323
Graco Inc. (GGG)	HOLT	NAÏVE	ESM_M3	HOLT3_M3	NAIVE3_M3	ESM3_M3	ESM	M15_M3	MEAN_M3	ARIMA_M3	MEAN	ARIMA
	0.0233	0.0234	0.0234	0.0234	0.0234	0.0235	0.0240	0.0241	0.0260	0.0267	0.0304	0.0445
Illinois Tools Works, Inc. (ITW)	ARIMA_M3	ESM3_M3	ESM_M3	ESM	HOLT	HOLT3_M3	NAIVE3_M3	NAÏVE	M15_M3	MEAN_M3	ARIMA	MEAN
	0.0714	0.0726	0.0727	0.0735	0.0740	0.0740	0.0741	0.0741	0.0759	0.0827	0.0941	0.0989
Parker-Hannifn Corporation (PH)	NAIVE3_M3	NAÏVE	HOLT3_M3	HOLT	ARIMA_M3	M15_M3	ARIMA	ESM_M3	MEAN_M3	ESM3_M3	ESM	MEAN
	0.1202	0.1204	0.1204	0.1208	0.1214	0.1219	0.1230	0.1231	0.1231	0.1250	0.1278	0.1448
Roper Technologies, Inc. (ROP)	ESM_M3	NAIVE3_M3	HOLT3_M3	NAÏVE	ESM3_M3	HOLT	M15_M3	MEAN_M3	ESM	ARIMA_M3	MEAN	ARIMA
	0.1165	0.1182	0.1182	0.1186	0.1187	0.1187	0.1197	0.1222	0.1234	0.1292	0.1392	0.2358
ELECTRICAL MACHINERY												
Emerson Electric Co (EMR)	HOLT3_M3	ARIMA_M3	ESM_M3	M15_M3	NAIVE3_M3	HOLT	ESM3_M3	NAÏVE	ARIMA	ESM	MEAN_M3	MEAN
	0.0561	0.0561	0.0562	0.0562	0.0562	0.0566	0.0567	0.0568	0.0571	0.0576	0.0584	0.0687
Eaton Corporation plc (ETN)	NAIVE3_M3	HOLT3_M3	NAÏVE	HOLT	ESM_M3	ARIMA_M3	ESM3_M3	M15_M3	ESM	MEAN_M3	ARIMA	MEAN
	0.0642	0.0642	0.0644	0.0644	0.0646	0.0646	0.0648	0.0652	0.0654	0.0668	0.0675	0.0776
Franlin Electric Co. Inc. (FELE)	ESM_M3	HOLT3_M3	NAIVE3_M3	ESM3_M3	HOLT	NAÏVE	M15_M3	ESM	MEAN_M3	ARIMA_M3	MEAN	ARIMA
	0.0341	0.0341	0.0342	0.0343	0.0344	0.0345	0.0347	0.0348	0.0357	0.0367	0.0415	0.0515
A. O. Smith Corporation (AOS)	ESM	ESM3_M3	HOLT	NAÏVE	HOLT3_M3	ESM_M3	NAIVE3_M3	M15_M3	MEAN_M3	ARIMA_M3	MEAN	ARIMA
	0.0268	0.0272	0.0275	0.0275	0.0280	0.0280	0.0280	0.0303	0.0343	0.0377	0.0409	0.1059
Regal Beloit Corporation (RBC)	HOLT3_M3	NAIVE3_M3	ARIMA_M3	ESM_M3	HOLT	NAÏVE	M15_M3	ESM3_M3	ARIMA	ESM	MEAN_M3	MEAN
	0.0749	0.0750	0.0751	0.0751	0.0756	0.0756	0.0757	0.0758	0.0763	0.0771	0.0771	0.0893
II-VI Incorporated (IIVI)	HOLT	NAÏVE	HOLT3_M3	NAIVE3_M3	ARIMA	ARIMA_M3	M15_M3	MEAN_M3	ESM_M3	ESM3_M3	ESM	MEAN
	0.0406	0.0406	0.0406	0.0406	0.0407	0.0410	0.0416	0.0421	0.0433	0.0445	0.0461	0.0471
ESCO Technologies, Inc. (ESE)	ESM	ESM3_M3	NAÏVE	HOLT	NAIVE3_M3	HOLT3_M3	ESM_M3	M15_M3	MEAN_M3	ARIMA_M3	MEAN	ARIMA
	0.0551	0.0553	0.0555	0.0555	0.0556	0.0556	0.0557	0.0574	0.0593	0.0654	0.0687	0.1081

APPENDIX C: M3 Forecast Study Results

SECURITY	RMSE #1	RMSE #2	RMSE #3	RMSE #4	RMSE #5	RMSE #6	RMSE #7	RMSE #8	RMSE #9	RMSE #10	RMSE #11	RMSE #12
PROTECTION—SAFETY EQUIPMENT & SERVICES												
MSA Safety Incorporated (MSA)	NAÏVE 0.0605	HOLT 0.0605	NAIVE3_M3 0.0609	HOLT3_M3 0.0610	ESM_M3 0.0625	ESM3_M3 0.0638	M15_M3 0.0642	ARIMA_M3 0.0643	ESM 0.0663	MEAN_M3 0.0669	MEAN 0.0782	ARIMA 0.1023
The Eastern Company (EML)	NAÏVE 0.0300	HOLT 0.0300	HOLT3_M3 0.0301	ESM 0.0301	NAIVE3_M3 0.0302	ESM3_M3 0.0302	ESM_M3 0.0304	M15_M3 0.0313	ARIMA_M3 0.0321	MEAN_M3 0.0327	ARIMA 0.0348	MEAN 0.0372
Napco Security Technologies, Inc. (NSSC)	NAÏVE 0.0096	NAIVE3_M3 0.0096	HOLT3_M3 0.0096	HOLT 0.0097	ESM 0.0097	ESM3_M3 0.0097	ESM_M3 0.0098	M15_M3 0.0101	MEAN_M3 0.0123	ARIMA_M3 0.0140	MEAN 0.0163	ARIMA 0.0253
Brady Corporation (BRC)	HOLT3_M3 0.0409	NAIVE3_M3 0.0410	ARIMA_M3 0.0412	HOLT 0.0412	ESM_M3 0.0412	M15_M3 0.0413	NAÏVE 0.0413	ESM3_M3 0.0415	ARIMA 0.0420	ESM 0.0421	MEAN_M3 0.0439	MEAN 0.0518
Johnson Controls Intl. plc (JCI)	MEAN_M3 0.1426	ESM3_M3 0.1432	ESM 0.1433	NAÏVE 0.1436	HOLT 0.1436	ESM_M3 0.1436	HOLT3_M3 0.1438	NAIVE3_M3 0.1438	ARIMA_M3 0.1446	ARIMA 0.1449	M15_M3 0.1462	MEAN 0.1567
Lakeland Industries, Inc. (LAKE)	ESM_M3 0.0232	NAIVE3_M3 0.0233	HOLT3_M3 0.0233	ESM3_M3 0.0233	NAÏVE 0.0235	M15_M3 0.0235	HOLT 0.0235	ESM 0.0236	MEAN_M3 0.0243	MEAN 0.0281	ARIMA_M3 0.0463	ARIMA 0.1051
Magal Security Systems Ltd. (MAGS)	NAIVE3_M3 0.0202	ESM_M3 0.0204	M15_M3 0.0205	HOLT3_M3 0.0206	NAÏVE 0.0207	ESM3_M3 0.0208	HOLT 0.0214	ESM 0.0215	MEAN_M3 0.0222	MEAN 0.0273	ARIMA_M3 0.0819	ARIMA 0.2167
POLLUTION CONTROL EQUIPMENT & SERVICES												
Donaldson Company, Inc. (DCI)	MEAN_M3 0.0329	NAIVE3_M3 0.0330	HOLT3_M3 0.0330	NAÏVE 0.0331	HOLT 0.0331	ESM_M3 0.0334	ESM3_M3 0.0335	ESM 0.0338	M15_M3 0.0339	ARIMA_M3 0.0341	MEAN 0.0380	ARIMA 0.0446
CECO Environmental Corp. (CECE)	ARIMA 0.0195	NAÏVE 0.0196	NAIVE3_M3 0.0196	HOLT 0.0197	HOLT3_M3 0.0197	ESM 0.0197	ESM3_M3 0.0197	ESM_M3 0.0199	ARIMA_M3 0.0200	M15_M3 0.0206	MEAN_M3 0.0212	MEAN 0.0247
Ecology & Environment, Inc. (EEI)	ESM3_M3 0.0140	HOLT3_M3 0.0140	ESM 0.0140	HOLT 0.0140	NAIVE3_M3 0.0140	NAÏVE 0.0140	ARIMA 0.0140	ESM_M3 0.0141	ARIMA_M3 0.0142	M15_M3 0.0146	MEAN_M3 0.0164	MEAN 0.0211
Tetra Tech, Inc. (TTEK)	NAÏVE 0.0468	HOLT 0.0469	ESM 0.0470	NAIVE3_M3 0.0471	HOLT3_M3 0.0472	ESM3_M3 0.0472	ARIMA 0.0474	ESM_M3 0.0476	ARIMA_M3 0.0482	M15_M3 0.0492	MEAN_M3 0.0492	MEAN 0.0549
Appliance Recycling Centers of America (ARCI)	HOLT3_M3 0.0122	ARIMA_M3 0.0122	NAIVE3_M3 0.0122	HOLT 0.0122	ARIMA 0.0122	NAÏVE 0.0123	M15_M3 0.0124	ESM_M3 0.0125	ESM3_M3 0.0127	MEAN_M3 0.0130	ESM 0.0131	MEAN 0.0153
Fuel Tech, Inc. (FTEK)	NAÏVE 0.0292	HOLT 0.0292	NAIVE3_M3 0.0293	HOLT3_M3 0.0293	ESM3_M3 0.0297	ESM 0.0297	ESM_M3 0.0298	M15_M3 0.0305	MEAN_M3 0.0310	ARIMA_M3 0.0314	MEAN 0.0365	ARIMA 0.0429

C.17. Medical Sector

Table C-37: Medical Sector M3 MAPE Accuracy Results

SECURITY	MAPE #1	MAPE #2	MAPE #3	MAPE #4	MAPE #5	MAPE #6	MAPE #7	MAPE #8	MAPE #9	MAPE #10	MAPE #11	MAPE #12
LARGE CAP PHARMACEUTICAL STOCKS												
Johnson & Johnson (JNJ)	ESM_M3	ESM3_M3	HOLT3_M3	NAIVE3_M3	HOLT	NAÏVE	ESM	M15_M3	MEAN_M3	ARIMA_M3	MEAN	ARIMA
	4.4387	4.4548	4.4568	4.4571	4.4926	4.4932	4.5077	4.5326	4.9049	5.7316	5.9657	8.6005
Pfizer Inc. (PFE)	NAIVE3_M3	HOLT3_M3	NAÏVE	HOLT	ESM_M3	ESM3_M3	M15_M3	ESM	MEAN_M3	ARIMA_M3	ARIMA	MEAN
	5.8127	5.8128	5.8151	5.8152	5.9481	6.0320	6.0429	6.1831	6.3121	6.3915	7.5626	7.5659
Merck & Co, Inc. (MRK)	HOLT	NAÏVE	ESM	HOLT3_M3	NAIVE3_M3	ESM3_M3	ARIMA	ESM_M3	ARIMA_M3	M15_M3	MEAN_M3	MEAN
	6.4003	6.4005	6.4908	6.5191	6.5193	6.5567	6.7005	6.7334	6.9536	7.2929	7.8882	9.3193
Eli Lilly and Company (LLY)	NAÏVE	HOLT	NAIVE3_M3	HOLT3_M3	ESM3_M3	ESM	ESM_M3	MEAN_M3	M15_M3	ARIMA_M3	MEAN	ARIMA
	6.2333	6.2335	6.2818	6.2819	6.3931	6.4221	6.4398	6.5277	6.7020	7.2951	7.3197	8.6850
Novo Nordisk A/S (NVO)	ESM3_M3	ESM	ESM_M3	NAIVE3_M3	NAÏVE	HOLT3_M3	HOLT	M15_M3	MEAN_M3	ARIMA_M3	MEAN	ARIMA
	6.6253	6.6636	6.6639	6.7695	6.7830	6.7917	6.8134	6.9948	8.1258	8.7733	10.1549	14.7281
Bristol-Myers Squibb Company (BMY)	NAIVE3_M3	HOLT3_M3	NAÏVE	HOLT	M15_M3	ESM_M3	ARIMA_M3	ESM3_M3	ESM	MEAN_M3	ARIMA	MEAN
	6.8873	6.8877	6.9362	6.9368	7.0015	7.1114	7.1840	7.2490	7.4328	7.5194	8.0529	9.2196
GlaxoSmithKline plc (GSK)	HOLT	NAÏVE	HOLT3_M3	NAIVE3_M3	ARIMA	ARIMA_M3	ESM_M3	ESM3_M3	M15_M3	MEAN_M3	ESM	MEAN
	5.1157	5.1182	5.1206	5.1209	5.1994	5.2319	5.2968	5.3698	5.3699	5.5012	5.5089	6.4856
BIOMEDICAL AND GENETICS STOCKS												
Amgen Inc. (AMGN)	HOLT3_M3	NAIVE3_M3	HOLT	NAÏVE	ESM_M3	M15_M3	ARIMA_M3	ESM3_M3	MEAN_M3	ESM	ARIMA	MEAN
	6.5194	6.5217	6.5226	6.5282	6.8179	6.8253	7.0742	7.1128	7.2472	7.5602	8.5789	8.7022
Celgene Corporation (CELG)	HOLT3_M3	HOLT	NAIVE3_M3	NAÏVE	ESM_M3	M15_M3	ESM3_M3	ESM	ARIMA_M3	MEAN_M3	MEAN	ARIMA
	11.1004	11.1161	11.1236	11.1425	11.4615	11.5830	11.6993	12.1320	12.8784	13.0419	16.6021	18.5271
Bio-Techne Corporation (TECH)	NAIVE3_M3	NAÏVE	M15_M3	HOLT3_M3	HOLT	ESM_M3	MEAN_M3	ESM3_M3	ARIMA_M3	ESM	MEAN	ARIMA
	7.7445	7.8399	7.9179	7.9239	8.0702	8.3009	8.4601	8.8384	9.1497	9.5918	10.3245	11.8157
Biogen Inc. (BIIB)	HOLT	NAÏVE	HOLT3_M3	NAIVE3_M3	ESM_M3	ESM3_M3	ARIMA_M3	ESM	M15_M3	MEAN_M3	ARIMA	MEAN
	10.3891	10.4168	10.5174	10.5391	10.9435	10.9671	11.0492	11.1447	11.3375	11.7477	13.7750	14.0290
Gilead Sciences, Inc. (GILD)	NAÏVE	HOLT	NAIVE3_M3	HOLT3_M3	ESM3_M3	ESM	ESM_M3	M15_M3	MEAN_M3	ARIMA_M3	MEAN	ARIMA
	8.4141	8.4175	8.4922	8.4962	8.5359	8.5570	8.6059	9.0720	9.5173	9.8627	11.3241	15.3191
Vertex Pharmaceuticals Inc. (VRTX)	NAÏVE	HOLT	NAIVE3_M3	ARIMA	HOLT3_M3	ARIMA_M3	ESM_M3	ESM3_M3	ESM	M15_M3	MEAN_M3	MEAN
	14.2582	14.3077	14.3519	14.3713	14.3898	14.5443	14.8604	14.8764	15.0751	15.1962	15.6352	19.4844
Regeneron Pharmaceuticals, Inc. (REGN)	HOLT	ESM	NAÏVE	HOLT3_M3	ESM3_M3	NAIVE3_M3	ESM_M3	M15_M3	MEAN_M3	ARIMA_M3	MEAN	ARIMA
	14.3914	14.4472	14.4862	14.6409	14.7171	14.7614	15.1777	16.5377	16.5551	17.0258	19.0453	22.3456

APPENDIX C: M3 Forecast Study Results

SECURITY	MAPE #1	MAPE #2	MAPE #3	MAPE #4	MAPE #5	MAPE #6	MAPE #7	MAPE #8	MAPE #9	MAPE #10	MAPE #11	MAPE #12
MEDICAL PRODUCTS MANUFACTURING												
Abbott Laboratories (ABT)	ESM 5.3968	ESM3_M3 5.3984	ESM_M3 5.4784	HOLT 5.4902	NAÏVE 5.4937	HOLT3_M3 5.5094	NAIVE3_M3 5.5105	M15_M3 5.8141	ARIMA_M3 5.8454	ARIMA 6.0480	MEAN_M3 6.1770	MEAN 7.4354
Stryker Corporation (SYK)	ESM3_M3 6.3778	ESM_M3 6.3853	HOLT 6.5232	NAÏVE 6.5331	HOLT3_M3 6.5478	NAIVE3_M3 6.5548	ESM 6.5564	M15_M3 6.8565	MEAN_M3 7.4558	ARIMA_M3 7.9244	MEAN 8.8462	ARIMA 13.2360
Bio-Rad Laboratories, Inc. (BIO)	ESM3_M3 8.0235	ESM 8.0530	ESM_M3 8.0778	NAIVE3_M3 8.1825	HOLT3_M3 8.1825	NAÏVE 8.2048	HOLT 8.2089	M15_M3 8.4293	ARIMA_M3 8.6936	MEAN_M3 8.8303	MEAN 10.4944	ARIMA 10.8066
Hill-Rom Holdings, Inc. (HRC)	ESM 7.8473	ESM3_M3 7.8759	NAÏVE 7.8911	HOLT 7.8921	ARIMA 7.9036	HOLT3_M3 7.9234	NAIVE3_M3 7.9242	ESM_M3 7.9552	ARIMA_M3 8.0547	M15_M3 8.2481	MEAN_M3 8.9084	MEAN 10.8332
Medtronic plc (MDT)	HOLT 6.0722	NAÏVE 6.0723	HOLT3_M3 6.0864	NAIVE3_M3 6.0864	ESM_M3 6.1189	ESM3_M3 6.1559	ESM 6.2940	M15_M3 6.3670	MEAN_M3 6.5710	ARIMA_M3 7.0982	MEAN 7.8621	ARIMA 9.2953
Baxter International Inc. (BAX)	ESM3_M3 6.2358	ESM 6.2498	ESM_M3 6.2697	NAIVE3_M3 6.2849	HOLT3_M3 6.2855	NAÏVE 6.3121	HOLT 6.3127	M15_M3 6.4603	ARIMA_M3 7.0197	MEAN_M3 7.1299	ARIMA 8.3735	MEAN 8.8698
Boston Scientific Corporation (BSX)	NAÏVE 9.3019	ESM 9.3026	HOLT 9.3028	ARIMA 9.3236	ESM3_M3 9.3556	NAIVE3_M3 9.3625	HOLT3_M3 9.3631	ESM_M3 9.4628	ARIMA_M3 9.5510	M15_M3 9.8325	MEAN_M3 10.7315	MEAN 13.3555
MEDICAL INSTRUMENTS MANUFACTURING												
Thermo Fisher Scientific Inc. (TMO)	ESM 7.4017	ESM3_M3 7.4373	ESM_M3 7.5402	NAÏVE 7.6494	HOLT 7.6594	NAIVE3_M3 7.6664	HOLT3_M3 7.6722	M15_M3 7.9264	ARIMA_M3 8.0384	MEAN_M3 8.2285	ARIMA 9.1425	MEAN 9.9160
Varian Medical Systems, Inc. (VAR)	NAÏVE 8.0680	NAIVE3_M3 8.0812	HOLT3_M3 8.1330	HOLT 8.1403	ESM_M3 8.1604	ESM3_M3 8.1607	ESM 8.2002	M15_M3 8.3039	MEAN_M3 8.7417	ARIMA_M3 9.1227	MEAN 10.6906	ARIMA 12.3175
ABIOMED, Inc. (ABMD)	NAIVE3_M3 17.0683	HOLT3_M3 17.0849	NAÏVE 17.1345	HOLT 17.1567	ESM_M3 17.1914	ESM3_M3 17.2048	ESM 17.3386	M15_M3 17.4668	ARIMA_M3 17.9083	MEAN_M3 18.0872	ARIMA 21.3786	MEAN 21.9347
Teleflex Inforporated (TFX)	ESM3_M3 7.0275	ESM_M3 7.0427	ESM 7.0899	NAIVE3_M3 7.1035	HOLT3_M3 7.1325	NAÏVE 7.1694	HOLT 7.2375	M15_M3 7.2475	ARIMA_M3 7.3044	MEAN_M3 7.8014	ARIMA 8.3606	MEAN 9.7989
Hologic, Inc. (HOLX)	NAIVE3_M3 12.3846	NAÏVE 12.3865	HOLT3_M3 12.4286	HOLT 12.4473	ESM_M3 12.5891	ESM3_M3 12.6127	ESM 12.7158	M15_M3 12.7315	MEAN_M3 13.0149	ARIMA_M3 13.6644	MEAN 14.8097	ARIMA 16.5273
IDEXX Laboratories (IDXX)	ESM3_M3 8.5380	ESM 8.5667	ESM_M3 8.6638	HOLT3_M3 9.0172	NAIVE3_M3 9.0178	HOLT 9.0677	NAÏVE 9.0692	M15_M3 9.2207	ARIMA_M3 9.3904	MEAN_M3 9.6449	ARIMA 10.7554	MEAN 11.0593
STERIS plc (STE)	ARIMA 6.8506	NAÏVE 6.8512	HOLT 6.8539	ESM3_M3 6.9124	NAIVE3_M3 6.9234	HOLT3_M3 6.9267	ESM 6.9464	ESM_M3 6.9960	ARIMA_M3 7.1656	M15_M3 7.5717	MEAN_M3 7.6084	MEAN 8.7946

Table C-38: Medical Sector M3 RMSE Accuracy Results

SECURITY	RMSE #1	RMSE #2	RMSE #3	RMSE #4	RMSE #5	RMSE #6	RMSE #7	RMSE #8	RMSE #9	RMSE #10	RMSE #11	RMSE #12
LARGE CAP PHARMACEUTICAL STOCKS												
Johnson & Johnson (JNJ)	ESM3_M3 0.0622	ESM_M3 0.0623	ESM 0.0625	HOLT3_M3 0.0630	NAIVE3_M3 0.0630	HOLT 0.0633	NAÏVE 0.0633	M15_M3 0.0641	MEAN_M3 0.0676	MEAN 0.0806	ARIMA_M3 0.0924	ARIMA 0.1966
Pfizer Inc. (PFE)	NAÏVE 0.0324	HOLT 0.0324	NAIVE3_M3 0.0325	HOLT3_M3 0.0325	ESM_M3 0.0332	ESM3_M3 0.0334	ESM 0.0340	M15_M3 0.0342	MEAN_M3 0.0345	MEAN 0.0396	ARIMA_M3 0.0434	ARIMA 0.0804
Merck & Co, Inc. (MRK)	HOLT 0.0665	NAÏVE 0.0665	HOLT3_M3 0.0669	NAIVE3_M3 0.0669	ESM3_M3 0.0687	ESM_M3 0.0688	ESM 0.0695	M15_M3 0.0718	ARIMA_M3 0.0726	MEAN_M3 0.0733	ARIMA 0.0796	MEAN 0.0836
Eli Lilly and Company (LLY)	NAIVE3_M3 0.0836	HOLT3_M3 0.0836	NAÏVE 0.0841	HOLT 0.0841	MEAN_M3 0.0851	M15_M3 0.0855	ESM_M3 0.0859	ESM3_M3 0.0876	ESM 0.0903	MEAN 0.0958	ARIMA_M3 0.1498	ARIMA 0.3222
Novo Nordisk A/S (NVO)	ESM 0.0326	ESM3_M3 0.0328	NAÏVE 0.0331	HOLT 0.0331	ESM_M3 0.0334	NAIVE3_M3 0.0334	HOLT3_M3 0.0334	M15_M3 0.0354	MEAN_M3 0.0409	MEAN 0.0502	ARIMA_M3 0.0517	ARIMA 0.1096
Bristol-Myers Squibb Company (BMY)	M15_M3 0.0602	NAIVE3_M3 0.0611	HOLT3_M3 0.0611	NAÏVE 0.0618	HOLT 0.0619	ESM_M3 0.0630	MEAN_M3 0.0640	ESM3_M3 0.0650	ESM 0.0673	ARIMA_M3 0.0720	MEAN 0.0764	ARIMA 0.1124
GlaxoSmithKlne plc (GSK)	HOLT3_M3 0.0419	NAIVE3_M3 0.0419	HOLT 0.0420	NAÏVE 0.0421	ARIMA_M3 0.0427	ARIMA 0.0432	M15_M3 0.0433	ESM_M3 0.0438	ESM3_M3 0.0451	MEAN_M3 0.0451	ESM 0.0471	MEAN 0.0520
BIOMEDICAL AND GENETICS STOCKS												
Amgen Inc. (AMGN)	NAIVE3_M3 0.1203	HOLT3_M3 0.1203	ESM_M3 0.1206	NAÏVE 0.1209	HOLT 0.1209	MEAN_M3 0.1216	M15_M3 0.1222	ESM3_M3 0.1234	ARIMA_M3 0.1258	ESM 0.1285	MEAN 0.1366	ARIMA 0.1677
Celgene Corporation (CELG)	M15_M3 0.1007	ESM_M3 0.1017	HOLT3_M3 0.1019	NAIVE3_M3 0.1020	MEAN_M3 0.1022	HOLT 0.1030	NAÏVE 0.1031	ESM3_M3 0.1037	ESM 0.1065	MEAN 0.1151	ARIMA_M3 0.1435	ARIMA 0.2742
Bio-Techne Corporation (TECH)	MEAN_M3 0.1121	M15_M3 0.1153	NAIVE3_M3 0.1190	NAÏVE 0.1210	HOLT3_M3 0.1225	ESM_M3 0.1256	HOLT 0.1259	MEAN 0.1277	ESM3_M3 0.1339	ESM 0.1437	ARIMA_M3 0.1633	ARIMA 0.2693
Biogen Inc. (BIIB)	HOLT 0.3067	NAÏVE 0.3068	HOLT3_M3 0.3085	NAIVE3_M3 0.3086	M15_M3 0.3199	MEAN_M3 0.3265	ESM_M3 0.3269	ESM3_M3 0.3365	ESM 0.3498	ARIMA_M3 0.3498	MEAN 0.3733	ARIMA 0.6870
Gilead Sciences, Inc. (GILD)	ESM 0.0666	ESM3_M3 0.0667	NAÏVE 0.0672	HOLT 0.0672	ESM_M3 0.0674	NAIVE3_M3 0.0675	HOLT3_M3 0.0676	M15_M3 0.0704	MEAN_M3 0.0716	MEAN 0.0837	ARIMA_M3 0.0847	ARIMA 0.1810
Vertex Pharmaceuticals Inc. (VRTX)	NAÏVE 0.1483	NAIVE3_M3 0.1485	ARIMA_M3 0.1489	HOLT 0.1492	HOLT3_M3 0.1492	ARIMA 0.1520	M15_M3 0.1556	ESM_M3 0.1558	ESM3_M3 0.1599	MEAN_M3 0.1647	ESM 0.1666	MEAN 0.1972
Regeneron Pharmaceuticals, Inc. (REGN)	ESM_M3 0.4255	ESM3_M3 0.4260	NAIVE3_M3 0.4276	HOLT3_M3 0.4281	ESM 0.4292	NAÏVE 0.4314	HOLT 0.4321	M15_M3 0.4324	MEAN_M3 0.4393	MEAN 0.5017	ARIMA_M3 0.6199	ARIMA 1.3729

SECURITY	RMSE #1	RMSE #2	RMSE #3	RMSE #4	RMSE #5	RMSE #6	RMSE #7	RMSE #8	RMSE #9	RMSE #10	RMSE #11	RMSE #12
MEDICAL PRODUCTS MANUFACTURING												
Abbott Laboratories (ABT)	ESM_M3 0.0312	ESM3_M3 0.0313	NAIVE3_M3 0.0315	HOLT3_M3 0.0315	HOLT 0.0317	NAÏVE 0.0317	ESM 0.0319	M15_M3 0.0322	MEAN_M3 0.0335	ARIMA_M3 0.0391	MEAN 0.0402	ARIMA 0.0660
Stryker Corporation (SYK)	ESM_M3 0.0749	HOLT3_M3 0.0760	NAIVE3_M3 0.0760	HOLT 0.0762	NAÏVE 0.0763	ESM3_M3 0.0765	M15_M3 0.0774	ESM 0.0799	MEAN_M3 0.0837	ARIMA_M3 0.0972	MEAN 0.0983	ARIMA 0.1795
Bio-Rad Laboratories, Inc. (BIO)	ESM_M3 0.1618	ESM3_M3 0.1631	M15_M3 0.1646	HOLT3_M3 0.1654	NAIVE3_M3 0.1654	ESM 0.1661	HOLT 0.1665	NAÏVE 0.1666	MEAN_M3 0.1742	MEAN 0.1982	ARIMA_M3 0.2298	ARIMA 0.3830
Hill-Rom Holdings, Inc. (HRC)	ESM3_M3 0.0548	ARIMA 0.0548	ESM 0.0549	NAÏVE 0.0550	NAIVE3_M3 0.0550	HOLT 0.0550	HOLT3_M3 0.0550	ESM_M3 0.0551	ARIMA_M3 0.0554	M15_M3 0.0565	MEAN_M3 0.0580	MEAN 0.0673
Medtronic plc (MDT)	HOLT3_M3 0.0651	NAIVE3_M3 0.0651	HOLT 0.0652	NAÏVE 0.0652	MEAN_M3 0.0657	ESM_M3 0.0661	M15_M3 0.0669	ESM3_M3 0.0671	ESM 0.0689	MEAN 0.0751	ARIMA_M3 0.0825	ARIMA 0.1510
Baxter International Inc. (BAX)	ESM3_M3 0.0400	ESM 0.0401	ESM_M3 0.0402	NAIVE3_M3 0.0402	HOLT3_M3 0.0402	HOLT 0.0403	NAÏVE 0.0403	M15_M3 0.0409	MEAN_M3 0.0430	ARIMA_M3 0.0457	MEAN 0.0503	ARIMA 0.0690
Boston Scientific Corporation (BSX)	ESM3_M3 0.0302	ESM 0.0302	ESM_M3 0.0303	NAIVE3_M3 0.0303	HOLT3_M3 0.0303	NAÏVE 0.0304	HOLT 0.0304	ARIMA_M3 0.0306	ARIMA 0.0306	M15_M3 0.0312	MEAN_M3 0.0340	MEAN 0.0412
MEDICAL INSTRUMENTS MANUFACTURING												
Thermo Fisher Scientific Inc. (TMO)	ESM3_M3 0.0998	ESM_M3 0.1001	ESM 0.1003	M15_M3 0.1030	NAIVE3_M3 0.1031	HOLT3_M3 0.1031	NAÏVE 0.1037	HOLT 0.1037	MEAN_M3 0.1059	ARIMA_M3 0.1141	MEAN 0.1205	ARIMA 0.1703
Varian Medical Systems, Inc. (VAR)	NAÏVE 0.0715	ESM3_M3 0.0715	HOLT 0.0715	NAIVE3_M3 0.0716	ESM 0.0717	HOLT3_M3 0.0717	ESM_M3 0.0718	M15_M3 0.0738	MEAN_M3 0.0761	ARIMA_M3 0.0768	MEAN 0.0885	ARIMA 0.0968
ABIOMED, Inc. (ABMD)	M15_M3 0.2706	ESM_M3 0.2767	HOLT3_M3 0.2806	ESM3_M3 0.2809	NAIVE3_M3 0.2809	HOLT 0.2854	ESM 0.2857	NAÏVE 0.2857	MEAN_M3 0.2885	MEAN 0.3377	ARIMA_M3 0.5021	ARIMA 1.0287
Teleflex Inforporated (TFX)	ESM 0.1123	ESM3_M3 0.1132	ESM_M3 0.1147	HOLT 0.1150	HOLT3_M3 0.1152	NAÏVE 0.1158	NAIVE3_M3 0.1160	M15_M3 0.1197	ARIMA_M3 0.1316	MEAN_M3 0.1334	MEAN 0.1628	ARIMA 0.2535
Hologic, Inc. (HOLX)	NAÏVE 0.0353	HOLT 0.0355	NAIVE3_M3 0.0356	HOLT3_M3 0.0357	ESM_M3 0.0369	ESM3_M3 0.0371	M15_M3 0.0375	ESM 0.0376	MEAN_M3 0.0388	MEAN 0.0452	ARIMA_M3 0.0489	ARIMA 0.0926
IDEXX Laboratories (IDXX)	ESM3_M3 0.1198	ESM 0.1199	ESM_M3 0.1208	M15_M3 0.1259	NAIVE3_M3 0.1280	HOLT3_M3 0.1280	NAÏVE 0.1291	HOLT 0.1291	MEAN_M3 0.1357	MEAN 0.1539	ARIMA_M3 0.1750	ARIMA 0.2875
STERIS plc (STE)	ESM 0.0488	ESM3_M3 0.0491	ESM_M3 0.0501	NAÏVE 0.0517	HOLT 0.0518	NAIVE3_M3 0.0519	HOLT3_M3 0.0519	ARIMA 0.0520	ARIMA_M3 0.0525	M15_M3 0.0540	MEAN_M3 0.0554	MEAN 0.0636

C.18. Oil and Energy Sector

Table C-39: Oil and Energy Sector M3 MAPE Accuracy Results

SECURITY	MAPE #1	MAPE #2	MAPE #3	MAPE #4	MAPE #5	MAPE #6	MAPE #7	MAPE #8	MAPE #9	MAPE #10	MAPE #11	MAPE #12
OIL & GAS												
Exxon Mobile (XOM)	MEAN_M3 5.1502	HOLT3_M3 5.1617	NAIVE3_M3 5.1696	ESM_M3 5.1774	HOLT 5.1820	NAÏVE 5.1978	ESM3_M3 5.2045	ESM 5.2993	M15_M3 5.3245	ARIMA_M3 5.4386	MEAN 5.8862	ARIMA 6.3991
Chevron Corporation (CVX)	HOLT 5.7559	NAÏVE 5.7608	HOLT3_M3 5.7871	NAIVE3_M3 5.7893	ESM3_M3 5.8778	ESM 5.8903	ESM_M3 5.9209	ARIMA 5.9545	ARIMA_M3 5.9775	M15_M3 6.1234	MEAN_M3 6.1972	MEAN 7.1342
BP p.l.c. (BP)	ARIMA 6.7754	NAÏVE 6.8959	HOLT 6.8968	NAIVE3_M3 6.9225	HOLT3_M3 6.9235	ESM 6.9518	ESM3_M3 6.9588	ARIMA_M3 7.0193	ESM_M3 7.0199	MEAN_M3 7.1213	M15_M3 7.2764	MEAN 8.4503
Royal Dutch Shell plc (RDS-B)	HOLT 6.1793	NAÏVE 6.1832	HOLT3_M3 6.1973	NAIVE3_M3 6.1995	ARIMA 6.2592	ARIMA_M3 6.3274	ESM_M3 6.4068	ESM3_M3 6.4420	M15_M3 6.4698	MEAN_M3 6.4748	ESM 6.5337	MEAN 7.6336
Sasol Limited (SSL)	ESM_M3 8.5093	ESM3_M3 8.5211	HOLT3_M3 8.5560	NAIVE3_M3 8.5676	ESM 8.5914	HOLT 8.6319	M15_M3 8.6371	NAÏVE 8.6487	MEAN_M3 9.4372	ARIMA_M3 9.6844	MEAN 11.4806	ARIMA 11.9193
TOTAL S.A. (TOT)	ARIMA 6.1846	HOLT3_M3 6.3088	NAIVE3_M3 6.3091	ARIMA_M3 6.3194	HOLT 6.3251	NAÏVE 6.3261	ESM_M3 6.4184	ESM3_M3 6.4275	ESM 6.5003	M15_M3 6.5529	MEAN_M3 6.5967	MEAN 7.6109
YPF Sociedad Anonima (YPF)	NAIVE3_M3 11.0656	ESM3_M3 11.0795	HOLT3_M3 11.0923	NAÏVE 11.0983	ESM 11.1224	ARIMA 11.1450	HOLT 11.1504	ESM_M3 11.1532	ARIMA_M3 11.2343	M15_M3 11.5587	MEAN_M3 13.1641	MEAN 17.4219
OIL & GAS EXPLORATION												
Apache Corporation (APA)	HOLT3_M3 9.2489	NAIVE3_M3 9.2569	HOLT 9.2787	NAÏVE 9.2911	ESM3_M3 9.2944	ESM_M3 9.3158	ESM 9.3466	M15_M3 9.5388	ARIMA_M3 9.9798	MEAN_M3 10.1727	ARIMA 12.4211	MEAN 12.5393
Noble Energy, Inc. (NBL)	HOLT 8.7319	ESM 8.7320	NAÏVE 8.7656	ESM3_M3 8.7835	HOLT3_M3 8.7893	NAIVE3_M3 8.8147	ESM_M3 8.9172	M15_M3 9.3788	MEAN_M3 10.3290	ARIMA_M3 10.3454	MEAN 12.5354	ARIMA 13.3009
Murphy Oil Corporation (MUR)	NAIVE3_M3 9.0115	HOLT3_M3 9.0181	NAÏVE 9.0386	ESM3_M3 9.0387	HOLT 9.0487	ESM_M3 9.0573	ESM 9.0755	M15_M3 9.2488	MEAN_M3 9.7049	ARIMA_M3 10.5318	MEAN 11.9599	ARIMA 14.1332
Devon Energy Corporation (DVN)	ESM 9.8857	HOLT 9.8923	ESM3_M3 9.8952	NAÏVE 9.8989	HOLT3_M3 9.9083	NAIVE3_M3 9.9146	ESM_M3 9.9716	M15_M3 10.3131	ARIMA_M3 10.3825	MEAN_M3 11.1943	ARIMA 11.2469	MEAN 14.0379
Anadarko Petrolium Corp. (APC)	ESM 9.3708	ESM3_M3 9.4322	HOLT 9.4627	NAÏVE 9.4664	HOLT3_M3 9.5051	NAIVE3_M3 9.5076	ESM_M3 9.5573	ARIMA 9.6114	ARIMA_M3 9.7398	M15_M3 9.9586	MEAN_M3 11.0958	MEAN 14.0837
EOG Resources, Inc. (EOG)	ESM 8.9947	ESM3_M3 9.0441	ESM_M3 9.1540	HOLT 9.1625	HOLT3_M3 9.1739	NAÏVE 9.1797	NAIVE3_M3 9.1865	M15_M3 9.5400	ARIMA_M3 10.0847	MEAN_M3 10.3593	ARIMA 12.2930	MEAN 12.8960
Cabot Oil & Gas Corporation (COG)	ESM 9.7076	ESM3_M3 9.7157	HOLT 9.7362	HOLT3_M3 9.7588	NAÏVE 9.7960	NAIVE3_M3 9.8098	ESM_M3 9.8233	M15_M3 10.3103	ARIMA_M3 11.1339	MEAN_M3 11.4099	MEAN 14.1170	ARIMA 14.6148

APPENDIX C: M3 Forecast Study Results

SECURITY	MAPE #1	MAPE #2	MAPE #3	MAPE #4	MAPE #5	MAPE #6	MAPE #7	MAPE #8	MAPE #9	MAPE #10	MAPE #11	MAPE #12
OIL & GAS DRILLING												
Helmerich & Payne, Inc. (HP)	ARIMA 11.1277	ESM 11.1445	NAÏVE 11.1480	ESM3_M3 11.1622	NAIVE3_M3 11.1644	HOLT3_M3 11.2456	HOLT 11.2513	ESM_M3 11.2666	ARIMA_M3 11.2982	M15_M3 11.7134	MEAN_M3 13.1750	MEAN 16.9647
Ensco plc (ESV)	HOLT3_M3 12.3743	ESM3_M3 12.3780	NAIVE3_M3 12.3870	HOLT 12.3991	ESM 12.3991	NAÏVE 12.4158	ESM_M3 12.4317	ARIMA_M3 12.5969	ARIMA 12.6636	M15_M3 12.7279	MEAN_M3 13.9805	MEAN 17.3330
Nabors Industries Ltd. (NBR)	HOLT 13.4641	HOLT3_M3 13.5604	NAÏVE 13.5651	NAIVE3_M3 13.6431	ESM 13.7697	ESM3_M3 13.7889	ESM_M3 13.9274	M15_M3 14.4724	ARIMA_M3 14.7362	ARIMA 16.3919	MEAN_M3 17.0061	MEAN 21.8887
Key Energy Services, Inc. (KEG)	ESM 23.3437	ESM3_M3 23.7098	HOLT 24.1263	ARIMA 24.1264	NAÏVE 24.1264	ESM_M3 24.1632	NAIVE3_M3 24.2808	HOLT3_M3 24.2812	ARIMA_M3 24.6997	M15_M3 25.2978	MEAN_M3 30.7380	MEAN 40.1484
Rowan Companies, plc (RDC)	ESM_M3 12.5208	ESM3_M3 12.5266	HOLT3_M3 12.5281	NAIVE3_M3 12.5369	ARIMA_M3 12.5417	ARIMA 12.5752	ESM 12.5810	HOLT 12.5842	NAÏVE 12.5963	M15_M3 12.6630	MEAN_M3 13.2778	MEAN 15.8393
Noble Corporation plc (NE)	HOLT3_M3 12.7371	ESM3_M3 12.7383	HOLT 12.7567	NAIVE3_M3 12.7570	ESM 12.7721	NAÏVE 12.7850	ESM_M3 12.7869	M15_M3 13.1143	ARIMA_M3 13.4172	MEAN_M3 13.8176	ARIMA 14.2772	MEAN 16.7283
Transocean Ltd. (RIG)	ARIMA 12.9505	HOLT 12.9669	ESM 12.9698	NAÏVE 12.9698	HOLT3_M3 13.0346	NAIVE3_M3 13.0364	ESM3_M3 13.0364	ESM_M3 13.1867	ARIMA_M3 13.3017	M15_M3 13.7264	MEAN_M3 15.0056	MEAN 19.1383
PRODUCTION & PIPELINE												
The Williams Companies, Inc. (WMB)	HOLT 13.2952	NAÏVE 13.2976	ESM 13.3283	HOLT3_M3 13.3770	NAIVE3_M3 13.3787	ESM3_M3 13.3868	ARIMA 13.4733	ESM_M3 13.5971	ARIMA_M3 13.8011	M15_M3 14.2602	MEAN_M3 19.0593	MEAN 26.0318
TransCanada Corporation (TRP)	ESM_M3 5.2842	NAIVE3_M3 5.2859	HOLT3_M3 5.2885	ESM3_M3 5.2906	ARIMA_M3 5.2994	NAÏVE 5.3254	HOLT 5.3291	ESM 5.3319	ARIMA 5.3479	M15_M3 5.3964	MEAN_M3 6.2700	MEAN 7.9742
Enbridge Inc. (ENB)	NAÏVE 5.2819	NAIVE3_M3 5.2865	ESM3_M3 5.2909	ESM_M3 5.3133	HOLT3_M3 5.3313	HOLT 5.3428	ESM 5.3586	M15_M3 5.5742	ARIMA_M3 5.9617	MEAN_M3 6.2502	MEAN 7.7321	ARIMA 11.2827
OIL FIELD MACHINERY & EQUIPMENT												
Weatherford International plc (WFT)	HOLT 17.8523	ESM3_M3 17.8549	NAÏVE 17.8585	ESM 17.8587	HOLT3_M3 17.8643	NAIVE3_M3 17.8692	ESM_M3 17.9285	M15_M3 18.2737	ARIMA_M3 18.5270	MEAN_M3 19.8700	ARIMA 20.2118	MEAN 24.0406
McDermott International, Inc. (MDR)	HOLT 18.3717	NAÏVE 18.5631	HOLT3_M3 18.6520	NAIVE3_M3 18.8358	ESM 18.8939	ESM3_M3 19.0618	ARIMA 19.2147	ARIMA_M3 19.2976	ESM_M3 19.3298	M15_M3 20.1335	MEAN_M3 23.6099	MEAN 30.3820
Matrix Service Company (MTRX)	NAÏVE 15.6604	HOLT 15.6867	ESM 15.7456	NAIVE3_M3 15.8646	HOLT3_M3 15.8854	ESM3_M3 15.9285	ESM_M3 16.2008	M15_M3 16.9442	ARIMA_M3 17.0774	ARIMA 17.6675	MEAN_M3 18.5200	MEAN 22.6869
ION Geophysical Corporation (IO)	NAIVE3_M3 21.7564	NAÏVE 21.7621	ARIMA 21.7676	HOLT3_M3 21.8032	HOLT 21.8264	ARIMA_M3 21.9771	M15_M3 22.4713	ESM_M3 23.2842	ESM3_M3 23.9294	ESM 24.7555	MEAN_M3 27.5835	MEAN 35.3918
Superior Energy Services, Inc. (SPN)	NAIVE3_M3 13.3684	HOLT3_M3 13.3855	NAÏVE 13.4128	HOLT 13.4355	ESM3_M3 13.4750	ESM 13.5373	ESM_M3 13.5474	M15_M3 14.0271	ARIMA_M3 14.3080	MEAN_M3 15.2415	ARIMA 15.5016	MEAN 18.9740

Table C-40: Oil and Energy Sector M3 RMSE Accuracy Results

SECURITY	RMSE #1	RMSE #2	RMSE #3	RMSE #4	RMSE #5	RMSE #6	RMSE #7	RMSE #8	RMSE #9	RMSE #10	RMSE #11	RMSE #12
OIL & GAS												
Exxon Mobile (XOM)	HOLT3_M3 0.0669	NAIVE3_M3 0.0670	HOLT 0.0671	MEAN_M3 0.0673	NAÏVE 0.0673	ESM_M3 0.0675	ESM3_M3 0.0683	M15_M3 0.0685	ESM 0.0699	ARIMA_M3 0.0726	MEAN 0.0760	ARIMA 0.0951
Chevron Corporation (CVX)	HOLT 0.0898	NAÏVE 0.0899	HOLT3_M3 0.0903	NAIVE3_M3 0.0903	ESM 0.0913	ESM3_M3 0.0913	ESM_M3 0.0919	MEAN_M3 0.0935	ARIMA_M3 0.0939	M15_M3 0.0945	ARIMA 0.0978	MEAN 0.1059
BP p.l.c. (BP)	ARIMA 0.0649	ARIMA_M3 0.0660	NAIVE3_M3 0.0661	HOLT3_M3 0.0661	NAÏVE 0.0662	HOLT 0.0662	MEAN_M3 0.0665	ESM_M3 0.0667	ESM3_M3 0.0667	ESM 0.0671	M15_M3 0.0678	MEAN 0.0763
Royal Dutch Shell plc (RDS-B)	HOLT3_M3 0.0723	NAIVE3_M3 0.0723	HOLT 0.0723	NAÏVE 0.0724	ARIMA_M3 0.0735	ARIMA 0.0736	ESM_M3 0.0743	M15_M3 0.0746	ESM3_M3 0.0751	MEAN_M3 0.0761	ESM 0.0765	MEAN 0.0877
Sasol Limited (SSL)	ESM_M3 0.0596	HOLT3_M3 0.0597	M15_M3 0.0597	NAIVE3_M3 0.0599	ESM3_M3 0.0599	HOLT 0.0602	NAÏVE 0.0605	ESM 0.0605	MEAN_M3 0.0672	ARIMA_M3 0.0748	MEAN 0.0807	ARIMA 0.1178
TOTAL S.A. (TOT)	ARIMA 0.0657	ARIMA_M3 0.0670	HOLT3_M3 0.0675	NAIVE3_M3 0.0675	HOLT 0.0677	NAÏVE 0.0678	M15_M3 0.0689	ESM_M3 0.0689	ESM3_M3 0.0697	MEAN_M3 0.0706	ESM 0.0709	MEAN 0.0803
YPF Sociedad Anonima (YPF)	NAIVE3_M3 0.0654	ESM3_M3 0.0655	NAÏVE 0.0655	HOLT3_M3 0.0656	ESM 0.0657	ARIMA 0.0657	ESM_M3 0.0657	HOLT 0.0657	ARIMA_M3 0.0660	M15_M3 0.0672	MEAN_M3 0.0707	MEAN 0.0857
OIL & GAS EXPLORATION												
Apache Corporation (APA)	HOLT 0.1304	HOLT3_M3 0.1305	NAÏVE 0.1309	NAIVE3_M3 0.1309	ESM3_M3 0.1313	ESM 0.1316	ESM_M3 0.1318	M15_M3 0.1345	MEAN_M3 0.1442	ARIMA_M3 0.1473	MEAN 0.1687	ARIMA 0.2420
Noble Energy, Inc. (NBL)	HOLT 0.0618	NAÏVE 0.0619	HOLT3_M3 0.0620	NAIVE3_M3 0.0621	ESM 0.0622	ESM3_M3 0.0623	ESM_M3 0.0627	M15_M3 0.0642	MEAN_M3 0.0673	MEAN 0.0782	ARIMA_M3 0.0877	ARIMA 0.1598
Murphy Oil Corporation (MUR)	NAIVE3_M3 0.0742	HOLT3_M3 0.0743	ESM_M3 0.0746	NAÏVE 0.0746	ESM3_M3 0.0748	HOLT 0.0748	ESM 0.0753	M15_M3 0.0754	MEAN_M3 0.0821	MEAN 0.0981	ARIMA_M3 0.1108	ARIMA 0.2153
Devon Energy Corporation (DVN)	HOLT 0.1036	NAÏVE 0.1038	HOLT3_M3 0.1041	ESM 0.1042	NAIVE3_M3 0.1043	ESM3_M3 0.1045	ESM_M3 0.1053	M15_M3 0.1084	MEAN_M3 0.1177	ARIMA_M3 0.1190	MEAN 0.1390	ARIMA 0.1601
Anadarko Petrolium Corp. (APC)	ESM 0.1026	ESM3_M3 0.1031	ESM_M3 0.1041	HOLT 0.1044	NAÏVE 0.1044	HOLT3_M3 0.1045	NAIVE3_M3 0.1045	ARIMA 0.1053	ARIMA_M3 0.1060	M15_M3 0.1079	MEAN_M3 0.1170	MEAN 0.1420
EOG Resources, Inc. (EOG)	ESM3_M3 0.0897	ESM 0.0899	ESM_M3 0.0900	HOLT3_M3 0.0910	NAIVE3_M3 0.0911	HOLT 0.0915	NAÏVE 0.0916	M15_M3 0.0922	ARIMA_M3 0.0984	MEAN_M3 0.0988	MEAN 0.1174	ARIMA 0.1313
Cabot Oil & Gas Corporation (COG)	ESM 0.0277	ESM3_M3 0.0278	HOLT 0.0281	ESM_M3 0.0282	NAÏVE 0.0282	HOLT3_M3 0.0283	NAIVE3_M3 0.0284	M15_M3 0.0297	ARIMA_M3 0.0328	MEAN_M3 0.0329	MEAN 0.0398	ARIMA 0.0502

APPENDIX C: M3 Forecast Study Results

SECURITY	RMSE #1	RMSE #2	RMSE #3	RMSE #4	RMSE #5	RMSE #6	RMSE #7	RMSE #8	RMSE #9	RMSE #10	RMSE #11	RMSE #12
OIL & GAS DRILLING												
Helmerich & Payne, Inc. (HP)	ESM 0.1014	NAÏVE 0.1014	ESM3_M3 0.1019	NAIVE3_M3 0.1019	ESM_M3 0.1029	HOLT3_M3 0.1032	HOLT 0.1032	ARIMA_M3 0.1043	M15_M3 0.1062	ARIMA 0.1117	MEAN_M3 0.1167	MEAN 0.1427
Ensco plc (ESV)	ESM 0.0789	HOLT 0.0792	NAÏVE 0.0793	ESM3_M3 0.0795	HOLT3_M3 0.0797	NAIVE3_M3 0.0797	ESM_M3 0.0803	M15_M3 0.0830	ARIMA_M3 0.0844	MEAN_M3 0.0896	ARIMA 0.0902	MEAN 0.1067
Nabors Industries Ltd. (NBR)	HOLT 0.0522	NAÏVE 0.0522	HOLT3_M3 0.0524	NAIVE3_M3 0.0525	ESM 0.0527	ESM3_M3 0.0528	ESM_M3 0.0532	M15_M3 0.0547	ARIMA_M3 0.0574	MEAN_M3 0.0608	MEAN 0.0748	ARIMA 0.0794
Key Energy Services, Inc. (KEG)	HOLT 5.5356	ARIMA 5.5367	NAÏVE 5.5367	ESM 5.5808	HOLT3_M3 5.5822	NAIVE3_M3 5.5829	ESM3_M3 5.6238	ESM_M3 5.6856	ARIMA_M3 5.6973	M15_M3 5.8629	MEAN_M3 6.1031	MEAN 7.2289
Rowan Companies, plc (RDC)	ESM_M3 0.0553	ESM3_M3 0.0553	HOLT3_M3 0.0554	NAIVE3_M3 0.0554	ARIMA_M3 0.0555	ARIMA 0.0556	ESM 0.0556	HOLT 0.0557	NAÏVE 0.0558	M15_M3 0.0562	MEAN_M3 0.0604	MEAN 0.0719
Noble Corporation plc (NE)	HOLT 0.0577	NAÏVE 0.0578	ESM 0.0580	HOLT3_M3 0.0581	NAIVE3_M3 0.0582	ESM3_M3 0.0583	ESM_M3 0.0589	M15_M3 0.0605	MEAN_M3 0.0622	ARIMA_M3 0.0662	MEAN 0.0703	ARIMA 0.0818
Transocean Ltd. (RIG)	HOLT 0.1457	ESM 0.1457	NAÏVE 0.1457	HOLT3_M3 0.1478	ESM3_M3 0.1478	NAIVE3_M3 0.1478	ARIMA 0.1480	ESM_M3 0.1507	ARIMA_M3 0.1533	M15_M3 0.1584	MEAN_M3 0.1738	MEAN 0.2100
PRODUCTION & PIPELINE												
The Williams Companies, Inc. (WMB)	HOLT 0.0514	NAÏVE 0.0514	HOLT3_M3 0.0516	ESM 0.0516	NAIVE3_M3 0.0516	ESM3_M3 0.0518	ARIMA 0.0522	ESM_M3 0.0523	ARIMA_M3 0.0530	M15_M3 0.0546	MEAN_M3 0.0608	MEAN 0.0773
TransCanada Corporation (TRP)	NAÏVE 0.0333	HOLT 0.0333	ESM 0.0334	NAIVE3_M3 0.0335	HOLT3_M3 0.0335	ARIMA 0.0335	ESM3_M3 0.0336	ESM_M3 0.0339	ARIMA_M3 0.0341	M15_M3 0.0350	MEAN_M3 0.0382	MEAN 0.0457
Enbridge Inc. (ENB)	NAÏVE 0.0260	HOLT 0.0261	NAIVE3_M3 0.0262	HOLT3_M3 0.0262	ESM_M3 0.0273	ESM3_M3 0.0276	M15_M3 0.0280	ESM 0.0284	MEAN_M3 0.0312	ARIMA_M3 0.0324	MEAN 0.0382	ARIMA 0.0747
OIL FIELD MACHINERY & EQUIPMENT												
Weatherford International plc (WFT)	HOLT 0.0462	NAÏVE 0.0462	HOLT3_M3 0.0463	NAIVE3_M3 0.0463	ESM3_M3 0.0470	ESM_M3 0.0470	ESM 0.0472	M15_M3 0.0478	MEAN_M3 0.0534	MEAN 0.0643	ARIMA_M3 0.0644	ARIMA 0.1156
McDermott International, Inc. (MDR)	ARIMA_M3 0.1000	HOLT 0.1029	NAÏVE 0.1043	HOLT3_M3 0.1043	NAIVE3_M3 0.1057	ESM 0.1062	ESM3_M3 0.1071	ESM_M3 0.1084	M15_M3 0.1120	MEAN_M3 0.1249	ARIMA 0.1268	MEAN 0.1501
Matrix Service Company (MTRX)	NAÏVE 0.0423	HOLT 0.0423	ESM 0.0424	NAIVE3_M3 0.0426	HOLT3_M3 0.0426	ESM3_M3 0.0427	ESM_M3 0.0431	M15_M3 0.0443	MEAN_M3 0.0472	ARIMA_M3 0.0475	MEAN 0.0552	ARIMA 0.0623
ION Geophysical Corporation (IO)	ARIMA 0.3652	NAÏVE 0.3655	HOLT 0.3663	NAIVE3_M3 0.3672	HOLT3_M3 0.3678	ARIMA_M3 0.3721	ESM3_M3 0.3739	ESM_M3 0.3745	ESM 0.3748	M15_M3 0.3799	MEAN_M3 0.4003	MEAN 0.4645
Superior Energy Services, Inc. (SPN)	NAÏVE 0.0572	ESM 0.0572	HOLT 0.0572	NAIVE3_M3 0.0575	ESM3_M3 0.0575	HOLT3_M3 0.0575	ESM_M3 0.0581	M15_M3 0.0603	MEAN_M3 0.0664	ARIMA_M3 0.0688	MEAN 0.0795	ARIMA 0.1134

C.19. Retail Sector

Table C-41: Retail Sector M3 MAPE Accuracy Results

SECURITY	MAPE #1	MAPE #2	MAPE #3	MAPE #4	MAPE #5	MAPE #6	MAPE #7	MAPE #8	MAPE #9	MAPE #10	MAPE #11	MAPE #12
FOOD & RESTAURANT												
McDonalds Corporation (MCD)	NAIVE3_M3 5.6185	HOLT3_M3 5.6191	ESM_M3 5.6289	NAÏVE 5.6420	HOLT 5.6425	ESM3_M3 5.6489	ESM 5.7277	M15_M3 5.7428	ARIMA_M3 6.2570	MEAN_M3 6.4944	MEAN 8.0219	ARIMA 8.5068
The Wendy's Company (WEN)	ESM 8.0171	ESM3_M3 8.0397	ARIMA 8.0656	NAÏVE 8.0656	HOLT 8.0669	NAIVE3_M3 8.0807	HOLT3_M3 8.0822	ESM_M3 8.1535	ARIMA_M3 8.2572	M15_M3 8.5871	MEAN_M3 10.2090	MEAN 13.2405
Cracker Barrel Old Country Store, Inc. (CBRL)	NAÏVE 8.5552	HOLT 8.5587	NAIVE3_M3 8.6277	HOLT3_M3 8.6306	ESM3_M3 8.7931	ESM_M3 8.8297	ESM 8.8712	ARIMA_M3 9.0495	M15_M3 9.2021	ARIMA 9.5184	MEAN_M3 9.8569	MEAN 12.2157
Jack in the Box Inc. (JACK)	ESM 7.6010	ESM3_M3 7.6203	NAÏVE 7.6282	HOLT 7.6359	NAIVE3_M3 7.6398	HOLT3_M3 7.6433	ESM_M3 7.6973	M15_M3 7.9752	ARIMA_M3 8.2040	MEAN_M3 9.1530	ARIMA 10.2792	MEAN 11.9245
Dine Brands Global, Inc. (DIN)	NAIVE3_M3 11.0489	HOLT3_M3 11.0634	ESM3_M3 11.0683	NAÏVE 11.0896	HOLT 11.1088	ESM_M3 11.1276	ESM 11.1288	ARIMA 11.1775	ARIMA_M3 11.2312	M15_M3 11.5309	MEAN_M3 13.2347	MEAN 16.9697
Starbucks Corporatino (SBUX)	ESM_M3 8.5003	ESM3_M3 8.6217	HOLT3_M3 8.6281	NAIVE3_M3 8.6282	M15_M3 8.6850	HOLT 8.7547	NAÏVE 8.7556	ESM 8.8917	ARIMA_M3 9.4681	MEAN_M3 9.6667	MEAN 12.1104	ARIMA 12.1675
The Cheescake Factory Inc. (CAKE)	NAIVE3_M3 8.6359	NAÏVE 8.6367	HOLT3_M3 8.6375	HOLT 8.6389	ESM3_M3 8.6935	ESM 8.7145	ESM_M3 8.7807	M15_M3 9.2338	ARIMA_M3 9.7359	MEAN_M3 9.9485	ARIMA 11.1805	MEAN 12.2231
APPAREL & SHOES												
Foot Locker, Inc. (FL)	ESM3_M3 10.0980	NAIVE3_M3 10.1201	ESM 10.1212	HOLT3_M3 10.1229	ESM_M3 10.1334	NAÏVE 10.1507	HOLT 10.1548	ARIMA 10.1669	ARIMA_M3 10.1909	M15_M3 10.4085	MEAN_M3 11.5930	MEAN 14.4031
Nordstrom, Inc. (JWN)	HOLT3_M3 10.3926	HOLT 10.3997	NAIVE3_M3 10.4022	ESM3_M3 10.4052	NAÏVE 10.4150	ESM 10.4250	ESM_M3 10.4745	ARIMA 10.5122	ARIMA_M3 10.5393	M15_M3 10.8810	MEAN_M3 12.1651	MEAN 14.8163
The Gap, Inc. (GPS)	NAIVE3_M3 10.2530	HOLT3_M3 10.2532	NAÏVE 10.3026	HOLT 10.3029	M15_M3 10.4046	ESM_M3 10.5757	ARIMA_M3 10.7386	ESM3_M3 10.8357	MEAN_M3 10.9836	ESM 11.1809	ARIMA 11.6711	MEAN 13.6382
Genesco Inc. (GCO)	ESM3_M3 12.0561	NAIVE3_M3 12.0600	HOLT3_M3 12.0607	ESM 12.0798	NAÏVE 12.0802	HOLT 12.0815	ESM_M3 12.1086	ARIMA_M3 12.4433	M15_M3 12.4464	ARIMA 12.8084	MEAN_M3 13.9405	MEAN 17.5594
L Brands, Inc. (LB)	NAIVE3_M3 11.3582	HOLT3_M3 11.3604	NAÏVE 11.4003	HOLT 11.4035	ESM_M3 11.4153	ARIMA_M3 11.4246	M15_M3 11.4575	ESM3_M3 11.4742	ESM 11.5882	ARIMA 12.0564	MEAN_M3 12.4277	MEAN 14.8944
Ascena Retail Group, Inc. (ASNA)	NAÏVE 12.1061	NAIVE3_M3 12.1278	HOLT 12.1549	HOLT3_M3 12.1659	ESM_M3 12.4448	ARIMA_M3 12.5269	ESM3_M3 12.5405	M15_M3 12.6589	ESM 12.7576	ARIMA 13.2720	MEAN_M3 14.0454	MEAN 17.2427
The Cato Corporation (CATO)	ESM3_M3 9.3076	ESM_M3 9.3428	HOLT3_M3 9.3463	NAIVE3_M3 9.3498	ARIMA 9.3700	ESM 9.3742	HOLT 9.3871	NAÏVE 9.3943	ARIMA_M3 9.4505	M15_M3 9.6921	MEAN_M3 9.9769	MEAN 11.6398

SECURITY	MAPE #1	MAPE #2	MAPE #3	MAPE #4	MAPE #5	MAPE #6	MAPE #7	MAPE #8	MAPE #9	MAPE #10	MAPE #11	MAPE #12
DISCOUNT & VARIETY												
Walmart Inc. (WMT)	HOLT3_M3 5.1425	NAIVE3_M3 5.1447	M15_M3 5.1493	ESM_M3 5.1980	HOLT 5.2027	NAÏVE 5.2061	ESM3_M3 5.3004	MEAN_M3 5.3522	ESM 5.4550	ARIMA_M3 5.7254	MEAN 6.2141	ARIMA 6.9363
Big Lots, Inc. (BIG)	HOLT3_M3 10.7305	NAIVE3_M3 10.7463	ESM3_M3 10.7561	ESM_M3 10.7713	HOLT 10.7854	NAÏVE 10.8011	ARIMA_M3 10.8160	ARIMA 10.8254	ESM 10.8298	M15_M3 11.0573	MEAN_M3 12.5996	MEAN 16.0077
Ross Stores, Inc. (ROST)	M15_M3 8.4346	ESM_M3 8.4871	HOLT3_M3 8.5757	NAIVE3_M3 8.5938	ESM3_M3 8.6177	HOLT 8.7067	NAÏVE 8.7300	ESM 8.8198	ARIMA_M3 9.4524	MEAN_M3 9.6526	MEAN 11.8985	ARIMA 14.2050
Costco Wholesale Corporation (COST)	ESM3_M3 6.0734	ESM 6.0774	HOLT3_M3 6.1184	NAIVE3_M3 6.1186	ESM_M3 6.1193	HOLT 6.1292	NAÏVE 6.1296	M15_M3 6.3382	ARIMA_M3 6.4814	MEAN_M3 6.7979	ARIMA 7.2504	MEAN 8.2985
The TJX Companies, Inc. (TJX)	ESM3_M3 6.5539	NAIVE3_M3 6.5554	ESM_M3 6.5716	NAÏVE 6.5732	ESM 6.5942	HOLT3_M3 6.5991	HOLT 6.6318	M15_M3 6.7628	ARIMA_M3 7.3431	MEAN_M3 7.6522	MEAN 9.4236	ARIMA 9.5903
Fred's Inc. (FRED)	HOLT 12.5920	ESM 12.6066	NAÏVE 12.6148	HOLT3_M3 12.6561	NAIVE3_M3 12.6725	ESM3_M3 12.6736	ESM_M3 12.7963	M15_M3 13.1845	ARIMA_M3 13.5631	MEAN_M3 14.7789	ARIMA 15.1309	MEAN 18.6206
Kohl's Corporation (KSS)	NAÏVE 7.9716	HOLT 7.9722	NAIVE3_M3 8.0004	HOLT3_M3 8.0008	ESM_M3 8.2924	ESM3_M3 8.3716	ARIMA_M3 8.4695	M15_M3 8.5279	ESM 8.5643	ARIMA 9.2061	MEAN_M3 9.2330	MEAN 11.1445

Table C-42: Retail Sector M3 RMSE Accuracy Results

SECURITY	RMSE #1	RMSE #2	RMSE #3	RMSE #4	RMSE #5	RMSE #6	RMSE #7	RMSE #8	RMSE #9	RMSE #10	RMSE #11	RMSE #12
FOOD & RESTAURANT												
McDonalds Corporation (MCD)	ESM3_M3 0.0720	ESM_M3 0.0721	ESM 0.0725	NAIVE3_M3 0.0726	HOLT3_M3 0.0726	NAÏVE 0.0727	HOLT 0.0727	M15_M3 0.0738	MEAN_M3 0.0803	ARIMA_M3 0.0909	MEAN 0.0959	ARIMA 0.1754
The Wendy's Company (WEN)	ARIMA_M3 0.0249	ESM_M3 0.0249	HOLT3_M3 0.0249	NAIVE3_M3 0.0249	ESM3_M3 0.0250	M15_M3 0.0250	HOLT 0.0251	ARIMA 0.0251	NAÏVE 0.0251	ESM 0.0252	MEAN_M3 0.0285	MEAN 0.0385
Cracker Barrel Old Country Store, Inc. (CBRL)	MEAN_M3 0.1092	NAIVE3_M3 0.1096	HOLT3_M3 0.1096	NAÏVE 0.1097	HOLT 0.1098	ESM_M3 0.1118	M15_M3 0.1122	ESM3_M3 0.1148	ESM 0.1198	ARIMA_M3 0.1206	MEAN 0.1297	ARIMA 0.1668
Jack in the Box Inc. (JACK)	ESM 0.0600	NAÏVE 0.0601	ESM3_M3 0.0603	NAIVE3_M3 0.0604	HOLT 0.0606	HOLT3_M3 0.0607	ESM_M3 0.0608	M15_M3 0.0627	MEAN_M3 0.0705	ARIMA_M3 0.0800	MEAN 0.0878	ARIMA 0.1737
Dine Brands Global, Inc. (DIN)	NAIVE3_M3 0.0931	ESM3_M3 0.0931	HOLT3_M3 0.0932	NAÏVE 0.0932	ESM 0.0933	HOLT 0.0933	ARIMA 0.0937	ESM_M3 0.0941	ARIMA_M3 0.0952	M15_M3 0.0994	MEAN_M3 0.1058	MEAN 0.1286
Starbucks Corporatino (SBUX)	ESM_M3 0.0298	ESM3_M3 0.0308	NAIVE3_M3 0.0308	HOLT3_M3 0.0308	M15_M3 0.0310	NAÏVE 0.0312	HOLT 0.0312	ESM 0.0327	MEAN_M3 0.0352	ARIMA_M3 0.0425	MEAN 0.0432	ARIMA 0.0871
The Cheescake Factory Inc. (CAKE)	NAÏVE 0.0461	HOLT 0.0461	NAIVE3_M3 0.0461	HOLT3_M3 0.0461	ESM3_M3 0.0462	ESM 0.0464	ESM_M3 0.0465	M15_M3 0.0481	MEAN_M3 0.0509	ARIMA_M3 0.0535	MEAN 0.0614	ARIMA 0.0834

SECURITY	RMSE #1	RMSE #2	RMSE #3	RMSE #4	RMSE #5	RMSE #6	RMSE #7	RMSE #8	RMSE #9	RMSE #10	RMSE #11	RMSE #12
APPAREL & SHOES												
Foot Locker, Inc. (FL)	ESM 0.0537	HOLT 0.0541	NAÏVE 0.0541	ESM3_M3 0.0542	ARIMA 0.0545	HOLT3_M3 0.0545	NAIVE3_M3 0.0545	ESM_M3 0.0550	ARIMA_M3 0.0558	M15_M3 0.0574	MEAN_M3 0.0631	MEAN 0.0796
Nordstrom, Inc. (JWN)	HOLT 0.0674	NAÏVE 0.0674	ESM 0.0678	HOLT3_M3 0.0680	NAIVE3_M3 0.0680	ESM3_M3 0.0682	ARIMA 0.0690	ESM_M3 0.0691	ARIMA_M3 0.0695	M15_M3 0.0718	MEAN_M3 0.0751	MEAN 0.0869
The Gap, Inc. (GPS)	NAIVE3_M3 0.0502	HOLT3_M3 0.0502	NAÏVE 0.0505	HOLT 0.0505	M15_M3 0.0506	ESM_M3 0.0517	MEAN_M3 0.0518	ESM3_M3 0.0530	ESM 0.0548	ARIMA_M3 0.0614	MEAN 0.0625	ARIMA 0.0897
Genesco Inc. (GCO)	NAÏVE 0.0784	HOLT 0.0784	NAIVE3_M3 0.0784	HOLT3_M3 0.0784	ESM3_M3 0.0787	ESM 0.0788	ESM_M3 0.0791	M15_M3 0.0812	ARIMA_M3 0.0816	MEAN_M3 0.0857	ARIMA 0.0861	MEAN 0.1039
L Brands, Inc. (LB)	NAIVE3_M3 0.0742	HOLT3_M3 0.0742	NAÏVE 0.0742	HOLT 0.0742	M15_M3 0.0754	ESM_M3 0.0767	ARIMA_M3 0.0771	MEAN_M3 0.0778	ESM3_M3 0.0783	ESM 0.0805	MEAN 0.0906	ARIMA 0.0948
Ascena Retail Group, Inc. (ASNA)	NAÏVE 0.0204	NAIVE3_M3 0.0205	HOLT3_M3 0.0206	HOLT 0.0207	ESM_M3 0.0210	M15_M3 0.0211	ESM3_M3 0.0213	ARIMA_M3 0.0216	ESM 0.0218	MEAN_M3 0.0221	MEAN 0.0263	ARIMA 0.0263
The Cato Corporation (CATO)	HOLT3_M3 0.0353	HOLT 0.0353	NAIVE3_M3 0.0353	NAÏVE 0.0353	ESM3_M3 0.0353	ESM 0.0354	ARIMA 0.0354	ESM_M3 0.0355	ARIMA_M3 0.0357	M15_M3 0.0365	MEAN_M3 0.0381	MEAN 0.0453
DISCOUNT & VARIETY												
Walmart Inc. (WMT)	M15_M3 0.0620	HOLT3_M3 0.0622	NAIVE3_M3 0.0623	HOLT 0.0629	NAÏVE 0.0630	ESM_M3 0.0638	MEAN_M3 0.0639	ESM3_M3 0.0656	ESM 0.0680	MEAN 0.0757	ARIMA_M3 0.0784	ARIMA 0.1285
Big Lots, Inc. (BIG)	HOLT 0.0547	HOLT3_M3 0.0548	NAÏVE 0.0548	NAIVE3_M3 0.0549	ARIMA 0.0549	ESM 0.0549	ESM3_M3 0.0549	ESM_M3 0.0552	ARIMA_M3 0.0556	M15_M3 0.0568	MEAN_M3 0.0609	MEAN 0.0754
Ross Stores, Inc. (ROST)	ESM_M3 0.0400	ESM3_M3 0.0402	HOLT 0.0404	HOLT3_M3 0.0405	NAÏVE 0.0406	NAIVE3_M3 0.0407	ESM 0.0412	M15_M3 0.0418	MEAN_M3 0.0458	ARIMA_M3 0.0551	MEAN 0.0571	ARIMA 0.1185
Costco Wholesale Corporation (COST)	ESM3_M3 0.0927	ESM 0.0928	ESM_M3 0.0935	HOLT 0.0952	NAÏVE 0.0952	HOLT3_M3 0.0952	NAIVE3_M3 0.0952	M15_M3 0.0977	MEAN_M3 0.1058	ARIMA_M3 0.1120	MEAN 0.1273	ARIMA 0.1886
The TJX Companies, Inc. (TJX)	ESM_M3 0.0222	ESM3_M3 0.0224	NAIVE3_M3 0.0224	M15_M3 0.0225	NAÏVE 0.0226	HOLT3_M3 0.0227	ESM 0.0228	HOLT 0.0230	MEAN_M3 0.0244	MEAN 0.0299	ARIMA_M3 0.0325	ARIMA 0.0607
Fred's Inc. (FRED)	HOLT 0.0330	NAÏVE 0.0330	HOLT3_M3 0.0331	NAIVE3_M3 0.0331	ESM3_M3 0.0333	ESM 0.0333	ESM_M3 0.0334	M15_M3 0.0340	MEAN_M3 0.0373	ARIMA_M3 0.0394	MEAN 0.0456	ARIMA 0.0630
Kohl's Corporation (KSS)	NAÏVE 0.0815	HOLT 0.0815	NAIVE3_M3 0.0816	HOLT3_M3 0.0816	ESM_M3 0.0838	M15_M3 0.0848	ESM3_M3 0.0849	ARIMA_M3 0.0860	ESM 0.0871	MEAN_M3 0.0911	MEAN 0.1089	ARIMA 0.1128

C.20. Transportation Sector

Table C-43: Transportation Sector M3 MAPE Accuracy Results

SECURITY	MAPE #1	MAPE #2	MAPE #3	MAPE #4	MAPE #5	MAPE #6	MAPE #7	MAPE #8	MAPE #9	MAPE #10	MAPE #11	MAPE #12
AIRLINE STOCKS												
Southwest Airlines Co. (LUV)	ESM_M3 9.4581	ESM3_M3 9.4918	HOLT3_M3 9.5126	NAIVE3_M3 9.5178	M15_M3 9.5534	HOLT 9.5832	ESM 9.5850	NAÏVE 9.5939	ARIMA_M3 9.6283	MEAN_M3 10.4690	ARIMA 10.7034	MEAN 12.7673
Alaska Air Group, Inc. (ALK)	NAIVE3_M3 10.6775	HOLT3_M3 10.6812	ESM_M3 10.6908	M15_M3 10.7284	ESM3_M3 10.7730	NAÏVE 10.7770	HOLT 10.7844	ESM 10.9217	ARIMA_M3 11.1108	MEAN_M3 11.2889	ARIMA 13.0936	MEAN 14.1162
PHI, Inc. (PHII)	NAIVE3_M3 11.2145	NAÏVE 11.2373	HOLT 11.2435	ESM 11.2435	ARIMA 11.2550	HOLT3_M3 11.2564	ESM3_M3 11.2641	ESM_M3 11.3626	ARIMA_M3 11.4690	M15_M3 11.7910	MEAN_M3 12.6384	MEAN 15.2428
Skywest, Inc. (SKYW)	NAIVE3_M3 12.5320	HOLT3_M3 12.5435	NAÏVE 12.5777	HOLT 12.5924	ESM3_M3 12.6906	ESM_M3 12.6923	ARIMA 12.7242	ARIMA_M3 12.7435	ESM 12.7926	M15_M3 13.0355	MEAN_M3 13.4825	MEAN 16.3400
Bristow Group Inc. (BRS)	HOLT3_M3 13.5396	NAIVE3_M3 13.5397	ARIMA 13.5830	HOLT 13.6140	NAÏVE 13.6143	ESM3_M3 13.6145	ARIMA_M3 13.6302	ESM_M3 13.6342	ESM 13.7028	M15_M3 13.9593	MEAN_M3 15.0421	MEAN 18.8122
Hawaiian Holdings, Inc. (HA)	NAÏVE 15.8612	NAIVE3_M3 15.8828	HOLT 15.8996	HOLT3_M3 15.9873	ESM_M3 16.9829	M15_M3 17.2681	ESM3_M3 17.3175	ARIMA_M3 17.4700	ESM 17.9162	MEAN_M3 18.0609	ARIMA 19.5870	MEAN 22.0332
RAIL STOCKS												
Union Pacific Corporation (UNP)	ESM3_M3 6.6037	ESM_M3 6.6079	ESM 6.6418	HOLT3_M3 6.6886	NAIVE3_M3 6.6917	M15_M3 6.7379	HOLT 6.7539	NAÏVE 6.7582	ARIMA_M3 7.3989	MEAN_M3 7.7080	MEAN 9.4632	ARIMA 10.9407
CSX Coporation (CSX)	ESM 8.4728	ESM3_M3 8.4834	HOLT 8.5170	NAÏVE 8.5242	HOLT3_M3 8.5273	NAIVE3_M3 8.5323	ESM_M3 8.5409	M15_M3 8.7782	ARIMA_M3 8.8254	ARIMA 9.2999	MEAN_M3 9.9633	MEAN 12.3156
Kansas City Southern (KSU)	M15_M3 33.4186	NAIVE3_M3 34.2095	HOLT3_M3 34.2121	ESM_M3 34.5594	NAÏVE 34.6321	HOLT 34.6385	ESM3_M3 35.3716	ESM 36.3483	MEAN_M3 37.9666	MEAN 44.2145	ARIMA_M3 44.6775	ARIMA 64.7951
Norfolk Southern Corporatino (NSC)	ESM3_M3 8.6394	ESM 8.6412	ESM_M3 8.6805	ARIMA_M3 8.7022	HOLT3_M3 8.7202	NAIVE3_M3 8.7240	HOLT 8.7396	NAÏVE 8.7459	ARIMA 8.7537	M15_M3 8.9269	MEAN_M3 9.3650	MEAN 11.1151
Canadian Pacific Railway Limited (CP)	HOLT3_M3 7.9358	NAIVE3_M3 7.9363	HOLT 7.9442	NAÏVE 7.9461	ESM_M3 7.9890	ESM3_M3 7.9914	ESM 8.0541	M15_M3 8.1616	MEAN_M3 9.1146	ARIMA_M3 9.2020	MEAN 11.1816	ARIMA 12.6472

SECURITY	MAPE #1	MAPE #2	MAPE #3	MAPE #4	MAPE #5	MAPE #6	MAPE #7	MAPE #8	MAPE #9	MAPE #10	MAPE #11	MAPE #12
TRUCK TRANSPORTATION STOCKS												
J.. B. Hunt Transport Services, Inc. (JBHT)	ESM3_M3 8.3531	ESM_M3 8.3694	ESM 8.3931	NAIVE3_M3 8.4218	HOLT3_M3 8.4274	NAÏVE 8.4746	HOLT 8.4819	M15_M3 8.5587	MEAN_M3 9.0158	ARIMA_M3 9.9386	MEAN 11.2105	ARIMA 14.8835
Werner Enterrises, Inc. (WERN)	MEAN_M3 7.3843	ARIMA_M3 7.4524	ESM_M3 7.4922	HOLT3_M3 7.5393	ESM3_M3 7.5430	ARIMA 7.5448	NAIVE3_M3 7.5508	M15_M3 7.5946	HOLT 7.6629	ESM 7.6736	NAÏVE 7.6812	MEAN 8.5275
Heartland Express, Inc. (HTLD)	ARIMA_M3 7.0318	ARIMA 7.0709	ESM_M3 7.0735	HOLT3_M3 7.1021	NAIVE3_M3 7.1292	ESM3_M3 7.1302	M15_M3 7.1543	HOLT 7.2143	NAÏVE 7.2593	ESM 7.2595	MEAN_M3 7.2951	MEAN 8.1874
Marten Transport, Ltd. (MRTN)	ESM_M3 8.5560	ESM3_M3 8.5908	NAIVE3_M3 8.6142	HOLT3_M3 8.6340	ESM 8.7045	M15_M3 8.7302	NAÏVE 8.7539	HOLT 8.7729	MEAN_M3 9.2523	ARIMA_M3 10.0263	MEAN 11.0542	ARIMA 13.7303
Old Dominion Freight Line, Inc. (ODFL)	ESM 8.6165	ESM3_M3 8.6787	ESM_M3 8.7992	NAIVE3_M3 8.8452	HOLT 8.8459	NAÏVE 8.8522	HOLT3_M3 8.8527	M15_M3 9.2077	ARIMA_M3 9.8140	MEAN_M3 10.1478	MEAN 12.3136	ARIMA 15.6395
Landstar Systems, Inc. (LSTR)	NAIVE3_M3 7.0047	ESM3_M3 7.0133	ESM_M3 7.0185	HOLT3_M3 7.0393	NAÏVE 7.0471	ESM 7.0889	HOLT 7.0918	M15_M3 7.3008	ARIMA_M3 7.6904	MEAN_M3 7.9430	MEAN 9.7340	ARIMA 10.2701
Forward Air Corporation (FWRD)	ESM_M3 8.3990	ESM3_M3 8.4613	NAIVE3_M3 8.5341	ARIMA_M3 8.5777	HOLT3_M3 8.6023	M15_M3 8.6306	ESM 8.6394	NAÏVE 8.6980	HOLT 8.7923	ARIMA 9.0052	MEAN_M3 9.0490	MEAN 11.0633

Table C-44: Transportation Sector M3 RMSE Accuracy Results

SECURITY	RMSE #1	RMSE #2	RMSE #3	RMSE #4	RMSE #5	RMSE #6	RMSE #7	RMSE #8	RMSE #9	RMSE #10	RMSE #11	RMSE #12
AIRLINE STOCKS												
Southwest Airlines Co. (LUV)	HOLT3_M3 0.0465	NAIVE3_M3 0.0465	HOLT 0.0466	NAÏVE 0.0467	MEAN_M3 0.0468	M15_M3 0.0469	ESM_M3 0.0477	ARIMA_M3 0.0478	ESM3_M3 0.0486	ESM 0.0498	MEAN 0.0539	ARIMA 0.0593
Alaska Air Group, Inc. (ALK)	NAIVE3_M3 0.0640	NAÏVE 0.0641	HOLT3_M3 0.0641	HOLT 0.0642	ESM_M3 0.0647	M15_M3 0.0650	ESM3_M3 0.0651	ESM 0.0658	MEAN_M3 0.0667	ARIMA_M3 0.0765	MEAN 0.0772	ARIMA 0.1263
PHI, Inc. (PHII)	NAIVE3_M3 0.0519	HOLT3_M3 0.0521	NAÏVE 0.0521	ESM3_M3 0.0521	ARIMA 0.0522	HOLT 0.0522	ESM 0.0522	ESM_M3 0.0523	ARIMA_M3 0.0525	M15_M3 0.0535	MEAN_M3 0.0551	MEAN 0.0637
Skywest, Inc. (SKYW)	NAIVE3_M3 0.0482	HOLT3_M3 0.0483	NAÏVE 0.0483	HOLT 0.0484	ESM_M3 0.0487	ESM3_M3 0.0488	ARIMA_M3 0.0490	ARIMA 0.0490	ESM 0.0492	M15_M3 0.0498	MEAN_M3 0.0507	MEAN 0.0600
Bristow Group Inc. (BRS)	ARIMA 0.0700	HOLT3_M3 0.0700	NAIVE3_M3 0.0700	HOLT 0.0702	NAÏVE 0.0702	ESM3_M3 0.0704	ESM_M3 0.0706	ARIMA_M3 0.0706	ESM 0.0708	M15_M3 0.0722	MEAN_M3 0.0770	MEAN 0.0934
Hawaiian Holdings, Inc. (HA)	MEAN_M3 0.0407	NAÏVE 0.0420	HOLT 0.0421	NAIVE3_M3 0.0422	HOLT3_M3 0.0422	ESM3_M3 0.0422	ESM 0.0422	ESM_M3 0.0425	M15_M3 0.0438	MEAN 0.0453	ARIMA_M3 0.0633	ARIMA 0.1322

APPENDIX C: M3 Forecast Study Results

SECURITY	RMSE #1	RMSE #2	RMSE #3	RMSE #4	RMSE #5	RMSE #6	RMSE #7	RMSE #8	RMSE #9	RMSE #10	RMSE #11	RMSE #12
RAIL STOCKS												
Union Pacific Corporation (UNP)	ESM 0.0641	ESM3_M3 0.0646	NAÏVE 0.0655	HOLT 0.0655	ESM_M3 0.0655	NAIVE3_M3 0.0657	HOLT3_M3 0.0657	M15_M3 0.0682	MEAN_M3 0.0776	ARIMA_M3 0.0810	MEAN 0.0975	ARIMA 0.1755
CSX Coporation (CSX)	HOLT3_M3 0.0385	NAIVE3_M3 0.0385	HOLT 0.0386	NAÏVE 0.0386	M15_M3 0.0386	ESM_M3 0.0389	ESM3_M3 0.0393	ESM 0.0399	ARIMA_M3 0.0433	MEAN_M3 0.0435	MEAN 0.0514	ARIMA 0.0614
Kansas City Southern (KSU)	MEAN_M3 0.2710	M15_M3 0.2727	HOLT3_M3 0.2815	NAIVE3_M3 0.2817	HOLT 0.2853	NAÏVE 0.2854	ESM_M3 0.2861	MEAN 0.2886	ESM3_M3 0.2943	ESM 0.3034	ARIMA_M3 0.4077	ARIMA 0.7372
Norfolk Southern Corporatino (NSC)	ARIMA_M3 0.0936	ARIMA 0.0946	HOLT 0.0965	NAÏVE 0.0965	HOLT3_M3 0.0970	NAIVE3_M3 0.0970	ESM3_M3 0.0972	ESM 0.0973	ESM_M3 0.0976	M15_M3 0.1003	MEAN_M3 0.1049	MEAN 0.1222
Canadian Pacific Railway Limited (CP)	ESM_M3 0.1211	HOLT3_M3 0.1215	HOLT 0.1215	NAIVE3_M3 0.1216	NAÏVE 0.1216	ESM3_M3 0.1217	ESM 0.1239	M15_M3 0.1244	MEAN_M3 0.1378	ARIMA_M3 0.1567	MEAN 0.1667	ARIMA 0.2802
TRUCK TRANSPORTATION STOCKS												
J.. B. Hunt Transport Services, Inc. (JBHT)	ESM 0.0602	ESM3_M3 0.0603	ESM_M3 0.0608	HOLT3_M3 0.0609	HOLT 0.0609	NAIVE3_M3 0.0610	NAÏVE 0.0610	M15_M3 0.0630	MEAN_M3 0.0677	MEAN 0.0823	ARIMA_M3 0.0899	ARIMA 0.2139
Werner Enterrises, Inc. (WERN)	ESM_M3 0.0279	HOLT3_M3 0.0280	ESM3_M3 0.0280	NAIVE3_M3 0.0280	ARIMA_M3 0.0280	HOLT 0.0282	MEAN_M3 0.0282	ESM 0.0282	NAÏVE 0.0282	ARIMA 0.0283	M15_M3 0.0285	MEAN 0.0327
Heartland Express, Inc. (HTLD)	ARIMA_M3 0.0190	M15_M3 0.0192	ARIMA 0.0193	ESM_M3 0.0193	HOLT3_M3 0.0194	NAIVE3_M3 0.0194	ESM3_M3 0.0197	HOLT 0.0197	NAÏVE 0.0198	MEAN_M3 0.0200	ESM 0.0203	MEAN 0.0234
Marten Transport, Ltd. (MRTN)	ESM_M3 0.0134	M15_M3 0.0134	ESM3_M3 0.0135	HOLT3_M3 0.0136	NAIVE3_M3 0.0136	ESM 0.0137	HOLT 0.0138	NAÏVE 0.0139	MEAN_M3 0.0148	ARIMA_M3 0.0164	MEAN 0.0180	ARIMA 0.0283
Old Dominion Freight Line, Inc. (ODFL)	ESM 0.0749	ESM3_M3 0.0754	HOLT 0.0762	NAÏVE 0.0763	ESM_M3 0.0763	HOLT3_M3 0.0765	NAIVE3_M3 0.0766	M15_M3 0.0789	MEAN_M3 0.0847	MEAN 0.0993	ARIMA_M3 0.1013	ARIMA 0.2029
Landstar Systems, Inc. (LSTR)	HOLT 0.0636	HOLT3_M3 0.0636	NAIVE3_M3 0.0637	NAÏVE 0.0637	ESM_M3 0.0643	ESM3_M3 0.0644	ESM 0.0651	M15_M3 0.0655	MEAN_M3 0.0691	ARIMA_M3 0.0710	MEAN 0.0815	ARIMA 0.1095
Forward Air Corporation (FWRD)	ESM_M3 0.0483	ARIMA_M3 0.0484	M15_M3 0.0487	ESM3_M3 0.0489	MEAN_M3 0.0490	NAIVE3_M3 0.0491	HOLT3_M3 0.0492	ESM 0.0500	NAÏVE 0.0500	HOLT 0.0502	ARIMA 0.0506	MEAN 0.0576

C.21. Utilities Sector

Table C-45: Utilities Sector M3 MAPE Accuracy Results

SECURITY	MAPE #1	MAPE #2	MAPE #3	MAPE #4	MAPE #5	MAPE #6	MAPE #7	MAPE #8	MAPE #9	MAPE #10	MAPE #11	MAPE #12
ELECTRICAL POWER DISTRIBUTION												
American Electric Power Co., Inc. (AEP)	HOLT	ESM	NAÏVE	ARIMA	HOLT3_M3	ESM3_M3	NAIVE3_M3	ESM_M3	ARIMA_M3	M15_M3	MEAN_M3	MEAN
	5.7588	5.7593	5.7602	5.7654	5.7709	5.7719	5.7724	5.8254	5.8770	6.0541	6.4847	7.8660
Exelon Corporation (EXC)	ESM	ESM3_M3	NAÏVE	HOLT	NAIVE3_M3	HOLT3_M3	ESM_M3	M15_M3	ARIMA_M3	MEAN_M3	ARIMA	MEAN
	5.7783	5.8275	5.8506	5.8723	5.8882	5.8993	5.9313	6.2459	6.4046	6.9758	7.5969	8.6097
NextEra Energy, Inc. (NEE)	ESM	ESM3_M3	ESM_M3	NAÏVE	HOLT	NAIVE3_M3	HOLT3_M3	M15_M3	MEAN_M3	ARIMA_M3	MEAN	ARIMA
	4.8489	4.8665	4.9267	4.9383	4.9411	4.9475	4.9499	5.1786	5.7224	5.7933	6.9459	7.6994
Duke Energy Corporation (DUK)	HOLT	NAÏVE	HOLT3_M3	NAIVE3_M3	ARIMA	ESM3_M3	ESM	ESM_M3	ARIMA_M3	M15_M3	MEAN_M3	MEAN
	5.5112	5.5121	5.5232	5.5257	5.5383	5.5553	5.5606	5.6052	5.6340	5.8115	6.3063	7.8802
Dominion Energy, Inc. (D)	ARIMA_M3	ARIMA	ESM_M3	HOLT3_M3	NAIVE3_M3	ESM3_M3	HOLT	NAÏVE	ESM	M15_M3	MEAN_M3	MEAN
	4.6916	4.6958	4.7126	4.7294	4.7298	4.7309	4.7765	4.7772	4.7909	4.8050	5.0113	6.2570
Public Service Enterprise Group Inc. (PEG)	HOLT3_M3	NAIVE3_M3	ESM3_M3	ARIMA	HOLT	ESM	NAÏVE	ESM_M3	ARIMA_M3	M15_M3	MEAN_M3	MEAN
	5.6962	5.7014	5.7017	5.7033	5.7062	5.7132	5.7134	5.7295	5.7573	5.8925	6.3468	7.8012
The Southern Comnpany (SO)	ARIMA	HOLT	NAÏVE	HOLT3_M3	NAIVE3_M3	ARIMA_M3	ESM3_M3	ESM_M3	ESM	M15_M3	MEAN_M3	MEAN
	4.0939	4.1082	4.1146	4.1182	4.1205	4.1918	4.1927	4.2003	4.2367	4.3301	4.3329	5.1516
NATURAL GAS DISTRIBUTION												
National Fuel Gas Company (NFG)	HOLT3_M3	NAIVE3_M3	HOLT	NAÏVE	ESM3_M3	ESM_M3	ESM	M15_M3	ARIMA_M3	MEAN_M3	ARIMA	MEAN
	6.5889	6.5897	6.5973	6.5994	6.6169	6.6244	6.6571	6.7876	7.0019	7.4528	7.5542	9.2197
ONEOK, Inc. (OKE)	HOLT	NAÏVE	ESM	HOLT3_M3	NAIVE3_M3	ESM3_M3	ESM_M3	M15_M3	MEAN_M3	ARIMA_M3	MEAN	ARIMA
	8.0865	8.0871	8.0895	8.1494	8.1511	8.1542	8.2541	8.5597	9.8224	10.1883	12.3799	15.3241
UGI Corporation (UGI)	ESM3_M3	ESM	NAÏVE	NAIVE3_M3	HOLT	HOLT3_M3	ESM_M3	M15_M3	MEAN_M3	ARIMA_M3	MEAN	ARIMA
	5.1207	5.1338	5.1378	5.1417	5.1481	5.1548	5.1630	5.3884	5.8654	6.5872	7.1179	10.8308
MDU Resources Group, Inc. (MDU)	NAÏVE	NAIVE3_M3	HOLT	HOLT3_M3	ESM_M3	ESM3_M3	ESM	M15_M3	ARIMA_M3	MEAN_M3	ARIMA	MEAN
	6.5384	6.5433	6.5552	6.5562	6.6267	6.6311	6.7113	6.8147	7.5145	7.7850	9.4346	10.0366
New Jersey Resources Co. (NJR)	ESM_M3	ESM3_M3	NAIVE3_M3	HOLT3_M3	ESM	NAÏVE	HOLT	M15_M3	ARIMA_M3	MEAN_M3	ARIMA	MEAN
	4.3098	4.3348	4.3879	4.3938	4.4302	4.4533	4.4565	4.5150	4.5602	4.6680	5.0958	5.4587
Atmos Energy Corporation (ATO)	ESM3_M3	ESM_M3	HOLT3_M3	ESM	NAIVE3_M3	HOLT	NAÏVE	M15_M3	MEAN_M3	ARIMA_M3	MEAN	ARIMA
	4.8873	4.8966	4.9210	4.9281	4.9287	4.9552	4.9673	5.0295	5.3845	5.5976	6.3561	7.1297
Southwest Gas Holdings, Inc. (SWX)	ESM_M3	HOLT3_M3	ESM3_M3	NAIVE3_M3	HOLT	ESM	NAÏVE	M15_M3	MEAN_M3	ARIMA_M3	ARIMA	MEAN
	5.3720	5.3840	5.3847	5.3953	5.4370	5.4383	5.4545	5.4766	5.8130	5.8652	6.9412	7.2077

SECURITY	MAPE #1	MAPE #2	MAPE #3	MAPE #4	MAPE #5	MAPE #6	MAPE #7	MAPE #8	MAPE #9	MAPE #10	MAPE #11	MAPE #12
WATER SUPPLY												
SJW Group (SJW)	NAIVE3_M3 6.3660	MEAN_M3 6.3714	HOLT3_M3 6.3809	NAÏVE 6.4048	HOLT 6.4270	ESM_M3 6.4943	ESM3_M3 6.5626	M15_M3 6.6080	ESM 6.7250	ARIMA_M3 7.2877	MEAN 7.3532	ARIMA 8.8771
Aqua America, Inc. (WTR)	HOLT3_M3 5.4394	NAIVE3_M3 5.4417	ESM3_M3 5.4610	ESM_M3 5.4657	HOLT 5.4869	NAÏVE 5.4917	ESM 5.5172	M15_M3 5.6301	ARIMA_M3 5.9588	MEAN_M3 6.1537	ARIMA 6.7899	MEAN 7.5481
American States Water Company (AWR)	ESM3_M3 5.5381	ESM_M3 5.5530	ESM 5.5804	HOLT3_M3 5.5835	NAIVE3_M3 5.5845	MEAN_M3 5.6021	HOLT 5.6082	NAÏVE 5.6157	M15_M3 5.7853	ARIMA_M3 5.9364	MEAN 6.4189	ARIMA 6.7138
California Water Service Group (CWT)	ARIMA_M3 5.1770	ARIMA 5.1869	MEAN_M3 5.1900	ESM_M3 5.2576	HOLT3_M3 5.2856	ESM3_M3 5.3004	NAIVE3_M3 5.3177	M15_M3 5.3664	HOLT 5.3748	ESM 5.4161	NAÏVE 5.4260	MEAN 6.0661
Middlesex Water Company (MSEX)	MEAN_M3 4.9294	HOLT3_M3 5.0618	ESM_M3 5.0923	M15_M3 5.1114	HOLT 5.1264	ESM3_M3 5.1578	NAIVE3_M3 5.1799	ESM 5.2735	NAÏVE 5.3148	ARIMA_M3 5.5044	MEAN 5.6445	ARIMA 6.3689
Connecticut Water Service, Inc. (CTWS)	ARIMA 4.9005	ARIMA_M3 4.9037	ESM_M3 4.9609	MEAN_M3 4.9736	HOLT3_M3 4.9970	ESM3_M3 5.0209	M15_M3 5.0242	HOLT 5.0681	NAIVE3_M3 5.0746	ESM 5.1425	NAÏVE 5.2045	MEAN 5.8428

Table C-46: Utilities Sector M3 RMSE Accuracy Results

SECURITY	RMSE #1	RMSE #2	RMSE #3	RMSE #4	RMSE #5	RMSE #6	RMSE #7	RMSE #8	RMSE #9	RMSE #10	RMSE #11	RMSE #12
ELECTRICAL POWER DISTRIBUTION												
American Electric Power Co., Inc. (AEP)	ESM3_M3 0.0452	HOLT3_M3 0.0452	NAIVE3_M3 0.0452	ESM 0.0453	HOLT 0.0453	NAÏVE 0.0453	ARIMA 0.0454	ESM_M3 0.0455	ARIMA_M3 0.0457	M15_M3 0.0467	MEAN_M3 0.0505	MEAN 0.0601
Exelon Corporation (EXC)	ESM 0.0516	ESM3_M3 0.0519	NAÏVE 0.0520	HOLT 0.0520	HOLT3_M3 0.0522	NAIVE3_M3 0.0522	ESM_M3 0.0524	M15_M3 0.0544	MEAN_M3 0.0617	ARIMA_M3 0.0642	MEAN 0.0750	ARIMA 0.1067
NextEra Energy, Inc. (NEE)	ESM 0.0586	ESM3_M3 0.0589	ESM_M3 0.0596	NAIVE3_M3 0.0612	HOLT3_M3 0.0612	NAÏVE 0.0615	HOLT 0.0615	M15_M3 0.0623	MEAN_M3 0.0701	MEAN 0.0857	ARIMA_M3 0.1002	ARIMA 0.2165
Duke Energy Corporation (DUK)	HOLT3_M3 0.0609	HOLT 0.0609	NAIVE3_M3 0.0609	NAÏVE 0.0610	ARIMA 0.0611	ESM3_M3 0.0614	ESM_M3 0.0615	ESM 0.0616	ARIMA_M3 0.0617	M15_M3 0.0630	MEAN_M3 0.0691	MEAN 0.0840
Dominion Energy, Inc. (D)	ARIMA_M3 0.0409	ARIMA 0.0410	ESM_M3 0.0411	NAIVE3_M3 0.0412	HOLT3_M3 0.0412	ESM3_M3 0.0412	HOLT 0.0415	NAÏVE 0.0415	M15_M3 0.0416	ESM 0.0417	MEAN_M3 0.0429	MEAN 0.0514
Public Service Enterprise Group Inc. (PEG)	HOLT3_M3 0.0340	ESM3_M3 0.0341	NAIVE3_M3 0.0341	ESM_M3 0.0341	ARIMA_M3 0.0342	HOLT 0.0342	ESM 0.0343	NAÏVE 0.0343	ARIMA 0.0343	M15_M3 0.0348	MEAN_M3 0.0372	MEAN 0.0457
The Southern Comnpany (SO)	HOLT 0.0268	ARIMA 0.0268	NAÏVE 0.0269	HOLT3_M3 0.0269	NAIVE3_M3 0.0269	ARIMA_M3 0.0273	ESM_M3 0.0275	ESM3_M3 0.0276	ESM 0.0280	M15_M3 0.0281	MEAN_M3 0.0284	MEAN 0.0332

SECURITY	RMSE #1	RMSE #2	RMSE #3	RMSE #4	RMSE #5	RMSE #6	RMSE #7	RMSE #8	RMSE #9	RMSE #10	RMSE #11	RMSE #12
NATURAL GAS DISTRIBUTION												
National Fuel Gas Company (NFG)	HOLT3_M3 0.0611	NAIVE3_M3 0.0612	HOLT 0.0613	NAÏVE 0.0613	ESM_M3 0.0620	M15_M3 0.0623	ESM3_M3 0.0626	ESM 0.0635	MEAN_M3 0.0681	ARIMA_M3 0.0693	MEAN 0.0818	ARIMA 0.0947
ONEOK, Inc. (OKE)	HOLT 0.0498	NAÏVE 0.0499	ESM 0.0499	HOLT3_M3 0.0503	NAIVE3_M3 0.0503	ESM3_M3 0.0504	ESM_M3 0.0510	M15_M3 0.0530	MEAN_M3 0.0591	ARIMA_M3 0.0633	MEAN 0.0722	ARIMA 0.1313
UGI Corporation (UGI)	ESM3_M3 0.0216	ESM 0.0216	ESM_M3 0.0218	NAÏVE 0.0219	HOLT 0.0219	NAIVE3_M3 0.0220	HOLT3_M3 0.0220	M15_M3 0.0229	MEAN_M3 0.0254	MEAN 0.0312	ARIMA_M3 0.0344	ARIMA 0.0846
MDU Resources Group, Inc. (MDU)	M15_M3 0.0298	HOLT3_M3 0.0301	NAIVE3_M3 0.0302	ESM_M3 0.0302	HOLT 0.0304	NAÏVE 0.0305	ESM3_M3 0.0307	ESM 0.0315	MEAN_M3 0.0334	ARIMA_M3 0.0395	MEAN 0.0403	ARIMA 0.0721
New Jersey Resources Co. (NJR)	ESM_M3 0.0191	ESM3_M3 0.0192	HOLT3_M3 0.0193	NAIVE3_M3 0.0194	HOLT 0.0195	NAÏVE 0.0196	ESM 0.0196	M15_M3 0.0197	MEAN_M3 0.0212	MEAN 0.0260	ARIMA_M3 0.0265	ARIMA 0.0486
Atmos Energy Corporation (ATO)	ESM3_M3 0.0344	ESM_M3 0.0346	ESM 0.0347	HOLT3_M3 0.0351	NAIVE3_M3 0.0352	HOLT 0.0354	NAÏVE 0.0355	M15_M3 0.0358	MEAN_M3 0.0399	MEAN 0.0483	ARIMA_M3 0.0610	ARIMA 0.1373
Southwest Gas Holdings, Inc. (SWX)	ESM_M3 0.0451	M15_M3 0.0452	ESM3_M3 0.0453	HOLT3_M3 0.0453	NAIVE3_M3 0.0454	ESM 0.0457	HOLT 0.0457	NAÏVE 0.0458	MEAN_M3 0.0465	MEAN 0.0556	ARIMA_M3 0.0752	ARIMA 0.1614
WATER SUPPLY												
SJW Group (SJW)	MEAN_M3 0.0435	HOLT3_M3 0.0447	ESM_M3 0.0447	NAIVE3_M3 0.0448	M15_M3 0.0450	HOLT 0.0451	ESM3_M3 0.0451	NAÏVE 0.0452	ESM 0.0459	MEAN 0.0488	ARIMA_M3 0.0540	ARIMA 0.0847
Aqua America, Inc. (WTR)	ESM_M3 0.0210	HOLT3_M3 0.0210	NAIVE3_M3 0.0210	M15_M3 0.0211	ESM3_M3 0.0211	HOLT 0.0212	NAÏVE 0.0212	ESM 0.0213	MEAN_M3 0.0218	MEAN 0.0253	ARIMA_M3 0.0259	ARIMA 0.0393
American States Water Company (AWR)	ESM3_M3 0.0253	ESM_M3 0.0254	ESM 0.0256	HOLT3_M3 0.0265	M15_M3 0.0265	NAIVE3_M3 0.0265	HOLT 0.0268	NAÏVE 0.0269	MEAN_M3 0.0274	MEAN 0.0320	ARIMA_M3 0.0342	ARIMA 0.0676
California Water Service Group (CWT)	MEAN_M3 0.0234	ARIMA_M3 0.0235	ARIMA 0.0236	M15_M3 0.0238	ESM_M3 0.0239	HOLT3_M3 0.0240	NAIVE3_M3 0.0241	ESM3_M3 0.0243	HOLT 0.0243	NAÏVE 0.0245	ESM 0.0248	MEAN 0.0268
Middlesex Water Company (MSEX)	M15_M3 0.0278	MEAN_M3 0.0279	ESM_M3 0.0283	HOLT3_M3 0.0286	ESM3_M3 0.0288	NAIVE3_M3 0.0290	HOLT 0.0290	ESM 0.0295	NAÏVE 0.0297	MEAN 0.0321	ARIMA_M3 0.0421	ARIMA 0.0795
Connecticut Water Service, Inc. (CTWS)	ARIMA 0.0308	ARIMA_M3 0.0308	ESM_M3 0.0309	ESM3_M3 0.0311	M15_M3 0.0313	HOLT3_M3 0.0314	ESM 0.0317	HOLT 0.0318	NAIVE3_M3 0.0319	NAÏVE 0.0326	MEAN_M3 0.0330	MEAN 0.0395

C.22. Currency Exchange Rates

Table C-47: Currency Exchange Rates M3 MAPE Accuracy Results

SECURITY	MAPE #1	MAPE #2	MAPE #3	MAPE #4	MAPE #5	MAPE #6	MAPE #7	MAPE #8	MAPE #9	MAPE #10	MAPE #11	MAPE #12
CURRENCY EXCHANGE RATES												
Australian Dollar/ US Dollar	HOLT 3.3233	ESM 3.3243	NAÏVE 3.3247	HOLT3_M3 3.3322	NAIVE3_M3 3.3326	ESM3_M3 3.3329	ESM_M3 3.3616	M15_M3 3.4745	ARIMA_M3 3.4935	ARIMA 3.7739	MEAN_M3 3.8306	MEAN 4.7465
Euro/US Dollar	ARIMA 2.5453	NAÏVE 2.5773	ESM 2.5777	NAIVE3_M3 2.5787	ESM3_M3 2.5792	HOLT3_M3 2.5960	ESM_M3 2.6058	ARIMA_M3 2.6068	HOLT 2.6068	M15_M3 2.7212	MEAN_M3 3.1202	MEAN 4.0742
Euro/Australian Dollar	NAÏVE 2.6272	ESM 2.6284	HOLT 2.6300	ARIMA 2.6312	NAIVE3_M3 2.6396	ESM3_M3 2.6405	HOLT3_M3 2.6421	ESM_M3 2.6800	ARIMA_M3 2.7095	M15_M3 2.8158	MEAN_M3 2.8337	MEAN 3.4187
British Pound/US Dollar	HOLT 2.2764	ARIMA 2.2783	NAÏVE 2.2789	ESM 2.2789	HOLT3_M3 2.3045	NAIVE3_M3 2.3082	ESM3_M3 2.3082	ESM_M3 2.3540	ARIMA_M3 2.3822	M15_M3 2.4890	MEAN_M3 2.8604	MEAN 3.5951
Euro/British Pound	NAÏVE 1.7864	ESM 1.7890	ARIMA 1.7952	NAIVE3_M3 1.8022	ESM3_M3 1.8058	HOLT 1.8199	HOLT3_M3 1.8282	ESM_M3 1.8458	ARIMA_M3 1.8761	M15_M3 1.9882	MEAN_M3 2.0697	MEAN 2.6722
US Dollar/ Canadian Dollar	NAIVE3_M3 2.3468	HOLT3_M3 2.3478	ESM3_M3 2.3479	NAÏVE 2.3485	ESM 2.3496	HOLT 2.3497	ESM_M3 2.3655	M15_M3 2.4544	ARIMA_M3 2.4736	MEAN_M3 2.6365	ARIMA 2.8207	MEAN 3.1668
US Dollar/Swiss Franc	ARIMA 2.5625	NAÏVE 2.5629	ESM 2.5631	HOLT 2.5632	NAIVE3_M3 2.5785	ESM3_M3 2.5788	HOLT3_M3 2.5788	ESM_M3 2.6148	ARIMA_M3 2.6411	M15_M3 2.7519	MEAN_M3 2.9535	MEAN 3.7531
US Dollar Currency Index	NAÏVE 2.1518	ESM 2.1528	HOLT 2.1555	NAIVE3_M3 2.1701	ESM3_M3 2.1711	HOLT3_M3 2.1729	ESM_M3 2.2050	M15_M3 2.3165	MEAN_M3 2.6496	ARIMA_M3 2.7913	MEAN 3.3887	ARIMA 3.8393
US Dollar/ Japanese Yen	ESM3_M3 2.6373	NAIVE3_M3 2.6382	HOLT3_M3 2.6422	ESM 2.6465	ARIMA 2.6481	NAÏVE 2.6481	ESM_M3 2.6481	HOLT 2.6539	ARIMA_M3 2.6618	M15_M3 2.7414	MEAN_M3 2.9545	MEAN 3.7211
British Pound/ Japanese Yen	NAÏVE 3.0400	NAIVE3_M3 3.0451	HOLT3_M3 3.0479	ESM3_M3 3.0499	ESM 3.0506	HOLT 3.0535	ESM_M3 3.0807	M15_M3 3.2162	MEAN_M3 3.5456	MEAN 4.5803	ARIMA_M3 5.1401	ARIMA 11.3804

Table C-48: Currency Exchange Rates M3 RMSE Accuracy Results

SECURITY	RMSE #1	RMSE #2	RMSE #3	RMSE #4	RMSE #5	RMSE #6	RMSE #7	RMSE #8	RMSE #9	RMSE #10	RMSE #11	RMSE #12
CURRENCY EXCHANGE RATES												
Australian Dollar/ US Dollar	ESM 0.0005	HOLT 0.0005	NAÏVE 0.0005	NAIVE3_M3 0.0005	ESM3_M3 0.0005	HOLT3_M3 0.0005	ESM_M3 0.0005	M15_M3 0.0005	ARIMA_M3 0.0006	MEAN_M3 0.0006	MEAN 0.0008	ARIMA 0.0008
Euro/US Dollar	ARIMA 0.0006	NAÏVE 0.0006	ESM 0.0006	NAIVE3_M3 0.0006	ESM3_M3 0.0006	HOLT 0.0006	HOLT3_M3 0.0006	ARIMA_M3 0.0006	ESM_M3 0.0006	M15_M3 0.0007	MEAN_M3 0.0007	MEAN 0.0009
Euro/Australian Dollar	NAÏVE 0.0008	HOLT 0.0008	ESM 0.0008	ARIMA 0.0008	NAIVE3_M3 0.0008	ESM3_M3 0.0008	HOLT3_M3 0.0008	ESM_M3 0.0009	ARIMA_M3 0.0009	M15_M3 0.0009	MEAN_M3 0.0009	MEAN 0.0011
British Pound/US Dollar	HOLT 0.0008	ARIMA 0.0008	NAÏVE 0.0008	ESM 0.0008	HOLT3_M3 0.0008	NAIVE3_M3 0.0008	ESM3_M3 0.0008	ESM_M3 0.0008	ARIMA_M3 0.0008	M15_M3 0.0008	MEAN_M3 0.0009	MEAN 0.0011
Euro/British Pound	NAÏVE 0.0003	ESM 0.0003	ARIMA 0.0003	NAIVE3_M3 0.0003	ESM3_M3 0.0003	ESM_M3 0.0003	HOLT3_M3 0.0003	HOLT 0.0003	ARIMA_M3 0.0003	M15_M3 0.0003	MEAN_M3 0.0003	MEAN 0.0004
US Dollar/ Canadian Dollar	NAÏVE 0.0006	ESM 0.0006	HOLT 0.0006	NAIVE3_M3 0.0006	ESM3_M3 0.0006	HOLT3_M3 0.0006	ESM_M3 0.0006	M15_M3 0.0006	ARIMA_M3 0.0006	MEAN_M3 0.0007	ARIMA 0.0007	MEAN 0.0008
US Dollar/Swiss Franc	ARIMA 0.0006	NAÏVE 0.0006	ESM 0.0006	HOLT 0.0006	NAIVE3_M3 0.0006	HOLT3_M3 0.0006	ESM3_M3 0.0006	ESM_M3 0.0006	ARIMA_M3 0.0006	M15_M3 0.0006	MEAN_M3 0.0007	MEAN 0.0008
US Dollar Currency Index	NAÏVE 0.0387	ESM 0.0387	HOLT 0.0387	NAIVE3_M3 0.0390	ESM3_M3 0.0391	HOLT3_M3 0.0391	ESM_M3 0.0396	M15_M3 0.0414	MEAN_M3 0.0459	MEAN 0.0560	ARIMA_M3 0.0620	ARIMA 0.1283
US Dollar/ Japanese Yen	ESM 0.0553	ARIMA 0.0553	NAÏVE 0.0553	HOLT 0.0554	ESM3_M3 0.0554	NAIVE3_M3 0.0554	HOLT3_M3 0.0554	ESM_M3 0.0558	ARIMA_M3 0.0562	M15_M3 0.0577	MEAN_M3 0.0616	MEAN 0.0757
British Pound/ Japanese Yen	HOLT 0.1121	NAÏVE 0.1123	ESM 0.1125	HOLT3_M3 0.1130	NAIVE3_M3 0.1133	ESM3_M3 0.1134	ESM_M3 0.1149	M15_M3 0.1197	MEAN_M3 0.1327	MEAN 0.1625	ARIMA_M3 0.1686	ARIMA 0.4105

CPSIA information can be obtained
at www.ICGtesting.com
Printed in the USA
BVHW020935290421
606136BV00009B/727

9 780986 449697